Martens, Sables, and Fishers

A juvenile American marten male in a lodgepole pine.
Drawing by Consie Powell.

Martens, Sables, and Fishers

BIOLOGY AND CONSERVATION

EDITED BY

Steven W. Buskirk

Department of Zoology and Physiology
University of Wyoming, Laramie

Alton S. Harestad

Department of Biological Sciences
Simon Fraser University, Burnaby, British Columbia

Martin G. Raphael

USDA Forest Service, Pacific Northwest Research Station
Olympia, Washington

Roger A. Powell

Department of Zoology
North Carolina State University, Raleigh

COMSTOCK PUBLISHING ASSOCIATES, *a division of*

CORNELL UNIVERSITY PRESS   *Ithaca and London*

Library of Congress Cataloging-in-Publication Data

Martens, sables, and fishers : biology and conservation / edited by Steven W. Buskirk . . . [et al.].
  p.   cm.
  Includes bibliographical references and index.
  ISBN 0-8014-2894-7
  1. Martes. 2. Martes—Ecology. 3. Wildlife conservation. I. Buskirk, Steven.
  QL735.C25M313   1994
  599.74'447—dc20   93-42022

Contents

v

Part VII   Physiology and Reproduction

Preface

Research on forest carnivores, traditionally an obscure area of study, blossomed in Eurasia and North America during the 1980s. This heightened interest grew out of concerns about the fate of forested habitats worldwide and about the vulnerability of many populations of Carnivora (Miller and Everett 1986). Forest carnivores seem likely to continue to receive close scrutiny in the coming years.

The literature on martens, sables, and fishers is fairly large, but it contains many descriptive reports that have been related to important biological ideas and conservation principles to varying degrees. We saw a need for a comprehensive review that would describe what is known about the biology of martens, sables, and fishers, and would also identify gaps in our knowledge and promising hypotheses for future study. Until now the only books that have treated members of the genus *Martes* are *The Sable,* by V. V. Timofeev and V. N. Nadeev, published in Russian (1955); *Quaternary Evolution of the Genus* Martes *(Carnivora, Mustelidae),* by Elaine Anderson (1970); and *The Fisher,* by Roger Powell (1982, 2d ed. 1993). The Soviets also produced several specialized monographs between 1947 and 1965 on population estimation and management of sables, some of which were translated into English by the Israel Program for Scientific Translations (e.g., Formozov 1967).

Although this book covers an entire genus, our treatment of individual species is inevitably uneven. The yellow-throated marten (*Martes flavigula*) (including the Nilgiri marten [*M. gwatkinsi*]), which is known almost solely from records of occurrence (e.g., Liat et al. 1980) and from a few behavioral anecdotes (e.g., Wemmer and Johnson 1976, Matyushkin 1987), necessarily receives only passing mention. In contrast, our reviews of North American

ix

and western Eurasian species are based on extensive studies; Martin (this volume) reviews 22 published diet studies for the American marten. Tremendous unevenness in our geographic knowledge of the genus is one clear message of this book. The literature on *Martes* is wealthy in some taxonomic and conceptual areas but impoverished in others. Studies that address how marten populations use and are affected by landscapes, especially the effects of tropical deforestation on the yellow-throated marten, are almost nonexistent and are badly needed.

We used the occasion of the Symposium on the Biology and Management of Martens and Fishers as a means of soliciting the chapters presented here. Eight authors were invited to submit review chapters, and others were encouraged, after we read their abstracts, to contribute reviews or reports of original research. Still, only about a third of the presentations at the symposium are represented here by chapters, and three chapters are by authors who did not participate in the symposium. This is not the proceedings of that meeting, then, but a more deliberate and complete account. Each chapter received at least two reviews by referees who could choose to be anonymous, and each received editorial guidance and revision from one or more of the editors.

The common nomenclature of the genus *Martes*, as for other taxa, is neither precise nor systematic. For example, *Martes foina* is variously referred to in English as the stone marten, beech marten, baumarten, and house marten. We have tried to reduce confusion for readers by limiting authors to the common names listed in Table I.1, although in doing so we may have sacrificed some of the regional linguistic flavor that common names convey.

We hope this book will be useful to anyone who is intellectually or professionally interested in martens, sables, and fishers. Our intent has been to provide a comprehensive treatment of the biology of a group of related animals and to stimulate new discussion about the biological phenomena they help reveal as well as the best means of assuring their survival in the face of human influences.

STEVEN W. BUSKIRK
ALTON S. HARESTAD
MARTIN G. RAPHAEL
ROGER A. POWELL

Acknowledgments

Preparation of the manuscript for this book was supported by grants from the following institutions and individuals with scientific, regulatory, or personal interest in martens and fishers:

Wyoming Game and Fish Department
University of Wyoming
Simon Fraser University
U.S. National Park Service, Alaska and Rocky Mountain Regions
USDA Forest Service, Regions 1, 2, 6, 9, and 10
British Columbia Trappers' Association
Minnesota Trappers' Association
USDA Forest Service, Forest Sciences Laboratory, Olympia, Washington
USDA Forest Service, Forest Sciences Laboratory, Wenatchee, Washington
Forestry Canada, Newfoundland Forestry Centre
Ontario Ministry of Natural Resources
Yukon Department of Renewable Resources, Fish and Wildlife Branch
Fur Institute of Canada, Furbearer Conservation Foundation
British Columbia Ministry of Environment, Lands and Parks, Wildlife Program
British Columbia Ministry of Forests, Research Branch
Oeko-Log Wildlife Research and Applications
The Wildlife Society
Central Mountains and Plains Section, and Northwest Section, The Wildlife Society
Wyoming Chapter, The Wildlife Society
Max Bass
William Berg

We appreciate the efforts of our contributors, whose generosity of time, talent, and data made this book possible. The book benefited from reviews by referees whose efforts must be acknowledged anonymously. Ann Mullens Boelter provided invaluable help. She organized the distribution of manuscripts to reviewers and authors, revised figures and tables, and prepared six maps (Figs. I.2, 7.3, 17.1, 18.1, 18.2, and 24.1). She also copy edited the entire manuscript for clarity, brevity, and consistency before it was submitted to Cornell University Press. Robin Petrosky assisted with typing. Robb Reavill and Helene Maddux at Cornell University Press provided keen insights and diligent editing, which improved both the content and the form of the book.

S. W. B.
A. S. H.
M. G. R.
R. A. P.

Contributors

ELAINE ANDERSON, Zoology Department, Denver Museum of Natural History, Denver, Colorado 80205, USA

STEPHEN M. ARTHUR, Alaska Fish and Wildlife Research Center, U.S. Fish and Wildlife Service, Anchorage, Alaska 99503, USA

MICHEAL J. BADRY, Department of Forestry, Alberta Research Council, Edmonton, Alberta, Canada T6H 5X2

NIKOLAI N. BAKEYEV, B. M. Zhitkov All-Union Research Institute of Game Management and Fur Farming, Kirov, 610601, Russia

YURI N. BAKEYEV, Zone Laboratory, All-Union Institute of Game Management and Fur Farming, Krasnodar, 350062, Russia

MORLEY W. BARRETT, Alberta Environmental Centre, Vegreville, Alberta, Canada T0B 4L0

WILLIAM E. BERG, Department of Natural Resources, Forest Wildlife Populations and Research Group, Grand Rapids, Minnesota 55744, USA

SCOTT M. BRAINERD, Department of Biology and Nature Conservation, Agricultural University of Norway, N-1432 Ås, Norway

SLADER G. BUCK, Environmental and Natural Resources Management Office, MCB Camp Pendleton, California 92055, USA

STEVEN W. BUSKIRK, Department of Zoology and Physiology, University of Wyoming, Laramie, Wyoming 82071, USA

MICHEL CANTIN, Ministère du Loisir, de la Chasse et de la Pêche, Direction régionale de Québec, Charlesbourg, Québec, Canada G1G 5H9

ANTHONY P. CLEVENGER, Departamento de Biología Animal, Universidad de León, 24071 León, Spain

PAM J. COLE, Department of Forestry, Alberta Research Council, Edmonton, Alberta, Canada T6H 5X2

CRAIG COOLAHAN, USDA APHIS, Animal Damage Control, Sacramento, California 95815, USA

LINDA M. DIX, Ontario Ministry of Natural Resources, Parry Sound, Ontario, Canada P2A 1S4

RANDY K. DRESCHER, Department of Forestry, Alberta Research Council, Edmonton, Alberta, Canada T6H 5X2

THOMAS D. DRUMMER, Department of Mathematical Sciences, Michigan Technological University, Houghton, Michigan 49931, USA

CLÉMENT FORTIN, Ministère du Loisir, de la Chasse et de la Pêche, Direction régionale de Québec, Charlesbourg, Québec, Canada G1G 5H9

EDWARD O. GARTON, Deptartment of Fish and Wildlife Resources, University of Idaho, Moscow, Idaho 83843, USA

CHARLES J. GIBILISCO, Department of Biology, Montana State University, Bozeman, Montana 59717, USA

RON P. GRAF, Northwest Territories Department of Renewable Resources, Fort Smith, Northwest Territories, Canada X0E 0P0

MARY ANN GRAHAM, Research and Collections Center, Illinois State Museum, Springfield, Illinois 62703, USA

RUSSELL W. GRAHAM, Research and Collections Center, Illinois State Museum, Springfield, Illinois 62703, USA

ALTON S. HARESTAD, Department of Biological Sciences, Simon Fraser University, Burnaby, British Columbia, Canada V5A 1S6

HENRY J. HARLOW, Department of Zoology and Physiology, University of Wyoming, Laramie, Wyoming 82071, USA

J.-O. HELLDIN, Department of Wildlife Ecology, Swedish University of Agricultural Sciences, Grimsö Wildlife Research Station, S-730 91 Riddarhyttan, Sweden

MATHIAS HERRMANN, Department of Ethology, University of Bielefeld, D-4800 Bielefeld, Germany

THOR HOLMES, Museum of Natural History, University of Kansas, Lawrence, Kansas 66045-2454, USA

JEFFREY L. JONES, USDA Forest Service, Beaverhead National Forest, Wisdom, Montana 59761, USA

ALFRED J. KOLENOSKY, Department of Forestry, Alberta Research Council, Edmonton, Alberta, Canada T6H 5X2

WILLIAM B. KROHN, Maine Cooperative Fish and Wildlife Research Unit, U.S. Fish and Wildlife Service, University of Maine, Orono, Maine 04469, USA

DAVID W. KUEHN, Forest Wildlife Populations and Research Group, Minnesota Department of Natural Resources, Grand Rapids, Minnesota 55744, USA

ERIK LINDSTRÖM, Department of Wildlife Ecology, Swedish University of Agricultural Sciences, Grimsö Wildlife Research Station, S-730 91 Riddarhyttan, Sweden

YIQING MA, Institute of Natural Resources, Harbin 150040, People's Republic of China

AUDREY J. MAGOUN, Alaska Department of Fish and Game, Fairbanks, Alaska 99701, USA

SANDRA K. MARTIN, Department of Natural Resource Sciences, Washington State University, Pullman, Washington 99164-6410, USA

GARY M. MATSON, Matson's Laboratory, Milltown, Montana 59851, USA

RODNEY A. MEAD, Department of Biological Sciences, University of Idaho, Moscow, Idaho 83843, USA

ARCHIE S. MOSSMAN, Department of Wildlife, Humboldt State University, Arcata, California 95521, USA

CURT MULLIS, USDA APHIS, Animal Damage Control, Albuquerque, New Mexico 87113, USA

DAVID W. NAGORSEN, Royal British Columbia Museum, Victoria, British Columbia, Canada V8V 1X4

THOMAS F. PARAGI, Koyukuk and Nowitna National Wildlife Refuges, U.S. Fish and Wildlife Service, Galena, Alaska 99741, USA

ROLF O. PETERSON, School of Forestry and Wood Products, Michigan Technological University, Houghton, Michigan 49931, USA

KIM G. POOLE, Wildlife Management Division, Northwest Territories Department of Renewable Resources, Yellowknife, Northwest Territories, Canada X1A 2L9

ROGER A. POWELL, Department of Zoology, North Carolina State University, Raleigh, North Carolina 27695-7617, USA

GILBERT PROULX, Department of Forestry, Alberta Research Council, Edmonton, Alberta, Canada T6H 5X2

MARTIN G. RAPHAEL, USDA Forest Service, Pacific Northwest Research Station, Olympia, Washington 98502, USA

JØRUND ROLSTAD, Norwegian Forest Research Institute, Agricultural University of Norway, N-1432 Ås, Norway

KEN SEIDEL, Department of Forestry, Alberta Research Council, Edmonton, Alberta, Canada T6H 5X2

IVAN SHOW, Southwest Research Associates, Carlsbad, California 92024, USA

ANDREI A. SINITSYN, B. M. Zhitkov All-Union Research Institute of Game Management and Fur Farming, Kirov, 610601, Russia

BRIAN G. SLOUGH, Fish and Wildlife Branch, Yukon Department of Renewable Resources, Whitehorse, Yukon Territory, Canada Y1A 2C6

MARJORIE A. STRICKLAND, Ontario Ministry of Natural Resources, Parry Sound, Ontario, Canada P2A 1S4

MASAYA TATARA, Laboratory of Ecology, Department of Biology, Faculty of Science, Kyushu University 33, Fukuoka 812, Japan

LINDA E. THOMASMA, College of Environmental Science and Forestry, State University of New York, Syracuse, New York 13210, USA

IAN D. THOMPSON, Forestry Canada, National Research Institute, Chalk River, Ontario, Canada K0J 1J0

LI XU, Institute of Natural Resources, Harbin 150040, People's Republic of China

Martens, Sables, and Fishers

Introduction to the
Genus *Martes*

Steven W. Buskirk

Studying the biology of martens, sables, and fishers is important for several reasons. First, this group includes some of the loveliest and shyest carnivores in the world. For many people, their interactions with these animals in the wild are rare and memorable. Satisfying their curiosity about how these animals live is an end in itself. Second, over much of temperate Eurasia and North America, animals in this group are the most wilderness-dependent mammals still remaining in forest ecosystems that have been altered by humans. Their presence symbolizes the natural character of our remaining forests, but it also reminds us of the vulnerability of these animals to future environmental changes. It is imperative that we understand how our treatment of forests affects the animals that live in them. Third, the fur of some species in this group is extraordinarily luxurious and valuable. These furs have been much sought after in the past and will continue to be in the future. We should know how our fur hunting affects the populations from which the animals are taken. Last, members of this group present some of the best available models of several biological phenomena. Studying these animals enables us to understand ecology, physiology, evolution, and other subjects better than we could otherwise.

Seven species of martens, sables, and fishers compose the genus *Martes* (Mustelidae, Carnivora; Table I.1). The mustelids usually are distinguished from other Carnivora by the loss, in living representatives, of the carnassial notch from the upper fourth premolar; by a delicate zygomatic arch (Holmes 1980); by five digits that make contact with the substrate during walking; and by the enlargement of the anal sacs or scent glands (Wozencraft 1989). The first appearance of mustelids is in fossils of the early Miocene Epoch, and the

1

family has found its way to all zoogeographic regions through natural dispersal or translocation.

Phylogeny and Morphology

In most phylogenetic arrangements, the genus *Martes* is grouped close to *Eira*, the Neotropical tayra, within the subfamily Mustelinae, and has three lineages that have been distinct since the late Pliocene (Anderson 1970). Subgenus *Charronia* is represented by the yellow-throated marten (including the Nilgiri marten *Martes gwatkinski*: Anderson, this volume). Subgenus *Pekania* includes the fisher, and subgenus *Martes* includes the stone marten and the four species of boreal forest martens (Table I.1). The boreal forest martens (*M. martes, M. zibellina, M. melampus, M. americana*) are distributed allopatrically or parapatrically across the boreal and taiga zones from Ireland eastward across Eurasia and the Bering Strait to Newfoundland. These species exhibit such close taxonomic and ecological similarities that Anderson (1970; this volume) has called them a "superspecies."

Members of the genus *Martes* are small. They range from the small female *M. americana*, weighing about 400 g (Strickland et al. 1982*a*) to the medium-sized male *M. pennanti*, weighing about 4800 g (Strickland et al. 1982*b*). They have conservative body forms, resembling early mustelids (Anderson 1970) more than other modern mustelids. As in other Carnivora,

Table I.1.   Nomenclature of martens, sables, and fishers

Scientific name	Common name[a]
Martes martes	Eurasian pine marten (European pine marten, pine marten)
M. zibellina	sable (Russian sable)
M. melampus	Japanese marten
M. americana	American marten (marten, pine marten, American pine marten, American sable)
M. foina	stone marten (beech marten, house marten)
M. flavigula (including *M. gwatkinsi*)	yellow-throated marten (including Nilgiri marten)
M. pennanti	fisher (pekan, Pennant's marten)

Notes: Scientific nomenclature follows Honacki et al. (1982) and Corbet and Hill (1986). The first four species are the boreal forest martens.

[a] The first name listed is the one used in this book. Alternative common names are shown in parentheses. The common name "pine marten" is ambiguous.

their sizes differ markedly by sex, the selective pressures for which have been debated before (Moors 1980) and are discussed in the chapters that follow. The skull has the facial shortening of most mustelids, but greater height, and the dentition is typically mustelid, except for the retention of four upper and four lower premolars. The sagittal crest becomes dramatically enlarged with age in male fishers, and the baculum is well developed and variable by species, age, and region.

Morphological Specializations

Martens and fishers show levels of locomotor specialization that are intermediate among carnivoran families (Fig. I.1). Holmes's (1980) study of locomotor adaptations in the Mustelidae showed that martens and fishers lie between skunks and canids in the degree of cursorial specialization of the appendicular skeleton. These specializations include lengthening of the distal limb bones and a digitigrade stance. Indeed, among the mustelids, Holmes considered the American marten to be second only to the wolverine in its degree of cursorial adaptation. This limb elongation also may confer energetic savings during locomotion in snow (Raine 1987), a factor that has been given more consideration for ungulates (Klein et al. 1987) than for carnivores. The head of the humerus of *Martes* suggests high shoulder mobility, and the claws are sharply recurved, both features that facilitate arboreal movement. The martens and their allies seem therefore to mix adaptations for moving through trees with those for traveling about large home ranges (Buskirk and McDonald 1989; Powell, this volume), either on soft snow or on ground.

Although typical mustelid anal sacs are present in *Martes*, they are poorly developed compared with those in other genera, and their secretions are not well known. Nonetheless, martens and fishers are adept chemical communicators, possessing scent sources of at least four other types: abdominal glands (Hall 1926), plantar glands (Buskirk et al. 1986), urine, and feces. The use of these scents involves specific behaviors that have been described anecdotally and are likely important in social organization of populations. We know little about the ecological importance of scent marking, however, because of the difficulty of investigating chemical signals in free-ranging *Martes*.

A generalization of dubious validity states that mustelids have high resting metabolic rates for their body sizes, because of the high energetic costs associated with the long, thin shape of some members (Brown and Lasiewski 1972). In fact, as Knudsen and Kilgore (1990) and Harlow (this volume) have shown, a surprising number of mustelids have resting metabolic rates that are

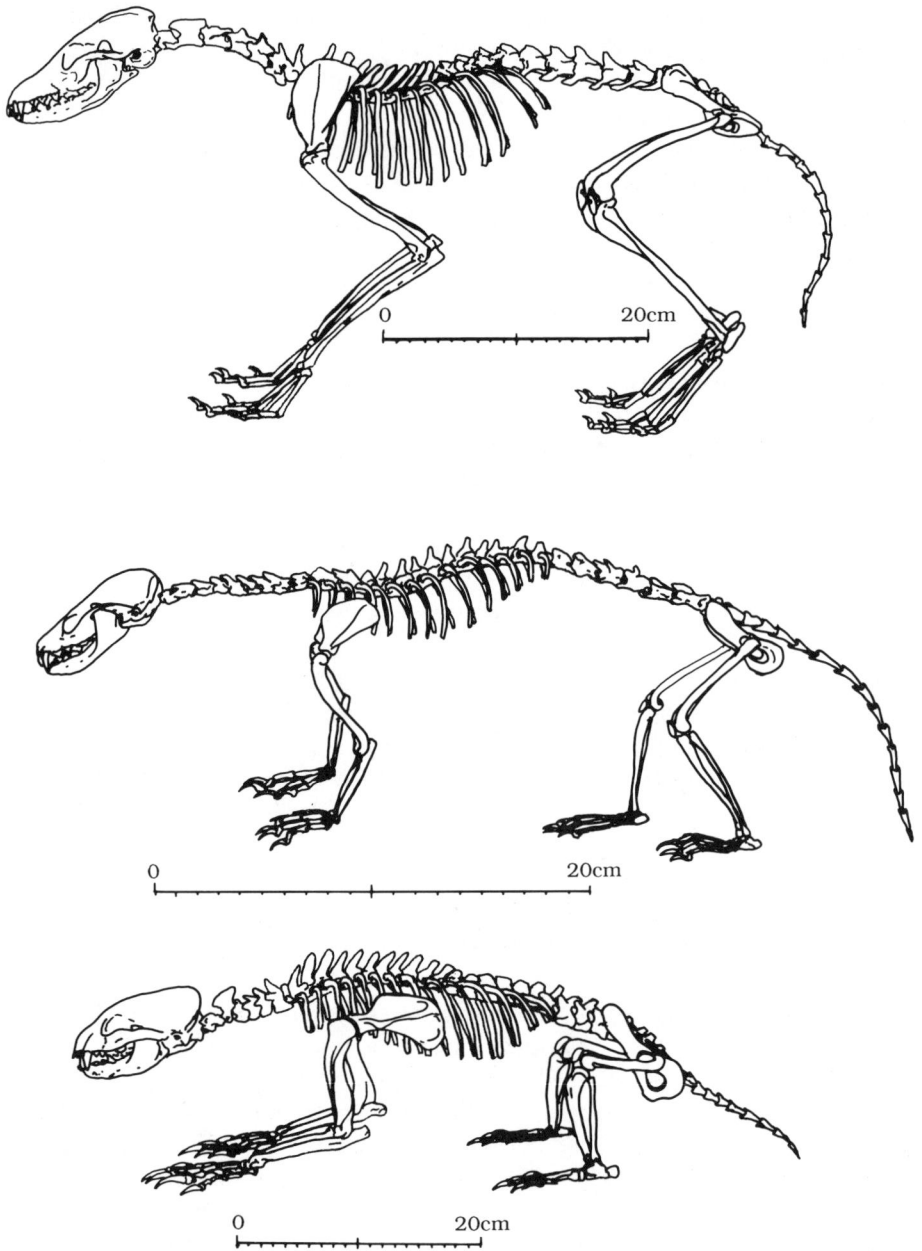

Figure 1.1. Skeleton of the American marten (*Martes americana; center*). Comparison with *Vulpes vulpes* (*top*) and *Taxidea taxus* (*bottom*) shows the intermediate degree of cursorial adaptation of the appendicular skeleton of *Martes*.

lower than predicted on the basis of body size. Further, some species are able to conserve energy through hypothermy, either shallow and daily (American martens: Buskirk et al. 1988) or torporous and prolonged (American badgers: Harlow 1981). Still, the relatively high lower critical temperature of *Martes* suggests that habitat use and behavior should be strongly influenced by thermal factors in cold environments. The image that emerges from the chapters in this book is that these animals integrate morphology, metabolic plasticity, foraging and thermoregulatory behaviors, and habitat specialization to balance energy budgets in cold climates.

Biogeography and Population Ecology

Anderson (1970) inferred a Palearctic origin for the genus *Martes*, based on the earliest reported (Miocene) fossils. As a result of eastward dispersal across Beringia, this genus is now distributed over the tropical, temperate, and boreal forest zones of the Old and New Worlds (Fig. I.2). In the Old World, the yellow-throated marten replaces the boreal forest martens in subtropical and tropical forest, extending southward to the islands of the Sunda Shelf (Borneo, Sumatra, and Java) that were connected to Asia during the Quaternary period (Anderson 1970). Although the Eurasian pine marten occupies extensive Mediterranean shrub habitats, the stone marten accounts for most of the nonforest distribution of *Martes*, occupying woodland and steppe habitats in central and western Eurasia (Fig. I.2). In the New World, American martens occur from the northern limit of trees to the isolated montane stands of mesic coniferous forest that extend as far south as New Mexico (Gibilisco, this volume). They also inhabit the larger oceanic islands (e.g., Vancouver Island, Newfoundland, and the Queen Charlotte Islands) that were once connected to the mainland.

Distributional dynamics are important to understanding both how marten populations have evolved and how they are affected by human activities today. As for other taxa, *Martes* distributions have responded to human actions mostly near their range boundaries and on islands. The chapters by Tatara, Gibilisco, and Thompson and Harestad show how insularization has aggravated the problems of population decline in response to human impact. In the southern part of the range of the American marten (Patterson 1984) and sable (Bakeyev and Sinitsyn, this volume), much can yet be learned about the processes of extinction by studying isolated *Martes* populations in relation to postglacial vegetation changes. Examination of past natural extinctions could reveal factors that affect persistence of *Martes* populations. Such increased understanding would be useful in conservation planning.

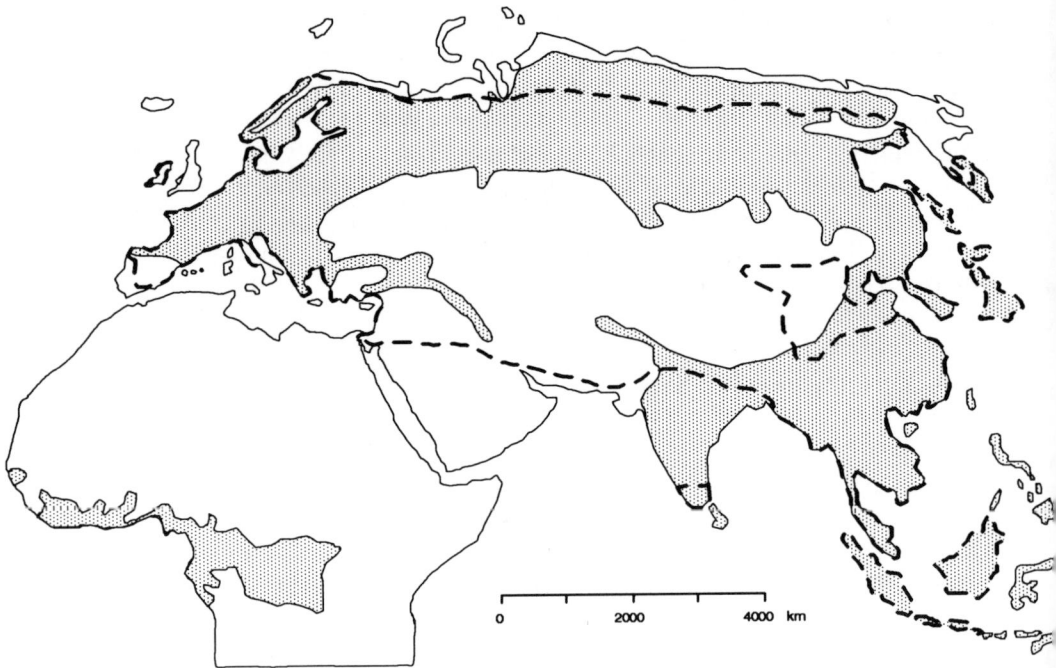

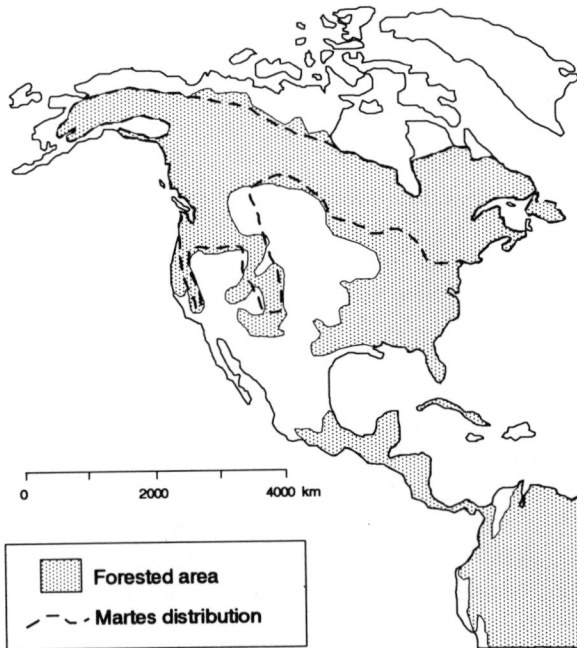

Figure 1.2. The worldwide distribution of the genus *Martes* and the distribution of forest vegetation. Major discrepancies occur in the distribution of the stone marten (*Martes foina*), which is not a forest obligate, and the Eurasian pine marten (*M. martes*), which occupies Mediterranean shrublands. Discrepancies on the North American map between the northern limit of trees and the limit of *Martes* are due to measurement error, as is the case for the Rocky Mountains. The most extensive forest in North America unoccupied by *Martes* is broad-leaved forest in the Mississippi Valley. The distribution of *Martes* follows Anderson 1970, as modified for North America by Gibilisco (this volume). Forest boundaries include all forest-woodland types (Anonymous 1981).

Forested area

Martes distribution

Like populations of most other mustelids (Powell 1979*a*), *Martes* populations are structured around the intrasexually territorial home ranges of resident adults. In addition, transient animals, often young, overlap with adults of the same or opposite sex (Powell, this volume). This spatial basis of population organization has spurred numerous researchers to describe the home ranges of various species, and these data have been used to test for the effects of sex, age, habitat quality, and prey abundance (Thompson and Colgan 1987*a*, Buskirk and McDonald 1989). The home ranges of most *Martes* populations are quite large, given the body size of the animals, and population densities are correspondingly low. This fact, combined with the late age of sexual maturity and small litters (Mead, this volume), makes these species vulnerable to overexploitation, a problem that has occurred to different degrees in the past. With appropriate management, however, *Martes* populations can produce sustained harvests and opportunities to observe them. Understanding spatial requirements can contribute to successful population management, including establishing conservation areas of the proper size. The more successful management schemes include conserving unharvested areas to retain as source populations from which dispersing animals can repopulate harvested areas (Strickland, this volume).

History, Research, and Management

The route to our current knowledge of the ecology and physiology of martens, sables, and fishers has been similar to that followed for other species of commercially important mammals. Management of carnivorous mammals can be thought of as a site- and taxon-specific mixture of several key elements: managing the fur harvest, controlling depredations on agriculture, protecting rare or endangered populations, and conserving populations by providing adequate habitat.

The boreal forest martens and the fisher were recognized early as lucrative fur-bearing animals and were important both in the eastward expansion of Russia under Peter the Great and Catherine II (Bakeyev and Sinitsyn, this volume) and in the westward and northward exploration of North America. This exploitation of furbearers led to scarcity or local extinctions in many areas of Eurasia and North America. Following these losses were programs that slowly developed the expertise, policies, and infrastructure to make the harvest of marten, sable, and fisher furs sustainable. This goal has been one of the most daunting in modern wildlife management, and in some respects it still eludes us. Because of the low densities and elusive behavior of martens, sables, and fishers, they simply cannot be enumerated like larger, more conspicuous species. Methods developed to monitor populations of secretive, low-density carnivores are more inferential, but nearly as reliable, as those for species that are more easily counted (Raphael, this volume). Strickland (this volume) describes a general strategy involving demographic and habitat monitoring that may apply to many harvested species with attributes similar to those of martens and fishers.

Economic losses, agricultural or otherwise, have never been important concerns in connection with martens and their allies, because this group mostly occupies cold areas with little nonforestry agriculture. With rare exceptions (Hall 1987), including the bewildering tendency of stone martens to chew on rubber parts of automobiles (Lower 1989), martens, sables, and fishers kill, eat, or injure almost nothing of value to humans. On the contrary, fishers were introduced to control porcupines (*Erethizon dorsatum*) in central and eastern North America in the 1950s and early 1960s, with apparent success (Powell 1982). I am aware of a few unpublished anecdotes that tell of martens raiding the nests or hack boxes of threatened or endangered raptors. Still, the question of their predation and its control is virtually absent from the management of martens and fishers, and therefore from this book.

Biological study of martens, sables, and fishers has had several distinct phases. Initially, research was a response to the overexploitation of *Martes* for fur and was driven by simple resource depletion. In the 1930s and 1940s, Soviet scientists launched major efforts to halt and reverse the vast geographical and numerical decline of the sable, while restoring fur production. The earliest North American field studies of martens and fishers, published in the 1940s and early 1950s (Marshall 1946, de Vos 1952), were mostly descriptive but had strong conservation overtones. Marshall's 1951*b* article "Pine Marten as a Forest Product" suggests his prescient concern that forestry consider wildlife as valuable in itself rather than as a constraint on wood production. Ways to achieve sustainable fur production were investigated a decade or two later in North America than in the Soviet Union, and to this emphasis was added concern about major environmental change, especially timber cutting and fire. Reproduction also was investigated on both continents at about the same time, in part because all *Martes* manifest the interesting phenomenon of embryonic diapause (Mead, this volume). One can read into these early reproductive studies the hope that improved knowledge would allow profitable fur farming. Except for the sable, which has an exceptionally valuable pelt, this hope has been largely unrealized because of the lengthy sexual maturation, delayed implantation, and small litter sizes of martens and fishers.

As production of ranch-raised fur has increased and public demand for natural furs has declined (Strickland, this volume), emphasis on managing wild carnivores for fur has diminished in many areas. Yet interest in the ecology of the group has increased, with a shift of emphasis from the utilitarian perspective to a broader conservation concern in which *Martes* are recognized as important components of forest communities, and even as indicators of habitat integrity (Buskirk 1992).

Although this trend toward appreciating *Martes* in a broad ecological context has refreshed our view of the genus, there is still much to learn. Research on the habitat ecology of martens, sables, and fishers, for example, has yet to consider the evolutionary importance of body form in determining communities and habitat associations. Interspecific differences in body size and shape are inversely related to the importance of behavior in structuring communities within families of Carnivora. In families such as Mustelidae that show a wide variation in body shape, there is greater tolerance of closely related species than there is in families such as the Canidae, in which interspecific tolerance is low, especially among related species of similar size (Chirkova 1967, Johnson and Sargeant 1977, Schamel and Tracy 1986). As a result, there is far greater diversity of mustelids than of other Carnivora in continental (Hall 1981) and locally sympatric associations. We also see greater species-level habitat specialization in families, including the Mustelidae, whose members have specialized shapes. Why mustelids radiated in size, shape, and habitat, while other carnivoran families radiated mostly in size, has not been explained. But the result is that mustelids tend to be habitat specialists, especially martens, sables, and fishers.

By and large, *Martes* associate with trees, the strongest exception being the stone marten (Fig. I.2). The forest-dwelling species have traditionally been regarded by scientists (Rosenszweig 1966, Eisenberg 1981:124) and nonscientists alike as arboreal hunters, but this notion was contradicted by the first field biologists to study martens and fishers (e.g., de Vos 1952). The arboreal habits of most *Martes* species have been overemphasized; martens, sables, and fishers generally spend most of their time on the ground and find their food on the forest floor or on snow. This is not to say, though, that they do not need trees; some species at some times rest almost exclusively in tree canopies (Buskirk and Powell, this volume). Several of the following chapters explore the link between *Martes* species and forests. Although these links are indirect and complex, mounting evidence indicates that they are specific, identifiable, and essential.

Conservation of Populations and Habitats

The association of the boreal forest martens and the fishers with late-successional forests has long been recognized; Ernest Thompson-Seton (1925) referred to the American marten's preference for the "glooms of firs" and to their adept use of "the brakes and tangles of this labyrinthine retreat." These mustelids specifically need overhead tree cover and physically complex structure at or near ground level (Buskirk and Powell, this volume). Old-growth forest provides both. In the habitat of American martens, for example, tree canopies provide cover, whereas fallen trees and branches, early seral growth, and the lower branches of living trees all contribute physical complexity at or near ground level. The importance of trees to martens, sables, and fishers lies not in their simple use of branches to hop and hunt along but in a highly complex need for prey and structure, interacting in several temporal and spatial ways.

Since 1980, the welfare of boreal forest martens and fishers has been tied to concerns about the maintenance of biotic diversity in north temperate and boreal coniferous forests. The apparent obligatory relationship between many populations of fishers and boreal forest martens and late-successional forests puts these species in the company of northern spotted owls (*Strix occidentalis caurina*), Sitka black-tailed deer (*Odocoileus hemionus sitkensis*), and red-cockaded woodpeckers (*Picoides borealis*) in the New World (Thomas et al. 1988, Thompson 1991) and of capercaillie (*Tetrao urogallus*) in the Old World (Rolstad and Wegge 1989). Whether we see these species as indicators of ecological conditions or as potentially threatened species, all these vertebrate inhabitants of late-successional forests have low population densities and reproductive rates and have become scarce or extinct in some places. These reductions and extirpations have stimulated heated scientific and public debate (Simberloff 1987, Thompson 1991) about the sustainability of biotas and ecosystems in the face of forestry practices that reduce or eliminate late-successional forests. The policies and practices that result from these debates have implications for sustaining biotic diversity and forested landscapes, on the one hand, and dealing with economic issues on the other. The most important unifying theme of the chapters that follow is the search for clearer understanding of the biological basis and variability, the obligatory nature, and the significance for conservation of the association of *Martes* with natural and managed forests. Improved knowledge of these factors must be translated into conservation actions if we are to have late-successional forest vertebrates, including martens, sables, and fishers, in future biotas.

PART I

Evolution and Biogeography

Introduction

Variation and the dynamics of evolution are obvious throughout this section. In her chapter, Anderson clarifies the evolution and prehistoric distributions within the genus *Martes*. Fossils are rare because martens have always been forest dwellers and small. Nonetheless, we know that *Martes* first appeared in the Miocene, shortly after the Mustelidae appeared. Early martens were distinct from numerous martenlike mustelids and diverged by the Pliocene into the three subgenera distinguished in living species: *Martes*, *Pekania*, and *Charronia*. The dynamics of evolution is seen today in the active divergence of the four sibling species known as boreal forest martens.

Anderson, Graham and Graham, and Gibilisco show the dynamic distributions of *Martes* from the Tertiary through the present. During the late Pleistocene in North America, fishers were restricted to the east, American martens were found across the continent, and noble martens were restricted to the west. Many communities containing these *Martes* species included species that no longer co-occur with *Martes*. As the postglacial climate warmed, *Martes* populations shifted to higher latitudes and altitudes, leading to present distributions and animal associations, which have not existed for long. Fishers remained restricted to eastern North America until the late Holocene, when noble martens became extinct, suggesting competitive exclusion. From data on past animal communities and habitats, Graham and Graham conclude that habitat selection limits the ranges of fishers and martens more than availability of prey.

American martens and fishers reached their nadir early in this century owing to overexploitation for fur and to habitat loss. Their ranges shrank to an all-time low. Both species have since recovered some of those ranges through protection, reintroduction, and habitat recovery, especially reforestation. Nonetheless, fisher distributions and population status in western North America are currently uncertain. American marten populations, naturally isolated in small, forested mountain ranges in the Rocky Mountains, have become even more isolated. Human land use prevents recolonization of some ranges. Gibilisco concludes that we can no longer "connect the dots" to outline species' ranges; we must look between the dots. As human populations grow, they change land use, fragment forests, and alter the habitat for martens and fishers.

Holmes and Powell and Nagorsen show that spatial and temporal variation in body size helps to explain the large sexual dimorphism in body size among martens and fishers. Males and females are isometrically the same except that the skulls, especially the teeth and jaws, of the two sexes are more similar in size than the postcranial skeleton. Selective pressures have kept food-catching structures similar in size in males and females. Sexual dimorphism in teeth, jaws, skulls, and postcranial bones does not vary with carnivoran sympatry; the diets of the two sexes do not diverge or converge depending on competition with other predators. Holmes and Powell conclude that niche differentiation has not been a force selecting for sexual dimorphism.

American martens in the Pacific Northwest exhibit considerable variation in body size but less sexual dimorphism. That patterns in body size do not match the patterns in sexual dimorphism indicates two different selective forces are at work. Presence or absence of slightly larger competitors best explains the variation in body size. Variation in male body size causes most of the variation in sexual dimorphism, and male body size appears to be best explained by abundance of food during growth.

ROGER A. POWELL

1   Evolution, Prehistoric Distribution, and Systematics of *Martes*

Elaine Anderson

The Holarctic and Oriental genus *Martes* includes the true martens, fishers, and the yellow-throated martens (Table 1.1). Formerly included in the genus *Mustela*, they are distinguished from that taxon in having four premolars in each jaw and a small but well-developed metaconid and a basined talonid on m1, the lower carnassial. *Mustela* has three premolars in both jaws, and on the m1 the metaconid is absent and the talonid trenchant. In addition, *Martes* has a longer face, an expanded inner lobe on the upper molar, a short auditory meatus, and distinct paraoccipital processes. Sexual dimorphism is pronounced (Holmes and Powell, this volume).

The geologic range of *Martes* extends back to the early Miocene (ca. 19 million years ago [mya]), a time when the Mustelidae make their first appearance. Tertiary martens are rare; their small body size and forested habitat were not conducive to fossil preservation. Many species have been described (Table 1.2) but most were not martens and only a few were ancestral to living species.

Evolution of Fishers

Two extinct species of fishers, *M. paleosinensis* and *M. anderssoni*, were described from Pontian (early Pliocene, ca. 4.8 mya) deposits in Shanxi and Gansu provinces, China. Both were about the size of a modern female fisher (*M. pennanti*). Characteristic is the presence of an external median rootlet on the upper carnassial (P4) and a shorter trigonid on the m1. Both species are rare, known from fewer than 20 specimens. The type skull of *M. anderssoni* has a well-developed sagittal crest and pronounced postorbital constriction,

13

Table 1.1.   Phylogenetic classification of *Martes*

Order Carnivora Bowditch, 1821
  Suborder Caniformia Kretzoi, 1945
    Superfamily Canoidea Fisher de Waldheim (Families Canidae, Procyonidae,
    Mustelidae, Phocidae)
      Family Mustelidae Fisher de Waldheim, 1817 (otters, badgers, skunks, weasels)
        Subfamily Mustelinae Fisher de Waldheim, 1817 (weasels, mink, ferrets,
        polecats, grison, wolverine, martens)
          Genus *Martes* Pinel 1792
            Subgenus *Martes* Pinel 1792 (true martens)
            Subgenus *Pekania* Gray 1865 (fishers)
            Subgenus *Charronia* Gray 1865 (yellow-throated martens)

Source: After Wozencraft 1989.

characters indicative of old age. Since specimens of both species are of the same geologic age and come from the same region, it is likely that only one species is represented. A larger sample would probably show that the supposed specific differences are due to age, sex, and individual variation.

The Chinese Pontian fishers are on the line leading to the American fishers, the extinct *M. diluviana* and the extant *M. pennanti*. Beginning with early Pleistocene (1.8 mya) deposits, fishers are no longer present in the Chinese faunas, but they appear in Irvingtonian faunas (early middle Pleistocene) in the United States. *M. diluviana* was described from three mandibular fragments and an isolated lower carnassial found in Port Kennedy Cave, Montgomery County, Pennsylvania (Cope 1899). The species was intermediate in size between *M. pennanti* and *M. americana* and had a better developed metaconid on m1, a large p4, and no posterior cusp on p3.

In 1933 J. W. Gidley and C. L. Gazin, paleontologists at the Smithsonian Institution, Washington, D.C., recognized a large *Martes* in the fauna of Cumberland Cave, Allegheny County, Maryland, and named it *M. parapennanti*. According to Gidley and Gazin (1938), the species, represented by 12 specimens including a partial skull, differed from *M. diluviana* in having a more robust jaw with teeth of comparable size and a smaller metaconid on m1. The upper carnassial (P4) was relatively shorter and the protocone was situated more anteriorly than in *M. pennanti*. Both species had an external median rootlet on P4, which is diagnostic of fishers.

In his study of Pleistocene Mustelidae, E. R. Hall (1936) determined that the characteristics separating *M. diluviana* and *M. parapennanti* were due to age, sex, and individual variation, and he recognized only *M. diluviana* as the early middle Pleistocene species. He also tentatively referred specimens from Conard Fissure, Newton County, Arkansas to *M. diluviana* (previously

Table 1.2. Tertiary *Martes*

Species (ref.)	Description	Country	Age	Comments	Reference
M. palaeosinensis (Zdansky 1924)	Size of *M. pennanti*, relatively short trigonid on m1; external median rootlet on P4	China	Pontian	Ancestral to *M. pennanti*	Zdansky 1924
M. anderssoni (Schlosser 1924)	Size, skull, and tooth morphology similar to *M. palaeosinensis*	China	Pontian	Probably conspecific with *M. palaeosinensis*	Anderson 1970
M. wenzensis (Stach 1959)	Intermediate in size between *M. palaeosinensis* and *M. martes*; no external median rootlet on P4	Poland	Early Villafranchian	True marten	Stach 1959
M. pentelici (Gaudry 1861)	Large size	Greece	Pontian		
M. woodwardi (Pilgrim 1931)	Large size	Greece	Pontian		
M. leporinum (Khomenko 1914)	Large size	Romania	?Pontian		
M. campestris (Gregory 1942)	Larger than *M. pennanti*	USA (S.D.)	Early Pliocene	Known from one jaw; believed to be related to *Sthenictis*, a Miocene lutrine	
M. kinseyi (Gidley 1927)	Slightly larger than *M. americana*; m1 long and straight with narrow talonid	USA (Mont.)	Miocene	Known from one jaw	
M. gazini (Hall 1931)	Size of *Mustela nigripes*; m1 has large metaconid and trenchant talonid	USA (Oreg.)	Late Tertiary	Not a marten	Hall 1931
M. laevidens (Dehm 1950)	Smaller than female *M. americana*; four premolars; m1 has semi-basined talonid	Germany	Early Miocene	Small true marten; probably on ancestral line	Anderson 1970

(continued)

Table 1.2. (Continued)

Species (ref.)	Description	Country	Age	Comments	Reference
M. stirtoni (Wilson 1968)	Martenlike mustelid	USA (Kans.)	Early Pliocene		Wilson 1968
M. pachygnatha (Teilhard & Piveteau 1930)	Short, heavy jaw; three premolars	China	Villafranchian	Not a marten	Teilhard & Leroy 1945
M. crassa (Teilhard & Leroy 1945)	Size of *M. palaeosinensis*; no p1; trenchant talonid on m1; folded enamel on canine	China	Middle Pliocene	Not a marten	Teilhard & Leroy 1945
M. zdanskyi (Teilhard & Leroy 1945)	Similar to living martens of the area	China	Villafranchian	Perhaps an early yellow-throated marten	Teilhard & Leroy 1945

identified as *M. pennanti* by Brown [1908]), noting that the fauna is the same age as that of Cumberland Cave and that Arkansas was far south of the extant fisher's range.

No other specimens of *M. diluviana* were reported until 1969–1970, when a crew from the Carnegie Museum, Pittsburgh, Pennsylvania, recovered some additional material from a reopened quarry in Cumberland Cave. These specimens showed that pronounced sexual dimorphism was present in *M. diluviana*. Then, in 1981, fossils were discovered in the Cheetah Room in Hamilton Cave, Pendleton County, West Virginia. Among the remains of small carnivores were two left lower carnassials (m1) of *M. diluviana*. Up until 1987 this extinct fisher had been found only in the eastern United States. During that summer, Darlene Emry, an amateur paleontologist, found a left mandible with p4–m1 in the Badger Room of Porcupine Cave, Park County, Colorado, at an elevation of 2900 m. This was the first western record of the species. Table 1.3 presents measurements of p4 and m1 of the specimens from Cumberland, Hamilton, and Porcupine caves

As of 1993, *M. diluviana* is known from about 25 specimens from five states. The age of these faunas is Irvingtonian (early middle Pleistocene). Repenning and Grady (1988) believed that the deposits in the Cheetah Room of Hamilton Cave were probably the oldest cave deposits in the eastern United States. The age of this fauna is between 820 and 850 thousand years ago (ka), based on the evolution of the microtine rodents found there. The other three eastern faunas have not been dated, but Port Kennedy has been considered to be Aftonian–early Kansan in age, Cumberland Cave has been considered pre-Illinoian, and Conard Fissure is considered Kansan–early Illinoian—in other words, middle Irvingtonian. Porcupine Cave, sealed until miners blasted it open in the late 1800s, is still being excavated and the fauna studied, but studies of the microtines from the Pit (Barnosky and Rasmussen 1988) indicate that the fauna is older than 450 ka.

The associated fauna at the five sites include coyotes and wolves (*Canis armbrusteri*, *C. priscolatrans*, *C. latrans*), bears (*Ursus americanus*, *Arctodus pristinus*), weasels (*Mustela frenata*, *M. erminea*), wolverines (*Gulo schlosseri*), river otters (*Lutra canadensis*), American badgers (*Taxidea taxus*), skunks (*Brachyprotoma obtusata*, *Mephitis mephitis*, *Spilogale putorius*), cheetahs (*Miracinonyx inexpectata*), jaguars (*Panthera onca*), and many species of rodents, lagomorphs, and large herbivores. The porcupine (*Erethizon dorsatum*) has been found at all five sites, leading to speculation that the extinct fisher may also have preyed on porcupines.

M. diluviana was probably close to the ancestry of *M. pennanti*. There are no other records of fishers until deposits of about 30 ka, which contain the

Table 1.3.   Measurements (mm) of lower p4 and m1 (carnassial) of *Martes palaeosinensis,*
M. anderssoni, M. diluviana, and *M. pennanti*

	p4		m1	m1 trigonid	
	Length	Width	Length	Length	Width
Martes palaeosinensis					
Uppsala Univ.					
3794	6.5	3.1	12.0	8.2	4.8
3795	6.9	3.6	12.0	8.3	4.6
M8	5.6	2.2	10.4	6.8	3.4
M11	6.0	3.2	11.5	7.5	4.3
M12	6.4	3.2	11.3	7.7	4.3
M14	—	—	12.6	8.3	4.3
M. anderssoni					
AMNH 50519	6.5	3.2	11.6	8.0	4.2
M. diluviana					
Cumberland Cave					
USNM 11877	7.2	3.3	12.5	8.2	4.5
11878	6.0	2.5	—	—	—
8129	7.0	3.0	11.4	7.9	4.3
12352	5.6	2.8	11.3	8.5	4.5
11876	5.7	2.5	10.1	7.0	3.5
12336	—	—	10.1	7.1	4.0
Hamilton Cave					
USNM No # 0-25 cm	—	—	10.7	7.1	4.0
No # 0-50 cm	—	—	10.8	7.5	4.5
Porcupine Cave					
CM 49108	5.8	2.7	10.9	7.4	4.1
M. pennanti[a]					
Recent					
Male			*N* = 26		
	8.2–8.9	3.5–3.8	11.6–14.0	8.6–10.1	4.5–5.3
Female			*N* = 39		
	6.8–7.3	3.0	10.4–12.0	7.6–9.1	4.1–4.5

Notes: AMNH = American Museum of Natural History, USNM = U.S. National Museum,
CM = The Carnegie Museum of Natural History. *N* = number of specimens. — = measurement not available.

   [a]p4 measurements from Anderson, unpubl. data. m1 measurements from Anderson 1970.

extant species. The oldest known fauna containing *M. pennanti* is Strait
Canyon, Highland County, Virginia, which is dated at 29,870 years bp
(Eshelman and Grady 1986). In the late Pleistocene it ranged as far south as
Georgia, Alabama, and Arkansas, throughout the Appalachian Mountains,
and west to Ohio and Missouri. It has not been identified in any western
faunas (Graham and Graham, this volume).

Evolution of Martens

The earliest known marten is *M. laevidens*, from lower Miocene beds in Germany. It was a small species with slender jaws and four premolars in each jaw half. The inner lobe of the upper molar is not greatly expanded, and the talonid of the lower carnassial is semibasined.

Then a long gap occurs in the fossil record. Not until the early Villafranchian (4 mya) are true martens again recognized. *M. wenzensis*, an animal the size of *M. paleosinensis* but morphologically similar to *M. martes* (Eurasian pine marten), was found in Poland. It had four premolars in each jaw, an expanded inner lobe on M1, no external median rootlet on P4, and a flat talonid on m1. Other large martens have been described from deposits of similar age in Greece and Romania. *M. wenzensis* may have been ancestral to *M. vetus* Kretzoi, an early middle Pleistocene species known from several cave deposits in Germany. Although the same size as *M. martes*, *M. wenzensis* showed characteristics of both the stone marten (*M. foina*) (proportions of P4 and M1, and jaw depth) and *M. martes* (short trigonid on m1, length of P4 equal to the width of M1 and mental foramina far apart). The associated fauna indicates that *M. vetus* was a forest dweller with habits similar to *M. martes*. It was ancestral to the *M. martes* line and perhaps to *M. foina* as well.

The fossil record of *M. foina* is incomplete. *M. vetus*, the early middle Pleistocene species, may have been ancestral inasmuch as specimens of both these species show several similarities in dentition. The stone marten is first recognized in middle Würm-age deposits in Lebanon and Israel and from the late Würm in northern Iraq and the Caucasus. Reports of its presence in western Europe during the last glacial are in error; these specimens belong to *M. martes* and in one case to *Mustela putorius*. In early Holocene times, *M. foina* is reported from the Middle East, Hungary, the Russian plains, the Crimea, and central Asia. It does not become a member of the southern and central European fauna until the middle–late Holocene, and it has never been a member of the British fauna.

The earliest record of *M. martes* (ca. 100 ka) is from the Riss-Würm interglacial travertine and cave deposits in central Europe, and it has been found in many late Pleistocene sites in central and western Europe. Found in association with other boreal animals, *M. martes* has been used as a paleontological indicator of coniferous forests and colder conditions (Anderson 1970). By the early Holocene it had spread to Scandinavia, where numerous specimens have been found in peat and Stone Age deposits. A trend of decreasing body size from the late Pleistocene to the Recent is evident.

Measurements of condylobasal length, width across the canines and carnassials, mandibular length and height, and dimensions of the cheek teeth show that specimens from the Würm are the largest, those from early Holocene sites are smaller, and Recent specimens from the same area are the smallest.

The sable (*M. zibellina*) and the Japanese marten (*M. melampus*) have been found in late Pleistocene faunas in Siberia and Japan, respectively. Both are closely related to *M. martes*, and like it, their ancestry goes back to *M. vetus*.

The American marten (*M. americana*) reached North America via Beringia in the early Wisconsinan (65–122 ka) and spread eastward. This population was apparently isolated in eastern North America by the Laurentide ice sheet, and not until the ice retreated was it able to reinvade western Canada and Alaska. A later immigration from Siberia populated the West Coast, the Sierra Nevada, and the Rocky Mountains. These martens show more similarities in cranial and dental characters to *M. zibellina* than to the American martens of the East.

The extinct species *M. nobilis* is now known from 14 faunas of late Pleistocene to perhaps late Holocene age in the western United States, compared with the four faunas Anderson (1970) reported. In 1970 the noble marten was thought to be boreal in its habitat associations, but it has now been reported from two sites in Nevada associated with nonboreal faunas (Grayson 1985); apparently it was adapted to a wider range of ecological conditions (Graham and Graham, this volume). It is a true marten (no external median rootlet on P4), intermediate in size between a female fisher and a male marten. The skull is broader, and P4 has a more robust appearance owing to a narrower protocone and a less distinct indentation between the paracone and the protocone. The trigonid on m1 is relatively short, and on a plot of length of m1 versus length of the trigonid, specimens of *M. nobilis* lie on a different trend axis than do those of *M. americana*, *M. diluviana*, and *M. pennanti* (Anderson 1970:82).

Many species of mammals were larger in the Pleistocene than their extant counterparts, just as northern races of warm-blooded species tend to be larger than races living farther south. *M. martes* shows a trend toward decreasing size from the late Pleistocene to the Recent. But measurements of *M. nobilis* from Holocene sites are as large as those of specimens from the late Pleistocene: there is no trend towards decreasing size. At three sites where *M. americana* and *M. nobilis* occur together, the American marten was the size of the extant members of the species from the same area, whereas the noble marten was much larger than the living species.

In a recent statistical study, Youngman and Schueller (1991) concluded that *M. nobilis* is a large representative of the "caurina" subspecies group rather than a distinct species. Several questions remain unanswered about the authors' conclusions and the validity of the species. Until they are addressed, I propose to call the large late Pleistocene–Holocene "caurina" marten *M. americana nobilis* to distinguish it from the smaller types of the same age (specimens the size of the extant marten from the same area). Subspecies recognition has been accorded to other Pleistocene species that, upon further study, have been shown to be identical, except for larger size, to an extant species. *Panthera atrox*, for example, is now called *P. leo atrox*, and *Ovis catclawensis* is now designated *O. canadensis catclawensis*.

The only fossil marten known from India, *M. lydekkeri*, was found in the Chinji Zone, Lower Siwaliks (Mio-Pliocene age). Colbert (1935) said it was the size of *M. martes* and had a basined talonid and a well-developed metaconid on the m1. It is considered to be on the line that led to *M. flavigula*, the yellow-throated marten. An extinct mid-Pleistocene subspecies, *M. f. tyrannus*, was recognized in the limestone fissure of Yenchinkou, Szechwan, China. It was larger than present-day male yellow-throated martens from the same area, but did not differ morphologically from them (Colbert and Hooijer 1953).

Systematics

Three subgenera of *Martes* are recognized: *Pekania* Gray, the fishers; *Charronia* Gray, the yellow-throated martens; and *Martes* Pinel, the true martens (Table 1.4 and Fig. 1.1).

The subgenus *Pekania* includes the extinct *M. palaeosinensis* (including *M. anderssoni*) and *M. diluviana*, and the extant fisher, *M. pennanti*. Characteristic of the subgenus is large size (*M. pennanti* is the largest living member of the genus) and the presence of an external median rootlet on P4, the upper carnassial.

Some authors (Ognev 1931, Simpson 1945) treated *Charronia* as a distinct genus, but the differences from *Martes* are slight and it is now considered a subgenus (Anderson 1970, Corbet 1980). The skull is similar to that of *M. foina* but is larger. Characteristic of the subgenus is the structure of the distal end of the baculum. It curves abruptly upward and backward to form a hook that is expanded distally and ends in four blunt processes that are quite variable in development. In other *Martes* the slightly expanded distal end of the baculum curves upward and is pierced by a slit or small hole. One, or perhaps two, living species are included in *Charronia*: the yellow-throated marten, *M. flavigula* (Corbet 1980); and perhaps the poorly known Nilgiri

Table 1.4. Subgenera of *Martes*: range, distinctive characters, habitat, and species

	Martes	*Pekania*	*Charronia*
Geologic range Prehistoric range	Early Miocene–Recent Europe, Middle East, Asia, North America	Early Pliocene–Recent China, North America	Middle Pleistocene–Recent India, China
Modern geographic range	Temperate Europe, Asia, North America	North American, from N. British Columbia across Canada S. of Hudson Bay to Nova Scotia, N.W. USA, Sierra Nevada, N. Rocky Mountains	E. Siberia, Manchuria, Korea, China, Taiwan, Tibet, Pakistan, N. India, S. India, Indonesia, Malaysia, Java, Sumatra, Borneo
Distinctive characters	Relatively small size; no external median rootlet on P4; distal end of baculum ends in rounded knob pierced by small slit	Large size; external median rootlet on P4 present; baculum similar to *Martes* but larger distal end spatulate with opening near middle	Medium size; no external median rootlet on P4; distal end of baculum curves sharply upward and backward to form a hook that ends in 4 blunt processes
Habitat	Mature coniferous and mixed forests, broad-leaved forests	Mature coniferous and mixed forests, broad-leaved forests	Temperate to tropical forests
Included species (* = extinct)	*M. laevidens, *M. wenzensis, *M. vetus, *M. nobilis, M. martes, M. zibellina, M. melampus, M. americana, M. foina*	*M. palaeosinensis including *M. anderssoni, *M. diluviana, M. pennanti*	*M. lydekkeri, M. flavigula, M. gwatkinsi*

	Subgenus Charronia	Subgenus Martes	Subgenus Pekania

Recent: M. flavigula   M. americana   M. melampus   M. zibellina   M. martes   M. foina   M. pennanti

Early Holocene: M. americana   M. martes   M. foina   M. pennanti

Late Pleistocene: M. americana   M. martes   M. foina   M. pennanti

M. melampus

M. zibellina

M. flavigula tyrannus   M. martes   M. diluviana

M. vetus

Early Pleistocene: M. wenzensis

Early Pliocene: M. lydekkeri   M. palaeosinensis

Miocene: ?   M. laevidens   ?

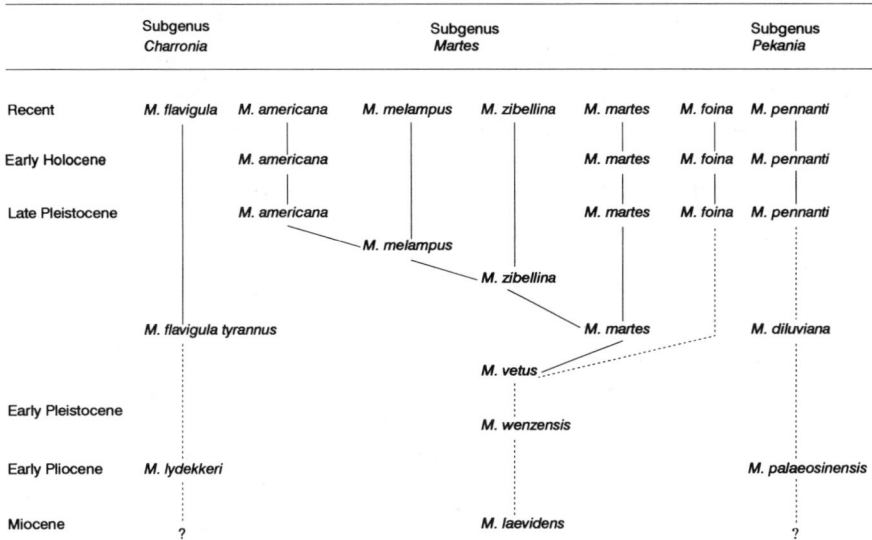

Figure 1.1. Phylogeny of *Martes*.

marten, *M. gwatkinsi* (Anderson 1970; Prasad, Geol. Surv. of India, pers. commun.). The latter is found only in the Nilgiri Hills and Tranvancore, southern India, completely isolated from other populations of *M. flavigula*. The few specimens available for study show several apparent cranial differences, but whether these are due to age, sex, geographic, or individual variation cannot be determined at this time. An analogous situation exists with the tahr (*Hemitragus jemilahicus*), which is found in the Himalayas, and *H. hylorius*, which has a restricted range in the hills of southern India. Perhaps these examples of discontinuous distribution resulted from populations shifting their range southward during the Pleistocene. When the climate ameliorated, many species returned to their former range, but those species that had reached the southern Indian hills persisted and became isolated from the main, northern populations.

True martens, subgenus *Martes*, are distinguished from the other two subgenera by their relatively small size and by the absence of an external median rootlet on P4. The extinct *M. laevidens*, *M. wenzensis*, *M. vetus*, and *M. nobilis* and the extant *M. martes*, *M. zibellina*, *M. melampus*, *M. americana*, and *M. foina* are true martens. They first appeared in Europe; later they spread across temperate Asia and into North America.

The four Holarctic species—*M. martes*, *M. zibellina*, *M. melampus*, and *M. americana*—replace each other geographically and have a combined range extending from Britain across Eurasia to eastern North America in mature coniferous and mixed forests. They show numerous similarities in morphology, habits, and habitat, and many authors have noted the close relationship between them. Of the four, *M. martes* is the largest and *M. americana* the smallest. All have long, relatively narrow skulls, long auditory bullae that are set rather close together, a well-developed protocone on P4, an expanded inner lobe on M1, and a relatively short trigonid on m1. For most characteristics there is a cline running from west to east: the characters are more pronounced in *M. martes*, intermediate in *M. zibellina* and *M. melampus*, and least pronounced in *M. americana* (Anderson 1970).

Hagmeier (1961:133) suggested that "on grounds of similarity of cranial characteristics, allopatricity of range and similarity of general ecological attributes, there is reason to suspect that *M. americana*, *M. martes*, *M. zibellina* and possibly *M. melampus* may one day be shown to be members of a single circumboreal species." Additional studies (Anderson 1970) show a close relationship among these four species, and they can probably be considered a "superspecies," defined by Mayr (1963:499) as "a monophyletic group of entirely allopatric species that are morphologically too different to be included in a single species."

Many subspecies of *Martes* have been named chiefly on the basis of size and coat color. Thus, subspecies were described as larger or smaller (Nagorsen, this volume), lighter or darker, with a small or extensive throat patch. Individual variation was seldom considered. Corbet (1980) listed the currently recognized subspecies of *M. foina*, *M. martes*, *M. zibellina*, *M. melampus*, and Palearctic *M. flavigula*.

Hall (1981) recognized three subspecies of *M. pennanti*: *M. p. pennanti*, found from southern Manitoba south of Hudson Bay to Nova Scotia, around the Great Lakes region, and from New England south to North Carolina; *M. p. columbiana*, found along the west coast from British Columbia east to northern Manitoba; and *M. p. pacifica*, found along the west coast from British Columbia to central California and east to the Sierra Nevada (Gibilisco, this volume).

Hall (1981) continued to recognize 14 subspecies of *M. americana*. On the other hand, Hagmeier (1958) believed that at most 6 subspecies could be distinguished. He noted that "the partitioning of the species into subspecies was completely arbitrary because most of the variation was clinal and because at least some of the characters varied discordantly." I agree with his view. American martens have been divided into 2 subspecies groups, "americana" and "caurina," distinguished from each other by the shape of the skull and auditory bullae and the relative size of the upper molar (Table 1.5). The "caurina" group shows more similarities to *M. zibellina* than to the "americana" group and was a later immigrant into North America (Anderson 1970).

Table 1.5. Subspecies groups of *Martes americana*: characters, range, and subspecies

	"americana"	"caurina"
Skull	Relatively high and narrow	Relatively low and broad
Auditory bullae	Long and narrow	Short and broad
Upper molar	Relatively small	Relatively large
Geographic range	Alaska, central Canada to Nova Scotia and Newfoundland, New England, Great Lakes region	West Coast of Canada, Sierra Nevada, Rocky Mountains
Included subspecies (Hall 1981)	*M. americana abieticola, abietinoides, actuosa, americana, atrata, brumalis, kenaiensis*	*M. americana caurina, humboldtensis, nesophila, origenes, sierrae, vancouverensis, vulpina*
Included subspecies (Hagmeier 1961)	*M. americana americana, brumalis, actuosa*	*M. americana caurina, humboldtensis, nesophila*

2 Late Quaternary Distribution of *Martes* in North America

Russell W. Graham and Mary Ann Graham

The North American species of *Martes*, fishers (*M. pennanti*) and American martens (*M. americana*), occupy a broad region in the middle latitudes of Canada (Gibilisco, this volume). Their range generally correlates with the boreal spruce forest. Both species extend southward in the boreomontane environments of the western United States, and American martens range farther south to northern New Mexico. Historically, both fishers (Powell 1982) and American martens (Clark et al. 1987) extended into the northeastern United States. These areas generally were covered with broad-leaved deciduous or mixed deciduous—evergreen forests. Another species, *M. nobilis*, became extinct sometime during the late Holocene, which began 4000 years ago (4 ka).

These distributional patterns are, in part, the result of complex environmental changes throughout the late Quaternary. Human impact, both historic and prehistoric, on *Martes* populations and habitats have also contributed to the shape of the modern ranges (Gibilisco, this volume). To fully understand the modern distributions of *Martes* species and their habitat preferences, we need to appreciate the history of their distribution during the late Quaternary (last 40 ka) and the possible environmental factors that influenced them.

Late Quaternary climates and environments have fluctuated rapidly and frequently. During glacial times in the Pleistocene, extensive continental ice sheets covered most of the area occupied by the modern boreomontane biota of North America (Fig. 2.1). With the onset of the last major glaciation (the Wisconsinan), glacial ice reached its maximum extent about 20 ka and physically displaced many boreomontane species. The cooler climate allowed these species to move to lower latitudes and altitudes. This trend was reversed beginning 14 ka when climates began to warm, glacial ice retreated

26

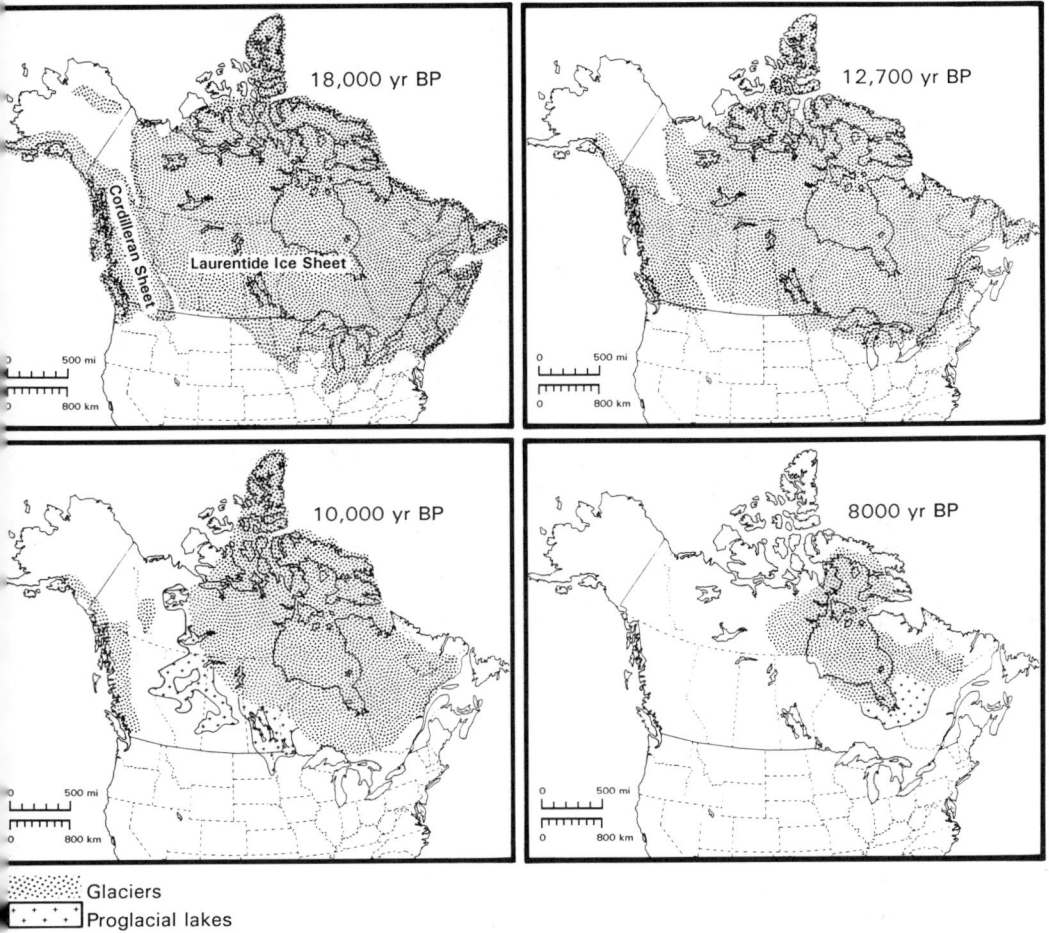

Glaciers
Proglacial lakes

Figure 2.1. Changing distribution of glacial ice and proglacial lakes in North America during the late Wisconsin. Adapted from Ritchie 1987.

(Fig. 2.1), and boreomontane species moved to higher latitudes and altitudes.

Paleobiological data suggest that range adjustments of plants and animals were species-specific rather than shifts of complete communities or ecosystems (Graham and Grimm 1990). Individual species responded to environmental changes at different rates and times and migrated in different directions. These reorganizations resulted in the formation of species associations that do not have modern analogs (Lundelius et al. 1983, Graham 1986, Graham and Grimm 1990). Even though *M. americana* and *M. pennanti* show a strong correlation in their modern distributions, they may not have done so in the past.

Evidence for the past distribution of *Martes* comes from bones and teeth found in paleontological and archeological sites. Clearly, taphonomy (the mechanisms governing the accumulation and preservation of fossil specimens) is critical in evaluating whether these specimens were local or came from some distance. Paleontological samples from cave sites, for example, generally reflect local conditions because they result from the activities of animals, primarily raptors, mammalian predators, and packrats, that do not roam far from caves. In some cases, pit caves trap animals that stumble into them, a very local sample indeed. Fossils in alluvial deposits, however, may represent either local accumulations or long-distance transport by streams, and samples from archeological sites may reflect long-distance transport by humans.

The distributions of *Martes* detailed here were generated by a geographic information system (GIS) from a computerized database of paleontological and archeological sites; localities and references are provided in Table 2.1, which starts on page 50. We did not search exhaustively for archeological site reports for the historical period, and so our maps (Figs. 2.8–2.9) for this time period may underrepresent such sites. We are confident, however, that we have not missed records that would significantly extend the historical range. The database included the species of *Martes* found at the site, site location, age of the deposits, methods of accumulation, and cultural associations. This collection of data allows comparison of sites of similar ages and taphonomic settings. Because of inadequate information, several sites in Table 2.1 are not shown on the maps.

Late Pleistocene Distribution of *Martes*

Martes americana

During the late Pleistocene *M. americana*, like many boreomontane species, extended to lower elevations and latitudes than it does today (Fig. 2.2). In the eastern United States, *M. americana* occurred as far south as northern Alabama, and many sites have been recorded from eastern and central Tennessee. In Cheek Bend Cave, Maury County, Tennessee (Klippel and Parmalee 1982) and Baker Bluff Cave, Sullivan County, Tennessee (Guilday et al. 1978) marten remains date to the full-glacial (ca. 17–20 ka), a time of maximum cold. American martens have been found as far south as Bell Cave, Colbert County, Alabama, in levels that presumably date to 11.8 ka (E. Anderson, Denver Mus. of Nat. Hist. pers. commun.).

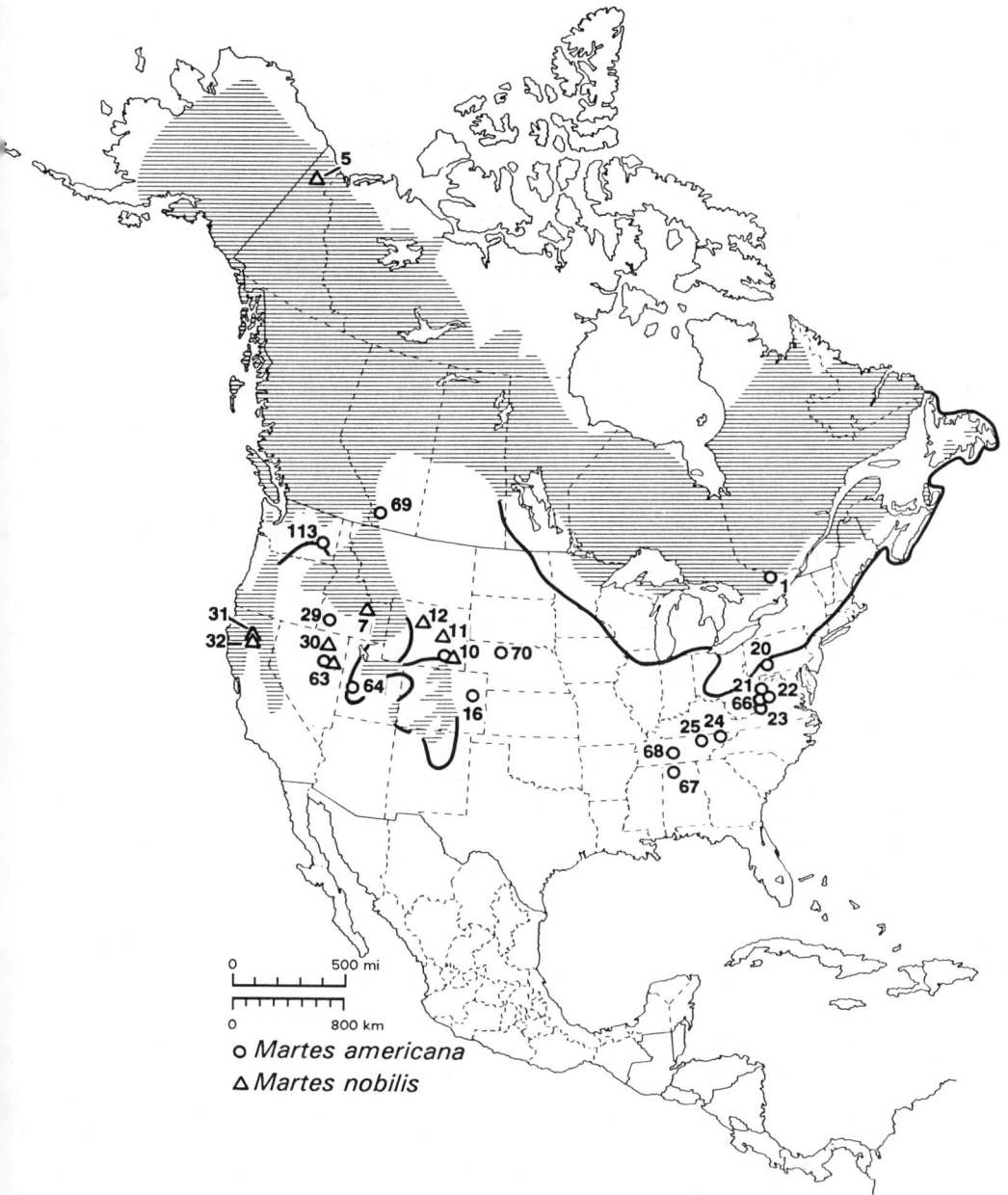

igure 2.2. Late Pleistocene localities of *Martes americana* and *M. nobilis*. Modern (*hatched area*) and istoric (*solid line*) distribution of *M. americana* are after Hall (1980) and Clark et al. (1980), respectively. lames of numbered sites and ages of fossils are in Table 2.1.

In all the eastern sites, *M. americana* is associated with other boreomontane taxa such as the yellow-cheeked vole (*Microtus xanthognathus*), southern red-backed vole (*Clethrionomys gapperi*), bog lemming (*Synaptomys* spp.), and frequently the heather vole (*Phenacomys intermedius*). These sites also contain the least shrew (*Cryptotis parva*), short-tailed shrew (*Blarina* spp.), and eastern chipmunk (*Tamias striatus*), which today occur primarily in deciduous forest. It is also not uncommon to find grassland species such as pocket gophers (*Geomys* spp.) and thirteen-lined ground squirrels (*Spermophilus tridecemlineatus*) associated with the boreomontane and deciduous species.

M. americana has not yet been recovered from sites in the central and upper Midwest nor from the central south; however, this marten did occur in late Pleistocene environments along the western plains. In Nebraska, *M. americana* has been recovered from a stratified site, Smith Springs, that dates to the late Wisconsinan (Voorhies 1990). Other boreomontane species from this site include montane vole (*Microtus montanus*), yellow-cheeked vole, and wolverine (*Gulo gulo*). As elsewhere in North America, these boreomontane species are intermixed with grassland and deciduous forest species. In contrast to southeastern sites, those in the western plains have a predominance of grassland rather than deciduous forest taxa.

In Colorado, *M. americana* has been tentatively identified from a Hell Gap (ca. 10 ka) cultural level at the Jones-Miller site, Yuma County (V. Rawn-Schatzinger, Bartelsville, Okla., pers. commun.), again associated with boreomontane and grassland species. Only a few species at this site could be characterized as having deciduous forest affinities, however, *M. americana* has also been found at the Chimney Rock Animal Trap in Larimer County, Colorado, but its age is uncertain, although assumed to be late Pleistocene (Hager 1972).

American martens occurred in southeastern Wyoming at the Bell Cave site, Albany County, around 12 ka (Zeimens and Walker 1974), in cave deposits not associated with any cultural materials. Other boreomontane species from the site include water shrew (*Sorex palustris*), pika (*Ochotona princeps*), and heather vole (Walker 1987).

In the western United States, *M. americana* is known from eastern Washington, eastern Nevada, and western Utah, which may have had American martens in historic times (Fig. 2.2). Only one site, Wilson Butte Cave in southwestern Idaho (Gruhn 1961), is outside the historic range of *M. americana*. The marten remains are from strata that date from between 9.95 and 14.5 ka (Lundelius et al. 1983). *Martes* remains have not yet been found in the southern Rocky Mountains (Arizona, New Mexico, and trans-Pecos Texas) or the southern plains (Kansas, Oklahoma, and northern and central Texas), even though these areas have abundant fossil-bearing sites.

Martes nobilis

M. nobilis, the extinct noble marten, is distinguished from *M. americana* by its larger size and morphological differences in the teeth and skull (Hall 1926; Anderson 1970; Kurtén and Anderson 1980; Anderson, this volume). Youngman and Schueler (1991) recently suggested that large size may not be a valid character for this taxon. Fluctuations in sexual dimorphism as a response to food abundance (Holmes and Powell, this volume) also suggest that size may be highly variable and may not be a good taxonomic character for *Martes*. As Nagorsen (this volume) indicates, however, size variation in modern *M. americana* in the Pacific Northwest is poorly understood, and aberrant geographic patterns may reflect phylogenetic relationships to some extent. For the purposes of this chapter, we have retained the name *M. nobilis* because of the morphological information that it conveys, but we realize that it may be relegated to lower taxonomic status later (Anderson, this volume).

The noble marten was widespread in the central Rocky Mountains and extended into northern California during the late Pleistocene (Fig. 2.2). Fossil specimens are especially abundant in Wyoming and northern Colorado. Generally, *M. nobilis* is found in Pleistocene deposits with such boreomontane taxa as pika, heather vole, southern red-backed vole, and water shrew (Anderson 1970).

M. nobilis has been found with *M. americana* in three sites; Chimney Rock Animal Trap, Colorado; Bell Cave, Wyoming; and Snake Creek Burial Cave, Nevada. The faunal remains from Chimney Rock are disturbed and mixed together (Hager 1972), so it is not possible to determine whether the two taxa were contemporaneous. The remains of *M. nobilis* and *M. americana* from Bell Cave (Walker 1987) and Snake Creek Burial Cave (Mead and Mead 1989) suggest, however, that these two species were sympatric and contemporaneous. In both sites, *Martes* remains are from noncultural deposits.

Martes pennanti

Like American martens, fishers (*M. pennanti*) existed at lower altitudes and latitudes during the late Pleistocene than they do today. But, unlike American martens, which were widespread over the northern United States, fishers have been found only in the eastern half of the country. They extended southward along the Appalachian Mountains into northern Alabama and northern Georgia (Fig. 2.3). Fishers are known from deposits in Bell Cave, Colbert County, northwestern Alabama, that date to 11.8 ka. They are associated with *M. americana* and numerous other boreomontane taxa in these deposits.

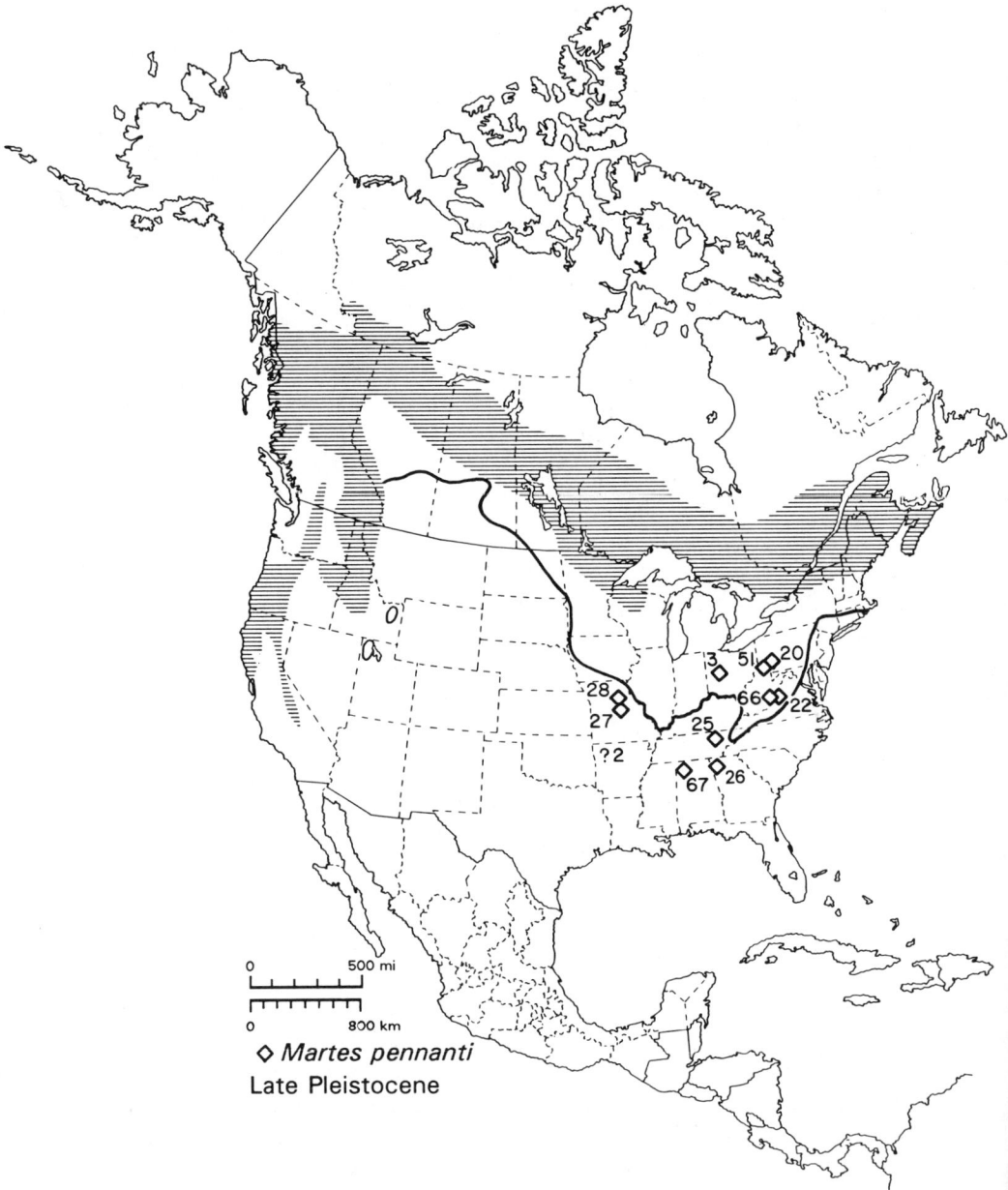

Figure 2.3. Late Pleistocene localities of *Martes pennanti*. Modern (*hatched area*) and historic (*solid line*) distribution of *M. pennanti* are after Hall (1981) and Powell (1982), respectively. Names of numbered sites and ages of fossils are in Table 2.1.

M. pennanti has also been recovered from Pleistocene deposits at Ladds, Bartow County, Georgia (Ray 1967). The Ladds limestone quarry contains many fossil-bearing fissures of varying ages (Ray 1967). Some of the fissure deposits have been dated between 10 and 11 ka (Holman 1985*a*), and Holman (1985*b*) believed that many of the Ladds localities are contemporaneous. Some taxa such as *Peromyscus cumberlandensis* and *Miracinonyx studeri*, however, represent an age of early Pleistocene land mammals (Irvingtonian). Unfortunately, neither the exact provenience nor the age (other than Pleistocene) of the *M. pennanti* specimens from Ladds can be determined. The Ladds local fauna contains several other boreomontane species (*Sorex cinereus*, *Sorex fumeus*, and *Zapus hudsonius*) that occur at higher elevations in Georgia today.

M. pennanti has also been found in late Pleistocene deposits in Robinson Cave, Overton County, north-central Tennessee (Guilday et al. 1969). Five extinct mammalian species are known from this cave, and of the 52 extant mammalian taxa, 91% occur in the Minnesota-Wisconsin area today (Kurtén and Anderson 1980). Most of the faunal remains from Robinson Cave accumulated as the result of a sinkhole trap or as the prey of predators that inhabited the cave.

Fisher remains have been recorded for two caves in the Ozark region of central and southern Missouri. Brynjulfson Cave No. 1, Boone County, contains a late Pleistocene assemblage of extinct and extant boreomontane taxa (Parmalee and Oesch 1972). Bat Cave, Pulaski County (Hawksley et al. 1973), contains a similar late Pleistocene fauna that reflects cooler and moister conditions than occur there today. *M. pennanti* has also been identified from Peccary Cave, Newton County, Arkansas (Kurtén and Anderson 1980). Peccary Cave contains a complex stratigraphic sequence that dates from the late Pleistocene (ca. 16–17 ka) to the late Holocene (ca. 2 ka) (Semken 1984, Stafford and Semken 1990). Unfortunately, the exact stratigraphic provenience of the *M. pennanti* specimen is not known, but we believe it reasonable to assume that it is from the late Pleistocene. This assumption is strongly supported by the similarity of other late Pleistocene boreomontane species from Brynjulfson Cave No. 1, Bat Cave, and Peccary Cave. *M. americana* has not yet been identified from any of the numerous fossil-bearing sites in the Ozark region.

Holocene Distribution of *Martes*

Martes americana

Only three localities for *M. americana* can be assigned to the early Holocene (8–10 ka). One occurs in peat deposits of an abandoned channel of the

Mississippi River in Whiteside County, northwestern Illinois (Graham and Graham 1990). This site is located at the margin of the known historic distribution of *M. americana* (Fig. 2.4), as mapped by Clark et al. (1987), and provides firm evidence of *M. americana* in this area in the early Holocene.

Another early Holocene record comes from Deer Creek Cave, Elko County, north-central Nevada (Ziegler 1963). This site is well outside the current and historic range of *M. americana* (Fig. 2.4), but the marten remains in this cave are associated with human artifacts. Ziegler (1963) made this identification before *M. nobilis* was recognized as a distinct taxon (Anderson 1970); therefore, it may be worth reexamining this material. *M. americana* is also known from early Holocene deposits (7–10.6 ka) at Baker Bluff Cave, Sullivan County, Tennessee (Guilday et al. 1978).

All three middle Holocene (4–8 ka) records of *M. americana* are within its historic range (Fig. 2.4). Raddatz Rockshelter, Sauk County, Wisconsin, contains a stratified faunal and cultural sequence that ranges from the early to late Holocene (Cleland 1966). *M. americana* has also been recovered from archeological deposits at Frontenac Island, Cayuga County, New York (Cleland 1966), and is associated with middle Archaic (middle Holocene) human artifacts at both sites. *M. americana* is also in middle Holocene cultural deposits at the Chief Joseph Dam site in Okanogan County, Washington (Livingston 1984).

Fifteen localities, excluding historic sites, document the late Holocene distribution of *M. americana*. Thirteen of these sites are within the historic range of the species, but three sites (Pabst, Dewitt County, Illinois; Phipps, Cherokee County, Iowa; and Harder, southern Saskatchewan) are well outside the range (Fig. 2.5). At the Pabst site both *M. americana* and *M. pennanti* are known from deposits that date between 3 and 4.3 ka (Lewis 1979). *M. americana* was tentatively identified on the basis of a second phalanx from the Harder site, which dates to about 3.4 ka (Dyck 1977). At Phipps, this marten has been recorded from 600 to 800-year-old deposits (Semken and Falk 1987). The marten remains from all three sites are associated with human materials.

Martes nobilis

Noble martens are known from only two early Holocene sites in the western United States. Both are near the modern distribution of *M. americana* in Idaho (Fig. 2.4). The remains of at least 25 individuals of *M. nobilis* accumulated in Moonshiner Cave in a natural pit trap and are not associated with human activity (White et al. 1984). The early Holocene date for Moon-

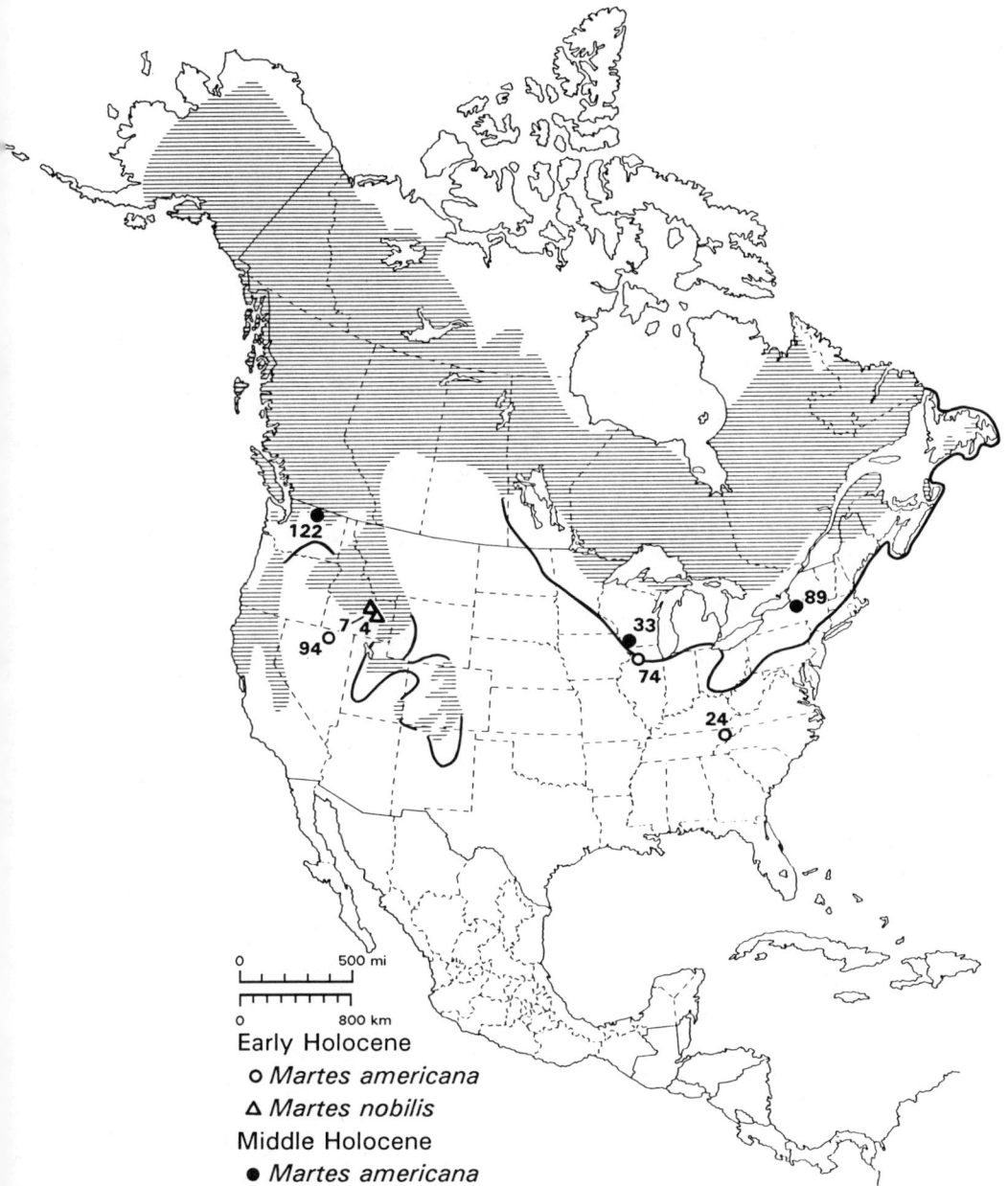

Figure 2.4. Early Holocene localities of *Martes americana* and *M. nobilis* and middle Holocene localities of *M. americana*, compared with modern (*hatched area*) and historic (*solid line*) distribution of *M. americana*. Names of numbered sites and ages of fossils are in Table 2.1.

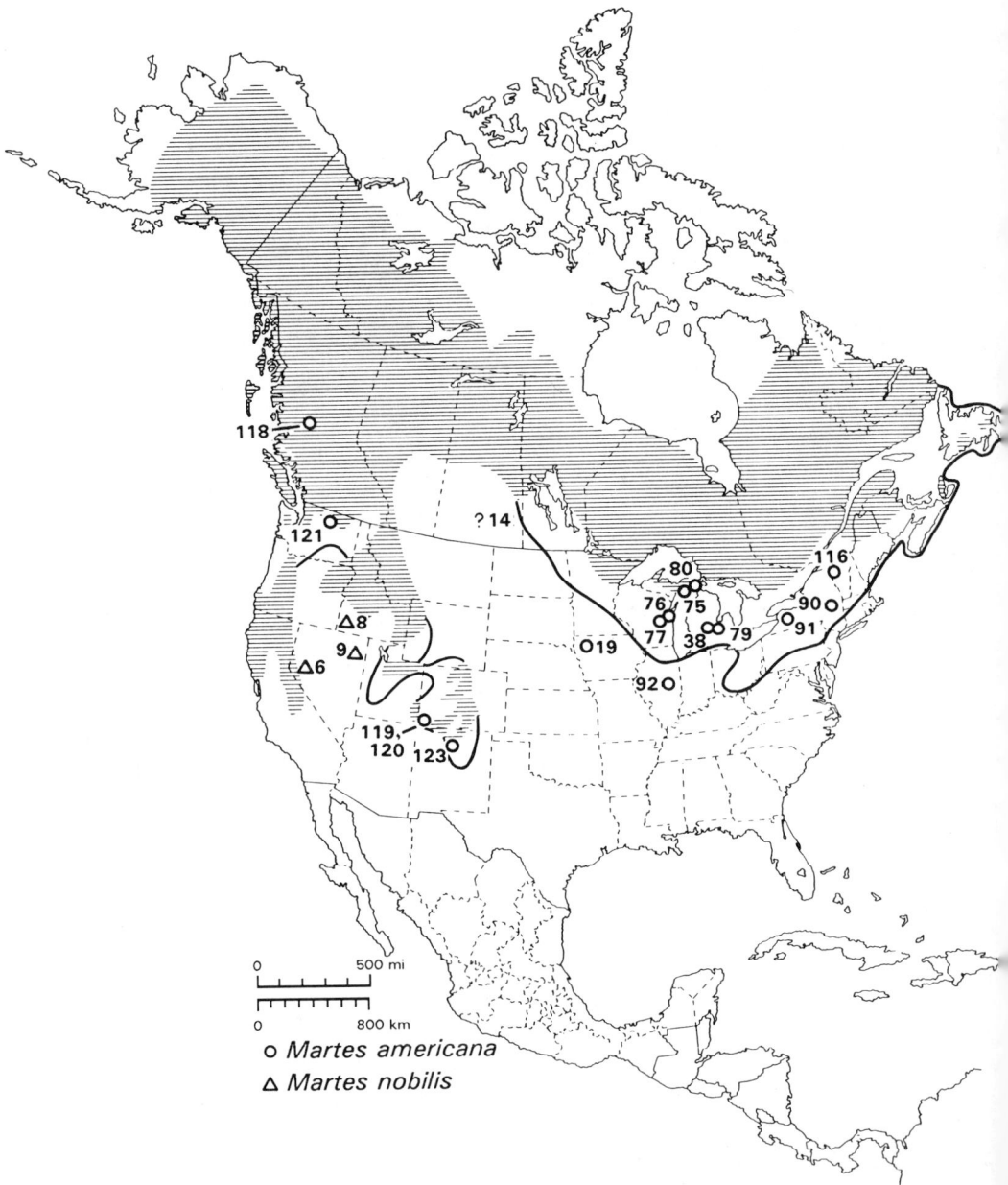

Figure 2.5. Late Holocene localities of *Martes americana* and *M. nobilis* compared with modern (*hatch* *area*) and historic (*solid line*) distribution of *M. americana*. Names of numbered sites and ages of fossils a in Table 2.1.

shiner Cave is based on faunal comparisons with a dated sequence from nearby Middle Butte Cave (White et al. 1984). The presence of other boreomontane species such as wolverine, short-tailed weasel (*Mustela erminea*), least weasel (*M. rixosa*), and lynx (*Felis canadensis*) suggests that the early Holocene climate was cooler than today (White et al. 1984). *M. nobilis* is associated with *M. americana* at this locality. At Jaguar Cave, *M. nobilis* is known from late Pleistocene and early Holocene strata. The early Holocene deposits are dated at 7 to 10.4 ka and contain cultural materials. *M. nobilis* is the only *Martes* identified from this site.

We have no middle Holocene records for *M. nobilis*. Three late Holocene sites (Dry Creek Rockshelter, Ada County, Idaho; Bronco Charlie Cave, Elko County, Nevada; and Hidden Cave, Churchill County, Nevada) have yielded the remains of *M. nobilis* (Fig. 2.5). All three sites are outside the historic limits of *M. americana* and contain late Holocene cultural materials. At Dry Creek Rockshelter, the remains of noble martens occur in stratigraphic horizons that have been securely dated at between 2.55 and 3.27 ka (Webster 1978). Also, these dated horizons sit directly on bedrock, so the noble marten remains could not have been redeposited from older sediments (Grayson 1984). Dry Creek Rockshelter offers the firmest evidence for the survival of this species into the late Holocene.

Bronco Charlie Cave sediments containing noble martens date from between 3.5 and 1.2 ka, based on correlations with other dated archeological sites that contain the same artifact types (Grayson 1984, 1987). The noble marten fossils from Hidden Cave are dated at about 3.5 ka (Grayson 1984, 1987). Grayson (1984:237) notes the possibility, however, that the Hidden Cave noble marten material may have been redeposited from older sediments. *M. nobilis* has also been found in sediments in Jaguar Cave that have been dated at 4 ka, but this date may be inaccurate or the fossils may have been redeposited (Grayson 1984:237). Therefore, we have not included Jaguar Cave as a late Holocene record for *M. nobilis*.

Martes pennanti

Only three records of *M. pennanti* date from the early Holocene. All are from the eastern United States (Fig. 2.6) and are within the historic range of *M. pennanti* (Powell 1982). Two of the sites, Baker Bluff Cave and Ripplemead Quarry, both in Virginia, occur in the central Appalachians near the southern boundary for the historic distribution of *M. pennanti*. Both sites are derived from natural accumulations and are not associated with any cultural complexes. The third record is a partial skeleton from peat deposits in an abandoned channel of the Mississippi River in northwestern Illinois. These

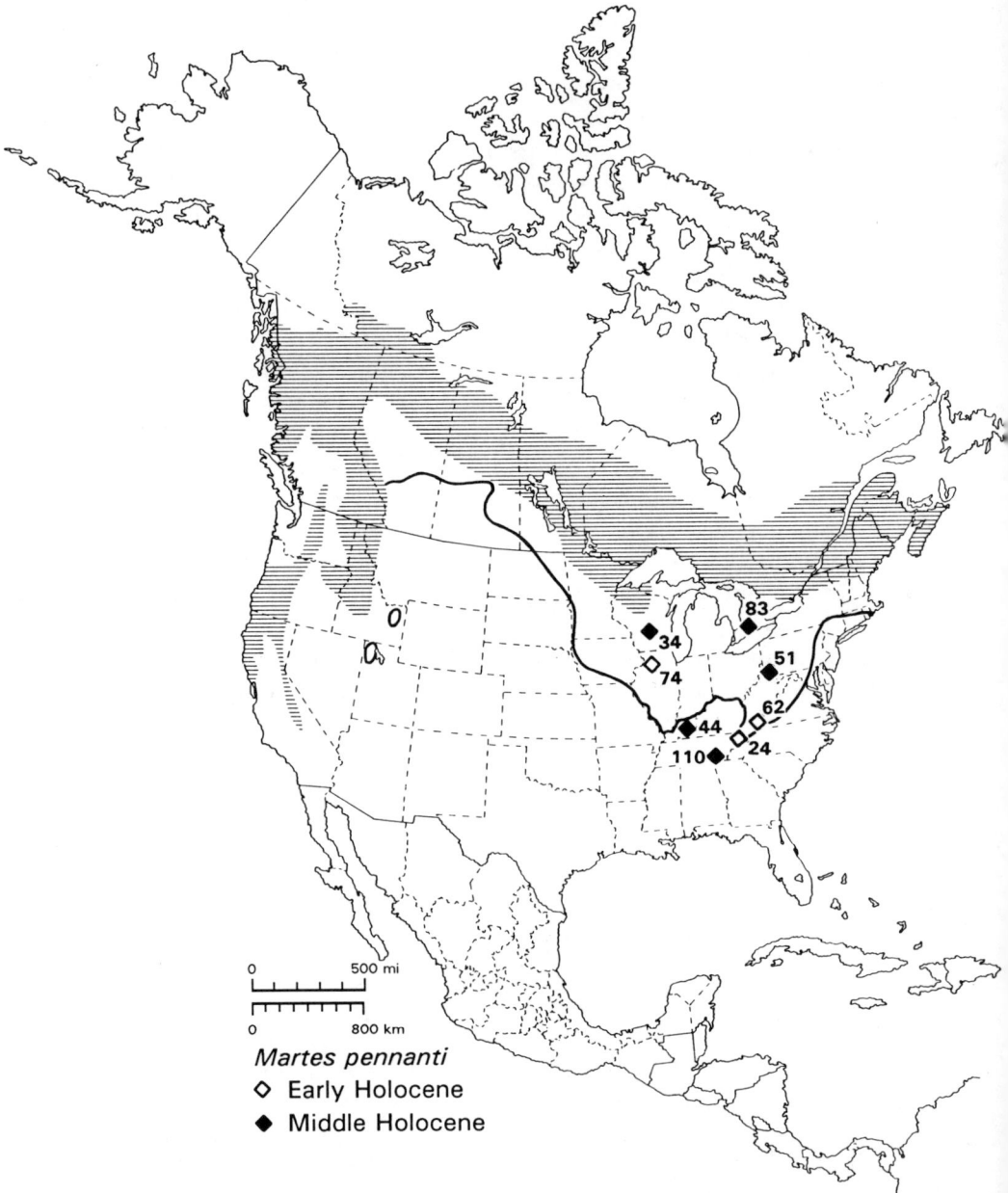

Figure 2.6. Early and middle Holocene localities of *Martes pennanti* compared with modern (*hatched are*
and historic (*solid line*) distribution of *M. pennanti*. Names of numbered sites and ages of fossils are in Tab
2.1.

skeletal elements occurred near the base of the peat, which dates between 10.13 and 8.86 ka (Graham and Graham 1990).

Five localities document the middle Holocene distribution of *M. pennanti* in the eastern United States (Fig. 2.6). Three of these sites are within the historic range, but are south of the modern range (hatched area in Fig. 2.6) of *M. pennanti*. Two localities are even south of the historic boundaries. Indian Knoll (Ohio County, Kentucky) is less than 50 km south of the historic boundary, and fisher remains at this site are associated with middle Archaic cultural materials that date between 4.5 and 6.1 ka. Westmoreland-Barber (Marion County, Tennessee) is also an archeological site, but is more than 100 km south of the historic boundary. The exact age of the *M. pennanti* material is not known; cultural materials from the site date between 1.15 and 5.0 ka (Faulkner and Graham 1966). Fisher remains from this site may, therefore, represent either middle or late Holocene times.

Twenty-three late Holocene sites have been documented for *M. pennanti*, and all but two fall within its historic range (Fig. 2.7). A lower jaw of a fisher, extensively modified by human workmanship, has been recovered from a late Woodland cultural level that dates between 1.2 and 1.6 ka at the Mason site, Franklin County, Tennessee (Parmalee 1968). The other locality, Etowah (Parmalee 1960*d*, Barkalow 1961), is an archeological site located in Bartow County, Georgia. The fisher remains were recovered from a mound and have been dated between 450 and 850 years before present (ybp) (Parmalee 1960*d*).

Historic Distribution of *Martes*

So far we have considered only records from archeological and paleontological sites. We have not incorporated information from written documents or from other records collected by early biological surveys. Therefore, our analysis differs considerably from that by Gibilisco (this volume) and other published accounts (Powell 1981*a*, Clark et al. 1987). Furthermore, our investigation of the historic archeological literature is not exhaustive. Still, the historic distribution of *Martes* offers some interesting perspectives on habitat requirements of this genus.

Martes americana

All the historic *M. americana* localities fall within the species' historic range (Clark et al. 1987), and, in fact, all but one of the localities occur near the limits of the historic distribution in southern Canada and northern Michi-

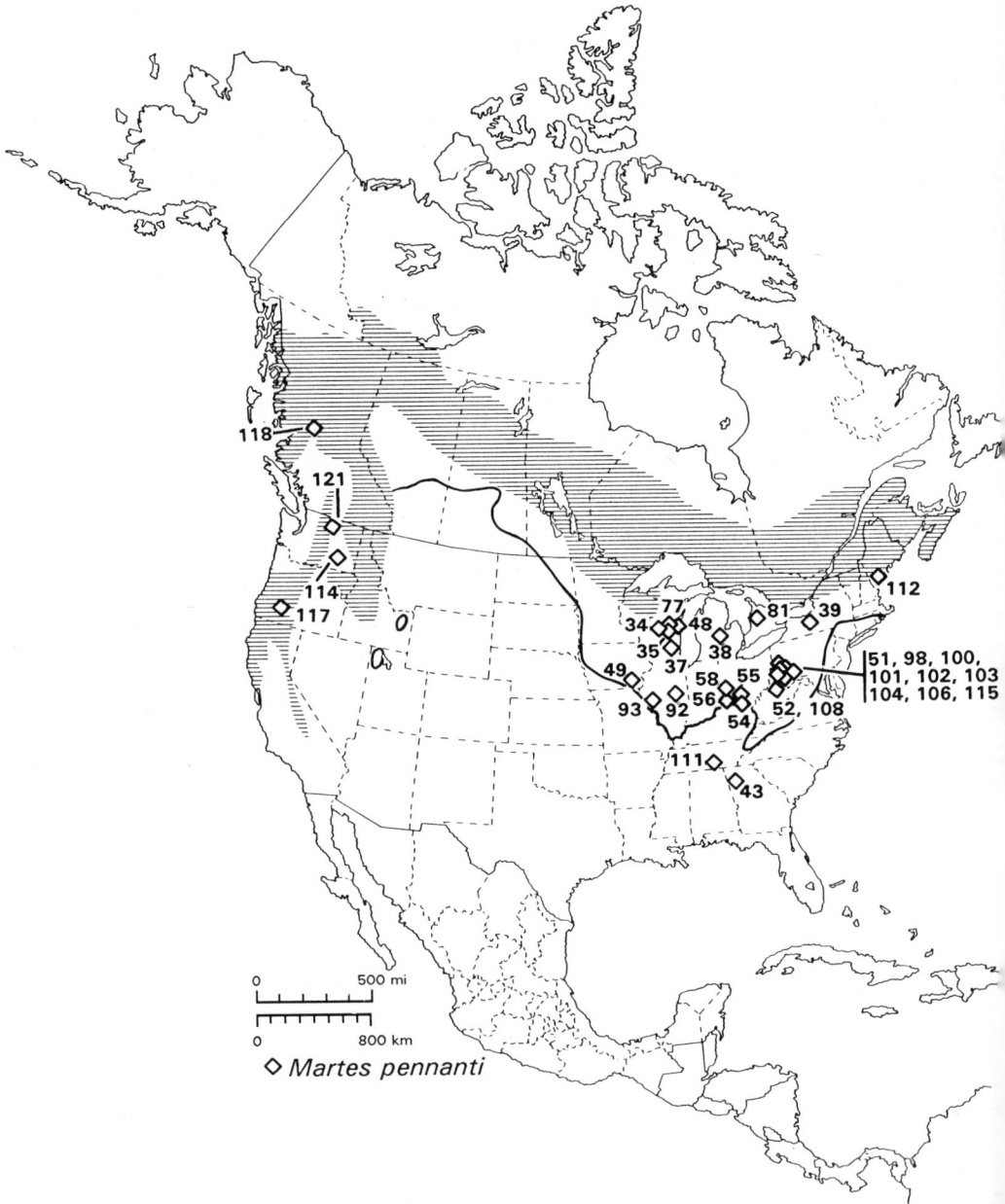

Figure 2.7. Late Holocene localities of *Martes pennanti* compared with modern (*hatched area*) and histc (*solid line*) distribution of *M. pennanti*. Names of numbered sites and ages of fossils are in Table 2.1.

gan (Fig. 2.8). The southernmost locality is an archeological site, Madison-ville, located in Hamilton County, Ohio. This site represents late Mississip-pian and Fort Ancient cultures that date between 280 and 310 ybp.

Martes nobilis

Because we have no historic or modern records of noble martens, it ap-pears that they became extinct sometime in the late Holocene (Grayson 1984). Youngman and Schueler (1991) suggest, however, that *M. nobilis* is merely a phenon of *M. americana*.

Martes pennanti

Sixteen sites document the historic distribution of *M. pennanti* (Fig. 2.9). All but one are located within or near the boundary of its known historic distribution. The single site well outside of this boundary is the Law's arche-ological site located in Marshall County, northeastern Alabama. A fisher maxillary fragment with several premolars was found associated with a hu-man burial that probably dates to 1700 A.D. (Barkalow 1961). This site is near the area in eastern Tennessee from which Audubon and Bachman pre-sumably examined skins of *M. pennanti* (Barkalow 1961:544).

Evolution of *Martes* Habitat

Late Wisconsinan Environments

Today North American *Martes* occur primarily in the boreal forest regions across Canada and the mountainous regions of the western United States (Gibilisco, this volume). But in historic times, both *M. pennanti* and *M. americana* occurred in mixed deciduous forest environments in the eastern United States (Figs. 2.2, 2.3).

In almost all late Wisconsinan sites that have noble martens, American martens, or fishers, they are associated with other boreomontane mammal species that today occur at higher latitudes and altitudes. In the eastern United States, both fishers and American martens extended as far south as northwestern Alabama (Figs. 2.2, 2.3). Fishers have also been found in Pleistocene deposits of uncertain age in northwestern Georgia. Cool and moist environments are also reflected in studies of pollen in these regions. Macrofossils (cones, needles, wood) of spruce (*Picea* spp.) dated at 21.3 ka have been found as far south as southeastern Georgia, but as yet none have been recovered from the numerous paleobotanical sites of Florida (Watts

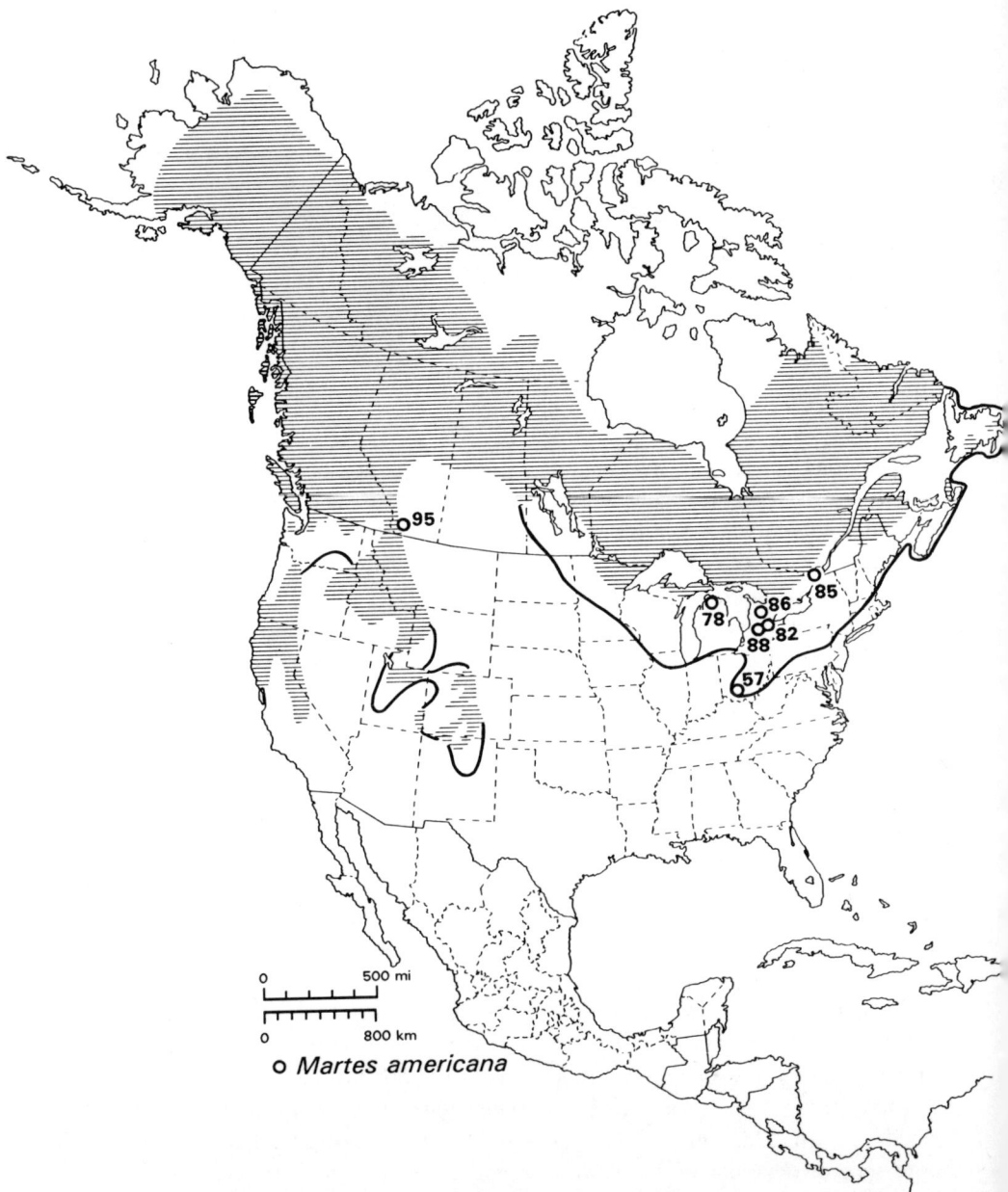

Figure 2.8. Historic localities of *Martes americana* compared with modern (*hatched area*) and histor (*solid line*) distribution of *M. americana*. Names of numbered sites and ages of fossils are in Table 2.

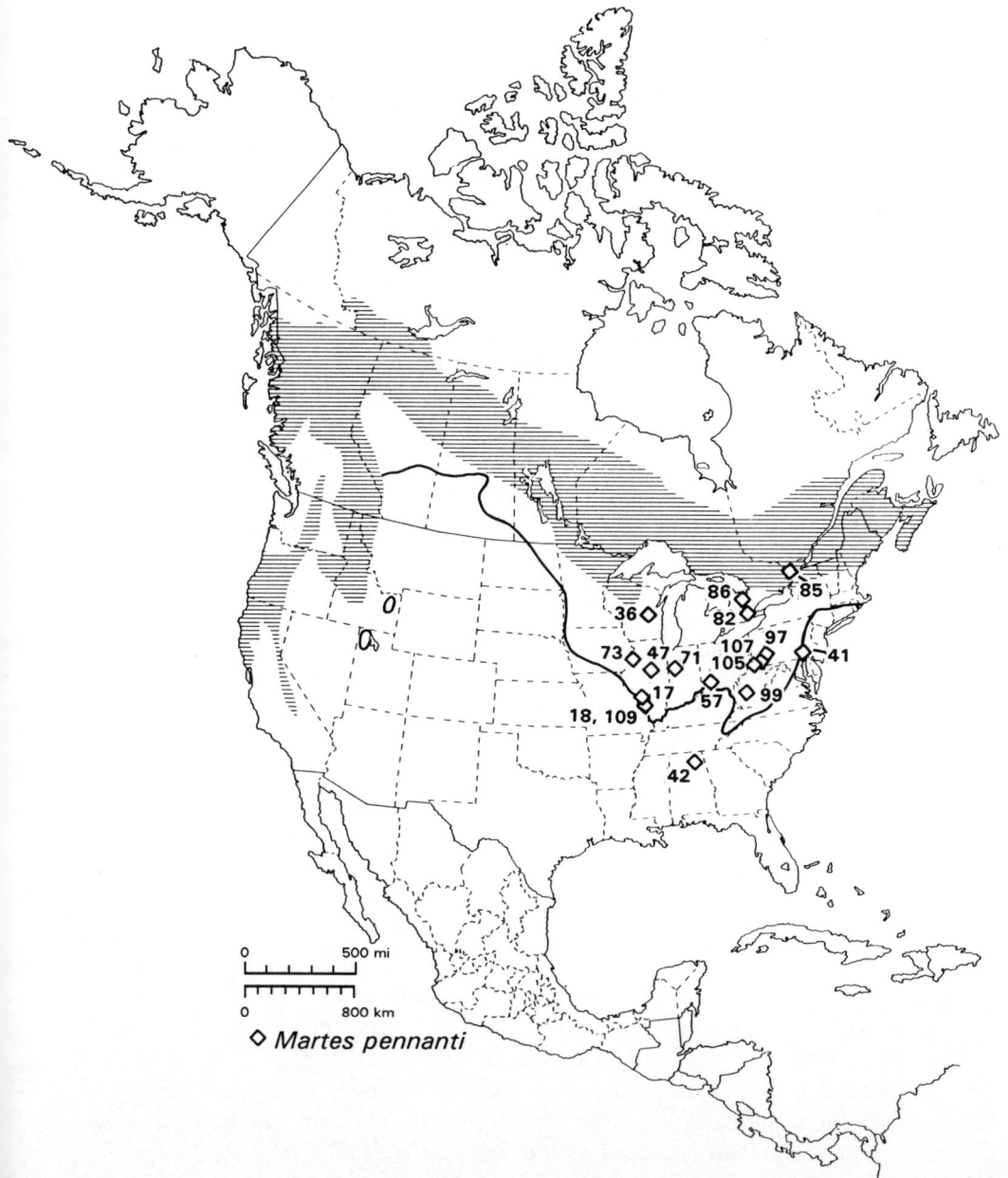

Figure 2.9. Historic localities of *Martes pennanti* compared with modern (*hatched area*) and historic (*solid line*) distribution of *M. pennanti*. Names of numbered sites and ages of fossils are in Table 2.1.

1983). Jack pine (*Pinus banksiana*) was the dominant tree of southeastern forests during the full-glacial.

Fishers are known from several late Pleistocene sites in the Ozarks of Missouri and Arkansas. Pollen studies in western Missouri indicate that a predominantly jack pine forest with diverse herbs covered the area from about 27.5 ka to 21 ka (King 1973). After 21 ka, jack pine was extirpated from the region and spruce became abundant (Watts 1983). The spruce forest became more diverse after 16 ka with the addition of broad-leaved trees, and spruce remained in the area until after 13.5 ka (King 1973). None of the fisher remains from the Ozarks are securely dated, but they appear to represent a time when spruce forest was prevalent. For instance, the boreomontane fauna from Peccary Cave (Newton County, Arkansas) appears to date to about 16 ka (Semken 1984, Stafford and Semken 1990).

Spruce forests dominated the upper midwestern United States (Van Zant et al. 1980) in what Rhodes (1984) has referred to as a boreal grassland, reflecting the open nature of the forest. American martens have been recorded from Pleistocene sites in north-central Nebraska and eastern Colorado. The Jones-Miller site (ca. 10 ka) in eastern Colorado is somewhat enigmatic, since it is generally assumed that boreal forest retreated from the plains region well before 12 ka (Watts 1983). Human transport of the American marten remains is a possibility, but inasmuch as it is not the only boreomontane mammal from the site, indigenous marten populations and boreomontane environments are also a possibility. Furthermore, Walker (1987) has shown that boreomontane mammals in Wyoming inhabited elevations lower than they do today well into the early Holocene. It therefore appears that *Martes* and other boreomontane species may have occupied parts of the western plains near the Rocky Mountains after they had left other areas of the central and eastern plains.

In the northwestern United States, the tree line was lowered by as much as 1000 m during the late Wisconsinan in areas such as the Olympic Mountains (Heusser 1977), Yellowstone Plateau, and the Snake River Plain (Baker 1983). Intermontane basins were not forested, however. The basin vegetation of the full-glacial was more like periglacial steppe because of the cold, dry environment at these lower elevations (Barnosky et al. 1987). Thus, the forested area in parts of the northwestern United States was significantly reduced during the full-glacial. Trees grew mostly along the base of mountains and in small protected refuges (Barnosky et al. 1987).

The few late Pleistocene records of *Martes* from the northwestern United States contrast with numerous records in the East and the northern Rocky Mountains (Figs. 2.2, 2.3). Both *M. americana* and *M. nobilis* occur in the western United States during the late Quaternary. *M. nobilis* is the most abundant fossil species, found primarily in environments in the northern Great Basin and central Rocky Mountains (Fig. 2.2). Unfortunately, there are no known sites along the Pacific Coast of British Columbia to provide insight into the relict hypothesis for body size on the Queen Charlotte Islands (Nagorsen, this volume).

Martes has not been found yet in the southwestern United States (southern California, southern Nevada, Arizona, or New Mexico), even though full-glacial woodlands and forests spread to lower elevations (Spaulding et al. 1983, Van Devender et al. 1987). However, these woodlands were variably composed of pinyon (*Pinus edulis*), juniper (*Juniperus* spp.), and oak (*Quercus* spp.) (Van Devender et al. 1987), none of which is preferred habitat for *Martes* today (Buskirk and Powell, this volume). Likewise, *Martes* has not been recovered from late Quaternary archeological or paleontological sites on the southern plains or the Gulf Coastal Plain.

As we have stated, the individualistic response of species to these environmental changes resulted in the creation of species associations without modern analogs (Graham 1979, Lundelius et al. 1983, Graham and Mead 1987). In the unglaciated regions of the northern United States, for example, arctic species (e.g., *Dicrostonyx torquatus*) and boreomontane species (e.g., *Microtus xanthognathus*) were integrated with more temperate species that today occupy grasslands (e.g., *Spermophilus tridecemlineatus*) and deciduous forest (*Tamias striatus*) (Graham 1976). Contemporaneity and sympatry of these nonanalog associations has been demonstrated by new radiocarbon dating techniques that isolate specific amino acids in rodent bones (Stafford and Semken 1990). Because *Martes* was a component of these nonanalog Pleistocene communities, its adaptations to modern environments may reflect only its tolerance for different habitat types.

Holocene Environments

In Holocene time (8–10 ka), forests developed throughout the eastern United States, but unlike during the late Pleistocene, spruce (*Picea* spp.) was not important, even in such areas as Maine where spruce is abundant today (Davis 1983). The individualistic northward shifts of tree species continually changed the composition of these eastern forests (Davis 1981, 1983; Jacobson et al. 1987). Changes in the mammalian fauna of the eastern United States throughout the Holocene were negligible, however (Semken 1983, Guilday 1967).

Early and middle Holocene records, like paleontological sites of these ages, are sparse for both *M. pennanti* and *M. americana* in the eastern United States (Figs. 2.4, 2.6). In early Holocene deposits, *M. americana* occurs as far south as eastern Tennessee (Fig. 2.4), but in later deposits it does not extend farther south in the southeastern United States than its historic limit as defined by Clark et al. (1987). Archeological sites in Kentucky and Tennessee (Fig. 2.6) suggest that *M. pennanti* may have extended into these states during the middle Holocene. In late Holocene and historic deposits in the southeast, there are no extralimital records for *M. americana*, but *M. pennanti* is found in deposits dated to both of these periods as far south as the southern Appalachian Mountains.

These sites may reflect the presence of *M. pennanti* habitat in the southern Appalachian Mountains (Parmalee 1960*d*). Long-distance transport by human populations was not likely an important factor in the occurrence of *Martes* in southern Appalachian archeological sites (see Prehistoric Human Use of *Martes*, below). The paleontological and archeological records suggest that with the reduction of boreal forest habitat, spruce forest in particular, at the end of the Pleistocene, *M. americana* was restricted to environments at higher latitudes, whereas *M. pennanti* was able to persist, perhaps in low numbers, in special, local environments in the central and southern Appalachians.

In the midwestern United States during the early Holocene (8–10 ka), a forest dominated by elm (*Ulmus* spp.) replaced the late Pleistocene spruce forest (Webb et al. 1983). In the middle Holocene (4–8 ka), the prairie peninsula, an extensive wedge-shaped area of grassland, extended eastward across northern Missouri and southern Minnesota and into central Illinois and Indiana (King 1981). The prairie peninsula was replaced in the late Holocene with a forest-prairie mosaic in northern and central Illinois, southern Wisconsin, southeastern Minnesota, eastern Iowa, and northern and western Missouri (Webb et al. 1983).

M. americana is known from only a few sites along its historic boundary during the early and middle Holocene (Fig. 2.4). Likewise, *M. pennanti* occurred in northern Illinois and southern Wisconsin during the early and middle Holocene, respectively. In the late Holocene, *M. americana* appears to have spread southward and westward with the expanding riparian forests in Illinois and Iowa (Fig. 2.5). *M. pennanti* also is known from numerous localities in central Illinois, eastern Iowa, and southern Wisconsin during this time. It appears, therefore, that the late Holocene deciduous forests along the prairie-forest border provided suitable habitat for both *M. americana* and *M. pennanti*.

In the northern plains, spruce forest was replaced by grassland without an intermediary forest type (Watts and Bright 1968, Ritchie 1976). The Harder site in southern Saskatchewan represents the only Holocene record for the northern plains. This record must be viewed cautiously, however, inasmuch as Dyck (1977) only tentatively referred a second phalanx to this taxon. It is possible that *M. americana* extended onto the plains along riparian woodlands, as seems to have been the case in the midwestern United States during the late Holocene. There are no late Quaternary records of *M. pennanti* from the plains.

The early Holocene was a time of increased summer temperatures and drought in the northwestern United States (Barnosky et al. 1987). In fact, the early Holocene appears to be the warmest time of the postglacial for this area (Baker 1983, Barnosky et al. 1987). These warmer temperatures allowed the expansion of spruce and alder (*Alnus* spp.) forest to higher elevations, expanding the geographic range and relative abundance of this forest compared with its Pleistocene distributions (Barnosky et al. 1987). In the middle and late Holocene the forests of the northwestern United States became generally more closed, because of the cooler and moister climate (Baker 1983). Although spruce forest expanded in the northwestern United States during the Holocene, records for *Martes* declined. This apparent paucity may actually be due to the lack of high-altitude paleontological and archeological sites.

Van Devender et al. (1987:332) summarized the late Quaternary environment of the southern Great Basin this way: "The vegetation sequence in the Snake Range in the Great Basin was a late Wisconsin Great Basin bristlecone pine forest, an early Holocene limber pine woodland, a middle Holocene Utah juniper–big sagebrush woodland, and a middle to late Holocene pinyon-juniper woodland." The only records of *M. americana* in the Great Basin are restricted to the late Pleistocene and early Holocene. *M. nobilis* is the dominant species throughout the late Pleistocene and Holocene. *M. pennanti* does not occur in the western United States until the late Holocene, when *M. nobilis* becomes extinct. During the late Pleistocene, *M. nobilis* occurs in faunas containing other boreomontane species, but late Holocene *M. nobilis* occurs in faunas that lack boreomontane species (Grayson 1984).

Most of the modern range of *Martes* in North America occurs in boreal forest in glaciated areas of Canada. The history of boreal forest development is complex and varies across Canada as a function of latitude and the chronology of deglaciation (Ritchie 1987). Glacial ice did not retreat from most of the area now occupied by the boreal forest until after 12 ka. The western area of the boreal forest was ice-free by 10 ka, but the eastern boreal forest area was not free of ice until about 8 ka (Prest 1970). The succession of the boreal forest in western Canada differed distinctly from that in eastern Canada (Ritchie 1987). Pollen studies show that the eastern Canadian early boreal forest changed to a mixed (temperate) forest in the middle Holocene and then evolved into the modern boreal forest in the late Holocene (Liu and Lam 1985). In addition, the northern and southern boundaries of the boreal forest have shifted during the Holocene, but the central part has remained forested throughout postglacial history (Ritchie 1987).

The modern boreal forest of this area is significantly different from the Pleistocene spruce forests that extended south of the ice sheet (Wright 1981). The boreal forest of today is a uniformly dense closed-crown coniferous forest, but the proportions of the dominant conifers vary greatly across the landscape (Ritchie 1987). By contrast, the Pleistocene spruce forest was much more open and contained a variety of understory species (e.g., *Artemisia*) that are not found in the modern spruce forest (Cushing 1965, Wright 1981, Jacobson et al. 1987). The small mammal faunas from these late Pleistocene sites also indicate a different composition at the microenvironmental scale (Graham and Mead 1987). Most of the habitat occupied by *Martes* today is, therefore, relatively new in evolutionary time; unfortunately, the late Quaternary history of *Martes* in this area is unknown.

Extinction of *Martes nobilis*

The cause of the extinction of *M. nobilis* is unknown but several explanations have been offered. Before the discovery of Holocene specimens, it was widely accepted that noble martens, along with many other mammalian species, had succumbed to environmental changes at the end of the Pleistocene (Kurtén and Anderson 1980). The discovery of Holocene specimens not only refuted this hypothesis but stimulated reconsideration of the environmental adaptations for this species. Many of the late Pleistocene sites with *M. nobilis* contained boreomontane species, which suggested that *M. nobilis* was adapted to cool moist environments (Anderson 1970, Kurtén and Anderson 1980). However, the lack of boreal species in late Holocene deposits that contain *M. nobilis* suggests that it was adapted to a much broader range of habitats (Grayson 1984). In fact, since *M. nobilis* survived further environmental degradation in the early and middle Holocene, Grayson (1984) wondered if environmental change could be invoked at all to explain the demise of this species. These late Holocene records could, however, also represent a time when *M. nobilis* was displaced from its preferred habitat by a competitor that ultimately caused its extinction.

Some investigators (Anderson 1970, Kurtén and Anderson 1980) have proposed that *M. nobilis* may have been forced into extinction by competition with *M. americana*. As stated by Grayson (1984), the late Holocene survival of *M. nobilis* seems to refute this hypothesis, which is further contradicted by the fact that *M. americana* was widespread in the late Pleistocene and found to be contemporaneous with *M. nobilis* at several western sites (Fig. 2.2).

But because *M. nobilis* is known only from the western United States during the late Holocene and is intermediate in size between American martens and fishers, *M. pennanti* may have effectively competed with *M. nobilis* in some areas. *M. pennanti* probably was not a factor in the extinction of *M. nobilis* in the Great Basin, since it appears that *M. pennanti* never inhabited this area and *M. nobilis* persists from the late Pleistocene to the late Holocene.

The extinction of *M. nobilis*, as with many other species, is probably the result of several factors acting in concert. The distribution of *M. nobilis* was reduced progressively during the late Quaternary (Figs. 2.2–2.5). By the late Holocene, sites are known only from the Great Basin (Fig. 2.5). These range reductions may have been caused by environmental changes at the end of the Pleistocene. Many other mammalian species in the arid west exhibit similar range reductions during this time (Grayson 1984, 1987). Reduction in the geographic range of a species significantly increases the probability of extinction by making the species more vulnerable to a host of other factors (Brown and Maurer 1989). The possible impact of humans on *M. nobilis* must also be considered. At several early Holocene sites and all late Holocene sites, *M. nobilis* is associated with cultural levels.

Prehistoric Human Use of *Martes*

Almost all Holocene records of *Martes* come from archeological sites (Table 2.1). The predominance of archeological records over paleontological reports results, in part, from the fact that few Holocene noncultural sites have been studied (Semken 1983). The large number of archeological sites with *Martes* suggests, however, that this taxon was exploited by people at different times. In fact, Kinietz (1940) and other ethnographers indicate that the historic Ottawa and other groups trapped martens in the Great Lakes area.

The actual uses of *Martes* by these people is uncertain. Skins of these animals were likely worn as clothing or used as medicine bags. Although few ethnographic studies describe the use of *Martes* for clothing, early paintings by Bodmer (Hunt and Gallagher 1984) demonstrate that mustelids, especially *Mustela erminea, M. vison*, and *Lutra canadensis*, were important decorative items for clothing. Because the heads and paws were frequently left on the skins, the occurrence of these bones and the absence of others in archeological sites have been cited as evidence for the presence of skins; however, taphonomic processes such as trampling and compaction might also account for the preferential preservation of these denser bones.

Worked bones at such sites as Mason, Franklin County, Tennessee (Parmalee 1968), indicate that some parts may have been worn as decorations or

Table 2.1. Paleontological and archeological sites containing *Martes*

Site no.	Name	Location	Species	Age (ybp) or period	Reference
1	Greens Creek	Ontario	*M. americana*	10,000	Harington 1978
2	Peccary Cave	Newton, Ark.	*M. pennanti*	2290–16,830	Kurtén & Anderson 1980, Semken 1984
3	Carter	Darke, Ohio	*M. pennanti*	10,230	Kurtén & Anderson 1980
4	Moonshiner	Bingham, Idaho	*M. nobilis*	Early Holocene	White et al. 1984
5	Old Crow River IIA	Yukon Territory	*M. nobilis*	Late Pleistocene	Kurtén & Anderson 1980
6	Hidden Cave	Churchill, Nev.	*M. nobilis*	3600–3700	Grayson 1984
7	Jaguar Cave	Lemhi, Idaho	*M. nobilis*	7000–11,580	Kurtén & Anderson 1972
8	Dry Creek Rockshelter	Ada, Idaho	*M. nobilis*	2550–3270	Webster 1978
9	Bronco Charlie Cave	Elko, Nev.	*Martes* sp., *M. nobilis*	3500	Mead & Mead 1989, Grayson 1984
10	Bell Cave	Albany, Wyo.	*M. americana, M. nobilis*	10,000–13,500	Ziemans & Walker 1974
11	Little Box Elder Cave	Converse, Wyo.	*M. nobilis*	10,000–24,000	Anderson 1968
12	Little Canyon Creek Cave	Washakie, Wyo.	*M. nobilis*	Late Pleistocene	Frison & Walker 1978, Walker 1987
13	Natural Trap Cave	Bighorn, Wyo.	*Martes* sp.	15,500–20,250	Chomko & Gilbert 1987
14	Harder	Saskatchewan	*M. americana*	3360–3425	Dyck 1977
15	Chimney Rock Animal Trap	Larimer, Colo.	*M. americana, M. nobilis*	Late Pleistocene–Holocene	Hager 1972
16	Jones-Miller	Yuma, Colo.	*M. americana*	10,020	Graham 1987
17	Cahokia	St. Clair, Ill.	*M. pennanti*	400–800	Purdue & Styles 1987
18	Laurens	Randolph, Ill.	*M. pennanti*	100–400	Jelks et al. 1989, Purdue & Styles 1987
19	Phipps	Cherokee, Iowa	*M. americana*	600–800	Semken & Falk 1987
20	New Paris No. 4	Bedford, Pa.	*M. americana, M. pennanti*	11,300	Guilday et al. 1964
21	Eagle Cave	Pendelton, W.Va.	*M. americana*	Late Pleistocene	Guilday & Hamilton 1973

22	Natural Chimneys	Augusta, Va.	*M. americana,* *M. pennanti*	Late Pleistocene	Guilday 1962
23	Clark's Cave	Bath, Va.	*M. americana*	Late Pleistocene	Guilday et al. 1977
24	Baker Bluff Cave	Sullivan, Tenn.	*M. pennanti,* *M. americana*	Late Pleistocene–early Holocene	Guilday et al. 1978
25	Robinson Cave	Overton, Tenn.	*M. americana,* *M. pennanti*	Late Pleistocene	Guilday et al. 1969
26	Ladds Quarry	Bartow, Ga.	*M. pennanti*	Late Pleistocene?	Ray 1967
27	Bat Cave	Pulaski, Mo.	*M. pennanti*	Late Pleistocene	Hawksley et al. 1973
28	Brynjulfson Cave No. 1	Boone, Mo.	*M. pennanti*	Late Pleistocene	Parmalee & Oesch 1972
29	Wilson Butte Cave	Jerome, Idaho	*M. americana*	9950–14,500	Gruhn 1961, Lundelius et al. 1983
30	Smith Creek Cave	White Pine, Nev.	*M. nobilis*	11,000–13,000	Mead et al. 1982
31	Samwell Cave	Shasta, Calif.	*M. nobilis*	Late Pleistocene	Kurtén & Anderson 1980
32	Potter Creek Cave	Shasta, Calif.	*M. nobilis*	Late Pleistocene	Kurtén & Anderson 1980
33	Raddatz Rockshelter	Sauk, Wisc.	*M. pennanti,* *M. americana*	1150–10,000	Parmalee 1959, Cleland 1966
34	Durst Rockshelter	Sauk, Wisc.	*M. pennanti*	1150–8000	Parmalee 1960c
35	Bornick	Marquette, Wisc.	*M. pennanti*	660	Gibbon 1971
36	Bell	Winnebago, Wisc.	*M. pennanti*	220–330	Parmalee 1963
37	Aztalan	Jefferson, Wisc.	*M. pennanti*	1150–3000	Parmalee 1960b
38	Schultz	Saginaw, Mich.	*M. pennanti,* *M. americana*	1050–2490	Cleland 1966
39	Kipp Island	Seneca, N.Y.	*M. pennanti*	1450–2250	Cleland 1966
40	Wilson Sand Hill	New York State	*M. pennanti*	<550	Cleland 1966
41	Eschelman	Lancaster, Pa.	*M. pennanti*	325–350	Guilday et al. 1962
42	Laws	Marshall, Ala.	*M. pennanti*	250	Barkalow 1961
43	Etowah	Bartow, Ga.	*M. pennanti*	450–850	Parmalee 1960d
44	Indian Knoll	Ohio, Ky.	*M. pennanti*	4508–6100	Webb 1974
45	Graham Cave	Montgomery, Mo.	*M. pennanti*	1000–7630	Parmalee 1971

(continued)

51

Table 2.1. (*Continued*)

Site no.	Name	Location	Species	Age (ybp) or period	Reference
46	Arnold Research Cave	Callaway, Mo.	*M. pennanti*	Late Pleistocene–Holocene	Parmalee 1971
47	Zimmerman	LaSalle, Ill.	*M. pennanti*	<500	Brown 1975
48	Heins Creek	Door, Wisc.	*M. pennanti*	1218	Cleland 1966
49	Rock Run Shelter	Cedar, Iowa	*M. pennanti*	1330–3660	Semken 1983
50	Bryan	Minnesota	*M. pennanti*	350–750	Semken 1983
51	Meadowcroft Rockshelter	Washington, Pa.	*M. pennanti*	2930–11,300	Semken 1983
52	Globe Hill	Hancock, W.Va.	*M. pennanti*	2950–3950	Guilday 1956
53	Johnston	West Virginia	*M. pennanti*	<550	Cleland 1966
54	Feurt	Scioto, Ohio	*M. pennanti*	550–1150	Cleland 1966
55	Cramer Village	Ross, Ohio	*M. pennanti*	550–950	Cleland 1966
56	Baum	Ross, Ohio	*M. pennanti*	550–950	Cleland 1966
57	Madisonville	Hamilton, Ohio	*M. pennanti*, *M. americana*	280–310	Cleland 1966
58	Anderson	Warren, Ohio	*M. pennanti*	550–950	Cleland 1966
59	Darty Cave	Scott, Va.	*M. americana*, *M. pennanti*	Late Pleistocene–early Holocene	Eshelman & Grady 1986
60	Loop Creek Quarry Cave	Russell, Va.	*M. americana*	Late Pleistocene–early Holocene	Eshelman & Grady 1986
61	Back Creek Cave No. 1	Bath, Va.	*M. pennanti*	Late Pleistocene–early Holocene	Eshelman & Grady 1986
62	Ripplemead Quarry	Giles, Va.	*M. pennanti*	Early Holocene	Weems & Higgins 1977
63	Snake Creek Burial Cave	White Pine, Nev.	*M. americana*, *M. nobilis*	Late Pleistocene	Mead & Mead 1989
64	Crystal Ball Cave	Millard, Utah	*M. americana*	Late Pleistocene	Heaton 1985
65	Middle Butte Cave	Bingham, Idaho	*M. americana*	<5175	White et al. 1984
66	Strait Canyon	Highland, Va.	*M. pennanti*, *M. americana*	29,870	Eshelman & Grady 1986

52

No.	Site	Location	Species	Date	Reference
67	Bell Cave	Colbert, Ala.	*M. pennanti, M. americana*	11,820	Churcher et al. 1989
68	Cheek Bend Cave	Maury, Tenn.	*M. americana*	16,500–18,000	Klippel & Parmalee 1982
69	January Cave	Alberta	*M. americana*	Late Pleistocene	Burns 1984
70	Smith Springs	Cherry, Nebr.	*M. americana*	Late Pleistocene	Voorhies 1990
71	Fort Ouiatenon	Tippecanoe, Ind.	*M. pennanti*	164–233	Martin 1986
72	Prairie Creek	Davies, Ind.	*M. pennanti*	Late Pleistocene–Holocene	Tomak 1974
73	Crawford Farm	Rock Island, Ill.	*M. pennanti*	140–160	Parmalee 1964
74	Anderson Peat Mine	Whiteside, Ill.	*M. pennanti, M. americana*	7000–10,000	Graham & Graham 1990
75	Ogontz Bay	Delta, Mich.	*M. americana*	1150–1650	Martin 1991
76	Mero	Door, Wisc.	*M. americana*	1150–1650	Cleland 1966
77	Lasley's Point	Winnebago, Wisc.	*M. americana, M. pennanti*	680–960	Cleland 1966
78	Ft. Michilimackinac	Emmet, Mich.	*M. americana*	169–235	Cleland 1966
79	Younge	Lapeer, Mich.	*M. americana*	550–950	Cleland 1966
80	Juntenen	Mackinac, Mich.	*M. pennanti*	630–1125	Cleland 1966
81	Inverhuron	Ontario	*M. pennanti*	1150–1250	Cleland 1966
82	Middleport	Ontario	*M. pennanti, M. americana*	<550	Cleland 1966
83	Lawson	Middlesex, Ont.	*M. pennanti*	3950–4950	Cleland 1966
84	York County Sites	Ontario	*M. pennanti, M. americana*	not given	Cleland 1966
85	Roebuck Village	Ontario	*M. pennanti, M. americana*	<550	Cleland 1966
86	Sidey-Mackey	Ontario	*M. pennanti, M. americana*	<550	Cleland 1966
87	Lander Shelter	Ontario	*M. americana*	not given	Cleland 1966
88	Uren	Ontario	*M. americana*	<550	Wintemberg 1928
89	Frontenac Island	Cayuga, N.Y.	*M. americana*	3950–4450	Cleland 1966

(*continued*)

Table 2.1. *(Continued)*

Site no.	Name	Location	Species	Age (ybp) or period	Reference
90	Snell	Montgomery, N.Y.	*M. americana*	794	Cleland 1966
91	Oakfield	Genessee, N.Y.	*M. americana*	750–850	Guilday 1963
92	Pabst	DeWitt, Ill.	*M. pennanti, M. americana*	3020–4300	Lewis 1979
93	Kuhlman	Cass, Ill.	*M. pennanti*	550–1150	Parmalee 1960*a*
94	Deer Creek Cave	Elko, Nev.	*M. americana*	9670–10,085	Ziegler 1963
95	Eagle Cave	Alberta	*M. americana, M. pennanti*	Late Pleistocene–Holocene	Burns 1984
96	North Cove	Harlan, Nebr.	*M. nobilis, M. pennanti*	Late Pleistocene–early Holocene	Stewart 1987
97	McKees Rocks	Allegheny, Pa.	*M. pennanti*	400–620	Lang 1968
98	Boyle	Washington, Pa.	*M. pennanti*	1050–2200	Nale 1963
99	Mt. Carbon	Fayette, W.Va.	*M. pennanti*	450–550	Guilday & Tanner 1965
100	Drew	Allegheny, Pa.	*M. pennanti*	590–950	Buker 1970
101	Wadding Shelter	Armstrong, Pa.	*M. pennanti*	350–1050	Guilday et al. n.d.
102	Murphy's Old House	Armstrong, Pa.	*M. pennanti*	350–1050	Guilday et al. n.d.
103	Bonnie Brook	Butler, Pa.	*M. pennanti*	350–1050	Guilday et al. n.d.
104	Mathies Mine Village	Washington, Pa.	*M. pennanti*	1650–2200	Guilday et al. n.d.
105	Hartley	Green, Pa.	*M. pennanti*	250	Guilday et al. n.d.
106	Janitor	Westmoreland, Pa.	*M. pennanti*	350–1050	Guilday et al. n.d.

No.	Site	Location	Species	Date	Reference
107	Campbell	Fayette, Pa.	*M. pennanti*	520	Guilday et al. n.d.
108	Watson Farm	Hancock, W.Va.	*M. pennanti*	1650–2200	Guilday et al. n.d.
109	Ft. DeChartres	Randolph, Ill.	*M. pennanti*	150–200	Martin & Masulis 1988
110	Westmoreland-Barber	Marion, Tenn.	*M. pennanti*	1150–5000	Guilday & Tanner 1966
111	Mason	Franklin, Tenn.	*M. pennanti*	1150–1650	Parmalee 1968
112	Turner Farm	Knox, Me.	*M. pennanti*	3515–3650	Bourque 1976
113	Marmes Rockshelter	Franklin, Wash.	*M. americana*	9000–12,310	Gustafson 1972
114	No Name	Ferry, Wash.	*M. pennanti*	1000–3000	Collier et al. 1942
115	Gnagey	Somerset, Pa.	*M. pennanti*	760–1030	George 1983
116	Boucher	Franklin, Vt.	*M. americana*	1845–2665	Heckenberger et al. 1990
117	Umpqua/Eden	Douglas, Ore.	*M. pennanti*	<3000	Lyman 1991
118	Charlie Lake Cave	British Columbia	*M. pennanti*, *M. americana*	1400–4270	Driver 1988
119	Lemoc Shelter	Montezuma, Colo.	*M. americana*	150–1550	Hogan 1986, Neusius 1986
120	Grass Mesa Village	Montezuma, Colo.	*M. americana*	1025–1350	Neusius & Gould 1988
121	Chief Joseph Dam (450K258)	Okanagan, Wash.	*M. pennanti*, *M. americana*	100–4000 2000–4000	Livingston 1985
122	Chief Joseph Dam (450K11)	Okanagan, Wash.	*M. americana*	4000–7000	Livingston 1984
123	Arroyo Hondo Pueblo (Site 1235)	Santa Fe, N.Mex.	*M. americana*	525–650	Lang & Harris 1984

Notes: Locations are plotted on Figures 2.2–2.9. Some sites are plotted on more than one figure because multiple levels contain fossils of *Martes*. Some sites are not plotted because there is insufficient information.

used for ceremonial purposes. The cut or broken posterior skull and the posterior lower jaw of a fisher from the historic Bell Site in Wisconsin have been interpreted as remnants of a medicine bag (Parmalee 1963:67). The inclusion of *Martes* material with human burials (e.g., at the Cahokia, Mason, and Etowah sites) also attests to their ceremonial use. Caution should be exercised in interpreting grave goods as possible trade items from other areas. In one case, Heckenberger et al. (1990:200) documented a hide medicine bag from an early Woodland (ca. 2 ka) burial at the Boucher site in Vermont. This bag contained the remains of American martens, mink, red foxes (*Vulpes* spp.), raccoons (*Procyon lotor*), coiled snakes and other vertebrates. All these species are local to the site. Also, Catlin (1973:36–37) stated that items for the medicine bag were frequently procured by hunting or trapping local animals.

Many records of *Martes* from archeological sites outside its normal limits probably resulted from trade. Ethnographic evidence suggests that tribes in the Great Lakes region traded furs to people who lived hundreds or thousands of kilometers away (Kinietz 1940:245). Our analysis, however, suggests that at least some archeological sites may document local occurrences. For example, for the late Holocene there are two archeological sites (Pabst, Dewitt County, Illinois; and Phipps, Cherokee County, Iowa) with *M. americana* specimens that are well outside the species' historic range. Several lines of evidence suggest that these specimens may in fact be local.

First, the presence of both fishers and martens in the same levels at the Pabst site indicates that *Martes* habitat may have been moderately widespread in central Illinois. This possibility is supported by fisher remains from other central Illinois sites of similar age (Fig. 2.7). Second, both the Pabst and Phipps sites contain many other species (*Sciurus* spp., *Marmota* spp., *Scalopus* spp., and *Tamias striatus*) indicative of woodland habitats suitable for *Martes*. Pollen studies (Webb et al. 1983) also indicate that both localities were dominated by grasslands with riparian woodlands during the late Holocene. Finally, no other faunal specimens or cultural materials from these sites suggest trade with people outside the central Midwest. We, therefore, believe that *M. americana* may have actually extended farther west along the prairie-forest boundary during the late Holocene.

The Etowah site in Bartow County, Georgia, is another example of an apparent extralimital record that may actually indicate the presence of indigenous populations. The persistent occurrence of *M. pennanti* in paleontological and archeological sites in the southern Appalachians during the late Pleistocene and throughout the Holocene (Figs. 2.6, 2.7) suggests that *M. pennanti* probably inhabited this area. Also, if trade were the primary reason for the occurrence of *M. pennanti* in the southern Appalachians, then specimens could be expected at other archeological sites of similar cultures outside the Appalachians. In fact, the prestige conferred by *M. pennanti* in these areas probably would have been greater than in the Appalachians because of its scarcity at lower elevations; however, *M. pennanti* has not been recovered from any sites in the southeast outside of the Appalachian region. We believe that *M. pennanti* probably occurred naturally in small populations throughout the southern Appalachians until historic times, when human activity reduced its distribution. Parmalee (1960*d*), Barkalow (1961:545), and Hagmeier (1956*a*) also thought it quite probable that *M. pennanti* inhabited the southern Appalachians

Conclusions

The modern distributions of North American fishers and martens have been influenced by climatic and environmental fluctuations throughout the late Quaternary as well as by prehistoric and historic human activities. During the late Pleistocene, three species of *Martes* inhabited North America. *M. americana* was widespread across the northern part of the United States, whereas *M. nobilis* and *M. pennanti* were restricted to the western and eastern halves of the country, respectively. All three species extended to lower latitudes and altitudes than do either *M. pennanti* or *M. americana* today. Pleistocene records of *Martes* have not yet been found in the Southwest, southern Plains, or Gulf Coastal Plain. All three species are generally associated with other boreomontane species such as yellow-cheeked vole, red-backed vole and heather vole, in the late Pleistocene.

Many of the Pleistocene faunas contain mixtures of species that are not found together in any one area today. These mixtures appear to represent genuine ancient communities, as shown by radiocarbon dating of rodent teeth and jaws. These nonanalogous environments may reflect a broader habitat tolerance for *Martes* than it would seem from their current restriction to boreomontane environments. As the climate warmed during the late Pleistocene, many boreomontane species, including *Martes*, moved to higher latitudes and altitudes. The reorganization of these communities was individualistic; current associations, such as the strong correlation between the modern distributions for *M. pennanti* and *M. americana*, may not extend very far back in time.

Early and middle Holocene records are sparse for all three species, but late Holocene localities are more abundant. In the eastern United States, it seems that *M. americana* retreated northward sometime during the Holocene, but *M. pennanti* persisted as far south as the southern Appalachians into historic times. Also, *M. pennanti* appears to have been restricted to the eastern United States until the late Holocene, at which time it invaded the west. *M. nobilis* became extinct sometime during the late Holocene. The cause of its extinction is unknown, but it may have resulted from a combination of factors such as competition, habitat restriction, reduction of range, and human activities.

Most of the modern habitat of *Martes*—the boreal spruce forest in the middle latitudes of Canada—evolved only in the Holocene. In fact, some forests may not have reached their present composition until the late Holocene. Unfortunately, little is known about the history of *Martes* in this area.

The wide range of habitats inhabited by *M. pennanti* and *M. americana* in the past may have important implications for current issues in habitat management for *Martes*. Such implications raise questions about (among other things) the type of habitats and habitat reconstruction for reintroduced *Martes* populations and the consequences of future climate changes, whether natural or manmade. Since *Martes* does not currently inhabit all forest types and did not in the past, habitat selection more than the availability of specific prey may be an important limiting factor.

Acknowledgments

Many people have been involved in the compilation and analysis of the distributional patterns of North American *Martes*. We thank P. Parmalee, E. Anderson, D. K. Grayson, H. Semken, R. Stearley, G. McDonald, A. Guilday, R. B. Lewis, J. Peterson, and D. George for information about sites and publications. R. B. McMillan, T. Martin, B. Styles, A. Harn, and D. Esarey provided us with great insight into the potential uses of *Martes* by prehistoric people. J. Snider drafted all the figures, and M. Aiello and J. Morris assisted with manuscript preparation. D. K. Grayson and A. D. Barnosky provided critical editorial comments on earlier versions of this chapter. We thank all these people for their assistance. This work was partially supported by NSF Grant BSR 9005144.

3   Distributional Dynamics of Modern *Martes* in North America

Charles J. Gibilisco

The distributions of the American marten (*Martes americana*) and the fisher (*M. pennanti*) in North America have changed dramatically since the 1600s. The most restricted distributions occurred during the early twentieth century, followed by variable regional expansions and contractions in recent decades. Several researchers have conducted distributional studies of martens and fishers; most notable was Hagmeier's (1956*a*) study, along with those of Hall (1981), Strickland and Douglas (1981, 1983, 1987), Powell (1982), Strickland et al. (1982*a,b*), Deems and Pursley (1983), Clark et al. (1987), and Douglas and Strickland (1987).

This chapter reviews the distributions of fishers and martens in North America, discusses patterns and possible causes of regional distributional changes since the 1950s, and suggests some implications of these changes for management and conservation of these species.

Methods

During fall and winter of 1990–1991, I developed and mailed a two-page survey to 34 state, provincial, and territorial agencies chosen on the basis of Deems and Pursley's book (1983). The survey requested information on the official status, current distribution, type of habitat supporting highest densities, factors influencing distribution since the 1950s, primary sources of data, knowledge of isolated populations, and short-term (10–20 years) and long-term (100–200 years) prognoses for American martens and fishers. I received a 100% response: all those polled either completed the survey in whole or in part or referred me to other sources. Follow-up communications

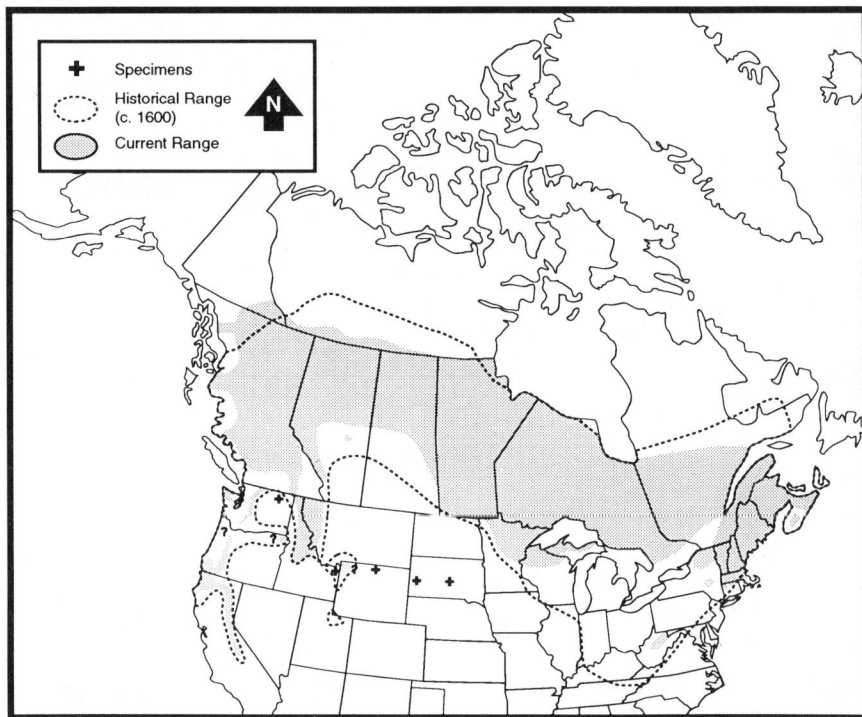

Figure 3.1. Historical and current distribution of fishers (*Martes pennanti*) in North America. Historical distribution derived from Hagmeier 1956*a*, Hall 1981, Powell 1982, and Douglas and Strickland 1987. Current distribution derived from maps and descriptions provided in the 1990–1991 survey, plus Hall 1981, Powell 1982, and Douglas and Strickland 1987.

were made as necessary. Selected survey responses are cited here as personal communications; a list of the individuals cited here and their affiliations is given at the end of the chapter.

Results

The current and historical distributions of fishers and American martens on a continental scale are shown in Figures 3.1 and 3.2. Compared with estimated historical distributions (ca. 1600; Hagmeier 1956*a*, Hall 1981), the current distributions of both species are smaller, particularly south of the Great Lakes region. Both species currently occupy much of their historical ranges, reflecting recoveries from heavy exploitation and habitat loss early in

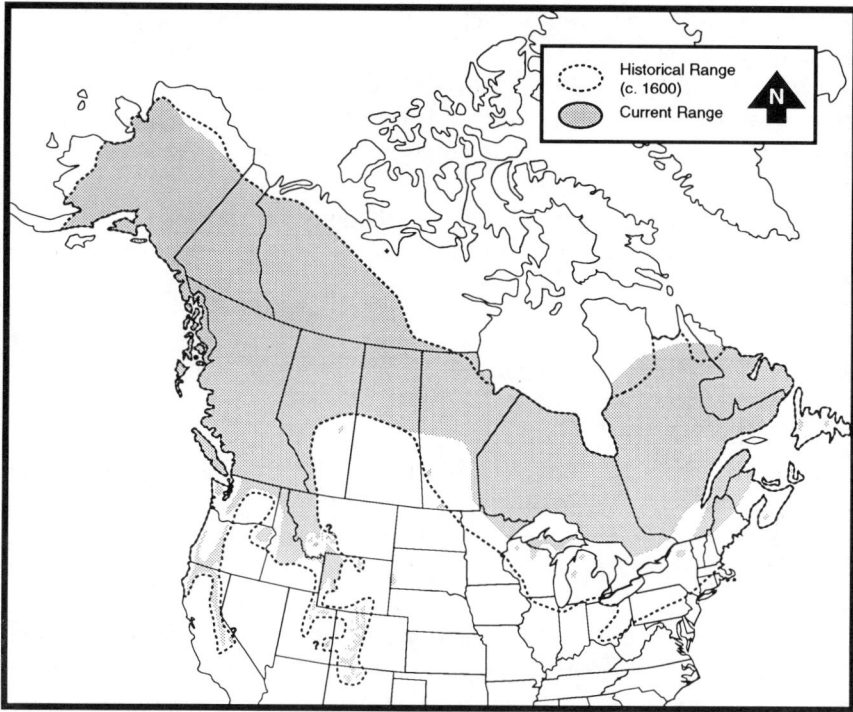

Figure 3.2. Historical and current distribution of American martens (*Martes americana*) in North America. Historical distribution derived from Hagmeier 1956*a*, Hall 1981, and Strickland and Douglas 1987. Current distribution derived from maps and descriptions provided in the 1990–1991 survey, plus Hall 1981, Clark et al. 1987, and Strickland and Douglas 1987.

the century (Anderson 1934, Allen 1942). Such recovery is particularly seen where continuous boreal forest still exists. Continent-wide distributional changes since the 1950s were, most survey respondents reported, relatively minor: of those who answered the question about factors influencing distributional change for American martens, 77% indicated little or no change in distribution since the 1950s; 67% of those answering the question with regard to fishers indicated stable or increasing distributions since the 1950s.

On the other hand, the regional distribution of each species reveals patterns of biogeographic interest, particularly where patchy or disjunct distributions exist. These include: for the American marten, Newfoundland, Nova Scotia, New York, Michigan, Montana, New Mexico, and Washington; and, for the fisher, Nova Scotia, New York, Connecticut, West Virginia, Alberta, Washington, Oregon, and California. Some of these disjunct populations are the result of relatively new reintroduction efforts (Berg 1982; Slough, this volume; B. G. Slough, pers. commun.). Some reflect natural extensions of former range (Berg and Kuehn, this volume; W. E. Berg, pers. commun.). Some may be range contractions related to habitat alteration by humans (Bissonette et al. 1989; J. A. Bissonette, pers. commun.). Others may be due to natural habitat changes (Hoffman and Knox Jones 1970; Patterson 1984; Graham and Graham, this volume) or to some combination of the above factors, as well as to a natural tendency toward disjunct populations at the edges of a species' range.

Regional Distributional Dynamics

Eastern North America

Fishers. During the mid-twentieth century fishers began to reinvade former range throughout many parts of northeastern North America, including Ontario, Quebec, Maine, and New York, as forest reinvaded abandoned farmland and as trapping restrictions and closures took effect (Strickland and Douglas 1981, Powell 1982, Douglas and Strickland 1987). Fishers have now spread out of central Massachusetts into Connecticut (P. Rego, pers. commun.); Massachusetts biologists reported fishers in the lightly forested agricultural region of southeastern Massachusetts (T. Decker, pers. commun.).

American martens. American martens in northeastern North America appear to be reoccupying much of their historical range. Toward the southern edge of their current range, several disjunct distributions are the result of reintroductions since the 1950s (Berg 1982; R. Earle, pers. commun.). But according to survey respondents, the future of the marten in the Atlantic provinces may be tenuous. Lands and Forests officials in Nova Scotia now consider American martens on Cape Breton "likely to disappear (if not already [gone])" (B. Sabean, pers. commun.), although they hope that reintroduced populations in western Nova Scotia will succeed. In Newfoundland, American martens are officially considered threatened, and provincial biologists are unsure about the future of this species (O. Forsey, pers. commun.). American martens are still gone from parts of their southeastern historical range, where trapping and habitat loss led to their extirpation (Clark et al. 1987, Strickland and Douglas 1987).

Central North America

Fishers. Fishers have reoccupied their former range throughout most of central North America, with increased sightings in parts of the historical range of North Dakota (S. H. Allen, pers. commun.) and a couple of incidental trappings in South Dakota (L. F. Fredrickson, pers. commun.). In Manitoba, fishers are now seen "at all times of the year in sparsely forested agricultural lands," according to Department of Natural Resources biologists (C. Johnson, pers. commun.).

American martens. American martens in central North America have refilled parts of their historical range, largely by natural expansions (Berg 1982, Berg and Kuehn, this volume). However, translocations in the Black Hills of South Dakota (Fredrickson 1989), the Cypress Hills in Saskatchewan (W. Runge, pers. commun.), and the Turtle Mountains in southern Manitoba (C. Johnson, pers. commun.) have resulted in disjunct distributions in these areas.

Western North America

Fishers. In western Canada, the distribution of fishers (Fig. 3.3) has remained stable or has increased. Range expansions have been noted in Alberta (Skinner and Todd 1988; A. W. Todd, pers. commun.). In Oregon, California, and Washington, the current distribution of fishers is in question (Gould 1987; G. Gould, pers. commun.; K. B. Aubry, pers. commun.; D. B. Marshall, pers. commun.; J. Thiebes, pers. commun.). Despite three reintroductions in various areas of Oregon, Harris et al. (1982) considered fishers extirpated from western parts of the state. Murie (1974) noted that by the 1950s fishers were no longer found in the coast range of Washington, and K. B. Aubry (pers. commun.) considered the prognosis for the fisher population on the Olympic Peninsula to be "poor." California Fish and Game biologists expected a loss of fishers on the west slopes of the Sierra Nevada over the next couple of decades, and expressed a concern about the total extirpation of fishers in the Sierra Nevada over the long term (G. Gould, pers. commun.). On the other hand, biologists involved in a continuing study to determine fisher and American marten presence on managed northern California timberlands believed that fishers and American martens may increase in the near future unless land-use changes interfere (Criss and Kerns 1990; S. J. Kerns, pers. commun.)

In central Idaho, fishers have expanded from a 1960s reintroduction (G. Will, pers. commun.). Similarly, in Montana, several reintroduction efforts have established fishers in the northwest portions of the state (Weckwerth and Wright 1968, Roy 1990). The precise distribution of fishers in Wyoming in the vicinity of Yellowstone National Park is questionable, inasmuch as only two specimens exist, both from outside the park (one on the east, one on the

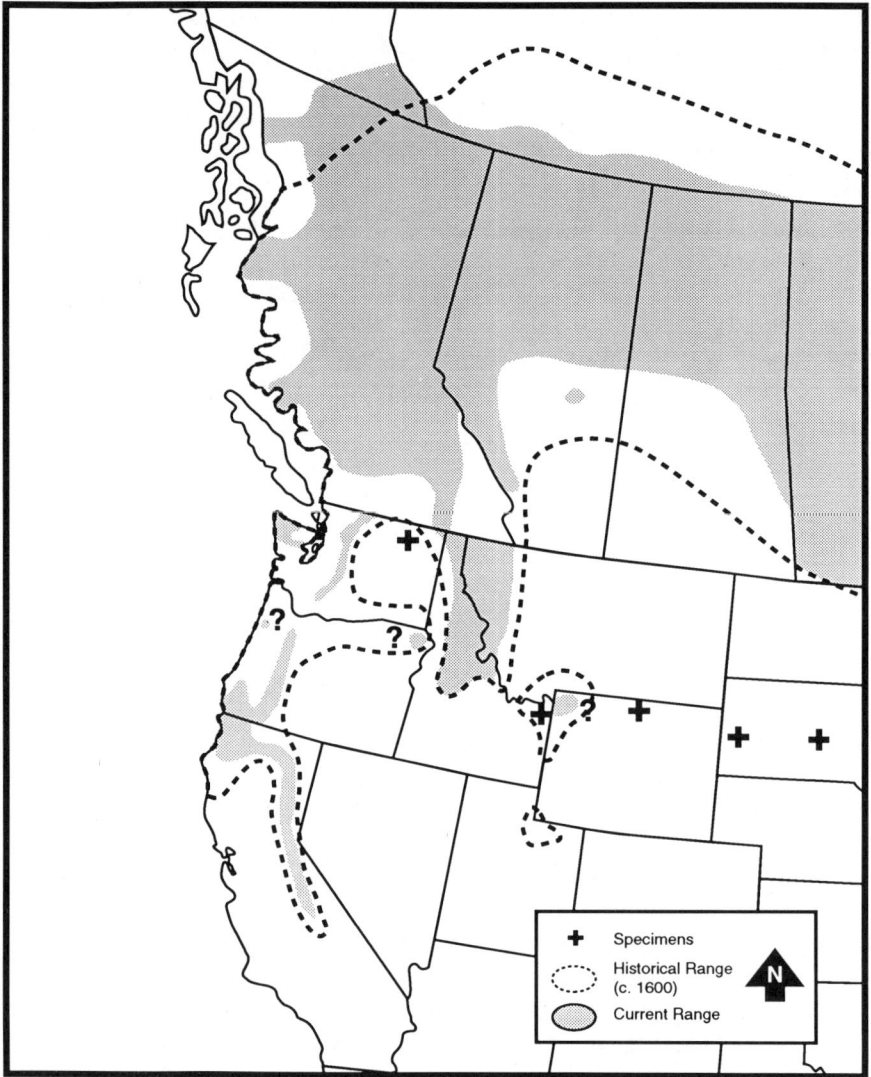

Figure 3.3. Historical and current distribution of fishers (*Martes pennanti*) in western North America. Historical distribution derived from Hagmeier 1956*a*, Hall 1981, Powell 1982, and Douglas and Strickland 1987. Current distribution derived from maps and descriptions provided in the 1990–1991 survey, plus Hall 1981, Powell 1982, and Douglas and Strickland 1987.

west) (Brown 1965; R. Rothwell, pers. commun.; Idaho Nat. Heritage Program, unpubl. data; S. Consolo-Murphy, pers. commun.).

American martens. American marten distribution in much of northwestern North America (Fig. 3.4) has remained relatively stable, including in Alaska (J. Whitman, pers. commun.), the Yukon (Slough, this volume), the Northwest Territories (K. G. Poole, pers. commun.), and British Columbia (J. D. Steventon, pers. commun.). Distribution in Alberta has begun to refill historical boundaries (Skinner and Todd 1988; see also Fig. 3.2).

Farther south in the western United States, American marten distribution is discontinuous. Hoffman and Knox Jones (1970), Patterson (1984), Graham and Graham (this volume), and others considered this pattern of distribution for American martens and other boreomontane species a reflection of forested montane islands, relics of a continuous coniferous forest during Pleistocene time. This landscape of mountainous forest patches continues across southwest Montana east into the Great Plains and south to New Mexico (Fig. 3.5).

Oregon Department of Fish and Wildlife biologists (J. Thiebes, pers. commun.) and an independent biologist working on status and distribution of American martens and fishers (D. B. Marshall, pers. commun.) consider the distribution of American marten east of the central Cascades to be disjunct. Oregon Fish and Wildlife also reported a disruption of historical coastal American marten distribution attributed to logging of coastal old growth (J. Thiebes, pers. commun.).

Distribution has not decreased overall in California, but the abundance of American marten in northern coastal California is being reduced by logging, which is leading to fragmentation of the distribution and possibly to the loss of the coastal distribution of American martens by the turn of the century (G. Gould, pers. commun.). In Washington, distribution does not appear to have changed significantly, but the prognosis for isolated American martens populations on the Olympic Peninsula is poor (K. B. Aubry, pers. commun.).

Discussion

Current patterns of distribution for American martens and fishers in western North America provide us with an opportunity to examine the implications of distributional dynamics for conservation and management of these species. From southwest Saskatchewan and Manitoba south through the intermountain west, forested islandlike mountain ranges rise from a more arid valley floor (Hoffman and Knox Jones 1970) (Fig. 3.5). The size of these

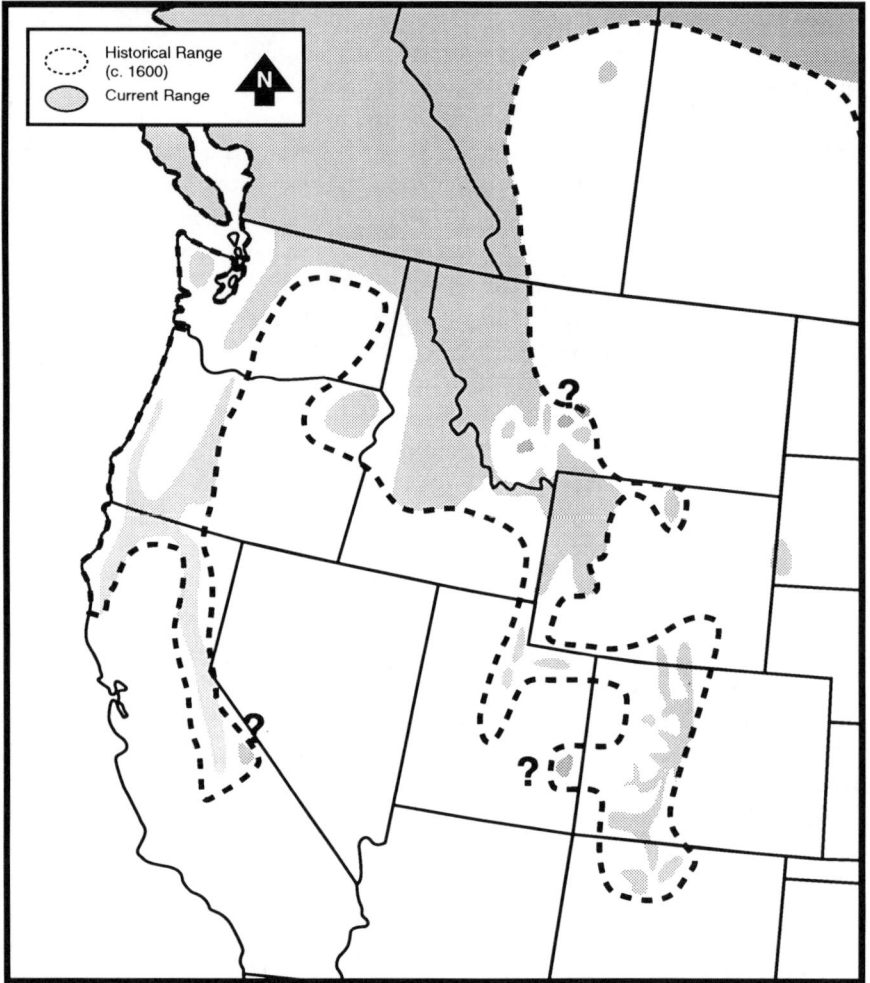

Figure 3.4. Historical and current distribution of American martens (*Martes americana*) in the western United States and southern Canada. Historical distribution derived from Hagmeier 1956*a*, Hall 1981, and Strickland and Douglas 1987. Current distribution derived from maps and descriptions provided in the 1990–1991 survey, plus Hall 1981, Clark et al. 1987, and Strickland and Douglas 1987.

mountain ranges varies from 29 km^2 to 11,600 km^2 (Picton 1979), with interisland distances ranging from less than 10 up to 50 km. Lower elevations surrounding these montane areas are frequently treeless and are dominated by sagebrush-grasslands, dry cropland, or desert.

These lowland environments likely are dispersal barriers for American

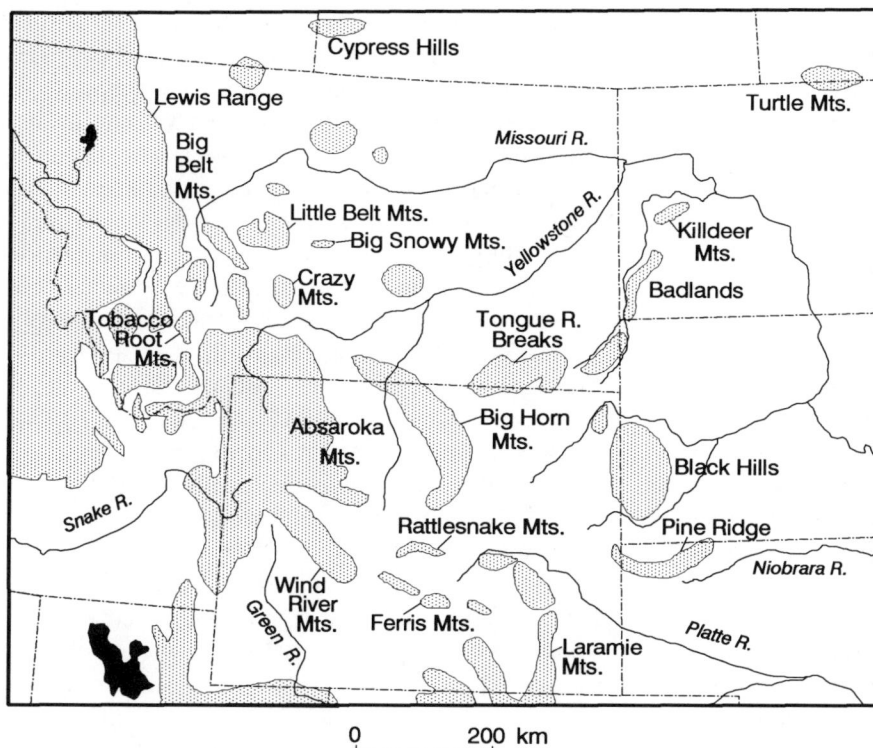

Figure 3.5. The northern Great Plains, central and northern Rocky Mountains, and Great Basin region, derived from Hoffman and Knox Jones 1970:358. Stippled areas indicate approximate distribution of montane areas.

martens. Marten biologists are nearly unanimous in believing that a distance of more than 5 km of unforested land below the conifer zone is a complete barrier to dispersal (Hawley and Newby 1957*a*, Burnett 1981). Unforested lands may serve as barriers for fishers as well; Hoffman and Pattie (1968), and Clark and Stromberg (1987) suggested that habitat patchiness may be one reason why fishers were never considered common in this region.

Possible consequences of this islandlike configuration are complex. As habitat area decreases, the ratio of interior to edge habitat decreases, population size decreases, and the probability of extinction increases unless counterbalanced by local immigration. Interisland distance and size, distance to mainland populations, and the nature of the intervening barrier are all variables that determine the probability of such immigration and population persistence (Wilcox 1978, Patterson 1984).

One might speculate, based on island biogeographic considerations, that the smaller, more isolated montane islands should have lost their American marten populations, whereas larger montane islands closer to a potential source of immigrants may still retain American martens. An example from Montana of three montane islands, similar in total area but varying in distance from the nearest American marten population, suggests that this area-distance relationship may not be enough by itself to predict population distributions (Fig. 3.5). The Snowy Mountains are about 733 km² in total area (Picton 1979), and at least 78 km from the nearest known American marten population; there are no records or sightings of martens in the Snowies. The Crazy Mountains, with an area of 932 km² (Picton 1979), are approximately 36 km from the nearest American marten population; the Crazies have a currently harvested American marten population (K. Aune, pers. commun.). The Tobacco Root Mountains, on the other hand, with an area of 1120 km² (Picton 1979) are only 18–20 km from the nearest known American marten population; although American martens were once found in the Tobacco Roots (Hawley and Newby 1957b), no American marten sightings or trappings have been documented for the range since the 1960s.

Many factors may help explain this incongruency. Picton (1979) found, for example, that topographic variation was more significantly predictive of numbers of big-game mammal species than was area alone. Vegetational and climatic changes associated with changes in elevation might also influence the occurrence of species such as American martens that are strongly habitat-selective (S. W. Buskirk, pers. commun.). Other factors affecting habitat quality and quantity within a given area might include current and historical human land-use patterns (e.g., the Tobacco Roots have an extensive and intensive mining and logging history, whereas the Crazy Mountains are much less disturbed); accuracy of American marten sightings and records; human access to American marten habitat; locations of American marten translocations; and the presence and nature of forested riparian ways that may serve as dispersal corridors linking the areas. Thus, not only the overall size and isolation of mountain ranges or other habitat islands but also the quality and quantity of the habitat and other factors may complicate the island biogeographic predictions for American martens and fishers.

Because of such complex and potentially synergistic factors, these naturally occurring isolated mountain ranges may serve as valuable study areas for population dynamics of American martens. Areas for further research include population persistence, dispersal potentials, genetic diversity within and among populations, and the possible effects of further habitat fragmentation from activities such as logging and road-building in an already patchy regional habitat. Many of these mountain islands are managed for multiple purposes (timber production, mining, recreation, oil and gas development, and other activities that affect American marten habitat), as are many larger, more continuous expanses of public forest (USDA Forest Service 1986, 1987). In the case of isolated mountain ranges, management practices may need to be adjusted to take into account a higher susceptibility to local extinction, if American martens are to persist in the long term (Schaffer 1983, Grayson 1987).

The western distribution of American martens provides a good case study of a scale phenomenon that may be at work when distribution lines are drawn. What seems to be a contiguous distribution at the continental scale breaks into isolated ranges at a local scale. The forests inhabited by American martens and fishers are subjected to habitat pressure from human activities, natural habitat changes, and even possible climate changes and other global phenomena, all of which will affect the ranges of both American martens and fishers in the future, as they have in the past (Graham and Graham, this volume). Similar factors, perhaps even more subtle ones, may well be occurring elsewhere on the continent.

Conclusions

Since the 1950s the distribution of fishers has recovered in some parts of its historical range thanks to natural and human-induced changes in forested habitats and to reintroduction efforts. This recovery has been concentrated in central and eastern portions of the fisher's former North American range; in western North America, on the other hand, current distribution is uncertain, and fishers are absent from their former range south and east of the Great Lakes region.

The distribution of American martens stretches continuously from Alaska to Quebec and Labrador and overall has changed little since the 1950s. In many parts of the American marten's southern range, however, populations are disjunct because of natural and human-caused factors. American martens are absent from most of their historical range south and east of the Great Lakes region.

While both American martens and fishers are refilling and expanding their historical distributions in several places, their distributional edges are unstable in some areas largely as a result of habitat alteration and fragmentation of forested environments; this fragmentation adds to the natural patchiness frequently found at distributional edges. In western areas, naturally isolated populations of American martens in particular are being further subjected to increasing human pressures and land-use changes.

If distributional patterns and dynamics serve as any kind of indicator of regional habitat quality and suitability, the patterns we are seeing for American martens and fishers in parts of North America may indicate a need for fine-tuned local management and research at a regional level that will keep an eye on habitat fragmentation and isolated populations.

Traditionally we have connected the dots, so to speak, to establish the perimeters of a species' distribution, even though it has always been understood that within such a boundary, animals are rarely equally distributed either in time or space. But it may be more appropriate now than ever before to look more carefully between the dots as growing human populations and resulting land-use changes affect the forests used by American martens and fishers in North America.

Acknowledgments

Special thanks are due all the individuals who completed my survey and spent time digging out additional materials and maps. I also thank Robert Moore, Harold Picton, Lynn Irby, and Katherine Hansen of Montana State University for their critiques. I am grateful to anonymous reviewers for their helpful comments and criticisms. Ed Madej of Great Divide Graphics, Helena, Montana, created the distribution maps.

Personal communications (i.e., survey responses) from the following individuals are cited in text:
S. H. Allen, North Dakota Fish and Game
K. B. Aubry, USDA Forest Service, Pacific Northwest Research Station
K. Aune, Montana Department of Fish, Wildlife, and Parks
W. E. Berg, Minnesota Department of Natural Resources
J. A. Bissonette, Utah State University
S. W. Buskirk, University of Wyoming
S. Consolo-Murphy, Yellowstone National Park
T. Decker, Massachusetts Department of Fish and Wildlife
R. Earle, Michigan Department of Natural Resources
O. Forsey, Government of Newfoundland and Labrador, Wildlife Division
L. F. Fredrickson, South Dakota Department of Game, Fish, and Parks
G. Gould, California Department of Fish and Game
C. Johnson, Manitoba Department of Natural Resources
S. J. Kerns, Wildland Resource Managers, Round Mountain, California
D. B. Marshall, Consulting Wildlife Biologist, Portland, Oregon
K. G. Poole, Northwest Territories, Department of Renewable Resources
P. Rego, Connecticut Department of Environmental Protection
R. Rothwell, Wyoming Game and Fish Department
W. Runge, Saskatchewan Department of Parks and Renewable Resources

B. Sabean, Nova Scotia Department of Lands and Forests
B. G. Slough, Yukon Department of Renewable Resources
J. D. Steventon, British Columbia Ministry of the Environment
J. Thiebes, Oregon Department of Fish and Wildlife
A. W. Todd, Alberta Department of Forestry, Lands, and Wildlife
J. Whitman, Alaska Department of Fish and Game
G. Will, Idaho Department of Fish and Game

4 Morphology, Ecology, and the Evolution of Sexual Dimorphism in North American *Martes*

Thor Holmes and Roger A. Powell

In a study of the arboreal adaptations of American martens (*Martes americana*), fishers (*M. pennanti*), and their close relatives in the subfamily Mustelinae, Holmes (1980) used ratios to minimize the effect of sexual size dimorphism in samples. This study revealed an unexpected sexual dimorphism in the ratios themselves. While male and female musteline postcranial skeletons are sexually dimorphic for size, they are proportionally the same. Female musteline postcranial skeletons are isometrically scaled-down models of males. This characteristic does not hold for male and female musteline skulls. The skulls of the sexes are also dimorphic, but females have larger skulls relative to their own body size than do males. In other words, male and female musteline skulls are closer to being the same size than are their bodies. This observation prompted our examination of sexual size dimorphism in *Martes*.

An extensive literature relates body size of predators to prey size. Competing species often partition resources along a size gradient (MacArthur 1972, Schoener 1974, Hespenheide 1975, Simberloff and Boecklen 1981). Although a diet of prey of different sizes should be reflected in differences in carnivore body size and in the sizes of their trophic structures, resource partitioning has not been verified among North American mustelids (Rosenzweig 1966, Powell 1979a, Powell and Zielinski 1983). Powell and Zielinski (1983) summarized the evidence for resource partitioning by mustelids and argued that some of the best evidence is for the genus *Martes*.

Sexual dimorphism is striking among mustelines; males may exceed females of the same population by 5–30% in length, most commonly by 10–20%, and may weigh twice as much. Brown and Lasiewski (1972) suggested that this sexual dimorphism permits resource partitioning by prey size, reduc-

72

ing competition between males and females of the same species (Fig. 4.1). This intuitively satisfying suggestion has been widely examined among vertebrates but has not always been supported (Selander 1966; Schoener 1967; Husar 1976; Snyder and Wiley 1976; Powell 1979*a*, 1981*a*). In contrast, Wilson (1975) argued convincingly that resource partitioning by size can occur only when the smaller species or sex has access to prey not available to the larger. Schoener (1974) argued that resource partitioning is so complex a problem that examination of a single "niche dimension" such as prey size may be too simple to be useful.

An older alternative hypothesis, traceable from Darwin (1871) to Trivers (1972), explains sexual dimorphism in vertebrates, and therefore in mustelids, as the result of selective pressures acting on male and female body size independently. Great sexual dimorphism in mammals, especially among carnivores, correlates with extreme polygyny and minimal male parental investment (Ralls 1976, Alexander et al. 1979). Moors (1974, 1980) and Erlinge (1979) hypothesized that sexual selection in weasels and other mustelines favors large male body size, whereas the bioenergetic constraints of reproduction dictate small female size (Fig. 4.1). Examination of fishers supported this hypothesis (Powell 1979*a*, 1981*a*; Powell and Leonard 1983).

No study has argued that resource partitioning is the sole driving force for sexual dimorphism. Once a carnivorous species is sexually dimorphic, however, selection should favor increased dimorphism when appropriately sized prey are limiting. Dietary overlap between the sexes in least weasels (*Mustela nivalis*) is most pronounced during critical times of low prey abundance and weasel life history (Erlinge 1975). The same is likely true for members of the genus *Martes*.

Thus, for vertebrates in general and mustelids in particular, hypotheses about sexual dimorphism have become complex and involve elements of both resource partitioning and sexual selection. Such complex hypotheses are difficult to test because of interacting variables, but the hypothesis that resource partitioning has been a contributing selective force for sexual dimorphism leads to straightforward predictions that can be tested easily.

We quantify sexual dimorphism here in order to test for a relationship between the level of sexual dimorphism and resource partitioning. If sexual dimorphism results from resource partitioning in any instance, it ought to be in the Mustelinae. In a variety of regions, two to three (sometimes more) congeners (*Martes, Mustela*) co-occur to form "hunting sets" in the sense of Rosenzweig (1966). As these carnivores exploit a finite resource, we might well expect to see some evidence of resource partitioning among these species. The hypothesis predicts that sexual dimorphism should be small when two or more competing species co-occur, compared with when they are

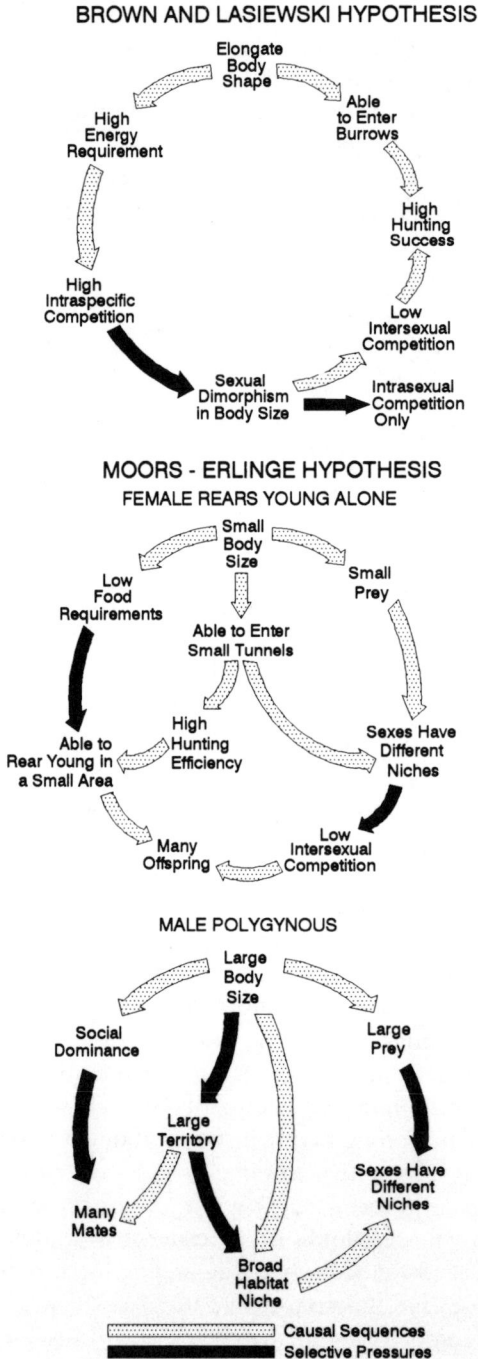

Figure 4.1. A simplified version of the Brown-Lasiewski model of sexual dimorphism in weasels (Moors 1980) and the Moors (1974) and Erlinge (1979) model of weasel sexual dimorphism.

allopatric. Ralls and Harvey (1985) examined geographic variation in sexual dimorphism in the three species of North American weasels and rejected the hypothesis that sexual dimorphism is a resource partitioning phenomenon. They found no pattern between sexual dimorphism and presence or absence of congeners, but their analyses were hindered by their computation of sexual dimorphism indices from condylobasal lengths of skulls.

The problem with using skull lengths to generate indices of sexual dimorphism derives from the complex structure of the mammalian skull whose size, shape, and proportions result from myriad selective pressures. Although condylobasal length is highly correlated with head and body length, Radinsky (1984) demonstrated that the facial skull of carnivores is more variable than the basicranium. The most appropriate way to resolve our question is with indices of sexual dimorphism based on trophic structures. This permits us to ask two essential questions. First, are sexual dimorphism indices based on trophic structures the same as such indices based on condylobasal length? Second, if they are not, do they vary in a pattern consistent with a resource partitioning hypothesis? Carnassials are evolutionarily the most dynamic trophic structures in mustelids (Butler 1946). A test of whether trophic-based indices of sexual dimorphism provide new information is whether they are consistent with indices based on condylobasal length.

Methods

To test the hypothesis of a relationship between sexual dimorphism and resource partitioning, we measured head and body length and a total of 13 cranial and dental measures that have trophic significance in 710 *Martes* specimens (*M. americana*: 302 males, 248 females; *M. pennanti*: 69 males, 91 females) using dial calipers accurate to 0.05 mm. These measures have been explained and justified for *Mustela* spp. by Holmes (1987).

Only adult specimens were used in these analyses (adult dentition, often with signs of wear, and closed facial sutures). Data were subdivided by subspecies (Hall 1981) and further divided by region (east of 84° longitude; central; or west of 100° longitude) because some subspecies have extensive east-west ranges. This division resulted in 13 subspecific transects for *M. americana* and 2 for *M. pennanti*. For specimens that were broken or missing teeth, values were computed with BMDPAM (Dixon 1983) if 5 or fewer of the 13 cranial variables were missing. This technique was important when sample sizes were small. Discriminant functions analyses (DFA) were carried out at the subspecies level with BMDP7M (Dixon 1983) to discriminate between males and females using the 13 skull and tooth variables. For specimens of unknown sex, an initial DFA assigned sex with high probabilities, and these assignments were accepted. Specimens that appeared to have been sexed incorrectly were noted during measurement. If such a specimen then fell well within the range for the opposite sex in the initial DFA, it was reassigned to the opposite sex. A second DFA was then performed on each subspecies, to be used in later analyses. This was justified only because the sexes are extremely different in mustelines and the probabilities associated with the assignments of unknowns to sex were very high ($P > 0.75$ in 99% of the cases, $P > 0.82$ in 95% of the cases).

Means, standard deviations, and coefficients of variation were computed for each sample. Correlation matrices were examined for measures with low correlations to other skull measures, particularly to condylobasal length of the skull. Principal components analyses (PCA) were carried out with BMDP4M (Dixon 1983). PCA using correlation matrices were run with sex entered as a variable for four samples of *Martes* to examine how sex loaded onto the principle axes with the other measures.

Sexual dimorphism indices were computed for each measure in each sample by dividing male means by female means. Raw sexual dimorphism indices and log-transformed indices were all examined for normality (Ryan et al. 1980). An analysis of variance (ANOVA) was carried out for the set of sexual dimorphism indices to test for differences between indices based on condylobasal length and indices based on trophic structures.

Results

Means ($\pm$ SD) for cranial and dental measures used to test the hypothesis of a relationship between sexual dimorphism and resource partitioning are shown in Figures 4.2 and 4.3, and means appear only in Figures 4.4 to 4.6. All skull measurements were highly intercorrelated, so that estimates for missing values were likely very good. Although correlations of tooth measurements with skull measurements were poorer, broken skulls were more problematic than were missing teeth. Thus, measurements for which correlations were poorest were seldom missing. Poorest correlations were found between condylobasal length and five tooth measures: molar width, molar depth, anterior carnassial width, posterior carnassial width, and carnassial length. These results encourage further examination of the data in the context of a resource partitioning hypothesis.

PCAs revealed that size was the major source of variation. Factor 1 had high positive loadings for all variables including sex. Subsequent factors were based principally on tooth and jaw measures (Table 4.1). Trophic struc-

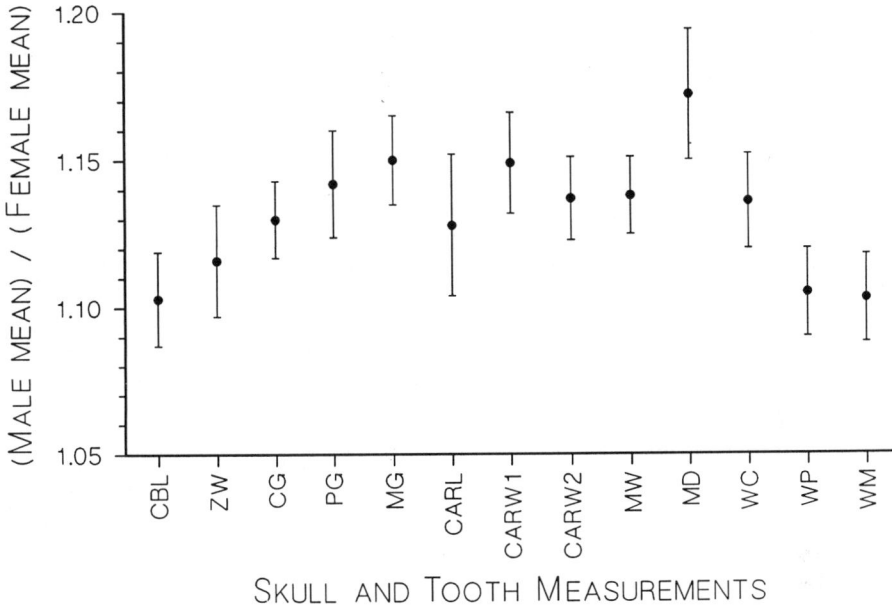

Figure 4.2. Means and 95% confidence intervals of sexual dimorphism indices for *Martes americana* (*n* = 13). CBL = condylobasal length of skull, ZW = zygomatic width, CG = distance from the base of the canine to the glenoid fossa, PG = distance from the front of the canine to the glenoid fossa, MG = distance from the back of the canine to the glenoid fossa, CARL = carnassial length, CARW1 = anterior width of the carnassial (at the protocone), CARW2 = posterior width of the carnassial (at the protocone), MW = molar width, MD = molar depth, WC = width at canines, WP = width at premolars, WM = width at molars. Summary statistics for all measurements used to make indices were reported by Holmes (1987).

tures appeared to contribute to overall variation. When PCAs were rerun without sex as a variable the results were surprisingly similar to the first iterations.

The pattern of sexual dimorphism indices shown in Figures 4.2–4.6 is consistent for all *Martes* samples. Distributions of all indices, log-transformed and nontransformed, did not differ from normal (*P* > 0.05).

ANOVAs on log-transformed indices revealed significant differences between indices based on different skull and tooth measures. In *M. americana* (Fig. 4.2; *n* = 13) there were low index values (range = 1.103–1.117) for condylobasilar length, zygomatic width, width at premolars, and width at molars. Other index values were high (range = 1.125–1.172) but variable. Only two sets of indices were obtained for *M. pennanti*. Thus, the pattern for

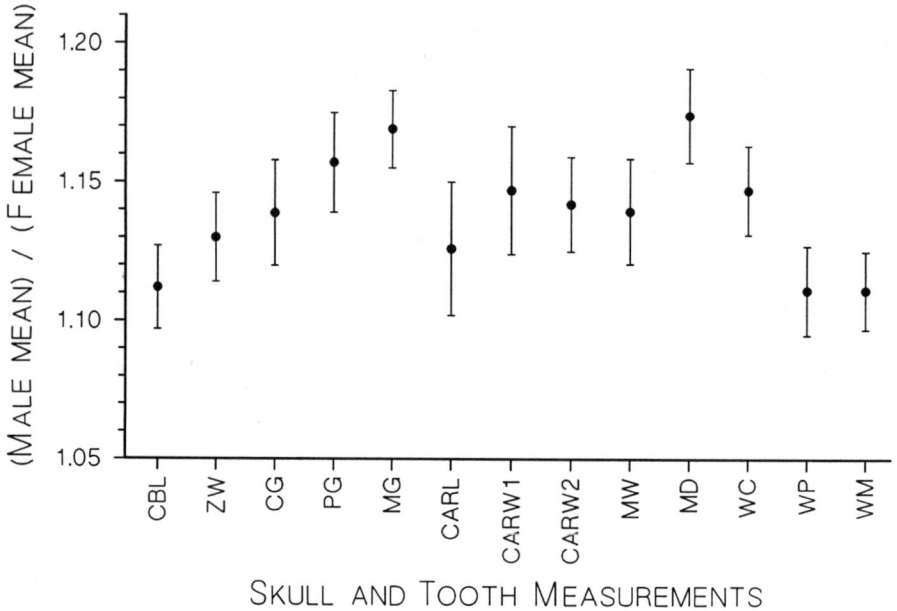

Figure 4.3. Means and 95% confidence intervals of sexual dimorphism indices for *Martes americana* and *M. pennanti* ($n = 15$). Abbreviations as for Figure 4.2.

all *Martes* ($n = 15$; Fig. 4.3) was mostly a reflection of that in *M. americana* except that *M. pennanti* is slightly more dimorphic than *M. americana*.

The sexual dimorphism indices for *M. americana* do not vary in response to sympatry with *Mustela* spp. (Fig. 4.4). The pattern of sexual dimorphism in *M. americana* is similar to that in weasels and does show some significant variation among indices. This variation is inconsistent with respect to sympatry or allopatry from *M. pennanti* (Figs. 4.4, 4.5). In three samples of island subspecies of *M. americana* allopatric from *M. pennanti*, the sexual dimorphism indices are lower than those of continental conspecifics. The differences between the sexual dimorphism indices for these three subspecies and the continental subspecies that co-occur with *M. pennanti* are very slight. In one small sample (four males, three females) of a continental subspecies (*M. a. origenes*) allopatric from *M. pennanti*, sexual dimorphism indices are dramatically lower than indices from *M. americana* subspecies sympatric with *M. pennanti*. Although this may be in part a problem of small sample size, the trend is exactly at variance with the predictions of a resource partitioning hypothesis. The indices for two subspecies of *M. pennanti* are virtually identical (Fig. 4.6). Fishers are more dimorphic than martens in

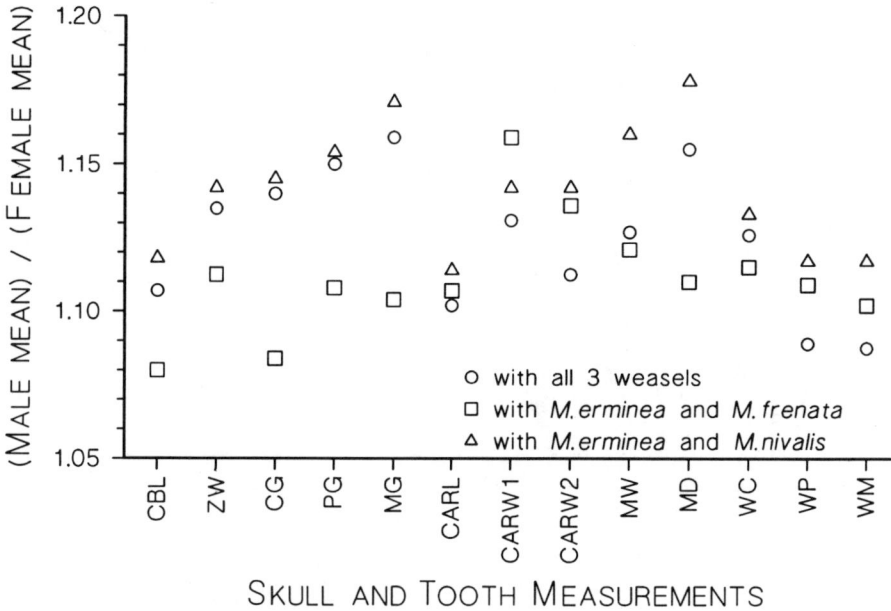

Figure 4.4. Sexual dimorphism indices for *Martes americana abietinoides* where it is sympatric with all three weasel species (*Mustela* spp.), sympatric with *M. erminea* and *M. nivalis* alone, and sympatric with *M. erminea* and *M. frenata* alone. Abbreviations as for Figure 4.2.

condylobasal length, zygomatic width, and tooth-row characters and about as sexually dimorphic as martens in their other characters.

Correlations between all trophic-based sexual dimorphism indices and condylobasal length indices in *Martes* were highly significant ($P < 0.01$). The patterns for *Mustela* were similar. Visual examination of plots of the sexual dimorphism indices of *Martes* corroborated the results of the correlation analysis (Figs. 4.4–4.6).

Discussion

Sexual dimorphism in *Martes* has not evolved to facilitate resource partitioning. Ratios of limb element measures are not sexually dimorphic; thus male and female mustelid bodies are built to the same scale (Holmes 1980). In contrast, ratios of limb element measures to condylobasal length of the skull are sexually dimorphic (Holmes 1980). This means that the skulls of

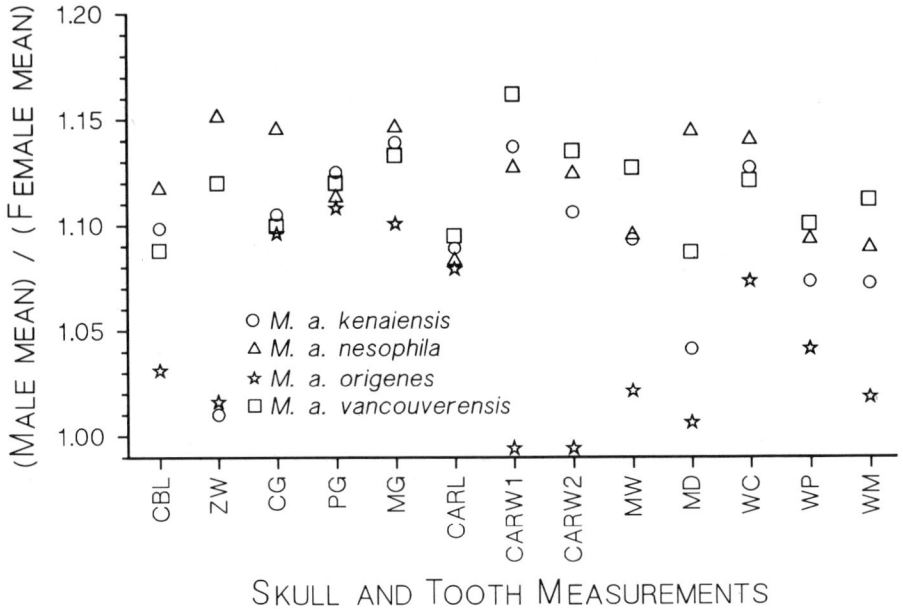

Figure 4.5. Sexual dimorphism indices for *Martes americana* subspecies that are allopatric from *M. pennanti*. Abbreviations as for Figure 4.2.

males and females are more similar in size than are their bodies, which exactly contradicts predictions of the resource partitioning hypothesis. Skulls are trophic structures in part and should be more different than the nontrophic limb elements if resource partitioning were responsible for sexual dimorphism in mustelid populations.

The proposition that sexual dimorphism in size among carnivores, particularly mustelids, is a resource partitioning phenomenon has been the focus of several studies that have examined both morphological variation and diet differentiation (Moors 1974, 1977; Erlinge 1979; Ralls and Harvey 1985; Dayan et al. 1989) and has been an incidental aspect of many others (e.g., Rosenzweig 1966, Ralls 1977, Alexander et al. 1979, King and Moors 1979, King and Moody 1982, Erlinge 1983). Most studies (Erlinge 1979, Moors 1980, Ralls and Harvey 1985) have also refuted the resource partitioning hypothesis for sexual dimorphism. Ralls and Harvey (1985) used condylobasal length of skulls to generate sexual dimorphism indices in their study, but our study reveals that such indices will mask sexual dimorphism because skulls are not as highly sexually dimorphic as the trophic region of the skull. Ralls and Harvey noted that the correlation between condylobasal

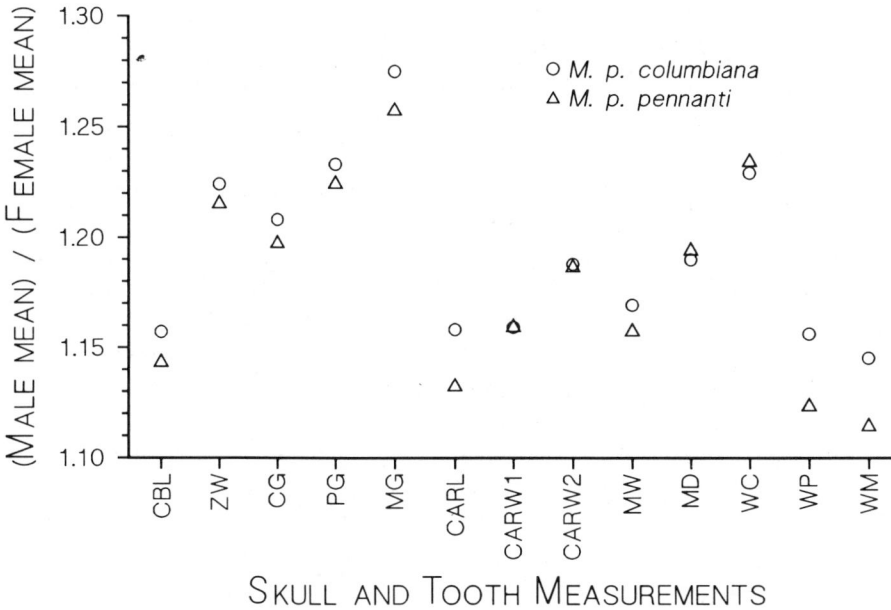

Figure 4.6. Sexual dimorphism indices for *Martes pennanti*. Abbreviations as for Figure 4.2.

Table 4.1. Results of principal components analyses with sex entered as a variable, showing the obvious relationship between sex and size in *Martes* and between sex and trophic structures

Species	Level	Factors	% Variance explained	Measurements loading
M. americana	Subspecies	1	80.4	All
		1–4	91.6	
		2–4		Teeth, sex, ZW, MG, WP
	Sample			
		1	80.6–82.5	All
		1–4	91.2–92.9	
		2–4		Teeth, sex, ZW, MG, WP
M. pennanti	Subspecies	1	84.7–88.1	All
		1–4	93.5–94.4	
		2–4		Teeth, sex, ZW

Note: ZW = zygomatic width, MG = distance from the back of the carnassial to the glenoid fossa, WP = width at premolars.

length of the skull and five "feeding measures" was quite high and eliminated them from further study. One of us also found high correlations between condylobasal length of skull and trophic measures, using an extensive sample of *Mustela erminea* ($n = 1360$; Holmes 1987). The lowest correlation coefficients in our sample (and in that of Ralls and Harvey) were between condylobasal length and tooth measures, suggesting that teeth (trophic structures) vary somewhat independently of condylobasal length of the skull (body size). This is consistent with the possibility that trophic structures are more responsive than condylobasal length to selective pressures for resource partitioning, leading to character displacement between the sexes. This observation could be taken as indirect support for a resource partitioning hypothesis and, if nothing else, encourages further investigation.

The PCAs of *Martes* species, subspecies, and subspecies transects produced results that were typical for morphological data sets (Table 4.1). Factor 1 was invariably a size axis and accounted for the majority of variation. In the PCAs with sex as a variable, sex loaded heavily on Factor 1. This finding points out the obvious relationship between sex and size in *Martes*. In these analyses, there is an interplay between sex and trophic structures, and among trophic structures on Factors 2–4.

Factors 2–4 involve the interplay of trophic structures, particularly tooth measures. In *M. americana*, Factors 2–4 involve almost exclusively tooth measures, although the molar-to-glenoid measure (MG), width at premolars (WP), and zygomatic width (ZW) contribute as well. In *M. pennanti*, Factors 2–4 again involve almost exclusively tooth measures, and only ZW as well. For both of these species, MG and WP are trophic measures and ZW is strongly associated with sex and size. Similar percentages of variation are explained by Factor 1 and Factors 1–4 for both *M. americana* and *M. pennanti*, although percentages are higher for *M. pennanti*. These PCAs suggest variation in trophic structures, especially the teeth, and again indirectly encourage further examination of a resource partitioning hypothesis.

Results of identical analyses of weasels (*Mustela frenata*, *M. erminea*, and *M. nivalis*) were very similar (Holmes 1987). Appendicular skeletal measurements and ratios revealed that both sexes of each species are built on the same body plan, but ratios that include skull length show that skulls are less dimorphic than body size. PCAs demonstrate that trophic structures are somewhat more dimorphic than skull length.

Do trophic considerations, specifically resource partitioning for food, mediate the level of sexual size dimorphism in mustelid populations? Ralls and Harvey (1985) have asserted that they do not, and their data tend to refute the hypothesis of a relationship between resource partitioning and sexual dimorphism.

The ANOVAs used in this study showed that different trophic structures produce different sexual dimorphism indices ($P < 0.005$) in *M. americana*, in all *Martes*, in various samples of *Mustela*, and in all *Mustela* plus all *Martes*. Thus the pattern of sexual dimorphism in indices based on condylobasal length is different from that in indices based on trophic structures throughout the Mustelidae.

Sexual dimorphism indices based on condylobasal length were consistently low for all species studied, especially for *M. erminea*. The pattern of sexual dimorphism index values for *Mustela* spp. was consistent across all species and geographic areas and did differ somewhat from the pattern documented for *Martes* (Holmes 1987). Zygomatic width in *M. americana* is not as dimorphic as in *Mustela* spp., whereas carnassial width at the protocone and molar width are more dimorphic.

Correlation analyses showed that the trophic-based sexual dimorphism indices parallel the sexual dimorphism indices based on condylobasal length. Thus, despite differences in indices of sexual dimorphism based on different skull structures and teeth, all indices exhibit similar patterns. In addition, the sexual dimorphism indices for *M. americana* do not vary in response to sympatry with *M. pennanti* or with *Mustela* spp. The refutation of a resource partitioning hypothesis is unequivocal. At the level of subspecies or subspecies transect, the sexual dimorphism indices for the three weasel species also refute the resource partitioning hypothesis.

Wilson (1975) noted that large predators capture prey unavailable to smaller predators, whereas the reverse is not true, giving a competitive advantage to large predators. *M. pennanti* is larger than *M. americana*. If *M. pennanti* has a competitive advantage over *M. americana*, it is reasonable to suggest that *M. pennanti* might make relatively few or small adjustments in the face of competition from *M. americana*, while *M. americana* would be much more responsive to the presence or absence of *M. pennanti*. A similar argument applies to *M. americana* and weasels. Such is distinctly not the case with respect to sexual dimorphism indices.

The distribution and abundance of mammalian body sizes is roughly log normal (Eisenberg 1981), and *Martes* spp. exploit mammalian prey in the range of sizes into which most mammals fall. Perhaps there are no resource-linked adjustments in sexual dimorphism, a considerable genetic or developmental feat, because of vast numbers of potential prey. Populations of *Martes* spp. may shift behavior, habitat, home range size, prey species, or some combination of these when competition becomes significant (Powell this volume). Kathy Ralls (pers. commun., Natl. Zool. Park, Smithsonian Inst.) has suggested that a morphological approach is unlikely to resolve further the relationship between sexual dimorphism and resources and has argued that long-term intricate food studies are needed for these species.

Abrams (1983) suggested that there is no universal limit to similarity.

Limiting similarity is more likely to explain differences in a species' number than in its overlap with another species. This suggestion implies that musteline populations are not so likely to undergo adjustments of levels of sexual dimorphism in the face of variable resource abundance as they are to adjust population numbers. Abrams's conclusions are consistent with the results of our study.

We ignored time in our study. Powell (1979a, 1981a) showed that the levels of sexual dimorphism in a population of *M. pennanti* varied as much temporally as do the geographic levels of sexual dimorphism found in museum collections. He and King and Moody (1982) showed that sexual dimorphism indices are highest in populations with large males and that variation in male body size is related to diet during the period of growth. Sexual dimorphism appears to track resource abundance.

Sexual dimorphism in mustelines is high and varies geographically. The pattern of this variation is not, however, what the resource partitioning hypothesis predicts. Some data show that male and female mustelines sometimes partition resources, and there are intersexual differences in the use of habitat, holes, and prey at least for parts of the year (Erlinge 1974, 1979, 1981; King and Moors 1979; King 1983). All the requisites to demonstrate adjustments in levels of sexual dimorphism in response to the presence or absence of a competitor seem to be present. The fact that the pattern of variation of sexual dimorphism is not that predicted by the resource partitioning hypothesis supports Moors's (1974, 1980) and Erlinge's (1979) suggestions that, while resource partitioning between sexes of musteline species exists, it is likely the result of sexual dimorphism, not the driving force leading to sexual dimorphism.

Acknowledgments

We thank C. Wozencraft and an anonymous reviewer for helpful comments on early versions of the chapter. Consie Powell drew the figures. For these analyses we used all the *Martes* specimens with skeletal material at the American Museum of Natural History, the Humboldt State University Museum of Zoology, the Los Angeles County Museum of Natural History, the National Museum of Natural History, the University of California at Berkeley Museum of Zoology, the University of Montana Museum of Zoology (only in part), and the University of Puget Sound Museum of Zoology.

5   Body Weight Variation among Insular and Mainland American Martens

David W. Nagorsen

Intraspecific geographic variation is an important phenomenon in evolutionary biology because of its implications for adaptation and speciation. Of particular interest are isolated islands, where strong selection pressures, founder effects, and genetic isolation interact to promote evolutionary divergence in phenotypic and genetic traits. In the mustelids, geographic variation in morphometric traits has been studied most intensively in short-tailed weasels (*Mustela erminea*). Continental and insular populations of this species demonstrate substantial variation in body weight and cranial size. This variation has provided a model for testing various hypotheses relating to the evolution of body size and sexual size dimorphism (Ralls and Harvey 1985, Erlinge 1987, King 1989*a*, Eger 1990).

Data on size and sexual variation among marten populations are scanty and mostly limited to taxonomic studies. Anderson (1970) summarized cranial size for the six Eurasian species, and Hagmeier (1955, 1958, 1961) mapped cranial size trends for American martens (*Martes americana*). Their research, however, focused primarily on the taxonomic implications of size variation rather than on its adaptive significance. Moreover, they did not evaluate size divergence in insular populations. Giannico and Nagorsen (1989) assessed multivariate patterns of variation in American marten skulls from three insular and two mainland areas in the Pacific Northwest. Their research suggests that size and sexual dimorphism are greater among insular than among mainland American martens. Using univariate ratios derived from cranial measurements, however, Holmes and Powell (this volume) conclude that three insular subspecies of martens are less dimorphic than continental subspecies.

Studies on body weight trends among marten populations are more limited. Shubin and Shubin (1975) tabulated body weight data for marten species in the Soviet Union, but their attempt to correlate the degree of size dimorphism with environmental severity was hampered by sampling deficiencies. With data on marten body weight available from only a few widely separated localities, there has been no attempt to evaluate weight variation across a broad geographic region in North America.

From 1975 to 1989 I collected data on carcass weights from American martens of known sex and age to evaluate patterns of geographic variation among three insular and five mainland areas in the Pacific Northwest. I was especially interested in divergence in body size and the degree of sexual size dimorphism in the insular populations. Because I had comparative weight data both from islands and from their presumed ancestral populations on the nearby mainland, I was able to test two hypotheses: first, that body size is larger in insular martens, as predicted by the island rule (Lomolino 1985); and second, that sexual size dimorphism is more pronounced in insular than mainland populations in accord with the trend demonstrated by Giannico and Nagorsen (1989) for cranial size.

Methods

I based my analyses on 1734 skinned carcasses collected during the winter trapping season (Table 5.1). Animals were sexed by anatomy; age was determined from counts of cementum annuli on lower canine teeth. Tooth sectioning and age determination were done by a commercial laboratory (Matson's

Table 5.1.   Collecting locations and dates for eight samples of American marten (*Martes americana*) carcasses used in analyses

Sample	Trapline locations	Collecting dates
Queen Charlotte Islands	Graham Island, Moresby Island, Louise Island	Jan–Feb 1984–1986
Vancouver Island	Entire island	Nov–Feb 1983–1986
Alexander Archipelago	Baranof Island, Chichagof Island	Dec–Feb 1975–1979
Southern B.C. coast	Kingcome Inlet, Loughborough Inlet, Powell Lake	Nov–Feb 1984–1989
Northern B.C. coast	Skeena River, Gitnadoix River	Dec–Feb 1986–1989
Alaska Panhandle	Aaron Creek, Bradfield River	Dec–Feb 1977–1979
Central B.C. interior	Ootsa Lake, Francois Lake, Bulkley Valley	Nov–Feb 1986–1989
Northern B.C. interior	Stikine River, Boya Lake, Spatzizi Plateau	Nov–Jan 1984–1987

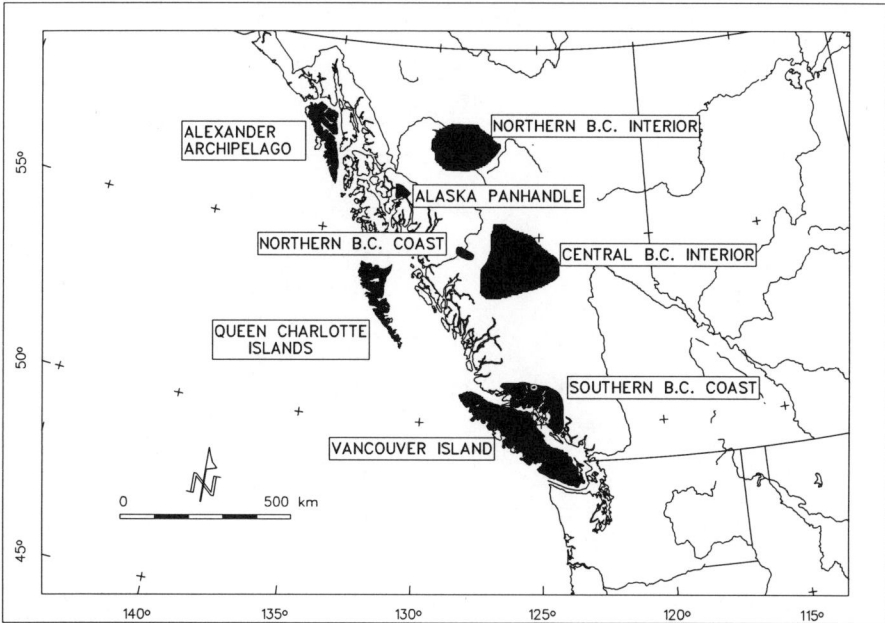

Figure 5.1. Locations of eight geographic samples of American martens (*Martes americana*) used in analyses.

Laboratory, Milltown, Montana). Carcass weights were measured to the nearest 0.1 g and were based on the entire carcass, including stomach and intestinal contents.

Because sample sizes were inadequate for use of individual traplines in any analyses of geogaphic variation, I pooled specimens from geographically proximate locations into eight groups (Fig. 5.1, Table 5.1). My samples represented geographically discrete areas, and I considered them valid for investigating variation on a broad scale. Movements of American martens between these areas have not been measured, but the coastal mountain ranges and water barriers that separate these areas would be expected to minimize gene flow. Specimens from each sample were subdivided into two age categories: juveniles and adults (Table 5.2). I defined juveniles as young of the year (cementum = 0) and adults as 1 year or older (cementum $\geq$ 1). Males and females were treated separately in all analyses of geographic variation except sexual variation. Sampling deficiencies prohibited analysis of among-month or among-year variation.

I determined the terrestrial carnivore species inhabiting islands and adjacent mainland areas from Cowan and Guiguet (1965) and Manville and

Table 5.2.   Skinned carcass weights (g) of winter-trapped American martens (*Martes americana*) from the Pacific Northwest

Group	Juveniles			Adults		
	$\bar{x}$[a]	SD	n	$\bar{x}$[a]	SD	n[b]
Males						
Queen Charlotte Islands	1138.3A	129.9	15	1253.8A	135.8	47 ***
Vancouver Island	913.3BC	110.6	137	980.0BC	103.5	264 **
Alexander Archipelago	838.0CD	115.5	86	963.5CD	137.6	39 ***
Southern B.C. coast	845.7CD	117.8	42	863.7E	89.5	45 ns
Northern B.C. coast	971.1B	181.3	34	1062.0B	149.5	26 *
Alaska Panhandle	822.6D	104.0	18	862.9E	157	15 ns
Central B.C. interior	842.5CD	127.1	142	907.7CDE	130.5	98 ***
Northern B.C. interior	806.3D	115.6	31	892.1DE	115.1	22 **
Females						
Queen Charlotte Islands	783.8A	85.2	17	785.5A	98.7	20 ns
Vancouver Island	627.6BC	81.6	136	626.8B	67.5	187 ns
Alexander Archipelago	575.6CD	75.3	51	600.0B	75.4	23 ns
Southern B.C. coast	570.5CD	57.7	15	570.5B	53.6	11 ns
Northern B.C. coast	657.9B	70.6	13	641.2B	100.6	9 ns
Alaska Panhandle	613.8BCD	80.3	14	578.0B	78.7	10 ns
Central B.C. interior	570.0CD	74.0	86	603.2B	78.2	49 *
Northern B.C. interior	546.1D	51.5	22	589.3B	37.2	10 **

[a]Means with the same letters are not different ($P > 0.05$) based on a comparison of eight means with Tukey's studentized range test.

[b]t-test comparing juveniles and adults: ***$P < 0.0001$, **$P < 0.01$, *$P < 0.05$. ns = not significant.

Young (1965). Based on mean weights of Gittleman (1985), I assigned species to five body size categories:

<0.5 kg: short-tailed weasels
0.5–1.5 kg: American martens, mink (*Mustela vison*)
1.6–5.0 kg: fishers (*Martes pennanti*), red foxes (*Vulpes vulpes*)
5.1–10.0 kg: bobcats (*Lynx rufus*), raccoons (*Procyon lotor*), river otters (*Lutra canadensis*)

>10.0 kg: black bears (*Ursus americanus*), cougars (*Felis concolor*), coyotes (*Canis latrans*), gray wolves (*Canis lupus*), grizzly bears (*Ursus arctos*), lynx (*Lynx canadensis*), wolverines (*Gulo gulo*).

Statistical Analyses

I used various univariate programs of SAS (SAS Institute 1985*a,b*) for my analyses. I assessed age and sexual variation within samples with the Student's *t*-test. Variation among the eight sample means was analyzed with a one-way analysis of variance and an a posteriori multiple comparison with Tukey's studentized range test using a general linear model (SAS PROC GLM). To express the degree of body size divergence in insular samples, I calculated the relative insular size index (S_R) of Lomolino (1985). S_R equals the mean weight of an insular sample divided by the mean weight of the nearest adjacent mainland sample. Means of insular and adjacent mainland American martens were compared using the Student *t*-test.

Body lengths (to the nearest millimeter) were available for five samples: Vancouver Island, the southern coast of British Columbia, the Queen Charlotte Islands, the Alexander Islands, and the Alaska Panhandle. I used these data to evaluate the correlation of body weight with linear size. Because weight is a curvilinear function of linear size, I transformed body weight to (weight)$^{1/3}$. I pooled sexes and age categories in this analysis to obtain the maximum range of variation.

I employed two widely used indices of sexual size dimorphism: the dimorphism ratio and the percent dimorphism. The dimorphism ratio equals male weight divided by female weight. I calculated percent dimorphism in weight from the formula of Rossolimo and Pavlinov (1974): $([\bar{x}_\male - \bar{x}_\female]/\bar{x}_\male)100$. To determine if sexual dimorphism increases with male size, I plotted $\log_{10}$ male body weight against $\log_{10}$ female body weight. If the sexes vary isometrically in body weight, then the slope (*b*) will equal 1; if *b* is greater than 1, dimorphism increases with male size (Ralls and Harvey 1985). Three techniques were used to estimate the best-fit isometric line: least squares regression, reduced major-axis, and major-axis (Pagel and Harvey 1988). The null hypothesis (*b* = 1) was tested with a one-tailed *t*-test. I also tested correlations between the dimorphism indices and $\log_{10}$ male body weight.

Results

Except for the two samples from interior British Columbia, females attained adult size by the first trapping season (Table 5.2). In contrast, adult males were larger than juveniles in all groups. Sexual dimorphism was pronounced, with males larger than females ($P < 0.0001$) for juveniles and adults in all groups. Carcass weight was correlated ($P \leq 0.0001$) with head and body length in the five samples (Vancouver Island: $r = 0.83$; Queen Charlotte Islands: $r = 0.84$; southern coast of British Columbia: $r = 0.77$; Alexander Archipelago: $r = 0.78$; Alaska Panhandle: $r = 0.71$).

Geographic Variation in Size

Analysis of variance revealed marked differences among means ($P \leq 0.0001$) for all sex-age groups. The most striking feature of the variation was the large size of American martens on the Queen Charlotte Islands (Table 5.2). Overall, there was a tendency toward greater body weight on islands and along the northern coast of British Columbia. American martens from other coastal mainland areas and interior British Columbia were smaller and demonstrated substantial overlap in size. This trend was consistent among juveniles and adults of both sexes. Body weight was correlated among sexes (juvenile females vs. juvenile males: $r = 0.897$, $P = 0.003$; adult females vs. adult males: $r = 0.975$, $P = 0.0001$) and among age categories within sex (juvenile females vs. adult females: $r = 0.930$, $P = 0.0008$; juvenile males vs. adult males: $r = 0.969$, $P = 0.0001$).

The insular-mainland comparisons (Table 5.3) supported the hypothesis of larger size on islands. The degree of insular gigantism, however, varied among the three island samples. It was most pronounced on the Queen Charlottte Islands; American martens from the Alexander Archipelago were weakly divergent in body weight from those of the Alaska Panhandle. For the Queen Charlotte Islands and Vancouver Island, divergence in size from mainland populations was evident even in the young of the year.

Geographic Variation in Sexual Dimorphism

Except for the small dimorphism index of American martens from the Alaska Panhandle, there was no geographic variation in the degree of sexual dimorphism among juveniles (Table 5.4). Adults demonstrated minor geographic variation, however, with dimorphism strongest in populations from the northern coast of British Columbia and islands. My results provide general support for the hypothesis of greater sexual dimorphism on islands. Mean dimorphism indices were larger for the island than mainland samples, but differences were not significant ($\bar{x}$ dimorphism ratio: $t = 1.33$, df $= 6$, $P > 0.10$; $\bar{x}$ percent dimorphism: $t = 1.60$, df $= 6$, $P > 0.10$). Small sample sizes, however, prohibit rigorous significance testing. American martens on

Table 5.3.   Measures of body weight divergence between three insular and adjacent mainland populations of American martens (*Martes americana*) in the Pacific Northwest

Comparison	S_R[a]	P[b]
Queen Charlotte Islands / northern B.C. coast		
Juvenile male	1.17	0.0008
Adult male	1.18	0.0001
Juvenile female	1.19	0.0001
Adult female	1.22	0.0026
Vancouver Island / southern B.C. coast		
Juvenile male	1.08	0.0016
Adult male	1.14	0.0001
Juvenile female	1.10	0.0024
Adult female	1.10	0.0060
Alexander Archipelago / Alaska Panhandle		
Juvenile male	1.02	0.5785
Adult male	1.12	0.0399
Juvenile female	0.94	0.1268
Adult female	1.04	0.4534

[a]$S_R = \bar{x}$ insular carcass wt./$\bar{x}$ adjacent mainland carcass wt.

[b]Paired *t*-test based on $\bar{x}$ carcass wt.

Table 5.4.   Variation in degree of sexual dimorphism in body weight among American martens (*Martes americana*) from the Pacific Northwest

Sample	Dimorphism ratio[a] Juv.	Adult	% Dimorphism[b] Juv.	Adult
Queen Charlotte Islands	1.45	1.60	45.2	59.6
Vancouver Island	1.46	1.56	45.5	56.3
Alexander Archipelago	1.46	1.60	45.6	60.5
Southern B.C. coast	1.48	1.51	48.2	50.9
Northern B.C. coast	1.48	1.66	47.6	65.6
Alaska Panhandle	1.34	1.49	34.0	49.3
Central B.C. interior	1.48	1.50	47.6	50.5
Northern B.C. interior	1.48	1.51	47.6	51.4

[a]$\bar{x}_\male/\bar{x}_\female$.

[b]$([\bar{x}_\male - \bar{x}_\female])/\bar{x}_\female)100$.

Vancouver Island and the Alexander Archipelago were more dimorphic than their mainland counterparts. But, although American martens on the Queen Charlotte Islands had larger dimorphism indices than four mainland samples, they were less dimorphic than the population from the northern coast of British Columbia. Variation in degree of sexual dimorphism among adults is largely a function of male size. Slopes (b) of the lines fitted to log male weight versus log female weight were greater than 1 (linear regression: $b = 1.198$, $t = 1.7095$, 6 df, $P < 0.10$; reduced major-axis: $b = 1.286$, $t = 2.4659$, $P < 0.05$; major-axis: $b = 1.253$, $t = 2.6431$, $P < 0.05$). The trend for increasing dimorphism with male size was also supported by the correlations of dimorphism indices with log male weight (dimorphism ratio: $r = 0.751$, $P < 0.05$; percent dimorphism: $r = 0.834$; $P < 0.01$).

Discussion

Because of its inherent variability, body weight may be less reliable than skeletal or linear body measurements as a measure of size. Nonetheless, my results demonstrate that carcass weight is a useful measure of overall body size in American martens. Although Giannico and Nagorsen (1989) analyzed skulls from only five of the eight samples used here, it is reassuring that the general trends in sexual and size variation appear similar for both size measures. The contradictory trend described by Holmes and Powell (this volume) for variation in sexual dimorphism among insular and mainland American martens from the Pacific Northwest may be attributed to sampling design. Their mainland samples were at the subspecies level and consequently encompass large geographic areas. Furthermore, one of their insular samples, *M. americana kenaiensis*, is actually a mainland race that has a distribution continuous with other mainland Alaskan populations (Gibilisco, this volume).

Another limitation of carcass weight as an indicator is that its heritability in martens has not been measured, and the contributions of genetic and environmental factors to variability of body size of a marten population are unknown. Most hypotheses relating to geographic variation in size represent a selectionist point of view (Gould and Johnston 1972) and assume that body size is inherited. On the other hand, changes in weight resulting from fluctuations in food abundance have been observed in several marten populations (Weckwerth and Hawley 1962, Thompson and Colgan 1987a), suggesting that some among-population differences may result from environmental effects. Body weight differences among Pacific Northwest American martens, especially between mainland and island populations, exceed those described by Weckwerth and Hawley (1962) and Thompson and Colgan (1987*a*), and I assume that this variation has an underlying genetic basis. Nonetheless, comparative studies on captive martens are required to estimate body weight heritability and assess the influence of diet on growth and adult size.

Geographic Variation in Size

The large body size I observed in insular American martens is consistent with the island rule, and the divergence values (S_R) of my insular samples are concordant with the mean S_R value calculated by Lomolino (1985) for terrestrial mustelids. Of the various hypotheses advanced to explain size differentiation in mustelids, those relating to competition and food resources may best account for marten insular gigantism. Climatic differences between island and coastal mainland areas in my study are minor, and it is unlikely that insular divergence reflects climatic adaptation such as Eger (1990) described for North American short-tailed weasels. Lomolino (1985) argued that decreased competition is a major factor responsible for the larger size of small and medium-sized insular mammals. According to this hypothesis, body size is constrained by the presence of larger competitors (McNab 1971). On islands where these competitors are absent, a greater mean body size may evolve in response to various selective pressures that favor a larger optimum body size. This argument is especially relevant to carnivore communities because competing species generally fall along a gradient of body sizes (Rosenzweig 1966).

The greater body size of insular American martens in the Pacific Northwest is congruent with a lack of competitors on these islands. My three island samples all support fewer large carnivores than their mainland counterparts (Table 5.5). Most significant is the absence of the 1.6–5.0-kg size class.

Table 5.5. Number of carnivore species inhabiting three insular and adjacent mainland areas in the Pacific Northwest

Sample	Size class (kg)					Total
	<0.5	0.5–1.5	1.6–5.0	5.1–10.0	>10.0	
Vancouver Island	1	2	0	2	4	9
Southern B.C. coast	1	2	2	3	5	13
Queen Charlotte Islands	1	1	0	2[a]	1	5
Northern B.C. coast	1	2	2	1	6	12
Alexander Archipelago	1	2	0	1	2	6
Alaska Panhandle	1	2	1	1	6	11

[a]Includes the raccoon, introduced in the 1940s.

Included in this group are the fisher and red fox, two potential competitors for martens. The carnivore fauna of the Queen Charlotte Islands is further reduced by the absence of mink, the only island group in my study that lacks this mustelid. Gigantism of American martens on the Queen Charlotte Islands is striking; females from this population, for example, are heavier than males from central Ontario (Strickland and Douglas 1987). This gigantism is concordant with the isolation and depauperate carnivore community of this archipelago.

Prey size and abundance may contribute also to insular gigantism (Case 1978). Among small mustelids, there is evidence that body size correlates to prey size (Erlinge 1987, King 1989*a*). Most small rodents and birds that martens prey on would be expected to be larger on islands (Foster 1963, Lomolino 1985). Greater prey abundance will also favor a large body size, and Case (1978) argued that food resources should be greater on islands for species with feeding territories. I suggest that the general shift to larger size among insular martens results from competition effects, but the degree of size divergence is determined by prey size or abundance.

Insufficient data, however, prohibit any critical assessment of the correlation between marten body size and food resources. Of my eight samples, diets are known only for martens on Vancouver Island (Nagorsen et al. 1989) and the Queen Charlotte Islands (Nagorsen et al. 1991), and these data are limited to winter. Diets of these insular martens are certainly more diverse than for martens inhabiting the continental interior because of the availability of small birds and marine resources such as salmon, in addition to small mammals. To test the effect of food resources on the body size of insular martens, I would have to obtain dietary data and measures of prey abundance from several mainland areas for critical mainland-island comparisons.

All these explanations assume that insular gigantism results from selection for increasing size on islands, with the insular population evolved from a small mainland ancestor. Gordon (1986), however, interpreted the large black bears on the Queen Charlotte Islands as a relict population derived from large forms that inhabited continental North America during the late Pleistocene and early Holocene. It is difficult to evaluate the relict hypothesis. With no fossil material available (Graham and Graham, this volume), the historical zoogeography of American martens on these islands is speculative. There is evidence (Heusser 1989) that ice-free areas existed on the Queen Charlotte Islands, Alexander Archipelago, and Vancouver Island during the Vashon Glaciation, but these glacial refugia were probably too small to support viable populations of martens. Moreover, as Case (1978) noted, even if these insular populations are relicts derived from glacial refugia or late Pleistocene immigrants, the relict hypothesis fails to explain how selection pressures have maintained the large body size of carnivores inhabiting these islands.

No obvious size clines are evident among mainland American marten populations in the Pacific Northwest. The most noteworthy aspect of the size variation is the aberrant large size of animals from the northern coast of British Columbia, which are much larger than those from other coastal mainland and interior areas. I cannot account for this aberrant size. It is not simply a function of a coastal environment, because American martens from coastal Alaska and the southern coast of British Columbia are small. To some extent it may reflect phylogeny; American martens from the Alaska Panhandle and interior British Columbia are associated with the *americana* morphotype, whereas coastal populations from British Columbia are members of the *caurina* morphotype (Giannico and Nagorsen 1989). Anderson (1970; this volume) hypothesized that these two morphotypes are different lineages derived from separate invasions from Eurasia. The large size of American martens on the northern coast of British Columbia may be one extreme in a size cline that occurs among coastal *caurina* populations. Samples from the central coast of British Columbia and along the Skeena River valley traversing the Coast Mountains are needed to evaluate the patterns of size variation in this region and their association with ecogeographic factors and taxonomic boundaries.

Geographic Variation in Sexual Dimorphism

According to Brassard and Bernard (1939) the weight disparity between male and female American martens first appears at about 3 weeks; my results demonstrate that by 7–10 months (i.e., juveniles) sexual variation in body weight is substantial. The general consensus (Holmes and Powell, this volume) is that sexual size dimorphism in mustelids has evolved in response to opposing selection pressures, with sexual selection favoring large males and energetics or prey size constraining female size. Among-population variation in the degree of dimorphism has been attributed generally to factors affecting male size rather than those that restrict female size (Ralls and Harvey 1985). Pacific Northwest American martens demonstrate a similar pattern. The most dimorphic populations have both large males and females, but males are disproportionately larger.

The strong dimorphism of American martens on the northern coast of British Columbia parallels the large body size of this population. Why males are disproportinately larger in this mainland population is not clear. There are several possible explanations for the occurrence of relatively larger males on islands. Sexual selection may be more intense on islands. With no comparative data available on territoriality, sex ratios, and home range sizes for insular and mainland martens, I cannot determine the degree of male competition among these populations. Character displacement to reduce intersexual competition has been invoked to explain the increased sexual dimorphism in feeding structures of insular vertebrates such as birds (Selander 1966). Giannico and Nagorsen (1989) speculated that the stronger dimorphism in cranial size shown by insular American martens was an adaptation for resource partitioning, but Holmes and Powell (this volume) found no relationship between sexual dimorphism in cranial traits associated with feeding and resource partitioning. Because body weight is generally correlated with prey size among carnivores (Rosenzweig 1966), enhanced intersexual differences in body size could promote distinct feeding niches. There are problems with this hypothesis. Variations in the degree of size dimorphism among American marten populations in the Pacific Northwest are slight, and it seems unlikely that such minor variation would contribute to distinct feeding niches. Moreover, the limited dietary data (Nagorsen et al. 1989, 1991) provide no evidence for intersexual resource partitioning. Ralls and Harvey (1985) concluded that differences in size dimorphism among weasels was related to the amount of food available to males, with males attaining maximum adult size in environments with productive food supplies. There is evidence that the size of adult male short-tailed weasels (Powell and King 1989) and fishers (Powell 1979*a*) is restricted by food resources. This hypothesis certainly is consistent with the pattern of size dimorphism in my study. Rich terrestrial and marine resources, together with a lack of competitors on islands, could produce relatively large adult males.

Insular Trends for Other Martens

The trend for increasing body size and sexual dimorphism on islands demonstrated herein is based on only three insular-mainland comparisons, and thus may not hold true for all insular martens. Foster (1963) reported contradictory size trends for island forms of Eurasian pine martens (*M. martes*) and stone martens (*M. foina*) in the Mediterranean. The island rule as formulated by Lomolino (1985) predicts that the small mustelid species should exhibit some degree of gigantism on islands. Nonetheless, from my study, it would seem that the extent of size divergence varies considerably on islands depending on isolation, community structure, and available prey. Similarly, if variation in the degree of sexual dimorphism results from food resources, then size dimorphism would be expected to vary among islands and even temporally within an island population. King's (1989*a*) observation that insular weasels show no consistent pattern in body size underscores the need to examine size trends for other insular martens before any generalizations can be made. The only other continental islands in North America with martens are along the Atlantic coast of Canada (Cameron 1958). Several of these populations have been extirpated, and these islands are unsuitable for a comparative study. The six Eurasian martens all have insular forms (Hagmeier 1955; Anderson 1970; Tatara, this volume), however, and these could be used to test the association of size and dimorphism divergence with insularity.

Acknowledgments

My research was funded by the Royal British Columbia Museum and the British Columbia Ministry of Environment. I thank K. Morrison and D. Steventon for their efforts in obtaining carcasses from trappers and providing various weight and age data. G. Jarrell of the University of Alaska loaned specimens, permitted tooth extractions, and provided weights for the Alaskan martens. Finally, I am indebted to many trappers from Alaska and British Columbia whose cooperation made this study possible.

Introduction

This section continues the theme of variation and diversity in *Martes*. Powell shows that population size, structure, and spacing fluctuate in natural populations, in some cases dramatically. As prey populations change over time, populations of *Martes* rise or fall as well, and preventing such shifts may place *Martes* populations in unnatural conditions. Just as population size fluctuates, the size and spacing of home ranges within populations also should vary. The norm is for the intrasexual territories of males to be far larger than those of females, but *Martes* are territorial only at intermediate levels of prey abundance. When prey are scarce, fishers, martens, and sables should be transient. As prey become more abundant, there should be a progression from individual territories, through intrasexual territories, to no territories when prey are so abundant that space can be shared. The cost of sharing a territory with one member of the opposite sex benefits males by increasing their probabilities of mating, but it appears not to benefit females.

Herrmann documents how habitat quality affected home range size in a stone marten population in Germany. The stone martens had a hierarchy of preferences for different habitat types. Home ranges within many preferred habitat patches were small, whereas home ranges encompassing only non-preferred habitat patches were large. Most preferred habitat patches were found in and around villages. Home ranges of males were far larger than those of females, but home ranges of both sexes showed the same pattern relative to habitat. Individual stone martens that shifted their home ranges from forest to village showed pronounced decreases in home range size. Herrmann's data suggest that all the stone martens he studied maintained intrasexual territories.

Krohn et al. report that mortalities of fishers varied over the year by age and sex class. For their heavily trapped population in Maine, the mortalities of juvenile males and females did not differ, but those of adults differed from those of juveniles and between sexes. The highest mortalities occurred for juveniles during the trapping season and the lowest were for adult males outside the trapping season. Adult females had mortalities distinctly lower than those of adult males during the trapping season, and Krohn et al. believe this difference explains their population sex ratio, which favors females. They highlight the importance of including differences in mortality between sex and age classes and among times of the year in models of population dynamics.

ROGER A. POWELL

6 Structure and Spacing of *Martes* Populations

Roger A. Powell

Understanding basic population characteristics of a species or group of species provides a critical foundation for understanding the ecology and behavior of the animals. In this review I show how population dynamics, age structure, sex ratio, and spacing patterns are interrelated for *Martes* spp., and I argue that a basic characteristic of natural *Martes* populations is lack of stability. Such instability means, in turn, that age structures and mortalities are constantly responding to changing population conditions.

The way that individuals within a population space themselves is an often-ignored aspect of population structure. Spacing is affected by the population's size, age structure, mortality, and sex ratio. It is also affected by the abundance and productivity of food; hence population size and spacing interact in response to food. I present a model for spacing within *Martes* populations and show how spacing, food abundance, and mating patterns are all integrally related. I apply the model to a fisher (*M. pennanti*) population.

Despite their importance, data on the basic population characteristics of unexploited populations of *Martes* spp. are limited. Most populations studied have been harvested during the period of study. Yet knowledge of the basic characteristics of unharvested populations is also critical for understanding population responses to management. My final argument is that harvesting affects more than population size. It affects population dynamics, age structure, sex ratio, spacing patterns, and probably mating patterns and foraging costs. All these changes must be considered in management programs.

Population Dynamics

Unharvested populations of *Martes* exhibit marked fluctuations in size, sometimes in excess of an order of magnitude, in response to fluctuations in prey populations. Such fluctuations are to be expected because the intrinsic rates of increase (*r*) for prey exceed those for *Martes* spp., all of which reproduce for the first time when age 2 or older and all of which have small litter sizes (Mead, this volume). Thus, population- and age-specific mortalities continually change in natural populations.

Weckwerth and Hawley (1962) documented a fourfold change in population density of American martens (*M. americana*) over five years, and Thompson and Colgan (1987*a*) reported a sixfold change in a population harvested at a level low enough that its numbers responded to changes in prey populations. Lockie (1964) reported a threefold change in Eurasian pine marten (*M. martes*) density in Scotland over the course of six years, and Marchesi (1989) reported an eightfold change in the density of Eurasian pine martens in Switzerland; the importance of harvest to dynamics of Marchesi's population was not reported. Several Russian authors (e.g., Naumov and Rukovsky 1972, Grakov 1978) have reported very large fluctuations in Eurasian pine marten populations, but these reports were for harvested populations with unknown, though probably high, levels of exploitation.

A common goal of managing furbearing wildlife, including *Martes* populations, is to stabilize population sizes (Strickland, this volume). Stable populations are easier to manage because small changes in numbers can be monitored and modestly understood, especially changes caused by harvesting. Such harvested populations obviously cannot exhibit natural population dynamics or population structure.

Bulmer (1974, 1975) analyzed the dynamics of all those species, including fishers and American martens, that apparently respond to the population cycles of snowshoe hares (*Lepus americanus*) in Canada. His analyses of fur harvest records through 1909 probably reflect natural population fluctuations because they do not include the substantial population declines caused by overtrapping in the twentieth century (Rand 1944, de Vos 1952, Dodds and Martell 1971*a*, Powell 1982). American marten populations ceased to rise and fall and fisher population cycles declined in magnitude during the twentieth century (Bulmer 1974) as a result of overharvesting (Powell 1982).

Bulmer (1974) showed that the fisher population cycle lagged three years behind that of snowshoe hares. This is the lag expected for population declines caused by juvenile or adult mortality rather than by declines in reproduction. American martens increased and decreased in phase with snowshoe hares. Alternate prey for hare predators should increase and decrease in phase with snowshoe hares; therefore Bulmer postulated that other predators preyed heavily on American martens when hare populations declined. This is highly unlikely. Because American martens depend less on hares for food than do fishers, the direct effect of hare population cycles on martens should have been less pronounced but in phase with the cycles of fishers. I conjecture that because hares overeat their food supplies when populations are high (Keith and Windberg 1978, Pease et al. 1979, Vaughn and Keith 1981), they reduce food and cover for other prey important to American martens. Consequently, populations of those prey should decline from lack of food and high predation pressure, and would thus lead to American marten population declines in phase with those of snowshoe hares.

Bulmer (1974) also reported on population cycles of some predators in Eurasia that depend on mountain hares (*L. timidus*) as major prey but did not include data on Eurasian pine martens or sables (*M. zibellina*). Red foxes (*Vulpes vulpes*) and lynx (*Lynx lynx*) cycled in Eurasia, as did their North American counterparts. I expect that Eurasian pine martens and sables cycled in response to mountain hares.

In many parts of Europe and Asia, stone martens (*M. foina*) have lived commensally with humans for long periods (Bakeyev, this volume; Herrmann, this volume). The dynamics of those populations must be dependent on human activities. Some have lived so long in habitat heavily modified by humans that understanding population responses to human influences is critical. Commensal martens have opportunistic diets, prey heavily on commensal rodents when they are available (Skirnisson 1986), and eat domestic fruits (Rasmussen and Madsen 1985). I expect that their population dynamics respond to conditions that change commensal rodent populations, or that influence agriculture, or that influence human food disposal.

Age Structure

Although age structure of a population is related to survivorship, age structure is directly proportional to age-specific survivorship only for populations with stable sizes and age structures. This is seldom, if ever, the case for *Martes* populations. The ranges of age structures and the manners in which age structures change provide only indices of age-specific survivorship in *Martes* populations. Natural *Martes* populations will never have characteristic age structures or age-specific survivorship rates. Age structures do, however, undergo characteristic types of changes.

Age structures of natural populations change with population density. Age structures of harvested populations differ from those of unharvested populations because few individuals are able to survive to old-age classes, leading to age structures biased toward young animals (Douglas and Strickland 1987, Strickland and Douglas 1987). Sex ratios of harvested populations also differ because trapping generally increases mortality of males more than that of females (Buskirk and Lindstedt 1989). Structures of harvested *Martes* populations change mostly in response to pressure and timing of harvest; they respond to changes in prey availability only when prey populations fall below those capable of supporting the artificially low population.

Soukkala (1983) reported age structures of harvested American martens in Maine whose populations were subjected to light and heavy trapping (Table 6.1). Because of trapping bias, these age structures probably do not accurately reflect population age structure, but they should reflect relative differences in age structure caused by heavy trapping. The population under heavy harvesting pressure had a large proportion of young individuals. Heavily trapped American marten and fisher populations in Ontario were biased even more heavily towards young animals (Strickland and Douglas 1987, Douglas and Strickland 1987). For a sample of 56 Eurasian pine martens, Marchesi (1989) reported an age structure similar to that reported by Soukkala (1983) for American martens trapped under light pressure (Table 6.1). Slightly more than half of Marchesi's sample were martens that were shot or trapped; the rest were road mortalities. He also reported a higher percentage of juveniles in this sample of dead than in his sample of 23 live-trapped martens (22%). Because harvest is usually biased towards juveniles (Douglas and Strickland 1987; Strickland and Douglas 1987; Strickland, this volume), Marchesi's live-trapping results probably indicate correctly that the true population age structure had a lower percentage of 1-year-old martens and greater percentages of older martens than shown in Table 6.1.

Extensive empirical research reviewed by King (1989*b*) and theoretical work by Powell and Zielinksi (1983) on *Mustela* have shown that the popula-

Table 6.1.   Age structures for American martens (*Martes americana*) harvested in Maine, 1979–1981, and Eurasian pine martens (*M. martes*) that died in Switzerland, 1984–1987

Age (years)	Americans martens (% harvested)		Pine martens (% dead)
	Light pressure	Heavy pressure	
1	34	40	30
2	18	24	16
3	15	16	14
4	8	8	21
5	10	8	7
6	14	4	12

Note: American martens were subjected to light (mean harvest = 0.05/km²) and heavy (mean harvest = 0.25/km²) trapping pressure (Soukkala 1983). Eurasian pine martens were harvested, died on highways, or died from other causes (Marchesi 1989).

tions, and thus the age structures, of these close relatives to *Martes* are also never stable. King showed for stoats *(Mustela erminea)* in New Zealand that high reproduction and high juvenile survivorship in years of high prey *(Mus musculus)* populations led to disproportional representation of these cohorts in future years. In fact, the most numerous cohort usually derived from a year with a high house mouse population and even outnumbered younger cohorts.

The same phenomenon should occur in *Martes* populations. Douglas and Strickland (1987) reported that during a period of high prey availability in Ontario, juvenile fishers composed a higher-than-average proportion of the population. When prey populations were low and fisher populations declined, cohorts of old fishers composed higher-than-average proportions of the population. Similarly, Thompson and Colgan (1987*a*) reported that as prey populations declined, juvenile American martens dropped from 63% of the harvest to 38%, 16%, and finally 4%. During the same period, cohorts from years with higher prey populations composed greater and greater proportions of the population. I expect that natural populations will not have such large proportions of juveniles when prey are abundant.

Thompson's and Colgan's (1987*a*) data also show that if prey populations remain low for a very long time, young animals will ultimately increase again as a proportion of the age structure. This is because old martens finally die and the marten population structure adjusts to an extended period of low prey availability and low population size.

Beyond the problems of sample bias, understanding the age structure of *Martes* populations is severely hampered by biases in population biology and demography research. Historically, such research has been oriented to understand population stability (e.g., May 1973; Łomnicki 1978, 1988) and thus contributes little to understanding unstable populations. Instability in populations and age structures comes predominantly from unstable prey populations and may be compounded by competition between *Martes* species (May 1973, Powell and Zielinski 1983) and habitat change. Most martens live in northern habitats that are subject to intermittent change by fire and wind. High variance in the frequency of natural fire and patchy distribution of fires (Heinselman 1973, Payette et al. 1989, Clark 1990) create habitat patches that are of variable importance to *Martes* (Koehler and Hornocker 1977). The effects of such natural habitat change on population structure have not been studied.

Unstable age structure causes variation in population responses to change. Populations biased toward individuals with low reproductive value (v_x: Fisher 1958; juveniles or animals near age of last reproduction) are unable to respond as rapidly to increases in prey populations as are populations with high proportions of individuals with high reproductive value. If declines in prey population lead to high juvenile mortality, age structure becomes biased toward old adults with low v_x and population response to prey increases will be slow. Trapping, in contrast, tends to bias populations toward young age classes. Thus, *Martes* populations that decline from overtrapping may be able to recover faster than populations that decline from prey shortage. Rapid response cannot occur, however, if age structure becomes highly biased toward juveniles at the same time that sex ratio becomes highly skewed toward females. Strickland and Douglas (1981) found that under these conditions a significant percentage of adult females in Ontario did not reproduce, which hampered population response. In addition, because *Martes* spp. delay implantation, all populations delay responses to increases in prey populations.

We must include juveniles and nonresident adults as important subjects for research because these segments of a population affect and respond to resident adults. Sandell's (1986) data for stoats indicate that residency patterns may be misunderstood. He found that dominant male stoats traveled long distances during the breeding season, often beyond the edges of the territories maintained during the rest of the year. Subordinate male stoats remained in small home ranges during the breeding season. Research of smaller scope than Sandell's might have led to the false conclusion that subordinate stoats were resident and dominant stoats transient.

Sex Ratio

Sex ratios of unharvested *Martes* populations are not well known because true population sex ratios are difficult to determine. There is a significant bias toward trapping males (live-trapping and kill-trapping: Buskirk and Lindstedt 1989). Archibald and Jessup (1984) determined through live-trapping and radiotelemetry that the true sex ratio of American martens in their Yukon, Canada, study area stayed at 50:50 during most of 1979–1981. Yet fur-trapping results from the same area showed a significant bias toward males (59:41; X^2, $P < 0.05$). Many studies reported sex ratios of live-trapped animals that were highly skewed yet not significantly different from 50:50 owing to small sample sizes (Clark and Campbell 1976, Powell 1982, Buck et al. 1983, Marchesi 1989).

Males have significantly larger home ranges than females in all *Martes* species (Table 6.2). *Martes* also consistently exhibit intrasexual territoriality (Table 6.2) and therefore would be expected to have sex ratios biased toward females if all areas of appropriate habitat were occupied. Because *Martes*, especially fishers, have such large home ranges, I conjecture that true sex ratios for natural populations are likely close to 50:50. Trapping bias for

Table 6.2. Home range sizes (km²) for *Martes* species

Species	Male Mean ± SD	Male N	Female Mean ± SD	Female N	Within-sex home range overlap	Location and comments	Source
M. americana	7.1 ± 1.5	3	5.6 ± 2.8	4	Adults and juveniles	Yukon; juveniles and adults, >30 locations	1
	8.7	1	6.6 ± 2.0	2		Yukon; adults only, >30 locations	1
	4.8	4	2.3	4	Less than between sexes	Vancouver Island, B.C.	2
	27	1	17	1		Newfoundland	3,4
	4.6 ± 1.9	5	2.4 ± 1.4	5		New York State	5
	2.0 ± 2.6	5	0.6	1		Montana	6
	7.1 ± 2.9	10	7.9 ± 8.9	4		Alaska	7
	3.6 ± 1.4	4	1.1 ± 0.9	4	Females no; males in 1 of 2 years	Ontario	8
	10 ± 9.1	6	4.3 ± 2.8	2	Females no; males somewhat	Northwest Territories	9
	6.1 ± 5.9	6	1.9 ± 1.8	2		Northwest Territories; 95% MCP	9
	8.2 ± 2.0	5	1.7 ± 0.8	3	Males no within same year	Maine	10,11
	16 ± 5	3	4.3	1		Minnesota	12
	3.4 ± 0.7	4	1.0 ± 0.2	2		Ontario; old-age forest, abundant prey	13
	6.8 ± 0.8	5	4.2 ± 0.3	4		Ontario; old-age forest, scarce prey	13
	5.0 ± 1.1	2	3.1 ± 0.6	4		Ontario; cut forest, abundant prey	13
	11 ± 2	3	13 ± 1	3		Ontario; cut forest, scarce prey	13
	10 ± 0.7	2	12	1	None	Manitoba	14
	3.9 ± 1.0	6	3.2 ± 1.7	5		California	15,16

(continued)

Table 6.2. (*Continued*)

Species	Male Mean ± SD	N	Female Mean ± SD	N	Within-sex home range overlap	Location and comments	Source
	5.6	2	2.9	3	None or little	Maine; summer only	17
					Males no; females somewhat	Wyoming	18
					None or little	Montana	19
MEAN	8.1		2.3				
M. foina	3.0 ± 0.1	2	2.8	1	None or little	Germany; rural	20
	1.8	1	0.1	1		Germany; urban	20
	0.4 ± 0.2	6	0.2 ± 0.1	6	None	Germany	21
MEAN	1.7		1.4				
M. martes	0.4 ± 0.6		2.0 ± 3.0		None or little	Scotland	22
	9.7 ± 3.5	6	2.9 ± 1.1	5	None or little	Switzerland	23
	9.2 ± 3.8	6	3.4 ± 1.3	5		Switzerland; harmonic mean, 95% area	23
	58 ± 34	2	14 ± 0	2	None or little	Finland; from tracks; excludes unused areas >4 km^2	24
MEAN	23		6.5				
M. pennanti	33 ± 25	7	19 ± 12	6	None or little	Maine	25
	27 ± 24	7	16 ± 12	6		Maine; harmonic mean, 90% area	25
	50 ± 40	7	31 ± 23	6		Maine; harmonic mean, 99% area	25
	20 ± 12	3	4.2	1	None or little	California; adults only, >20 locations	26
	23 ± 12	4	6.8	2		California; adults and juveniles, >20 locations	26

108

19 ± 17	3	15 ± 3	2	None or little	New Hampshire; adults	27
26 ± 17	3	15 ± 6	3		New Hampshire; juveniles	27
23 ± 16	6	15 ± 5	5	Adults and juveniles	New Hampshire; adults and juveniles	27
35	1	15	1		Michigan	28
79 ± 35	6	32 ± 23	4		Idaho; adults and juveniles; harmonic mean, 90% area	29
49 ± 37	2	7.5 ± 4.2	5	None or little	Wisconsin; adults, >25 locations	30
39 ± 27	4	8.3 ± 3.7	7		Wisconsin; adults and juveniles, >25 locations	30
MEAN 38		15				

Notes: Areas were calculated from minimum convex polygons (MCP) unless otherwise stated. Means were calculated using only one figure for each sex in each study. Blank cells indicate data were not provided.

Sources:

1. Archibald and Jessup 1984
2. J. M. Baker, unpubl. data
3. Bateman 1986*
4. Bissonette et al. 1988*
5. M. K. Brown*
6. Burnett 1981*
7. Buskirk 1983*
8. Francis & Stephenson 1972
9. P. Latour et al., unpubl. data
10. Major 1979*
11. Steventon 1979*
12. Mech and Rogers 1977
13. Thompson and Colgan 1987*a*
14. Raine 1981*
15. Simon 1980*
16. Spencer 1981*
17. Wynne and Sherburne 1984
18. Clark and Campbell 1976
19. Hawley and Newby 1957*a*
20. Skirnisson 1986
21. Herrmann, this volume
22. D. Balharry, unpubl. data
23. Marchesi 1989
24. Pulliainen 1984
25. Arthur et al. 1989
26. Buck et al. 1983
27. Kelly 1977
28. Powell 1982
29. Jones 1991
30. Johnson 1984

*Data summary taken from Buskirk & McDonald 1989

109

males probably leaves more live females in harvested populations. If this skew is extreme, it can hamper the ability of a harvested population to replace harvested individuals.

Home Range

Home range sizes vary greatly within and between species of *Martes* (Table 6.2). This variation is due in part to lack of consistency in methods of calculating home range size, in part to minimum convex polygon methods, which include much area not actually used by animals, and in part to true variation. Buskirk and McDonald (1989) analyzed home range data for American martens from nine different geographical locations and found no obvious geographic patterns. They tested particularly for correlations between home range area and latitude, which should be an index of prey productivity, and found no relationship. In contrast, Thompson and Colgan (1987a) did find that American marten home range size responded to prey availability; home ranges were smallest in old-age forest when prey populations were high and largest in recently logged forest when prey populations were low. The failure of Buskirk and McDonald (1989) to find patterns was probably due to the lack of specific information on prey availability at each of the study sites in their analyses. They did find a positive relationship between body size of females and home range size, but they did not find such a pattern for males.

For each sex of each *Martes* species in Table 6.2, I calculated mean home range area using a single figure for each sex in each study. Because methods used were not consistent among studies or species, this figure can be used only for crude comparisons and no measures of variation are given. Nonetheless, two broad generalizations can be made for *Martes* home ranges. First, males' home ranges tend to be larger than females' of the same species in the same area. Second, home range sizes appear positively correlated with body size. Both generalizations are consistent with data on other carnivores (McNab 1986) and can be explained at least partially by the positive relationship between body size and metabolic requirements in mammals (McNab 1986). Because home ranges of males tend to be larger than those of females by a factor greater than expected from metabolic requirements, body size appears not to explain home range size completely. Buskirk and McDonald (1989) found the following for American martens: males averaged 1.5 times heavier than females, but males' home ranges averaged 1.9 times larger.

There are several possible explanations (not mutually exclusive) for disproportionate sizes of male's and female's home ranges. First, males may have energy requirements greater than expected from body size and therefore need disproportionately large home ranges. No field metabolic measurements have been made for *Martes* spp., but laboratory studies on *Martes* (Powell 1979*b*, 1981*b*; Worthen and Kilgore 1981; Buskirk et al. 1988) and other mustelines (Moors 1977, Casey and Casey 1979) indicate no unexpected sexual differences in metabolism and energy requirements. Second, the actual area used by males and females may be proportional to body size, though total home range is not. Home ranges of male and female *Martes* do overlap extensively, even when there is no overlap for members of the same sex (Table 6.2). Nonetheless, in some mustelines, males spend minimal time in the home ranges of females that lie within their own home ranges (Gerell 1970, Erlinge 1977). No published data quantify intensity of home range use of *Martes* spp. Third, males and females may space themselves to gain access to different resources: females may need access to food, whereas males may need access to females. Sandell (1986, Erlinge and Sandell 1986) found that among male stoats (*Mustela erminea*), home ranges and spacing depended on the distribution of females as well as on the dominance hierarchy of the males. Ims (1987; 1988*a,b*; 1990) has shown in theory and empirically that the spacing of male grey-sided voles (*Clethrionomys rufocanus*) depends on the spacing of females, whereas the spacing of females appears to depend on food distribution and productivity. Sandell (1989) has hypothesized this relation to be the case for solitary carnivores, such as *Martes*.

Population Spacing Patterns

Martes populations consistently but not universally exhibit intrasexual territoriality, where individuals maintain territories only with respect to members of the same sex. I have reviewed spacing patterns described for the Mustelidae (Powell 1979*a*) and for fishers specifically (Powell 1982) and found a correlation between intrasexual territoriality, large sexual dimorphism in body size, elongate shape, and high degree of carnivory. Recent research has generally supported this spacing pattern in *Martes* (Table 6.2) but has also provided some new insight. Research by Pulliainen (1984), Marchesi (1989), and Schröpfer et al. (1989) on Eurasian pine martens, by Wynne and Sherburne (1984) and Thompson and Colgan (1987*a*) on American martens, by Arthur et al. (1989*a*) on fishers, and by Skirnisson (1983, 1986) and Herrmann (this volume) on stone martens is consistent with intrasexual territoriality. Some other studies cited in Table 6.2, and possibly that of Tatara (this volume), document variation from intrasexual territories.

A species' spacing behavior varies across its range and through time (Bekoff and Wells 1981, Macdonald 1983, Moehlman 1987), and variation in spacing has been documented in *Martes*. Hawley and Newby (1957*a*), Francis and Stephenson (1972), Mech and Rogers (1977), Taylor and Abrey (1982), Archibald and Jessup (1984), and Wynne and Sherburne (1984) noted overlap of home ranges for American martens of the same sex. Archibald and Jessup (1984) demonstrated that juveniles and adults of the same sex sometimes had overlapping home ranges. Clark and Campbell (1976) documented lack of home range overlap for resident adult American martens of the same sex, but many martens in their study areas were transient or juvenile.

Intrasexual territoriality allows the overlap of home ranges between a male and female in the same area, although they may compete somewhat for limiting resources within their overlapping territories. In many territorial mammals, males and females have exclusive territories and thus no competitors of the same species (e.g., North American red squirrel, *Tamiasciurus hudsonicus*: Smith 1968). Intrasexual territoriality may be possible in mustelids, including *Martes*, because large sexual dimorphism allows two competing individuals that share space to have different diets. Research on mustelid diets (Coulter 1966; Erlinge 1975; Tapper 1976, 1979; Clem 1977*a*; Kelly 1977; Powell 1982; King 1989*b*), however, has consistently refuted resource partitioning among sexes, and the classically cited example in *Martes* does not stand up to statistical scrutiny (Yurgensen 1947, Powell 1979*a*). Holmes (1980, 1987; Holmes and Powell, this volume) has shown that sexual dimorphism in skulls and teeth—parts of the body particularly important for carnivory—is significantly smaller than sexual dimorphism in total body size in *Martes* and other mustelines. This finding likewise refutes resource partitioning as an explanation of intrasexual territoriality. I propose that intrasexual territoriality in *Martes* exists because the distribution and availability of the most likely limiting resource for these species, food, allows a trade-off not possible for species with individual territories, such as North American red squirrels. This trade-off allows male spacing to be affected by female spacing, as well as by the distribution of available prey. The effects of intrasexual territoriality on mating patterns have been modeled by Hixon (1987), but his model assumed particular patterns of territoriality a priori and did not predict when patterns of territoriality should change.

Model of Intrasexual Territoriality

Economic analyses have shown that territorial behavior occurs only when there is a limiting resource and when the cost of territorial defense is less than the benefit accrued from territoriality (Brown 1969, Carpenter and Mac-

Millen 1976, Kodric-Brown and Brown 1978, Hixon et al. 1983). An implication is that sharing a territory must have an economic benefit that exceeds the cost of intraspecific competition. Using cost-benefit analyses, Smith (1968) showed that North American red squirrels cannot share territories economically: they therefore maintain individual territories.

When a red squirrel harvests a cone (*Pinus* spp., other Pinaceae), that cone is no longer available for other squirrels. During most years, each red squirrel harvests a large proportion of the available cones, leaving none for competing squirrels that might try to harvest cones within its territory (Smith 1968). Harvesting of cones is thus limited by the number of cones. In contrast, individual martens (any *Martes* sp.) are seldom able to kill a significant proportion of their prey populations; when they do, this leads to prey shortage and to the decline or local extinction of the marten population. Marten populations can be maintained only when there are enough prey to allow martens to kill at most a sustainable surplus. In addition, the ability of martens to capture and to kill prey is limited not only by the size of their prey populations but also by the vulnerability of prey to capture.

Vulnerability of prey varies over space and time. The effects of this variability on territorial behavior have not been specifically studied for martens but can be modeled to develop predictions for spacing behavior. Prey populations are generally patchily distributed because their habitats are patchily distributed. The fishers I studied in the Upper Peninsula of Michigan preyed predominantly on snowshoe hares and porcupines (*Erethizon dorsatum*), which were found in mutually distinct habitat types that were intermixed throughout my study area (Powell 1979*b*, 1981*b*). Prey for other marten species also are patchily distributed, though not always so distinctly.

In addition, marten hunting behavior must affect prey vulnerability by affecting prey behavior. Jedrzejewski and Jedrzejewski (1990) and Ylönen (1989) showed that behavior of captive bank voles (*Clethrionomys glareolis*) and grey-sided voles changed when a stoat or least weasel (*Mustela nivalis*) entered a local habitat patch; the voles moved less, were more alert, and shifted to different habitat patches, if possible. Vulnerability of these captive voles was greatest before the stoat was perceived, decreased thereafter, and remained depressed for a day or more. This change in prey vulnerability can be called resource depression, in contrast to resource depletion, which occurs when a resource is permanently removed (Charnov et al. 1976). Prey vulnerability is depressed for a comparatively short time by the presence of a marten and then returns to its previous level; prey populations are not depleted by marten foraging because few prey are eaten.

Territorial behavior is economical only at intermediate levels of productivity for the limiting resource (Carpenter and MacMillen 1976). The limits for resource productivity (P) that permit territorial behavior in a patchy habitat can be modeled as

$$\frac{E + T}{a + b} < P < \frac{E}{a} \tag{6.1}$$

where E represents an animal's energy requirements per unit time, T is the cost of territorial behavior, a is the proportion of resource productivity P available to the animal when territorial behavior is not exhibited, and b is the additional proportion of resource productivity P that becomes available through territorial behavior (Carpenter and MacMillen 1976). When resource productivity exceeds E/a, food is so abundant that animals can share home ranges and still meet energy requirements. When resource productivity is less than $(E + T)/(a + b)$, food is too scarce to maintain a resident population and animals will be transient. This model predicts territorial behavior in nectarivorous birds (Carpenter and MacMillen 1976) and appears to predict territorial behavior in black bears (*Ursus americanus*: Powell 1986); similar economic models predict territorial behavior in other small vertebrates (Ebersole 1980, Hixon et al. 1983, Stamps and Tollestrup 1984). Therefore this model should apply to marten species. For a marten, maintaining an intrasexual territory requires sharing each part of its territory with one conspecific. (For simplicity, we will assume that each suitable habitat patch is included within the territories of one male and one female marten.) This decreases b below its value in unshared territories and consequently increases the lower limit of prey productivity that can support a resident marten population.

To quantify this effect: Once one marten has foraged in a habitat patch, prey vulnerability (P in the model) in that patch is depressed, and there are fewer or no vulnerable prey. If a marten is able to maintain a home range with x patches (i.e., acquires the necessary information to be familiar with x patches) and uses an average of y patches per day ($y < x$), then it will use the proportion y/x of its patches each day. If the marten can share each section of its home range with z other martens that use patches at random with respect to each other (i.e., are unable to know before entering a patch whether it has been depressed by another marten), then the probability that a particular patch entered by a marten has not been visited by another marten in the previous 24 hours is $(1 - y/x)^z$. If each patch is depressed for a proportion d of a day (d = [hours depressed]/24), then the probability (p) that a particular patch entered by the marten has not been depressed by a recent visit from another marten is $(1 - yd/x)^z$; y, d, and z must have values such that $0 < p < 1$. This expression is also the expected proportion of patches visited per day that have not been depressed recently by another marten and thus is the proportion of the vulnerable prey in the marten's home range that are available when the home range is shared, which is a.

Let $p = (1 - yd/x)$. Then

$$a = p^z \qquad (6.2)$$

In fact, a marten's ability to travel through its home range is limited and its reproductive output is limited. Therefore any variance around p that it experiences will inflate the negative effects of sharing the home range (Real 1980). This means that costs of sharing a territory are actually underestimated in the discussion that follows.

If the marten does not share its home range with any other martens, then the proportion of vulnerable prey available to it can be set to 1 (i.e., $a + b = 1$). Therefore the conditions under which a marten can maintain an individual territory can be specified by inserting equation 6.2 into equation 6.1:

$$b = 1 - p^z \qquad (6.3)$$

and

$$E + T < P < \frac{E}{p^z} \qquad (6.4)$$

If the marten shares each part of its home range with one other marten, then $b = (1 - p)^z - (1 - p)$, and

$$a + b = p \qquad (6.5)$$

The conditions under which a marten can tolerate this territorial overlap can be specified by inserting equations 6.2 and 6.5 into equation 6.1:

$$\frac{E + T}{p} < P < \frac{E}{p^z} \qquad (6.6)$$

Thus, the marten loses the proportion $1 - p = yd/x$ of the vulnerable prey to which it would have access were it to maintain strict territoriality. It also means that the lower limit of prey availability that allows martens to be resident is raised by the same proportion. If a marten must visit on the average y' undepressed patches per day in order to catch sufficient prey to survive, then if it shares its territory, it must actually visit a total of y^* patches, where

$$y^* = \frac{y'}{p} \qquad (6.7)$$

The ratio $(y^* - y')/y'$ is the relative cost to a marten of allowing one other marten to share its territory; it represents the proportional increase in number of prey habitat patches that must be visited to meet daily energy requirements. Thus the ratio is an increase in daily energy requirements and can be calculated from estimates of energy budgets, such as those given by Buskirk et al. (1988), by Powell (1979*b*, 1981*b*); and by Powell and Leonard (1983).

The ratio $(y^* - y')/y'$ also represents the increase in time or energy that is devoted to foraging and thus cannot be devoted to other activities such as reproduction. Alternatively, if a decrease in food availability does not directly decrease the time spent in reproductive effort because of increased time spent foraging, then $(y^* - y')/y'$ might represent a decrease in the quality of reproductive effort. This poorer reproductive effort could be manifested as a lower competitive ability among males or decreased attractiveness to females.

Given that most *Martes* populations do share their territories, some benefit must outweigh the cost $(y^* - y')/y'$. I hypothesize that this benefit is an increase in reproductive success due to familiarity with the resident members of the opposite sex. There should be no benefit to sharing a territory with a member of the same sex. For a male, the benefit could be decreased cost of finding a female, increased probability of fertilizing the females in his territory, or decreased probability of failing to reproduce. These latter two options can be explored with simple models.

Consider m male and f female martens, each of which maintains a non-overlapping territory. If mating is random, then each male has a probability of $1/m$ of mating with each female and is expected to fertilize f/m females. Each male's probability of *not* fertilizing any females is $[(m - 1)/m]^f$. If the same number of male and female martens maintain intrasexual territories, then each male is expected to have f/m females within his territory. If each male has a probability q $(q > [1/m])$ of fertilizing each female within his territory and a probability $(1 - q)/(m - 1)$ of fertilizing each female not in his territory, then each male is again expected to fertilize f/m females. Each male's probability of not fertilizing any females, however, is now

$$(1 - q)^{f/m}\left[1 - \frac{(1 - q)}{(m - 1)} \right] \qquad (6.8)$$

This probability is always less than $[(m - 1)/m]^f$, which means a male with an intrasexual territory has a smaller probability of failing to reproduce than does a male that does not share his territory with a female.

I find no benefit to a female from sharing a territory with a male. Her access to males is unlikely to be affected by whether male territories are adjacent to or overlap with hers. Similarly, her energy expenditure to find a mate will be unaffected because males will come to her. Thus this model predicts that intrasexual territoriality found in martens is imposed on females by males, which are larger than and dominant to females.

From equations 6.1, 6.4, and 6.6 it is possible to predict how territorial behavior should change in martens as prey populations, prey availability, and prey vulnerability change. A marten population should exhibit intrasexual territoriality when conditions meet equation 6.6. As prey populations increase, marten territories should decrease in size until further decreases eliminate critical patches of prey habitat. At this point the martens should cease to maintain territories and should tolerate considerable home range overlap with members of both sexes. As prey populations decrease such that the left-hand inequality in equation 6.6 is no longer met, marten spacing should change to exclusive territoriality. If prey populations decrease further such that the left-hand inequality in equation 6.4 is no longer met, the martens should cease to maintain stable home ranges but should become transient until prey populations again meet the conditions of the left-hand inequality in equation 6.4. Thus, from very low to very high prey population densities, the following pattern of change in marten spacing is predicted:

transient $\rightarrow$ exclusive territories, decreasing in size $\rightarrow$
intrasexual territories, decreasing in size $\rightarrow$
extensive home range overlap

I believe that this pattern—one that has been observed in the spacing of European badgers (*Meles* spp.; Pigozzi 1990)—may explain much of the variation home range overlap that has been reported for *Martes*. Other explanations are possible, of course, but these will usually make different predictions than are made here. Caro et al. (1989) documented two spacing strategies for adult male cheetahs (*Acinonyx jubatus*): intrasexual territoriality where female density was high, and transiency. Transient cheetahs were in poor health. Transient and temporary residents can be of any age or either sex in American martens and can make up a high percentage of a population (Hawley and Newby 1957a, Weckwerth and Hawley 1962, Soutiere 1978). No data are available on ages of other marten species with respect to residency status or on health of members of any *Martes* spp.

The model I have presented is not limited in application to *Martes* spp., but predicts spacing behavior for any solitary mammal. It also predicts spacing of social groups (e.g., wolf [*Canis lupus*] packs), but it cannot predict the formation of social groups beyond the prediction of extensive home range overlap. It can be used in conjunction with models that predict group formation (Macdonald 1983, Powell 1989).

Case History

The habitat in my fisher study area in the Upper Peninsula of Michigan was distinctly patchy with two distinct habitat (patch) types, each with a different predominant prey species (Fig. 6.1; Powell and Brander 1977; Powell 1978, 1979b, 1980a,b, 1981b, 1982). Porcupines were found almost exclusively in open, upland northern hardwood habitats that had interspersed stands of hemlock (*Tsuga canadensis*) and white pine (*Pinus strobus*). Each porcupine denned during winter in one to a few hollow trees or logs and traveled nightly to nearby hemlocks or white pines to eat. Porcupines tend to maintain fairly even spacing of home ranges (Roze 1989), and porcupine dens in my study area were spaced widely enough that individual porcupines could not warn each other of the presence of a fisher. Thus in winter each porcupine den created a porcupine habitat patch. During the winters of 1973–1974 through 1975–1976, I followed approximately 150 km of fisher tracks in the central part of my study area and counted porcupines in three subareas (Powell 1980a). During those three winters, at least one fisher track approached each known porcupine den in each porcupine census area. I therefore believe that most porcupine dens in the central study area were found by following fisher tracks. The average density of porcupine dens in the central study area was 1.23 dens/km^2. If 25 km^2 is taken as a representative home range size for a fisher (Table 6.2), then such a home range contained approximately 30 porcupine dens.

Snowshoe hares were found almost exclusively in dense, lowland habitats characterized by black spruce (*Picea mariana*), white cedar (*Thuja occidentalis*), and alder (*Alnus* spp.). In some parts of my study area, hare habitat was so extensive that it is inconceivable that a single fisher could depress vulnerability of hares throughout an entire patch. Some fisher home ranges, however, encompassed patches of lowland hare habitat, each of which was small enough that the fishers never traveled more than several hundred meters in each patch and rested no more than once in each patch. In these areas, average density of hare habitat patches was 1.86 patches/km^2 (Fig. 6.1). With 25 km^2 again a representative home range size for a fisher, such a fisher home range contained approximately 45 hare habitat patches.

Thus each fisher home range in my study area contained approximately 30 + 45 = 75 = x habitat patches. From over 150 km of fisher tracks and from telemetry data, I calculated that fishers averaged two active periods per day and traveled a minimum of 2.5 km per active period (Powell 1979b, 1982). From this, from data provided in earlier publications (Powell 1977, 1979b, 1982), and from unpublished data, I calculate that fishers visited a mean ($\pm$ SD) of 2 $\pm$ 2 porcupine dens and 4 $\pm$ 4 hare habitat patches each day, for a total of $y^* = 6$ patches visited per day.

If we assume that a fisher is equally likely to enter each of the two patch types per visit, then we can calculate y', the daily requirement for undepressed habitat patches, and $(y^* - y')/y'$, the cost of maintaining an intra-

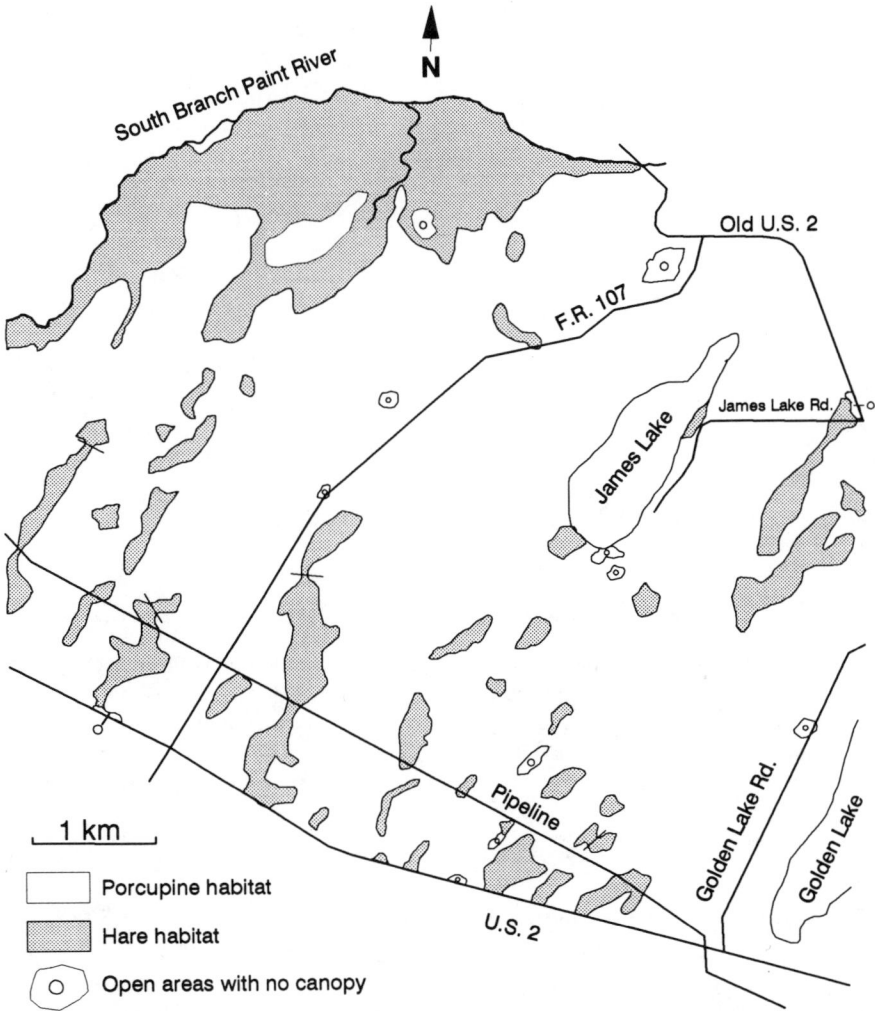

Figure 6.1. The central part of my fisher (*Martes pennanti*) study area in the Upper Peninsula of Michigan, which contained snowshoe hare habitat, porcupine habitat, and open habitats used by neither. The area used to calculate density of hare habitat patches was bounded by Golden Lake Road to the east, U.S. 2 to the south, the linear hare habitat patch to the west, and a line parallel to U.S. 2 intersecting the southern edge of James Lake to the north. Lines subdividing hare habitat patches show where detailed forest type maps (USDA Forest Service) showed separate patches. The pipeline subdivided all hare habitat patches that spanned it.

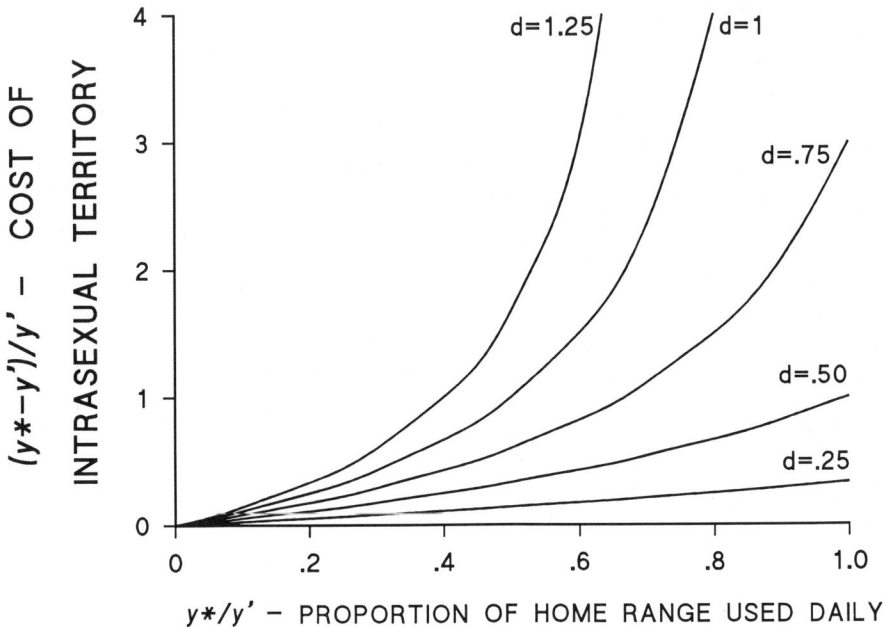

Figure 6.2. Cost of intrasexual territoriality relative to the proportion of habitat patches in a fisher's (*Martes pennanti*) home range that are visited each day and relative to the length of time, *d*, that prey remain depressed or invulnerable following a visit. Values used for calculations were $E = 1$, $T = E/10$, and $z = 5$. *E* can be considered the proportion of nonterritorial daily energy requirements that a fisher is expected to catch. The effects of varying these initial values is discussed in the text.

sexual territory over maintaining an individual territory (Fig. 6.2). For fishers, which have many more habitat patches in their home ranges than they visit in a day, the cost of sharing a territory with one other fisher is small (less than $(y^* - y')/y' = 0.1$, or 10%). If an animal visits most of the patches in its home range each day, the cost of intrasexual territoriality can be very high, especially if $d \geq 0.5$ day.

For fishers the range of food abundance or productivity for which intrasexual territoriality has the greatest benefit is large compared to that for unshared territories (Fig. 6.3). For an animal that visits most of its home range each day, however, there is less range over which intrasexual territoriality should be exhibited; the range completely disappears as *d* increases above 0.5.

The upper and lower limits of food availability that define when territorial behavior should be exhibited are given in equations 6.4 and 6.6. Note that they do not depend on the proportion of patches an animal visits per day nor on the length of resource depression. If the cost of maintaining a territory (T)

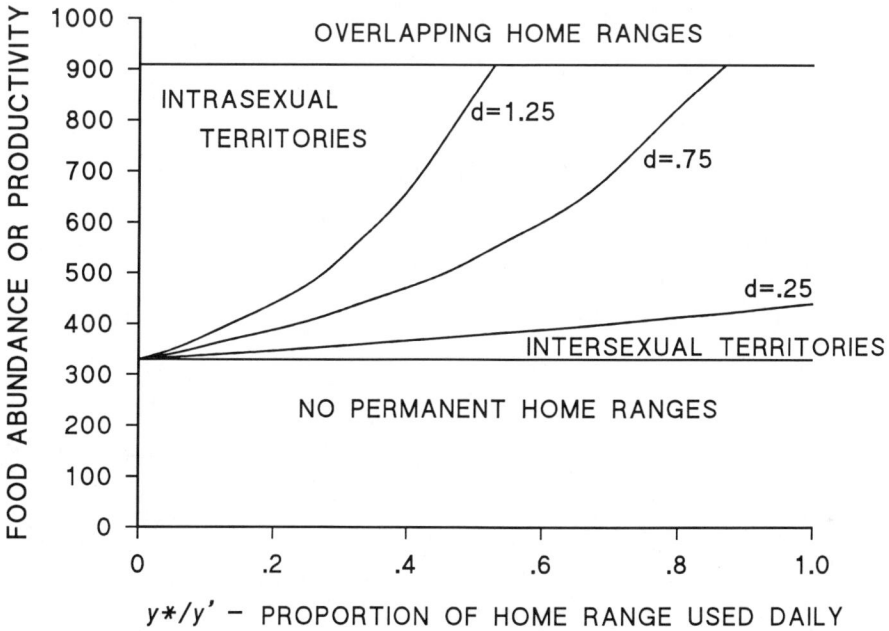

Figure 6.3. Home range and territory patterns, relative to the proportion of habitat patches in a fisher's (*Martes pennanti*) home range that are visited each day, to the length of time that prey remain depressed or invulnerable following a visit, and to the availability or productivity of food in the home range. Values used for calculations are as for Figure 6.2.

is high, the lower limit is raised. If T is high enough, animals should never be territorial. Likewise, if the number of conspecifics with which a home range can be shared is low (z is low), then the upper limit is lowered and can be lowered sufficiently to eliminate territoriality. These characteristics of Fig. 6.3 make sense. If the cost of maintaining a territory is very high, that cost can exceed all benefits and thus preclude territorial behavior. If the number of conspecifics that will share an undefended home range is very low, there is little benefit from territorial behavior and thus it should not be exhibited.

Acknowledgments

Mark Boyce and an anonymous reviewer made helpful comments on early versions of this chapter. Consie Powell drew Figure 6.1. The Department of Zoology and Physiology, University of Wyoming, graciously provided space and logistic support that allowed the chapter to be written.

7 Habitat Use and Spatial Organization by the Stone Marten

Mathias Herrmann

Home range size is related to energy requirements, foraging behavior, body size, and population density (McNab 1963, Jewell 1966, Lockie 1966, Brown and Lasiewski 1972, Harestad and Bunnell 1979, Buskirk and Mac-Donald 1989). Moreover, among mustelid species, spacing and mating patterns are related (Moors 1980, Hixon 1987, Sandell 1989*a*), and social status, determined by sex and age, and spacing patterns affect home range size.

Sandell (1989*a*) predicted that solitary carnivores should maintain exclusive territories when resources are relatively stable. He predicted that female home range size and spacing patterns are determined by the abundance and dispersion of food, especially during the most critical period of the year, whereas male spatial patterns are determined by the distribution of females. Because males can maximize reproductive success by defending territories that overlap the territories of several females, male home ranges should be larger than predicted by energy requirements.

Adult mustelids commonly show intrasexual territoriality (Ewer 1973, Powell 1979*a*), and males adopt one or both of two alternative ways to achieve mating: either they remain in their home ranges and try to monopolize several females (Eurasian pine martens, *Martes martes*: Marchesi 1989, Balharry 1991; Marchesi 1989; Tsushima martens, *M. melampus*: M. Tatara, Kyushu Univ., pers. commun.), or they roam and compete for access to each female that enters estrus (stoats, *Mustela erminea*: Erlinge 1977, Debrot and Mermot 1983, Erlinge and Sandell 1986, Sandell 1986; least weasels, *M. nivalis*: Lockie 1966, Erlinge 1974, Moors 1974; polecats, *M. putorius*: Weber 1987; minks, *M. vison*: Gerell 1970; fishers, *Martes pennanti*: Arthur et al. 1989*a*). Siberian weasels (*Mustela sibirica*) switch strategies (Sasaki

122

and Ono 1989). The maintenance of exclusive male home ranges should be the best tactic when females are densely and evenly distributed. Under these circumstances, male home range size should show little variation during the year (Sandell 1989*a*).

Reports that focus only on food availability, summarizing home ranges without discriminating seasons, habitats, or social status, are inadequate to test Sandell's (1989*a*) predictions. This chapter presents data on stone martens (*Martes foina*) that bear on those predictions. Three major questions are considered: To what extent can space use be explained by habitat selection? How do individuals structure their use of space and habitats? And, how do social factors, such as mating system or parental care, influence space use?

Study Area

The study area was located in the Saarland in southwest Germany, near the French border (49°12′N, 7°11′E). The climate is Atlantic; temperatures during the study period (1980–1985) averaged 8.9°C (January 0.4°C, July 18.4°C) and rainfall averaged 900 mm/year. The hilly landscape (200–400 m above sea level) is densely populated (342 people/km²) and intensively cultivated: the southern part of the study area (59 km²) is agricultural. The floodplain of the central valley is mostly grassland with villages of 500–3000 people along the floodplain border. Arable land with shell-limestone soils occurs on the moderate slopes and plateaus. Many slopes have traditional orchards, and some slopes and the tops of the hills have large hedges and forests. The northern part of my study area is a large forest (30 km²) on new red sandstone.

It was not practicable to quantify food availability because foods were highly diverse and fluctuated both in time and space, especially in the village (e.g., foods on composts: Herrmann 1987, Labhardt 1991). Fruits (*Prunus* spp., *Malus* spp., *Pyrus* spp., *Crataegus* spp., *Sorbus* spp., *Rosa* spp., *Rubus* spp., *Taxus baccata*, *Ribes* spp.) were available from the end of June until winter. In or near the villages, fruit trees were most common in orchards and gardens. In agricultural land they were found in hedges, in seminatural woodland, and in traditional orchards. Fruit trees were rare in the forest. Human food scraps and chicken eggs were available only in gardens or near buildings in the village and on farms. Passerine birds (e.g., house sparrow, *Passer domesticus*; starling, *Sturnus vulgaris*; *Turdus* spp.; *Parus* spp.) were most common in villages, in small forests, and along river banks and hedges. The most abundant small mammals were bank vole (*Clethrionomys glareolus*), common vole (*Microtus arvalis*), field vole (*M. agrestis*), long-tailed field vole (*Apodemus sylvaticus*), yellow-necked field vole (*A. fla-vicollis*), house mouse (*Mus domesticus*), and brown rat (*Rattus norvegicus*). After harvest (August–September), *Apodemus* spp. and other species migrated into fallow fields, gardens, river banks, hedges, and seminatural woodland, making them more abundant there than in the forest or on open fields. Earthworms (*Lumbricus* spp.) were most easily found in plowed fields on shell-limestone soils.

In wooded areas, stone martens coexisted with Eurasian pine martens. Red foxes (*Vulpes vulpes*), badgers (*Meles meles*), polecats, stoats, and least weasels were also in the study area (Herrmann 1991), and in and near villages the most frequent predators were domestic and feral cats (*Felis catus*). I assumed that the entire study area was populated by stone martens; they were abundant, and deserted home ranges were rapidly recolonized (Skirnisson 1986, Broekhuizen et al. 1989, hunters' reports, per. observ.).

Methods

Trapping and Radiotelemetry

Wooden livetraps (100 × 40 × 40 cm) were baited with white chicken eggs. Traps were placed and prebaited 100 days before they were set to capture martens. Sixteen stone martens (6 males, 10 females) were trapped. They were anesthezised, measured, and sexed, and their age was estimated by visual examination of the teeth. They were classified as juvenile (less than 1 year), subadult (between 1 and 2 years), or adult (2 years or older). These three ages corresponded to social status: young animals living in their mothers' home ranges, prereproductive animals after dispersal, and mature animals. Age estimates from cementum annuli of four animals were consistent with the visual estimates. A transmitter collar (37 g, 148–151 mHz, "Wagener, Köln") was put around the neck of each, and transmitter signals could be received from up to 1 km away. After three months, when transmitter batteries needed to be replaced, the stone martens were recaptured in livetraps. In 1985 I also surrounded rest sites with up to 400 m of 4-m-high fish nets and chased martens into the nets. This worked on seven of eight attempts.

Fourteen stone martens (five males, nine females) were located nightly for a minimum of 42 days and a maximum of 676 days. A total of 7344 locations were made between August 1980 and June 1985. During continuous observations (every 15 minutes for an entire or half night) I recorded the martens' location, activity (active or not during 30 seconds of monitoring), and habitat, as well as weather conditions. In villages I approached stone martens closely (≤50 m); here sight observations were frequent (about six a night). On farmland and forest, I mostly used triangulation. I located resting sites by approaching on foot. The coordinates of each location were recorded using a 50-m grid system, and for every location an estimate of accuracy was made in the field.

Convex polygon home ranges (Mohr and Stumpf 1966) were used to quantify habitat availability for each individual marten for several reasons. First, the territorial behavior of the martens restricted their movements to areas that can be accurately mapped as polygons. Second, polygons include rarely visited parts of home ranges. Third, polygon area is not strongly influenced by the number of locations above a sample size of 100 (Buskirk and MacDonald 1989). Home ranges were plotted for periods during which martens showed no major home range shifts for at least one month and in some cases up to a year and a half excluding excursions. Successive locations could be used for home range determination because all animals were able to move within 15 minutes from one end of their home ranges to another. Territoriality and home range overlap were quantified with a cell model (Herrmann 1987). Each 50-by-50-m cell visited at least once by a given marten or surrounded by at least three used cells was included in its home range.

Habitat

Habitats were classified on two scales. On a coarse scale (>10 ha), three "biotopes" were distinguished: village, farmland, and forest. On a fine scale, 19 "habitats" were classified within the biotopes: clearing, thicket, coniferous forest, deciduous forest, mixed forest, seminatural woodland, hedge, fallow field, garden, grassland, arable land, riverbank, orchard, road, house, commercial or public building, farm, ruin, or chicken yard. Habitat patches were small in the village biotope ($\bar{x}$ [± SD] = 0.12 ± 0.03 ha) and large in forest and farmland (2.45 ± 0.85 ha).

All habitats were mapped and overlaid with the 50-m grid system. Habitats in each cell were indexed in units according to their total area:

    1 unit =   1–25% of the cell area
    2 units = 26–50% of the cell area
    3 units = 51–75% of the cell area
    4 units = 76–100% of the cell area

To avoid underestimation of biologically important structures, I indexed very small habitats (chicken yards, hedges) as equal to habitats that covered up to

25% of a cell's area. Habitat types were considered "available" to a particular marten if they occurred within its convex polygon home range.

Because mean patch size of many habitats was small relative to mean radio location error (50 m), habitat use was estimated as follows. For the four 50-by-50-m cells surrounding the grid intersection closest to a telemetry location estimate, the number of units in each habitat was summed. These habitat data for all locations were then summed for each individual marten during each season to give a final habitat use index for that individual. Locations during daytime were excluded because human disturbance inhibited martens' movements. Locations of resting martens were excluded because I assumed that resting sites were not important for spatial requirements. And locations of low accuracy (error > 125 m) were excluded.

Because habitat availability differed for the animals, Johnson's (1980) hierarchical method was modified to examine habitat selection. The habitat use index of each marten was divided by the sum of the habitat units in that marten's convex polygon home range (availability) to index "relative preference." Indices of relative preferences were then ranked. Habitats with high values were more attractive than those with lower values. A hierarchy was developed for each individual during each season ($n = 34$). Using these data, I could determine only a relative preference for the habitats (Johnson 1980).

To get an overall picture of the habitat preferences of all stone martens during the whole year, I calculated a quotient of preference for each habitat type. This quotient equals the number of cases when a given habitat was preferred over all other habitats, divided by the number of cases when it was not preferred. The quotients for each habitat type were then ranked.

Home Range Size

To determine the influences on the home range size of biotope, age, sex, and season, I compared pairs of individual animals for which all other variables were equal using Wilcoxon paired-rank tests. Because the number of pairs was limited by the number of individuals living under comparable circumstances, I included in the comparisons home range sizes of stone martens published by Skirnisson (1986) and Müskens et al. (1989) for which biotope, season, sex, and age were stated.

Results

Habitat Preference

Over all martens, habitat preferences were structured significantly (Bowker symmetric index: $X^2 = 772$, $P < 0.001$; Lundberg 1987). The quotient of preference from most to least preferred was as follows: chicken yard, farm, orchard, ruin, house, thicket, seminatural forest, garden, commercial or public building, road, coniferous forest, hedge, riverbank, fallow field, grassland, mixed forest, deciduous forest, arable land, clearings. Presumably, this order represents a hierarchy of descending habitat quality.

Availability of habitats for stone martens that lived in villages was not the same as for those that lived in farmland and forest (Fig. 7.1). The five most preferred habitats (chicken yard, farm, orchard, ruin, and house) were 5.1 times more available to the martens that had the most important parts of their home ranges in villages (>1/3) than to those in the other biotopes. The five least preferred habitats (clearing, arable land, deciduous forest, mixed forest, grassland) were 2.6 times more available to the animals that lived predominantly in forest and farmland (>2/3) than to those in villages.

Home Ranges

Four factors appear to affect home range size: biotope, sex, age (approximate social status), and season (Herrmann 1989*a*). Animals of the same sex and age class were compared during similar seasons to examine the influence of biotope on home range size (Table 7.1). Home ranges in villages were significantly smaller than home ranges in farmland and forests (Wilcoxon paired rank test, $P < 0.005$, $n = 10$).

To test the importance of sex for home range size, I compared four female and four male stone martens of the same age during identical seasons in the same habitat (Table 7.2). Skirnisson's (1986) data were included in the statistical test. Home range sizes of males were significantly larger than for females (Wilcoxon paired-rank test, $P < 0.01$, $n = 9$).

In six of seven comparisons of adult versus juvenile or subadult stone martens matched for sex, habitat, and season (Table 7.3), adults had larger home ranges. No comparable data are available in the literature to enlarge the sample size for this test. Four comparisons were made of home range size for martens of the same age, sex, season, and biotope (Table 7.4). In all cases the martens' home range sizes did not differ. The home ranges of most animals were stable. Three females (designated TH, NN, and SL) and one male (KA) did not shift their home ranges during 20, 12, 7, and 12 months of observations, although their home range sizes did change with season.

Seven stone martens were observed for more than six months (Fig. 7.2). Two adult martens (KA and TH) were tracked during and entire year. During summer (the breeding season: Jun–Aug) the home range size of the adult male (KA, ▲) was largest (56 ha). During fall (Sep–Nov) he reduced his home range size slightly (50 ha). During winter (Dec–Feb) his home range was smallest (10 ha), and he used only 18% of his summer home range.

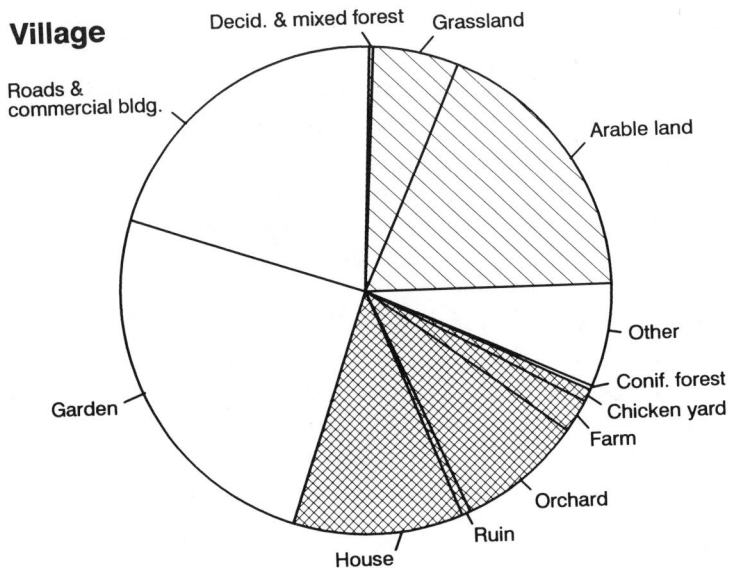

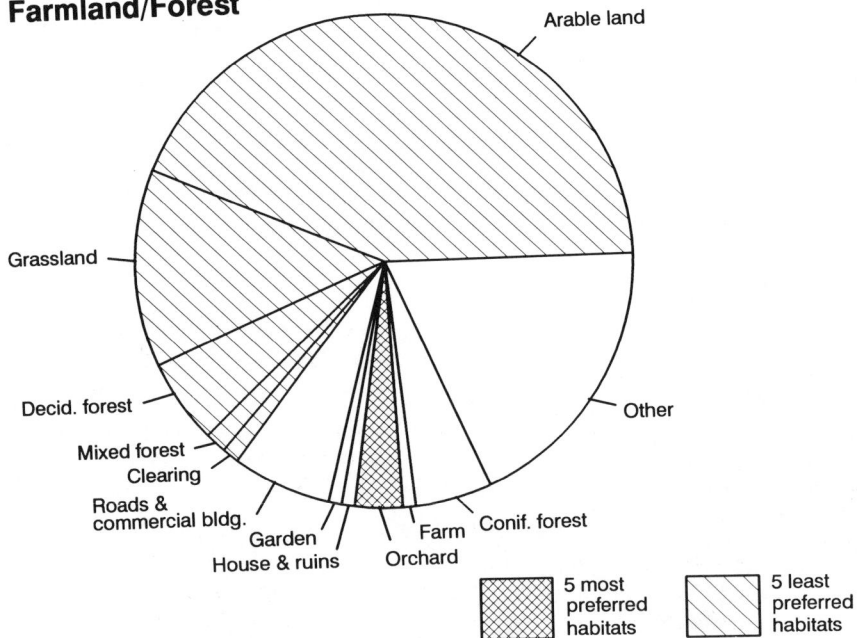

Figure 7.1. Relative availability of habitats in the home ranges of stone martens (*Martes foina*) in villages and in farmland and forests. The five most preferred habitats were most common in villages, whereas the five least preferred habitats were most common in farmland and forests. Other = riverbank, seminatural forest, fallow, hedge, and thicket.

Table 7.1. Home range size of stone martens (*Martes foina*) of the same sex and age class during comparable seasons in villages and in farmland and forests

	Village				Farmland/forest		
Age and sex class	Home range (ha)	Season	*n*	Age and sex	Home range (ha)	Season	*n*
Juv. ♂	30	Aug–Oct	208	Juv. ♂	87	Aug–Sep	63
Juv. ♀	19	Oct–Dec	152	Juv. ♀	90	Oct–Dec	82
Juv. ♀	16	Nov–Feb	104	Juv. ♀	110	Jan–Feb	287
Adult ♀	24	Mar–May	155	Adult ♀	78	Mar–May	872
Adult ♀	27	Jun–Aug	163	Adult ♀	37	Jun–Aug	179
Adult ♀	17	Sep–Nov	155	Adult ♀	74	Sep–Nov	169
Adult ♂	44	Oct–Feb	169	Subad. ♂	57	Oct–Feb	207
Subad. ♀	41	Jun–Mar	999	Subad. ♀	42	Apr–Mar	318
Adult ♀	27	Dec–Apr	516	Adult ♀	57	Dec–Apr	193
Juv. ♀	16	Nov–Feb	104	Juv. ♀	98	Jan–Feb	84

Note: Home ranges in villages were significantly smaller than those in farmland and forests ($P < 0.005$, $n = 10$, Wilcoxon paired-rank test).

Table 7.2. Home range sizes of stone martens (*Martes foina*) of different sex but in the same age class in comparable seasons and biotopes

			Female		Male	
Age class	Season	Biotope	Home range (ha)	*n*	Home range (ha)	*n*
Adult	May–Jul	Village	22	217	61	217
Adult	Mar–May	Village	30	525	41	196
Adult	Jun–Aug	Village	27	244	56	289
Adult	Sep–Nov	Village	19	200	50	218
Adult	Dec–Feb	Village	16	198	10	83
Juv.	Aug–Oct	Village	21	198	30	208
Adult	May–Jul	Forest/farm	185	>300	310	>300
Adult	Jan–Mar	Forest/farm	185	>300	292	>300
Adult	Mar–Jun	Village	72	>300	88[a]	>300

Notes: Males had larger home ranges than did females ($P < 0.01$, $n = 9$, Wilcoxon paired-rank test). The first six rows contain data for two females and three males; data for the last three rows are from Skirnisson 1986 for two females and two males.

[a]Mar–May.

Table 7.3. Home ranges sizes of stone martens (*Martes foina*) of different age classes but of the same sex in comparable biotopes and seasons

		Juvenile/subadult			Adult		
Sex	Biotope	Home range (ha)	Season	*n*	Home range (ha)	Season	*n*
♂	Village	30	Aug–Oct	208	57	Aug–Oct	311
♀	Village	21	Aug–Oct	198	26	Aug–Oct	262
♀	Village	16	Nov–Feb	104	17	Nov–Feb	127
♀	Farmland	28	Dec–Mar	151	54	Dec–Mar	159
♀	Farmland	30	Apr–Jun	79	62	Apr–Jun	40
♂	Farmland	72	Dec–Mar	142	42	Dec–Feb	145
♀	Village	16	Nov–Feb	104	18	Nov–Feb	163

Note: $P > 0.05$, $n = 7$, Wilcoxon paired-rank test.

During spring (Mar–May) he again used a great proportion of his home range from the previous summer (41 ha). The seasonal variation of the adult female (TH, ○) was similar but less extensive. During spring, when she had to provide for young, her home range was larger (30 ha) than during summer (27 ha), when her young followed her. During fall (19 ha) and winter (16 ha) her home range size was reduced to 59% of the summer range. Most martens that were tracked for less than a year showed similar patterns (Fig. 7.2).

Despite extensive trapping, a second adult stone marten of the same sex was never trapped within an adult marten's home range. Twice, neighboring females were radio-tracked simultaneously. In the farmland study area, fe-

Table 7.4. Home ranges sizes of stone martens (*Martes foina*) of the same sex and age class in comparable isotopes and seasons (note the striking similarities)

Age and sex class	Biotope	Home range (ha)	Season	*n*
Juv. ♀	Village	16	Nov–Jan	104
Juv. ♀	Village	16	Nov–Jan	144
Adult ♂	Village	56	Jun–Aug	289
Adult ♂	Village	61	May–Jul	217
Juv. ♀	Forest	90	Oct–Dec	82
Juv. ♀	Forest	110	Jan–Feb	297
Subad. ♂	Farmland	48	Aug–Oct	86
Subad. ♂	Farmland	36	Jul–Oct	208

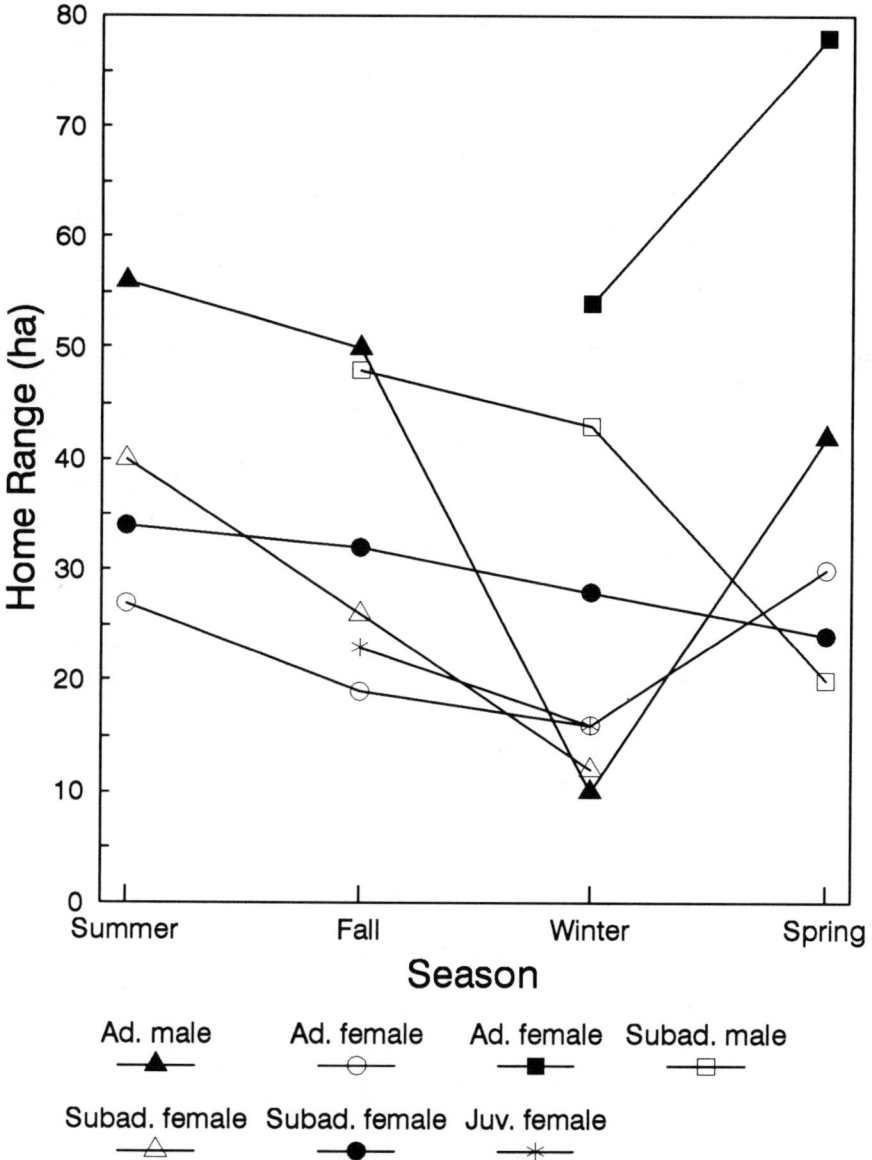

Figure 7.2. Seasonal home range areas for seven stone martens (*Martes foina*) radio-tracked for at least six months.

males NN and SL had a long common border but little overlap of their home ranges from December 1983 to March 1984. Female NN died in April, and during the next three months female SL was radio-tracked in her old range as well as in the deserted home range of NN. In the village study area, females SN and TH had exclusive home ranges in July–October 1983. Neighboring adult males could not be caught. Stability of home ranges and aggressive interactions observed near home range borders support the idea of exclusive male territories.

In the village, home ranges of adult male BP and adult female TH over-lapped completely during May–July 1983. During the mating season the male and the female were tracked several times at the same location, and the female visited the male at his preferred resting site.

A mother and her cubs were radio-tracked simultaneously. In a village a son (TR) and a daughter (SB) of female TH were caught at age 5 months in 1983. The home ranges of all three overlapped completely from August to October (Fig. 7.3). The home range of SB was nearly identical with that of her mother. TR undertook a few excursions out of his mother's seasonal home range, but all excursions lay within the home range his mother used throughout the year.

The mother was tracked together with her son in 26% of all active loca-tions ($n = 92$) and together with her daughter in 16% of all active locations ($n = 92$). In 33% of all cases ($n = 30$) she was found at the same resting site as her son, and in 13% she was found resting with her daughter. Several times one of the young was seen closely following the mother, and all three martens

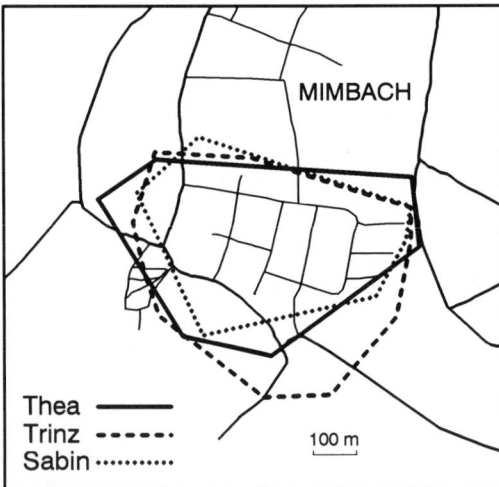

Thea ———
Trinz – – – – – ·
Sabin ··········

100 m

Figure 7.3. Home ranges of adult female Thea (26 ha, $n = 262$), her son Trinz (30 ha, $n = 208$), and her daughter Sabin (21 ha, $n = 198$) from 6 August to 31 October 1983 in the village of Mimbach.

were sighted playing, chasing each other, and running together. The son's transmitter failed on 30 October, but he was seen next June and caught in July within the home range used the previous fall. A few days after release he shifted his range to the next village. The daughter disappeared in mid-January.

Discussion

Home range sizes of the stone martens studied varied considerably. Seasonal home range sizes varied from 12 to 211 ha. Skirnisson (1986) found a range of 20–310 ha, Broekhuizen (1983) 78–777 ha, and Krüger (1989) 145–675 ha. One objective of my study was to determine how habitat quality and social organization affect this variation in home range sizes.

My results indicate that four factors affect home range size of stone martens:

1. *Habitat quality*. The higher habitat quality in villages was correlated with significantly smaller home range sizes.
2. *Sex*. Male home ranges were significantly larger than female home ranges.
3. *Season*. The home range sizes of stone martens appeared to vary seasonally. Home ranges of adult males in particular were largest during summer and smallest during winter.
4. *Age or social status*. Size of home range and seasonal variation in size appeared related to an animal's age. The home ranges of adult stone martens were larger than those of juveniles or subadults, especially during summer (mating sesason).

Habitats used by the stone martens differ from those used by other boreal forest martens (Buskirk and Powell, this volume; Bakeyev and Sinitsyn, this volume; Tatara, this volume). Stone martens are synanthropic, living even in the centers of European cities (Nicht 1969, Herrmann and Knapp 1984, Tester 1986, Klausnitzer 1987, Broekhuizen et al. 1989, Romanowski 1989, Libois and Waechter 1991). Consequently, a broad variety of urban, rural, and forest habitats is available to stone martens. My results corroborate patterns of habitat selection reported by Broekhuizen (1983) and Skirnisson (1986) for individuals. Both found arable land and grassland to be used less than expected, whereas hedges, groves, ditches, and village habitats were used more than expected. In my study area high-ranking habitats were most abundant in villages, whereas low-ranking habitats were most abundant in farmland and forest, and home range size of stone martens was significantly smaller in the biotopes with high-ranking habitats than in those with low-ranking habitats. But were resources more abundant in high-ranking habitats?

Stone martens are omnivorous food generalists (Schoener 1968). Throughout the year they eat small mammals, birds, eggs, and human scraps. During summer, fall, and early winter, fruits are i. .portant, but during winter and spring earthworms often are important. In villages fruits and human scraps compose much of the diet, whereas in other habitats small mammals and earthworms are more important (Waechter 1975, Goszczynski 1976, Delibes 1978, Amores 1980, Chotolchu et al. 1980, Kalpers 1983, Rasmussen and Madsen 1985, Skirnisson 1986, Tester 1986, Ansorge 1989a, Marchesi et al. 1989, Romanowski 1989).

Food is abundant in villages year-round. Favorite foods (Waechter 1975, Herrmann 1987), such as cake, meat, cheese, eggs, and food for domestic animals, were found nearly every morning on composts and dung hills or in farms or chicken yards within the stone martens' home ranges. Such food was not completely consumed by the martens during the night. I hypothesize that food shortage occurs only in the forest and farmland habitats, and only during late winter and spring.

Warm, dry resting sites are another resource important to stone martens (Skirnisson 1986, Foehrenbach 1987), especially during winter (Waechter 1975). In villages, stone martens sleep in straw, hay, and roof insulation during winter. Resting sites for stone martens in farmland and woodland areas are not as good (Waechter 1975). Only holes in the earth and trees or resting sites in dense vegetation are available. Stone martens in fields or woodlands that contain buildings in their home ranges use these intensively.

Cover and vertical structures provide protection against predators and facilitate foraging (Waechter 1975, Skirnisson 1986). The best cover was available in or near buildings, in gardens, along riverbanks, in hedges and thickets, and in seminatural forest. The assessments of food, cover, and resting sites in the habitats agreed with the habitat selection hierarchy I found.

The size of a stone marten's home range appears to be influenced by habitat quality, as hypothesized by Buskirk and McDonald (1989) for American martens (*M. americana*). Stone martens that live predominantly in villages have smaller home ranges because there are more high-quality habitats in this biotope than in farmland or forests. Broekhuizen (1983) first discussed an inverse relationship between the quality of the habitat and the size of the home range on farmland. Skirnisson (1986) found smaller home ranges for stone martens that lived predominantly in villages and discussed diet and resting sites as possible causes.

Availability of resources does not explain seasonal variation in home range size, however. Four stone martens tracked for more than a year by Skirnisson (1986) and Müskens et al. (1989) exhibited the same seasonal fluctuations in home range size as I found. During summer and fall, food and cover exceed what is needed. Resting sites are not so important during this time of the year. Resource scarcity may occur in late winter and spring, however, particularily in forest and farmland. In contrast to the predictions of Harestad and Bunnell (1979) relating home range size to resources, the home ranges of stone martens were smallest during winter and largest during summer.

Resources also did not appear to account for home range size differences between sex and age classes. Male stone martens are 1.3 times larger than females (Schmidt 1943; Koenig and Mueller 1986*a,b*). Sandell (1989*a*) predicted that the home range size of such males should be 1.2 times that of females based on energetic requirements. Observed male home ranges were on average 1.8 times larger than female home ranges, results similar to data collected by Foehrenbach (1987), Krüger (1989), and Müskens et al. (1989), and larger than predicted on the basis of energetic requirements. No sex difference in diet, as postulated by Brown and Lasiewski (1972), has been found to explain the large home range size of adult males (Skirnisson 1986, Ansorge 1989*a*). Thus, energetic needs and resources are not the primary cause of sex and age differences in home range size.

The major factor causing the disproportionate home range sizes of males seems to be related to mating; the largest home ranges were found during the mating season, not in winter when resources should be scarcest. As predicted by Sandell (1989*a*) for solitary carnivores, when resources are relatively stable stone martens appear to have exclusive intrasexual territories (Powell 1979*a*; this volume). Territoriality allows the spatial distribution of adult stone martens to be stable throughout the year, even during mating season (Herrmann and Hendrichs 1989, Hovens and Janss 1990). By maintaining territories that are as large as they can defend, male stone martens increase the chances that their territories will overlap with more than one female territory (Skirnisson 1986, Krüger 1989, Müskens et al. 1989). Maximizing matings should increase a male's fitness. In addition, the intersexual competition will be minimized the more female territories are covered by one male territory.

Since the sex ratio of stone martens is close to 1:1 (Powell, this volume) and all territories are occupied, what happens to the surplus males? In captivity, males mature socially more slowly than females (Schmidt 1943, Kugelschafter 1988). A 1½-year-old subadult male tracked by Müskens et al. (1989) was tolerated in his natal home range by his likely father. Juvenile males in my study stayed for at least 1 year (GR) and 1¼ years (TR) in their natal home ranges, whereas juvenile female MR dispersed at about age 9 months and SB disappeared for unknown reasons at about age 10 months. Subdominant males may be excluded from reproduction and be forced onto small home ranges in poor habitats (Erlinge 1967, 1968; Gerell 1970; Pulliam 1988). The two subadult males I tracked fit this pattern and reduced home range size before the mating season began.

In contrast with Sandell's (1989a) hypothesis, the home ranges of adult male stone martens became small during winter. The energetic costs of territorial defense may be too large to maintain large territories all year. In March, long before the next mating season but as testes grow larger (Madsen and Rasmussen 1985, Hesse 1987), territorial behavior and size of male home ranges increases.

For females, Sandell (1989a) predicted that home ranges should be just large enough to provide support during the most critical period of the year. Like Sandell (1989a), I hypothesize that females increase their home range sizes to support their young. Kugelschafter (1989) and I (1989b) hypothesized that a home range that enables the young to stay longer within the familiar territory of the mother (Müskens et al. 1989, Lucas 1989) gives the young time to learn. Large home ranges may also allow daughters to take over parts of their mothers' home ranges. Clearly, the relationship between spacing patterns, social dominance, resource abundance, and fitness will bear much close examination in the case of stone martens.

Acknowledgments

I thank H. Hendrichs for supervising my study, and I thank all those who helped me with my fieldwork, especially K. U. Goss, R. Roechert, and M. Trinzen. I thank T. Caro, W. Dressen, K. Kugelschafter, F. Trillmich, and E. Zimen for helpful discussions on earlier versions of this chapter.

8 Mortality and Vulnerability of a Heavily Trapped Fisher Population

William B. Krohn, Stephen M. Arthur, and Thomas F. Paragi

The fisher (*Martes pennanti*) was eliminated from much of North America south of Canada during the late 1800s (Strickland and Douglas 1981). In Maine, fisher populations reached a low point during the 1930s, when they were restricted to the extensive forests of western Maine (Coulter 1960, 1966). Except for an experimental season in January 1950, hunting and trapping of fishers in Maine was prohibited from 1937 to 1954. Trapping was legalized in 1955, and today the species is common statewide except in the eastern tip of Maine (Clark 1986).

Coulter (1960), Powell (1979c), and others have commented on the susceptibility of fishers to overtrapping. Based on an analysis of five predator-prey models, Powell (1979c) predicted that annual harvests in excess of 4.0 fishers per 100 km^2 may lead to extermination of local populations. In contrast, the average annual harvest in south-central Maine was 6.6 animals per 100 km^2 during 1976–1982 (Clark 1986). Powell's models did not use empirically estimated mortality rates, nor did he account for the possibility of differential vulnerability among age and sex classes. Strickland and Douglas (1981) observed that age and sex ratios of harvested fishers differed in populations that were presumed to be declining rather than increasing. They concluded that juveniles were more vulnerable than adults to trapping, and that the ratio of trapped juveniles to adult females (≥ 2 years old) could be used to monitor the intensity of the fisher harvest (see also Strickland, this volume). Strickland and Douglas (1981) also concluded that adult females outnumbered adult males in the population. In a review of fisher ecology and management, Douglas and Strickland (1987:526–527) noted that age- and sex-specific mortality rates and the relations between trapping and nontrapping mortalities need clarification. In addition, measures of differential trapping vulnerability are needed to correct for biases in trapped samples and may provide insight into patterns of natural mortality in untrapped populations.

As part of a long-term study on spatial patterns, habitat use, food habits, and reproduction of fishers in south-central Maine (Arthur et al. 1989*a,b*; Arthur and Krohn 1991; Paragi 1990), we collected data on trapping mortality and differential vulnerability. Our objectives were threefold: first, to document the causes of death in a sample of radio-collared fishers; second, to estimate mortality rates during trapping and nontrapping intervals for four age and sex classes; and third, to use these data to test for differential vulnerability to fur trapping among the age and sex classes.

Study Area and Methods

The 500-km² area used for our telemetry studies extends approximately 50 km inland from Penobscot Bay in south-central Maine. The area is covered with second-growth forests interspersed with small farms. More details on the land use, topography, and climate of the study area were presented in previous works (Arthur et al. 1989*a,b*).

Our study was conducted from 1984 to 1989. Arthur (1988) documented the techniques used to capture, handle, and equip fishers with radio collars. Live-trapping was conducted from September through May during the first two years of the study. Beginning in 1986, live-trapping efforts were mainly conducted in September and October, and to a lesser degree in February and March. Transmitters had a life expectancy of 12 months, and we recaptured and recollared animals that survived the fall trapping season.

From 1984 through 1989, the fur-trapping season averaged 39 days (range: 36–42) and occurred from late October to early December. There were no limits on the number of trappers, number of traps used, or harvest per trapper, although trappers were required to check traps at least once every 24 hours. Because radio collars caused wear of fur on some animals, trappers were compensated US$20 in 1988 and 1989 when they captured a radio-collared animal.

We attempted to locate each radio-collared animal at least twice each week during the trapping season and at least twice each month for the rest of the year. Dates of death were determined by reports from fur trappers or by using the midpoint between the last date a fisher was known to be alive and the first date the animal was known to be dead. Animals were necropsied when cause of death was questionable. When contact was lost with a radio-collared animal (i.e., the collar slipped off or the radio failed), we assumed it was alive up to the last day of contact. When an animal with a failed or slipped collar was recaptured and recollared, we excluded the radio-days during the period of no contact to avoid possibly underestimating nontrapping mortality (i.e., the only animals we could possibly recontact were those that lived).

Juveniles were defined as fishers less than 1 year of age; adults were 1 year old or more. Fisher ages were estimated by counting cementum annuli of first premolars (Arthur et al. 1992; Strickland et al. 1982). Sample sizes were too small to distinguish the survival of yearlings (= 1 year) from older animals (≥ 2 years). The birthdate was set at 15 March because most natal dens were initiated in mid-March (Paragi 1990:16). Some juveniles became independent of their mothers during late August or early September (Arthur and Krohn 1991). Thus, the nontrapping interval for juveniles was defined as 1 September to the start of trapping, plus the period from the end of trapping to 15 March for each year (a total of 156 days). The nontrapping interval for adults went from 15 March of one year to 15 March of the next, excluding the trapping season (a total of 326 days).

The MICROMORT program calculates survival as the number of radio-days that animals survived divided by the total number of radio-days that animals were monitored (Heisey and Fuller 1985). Thus, these rates are the proportions of days survived. We compared survival rates among all age and sex classes during the trapping interval, using the multiple comparison procedure recommended by Zar (1984:402), with the arc-sin transformation of proportion of days survived. During the nontrapping interval, we used different interval lengths for adults and juveniles, so we made no comparisons between age groups. Instead, we compared interval survival rates between juvenile males and juvenile females and between adult males and adult females (Zar 1984:396). MICROMORT assumes that survival is constant throughout the interval over which it is estimated (Heisey and Fuller 1985), and no evidence was found to reject this assumption (see Paragi 1990:46).

The vulnerability of age and sex classes to fur trapping was examined by comparing trapping mortality rates among classes. The fundamental assumption of the vulnerability analysis is that radio-collared and uncollared fishers were harvested at the same rate. We had no way to test this assumption, but believe that most resident adults were radio-collared (see Arthur et al. 1989a).

Results

Human-related causes accounted for 94% ($n = 47$) of the 50 deaths recorded during the study (Table 8.1). Of the human-related causes of death,

Table 8.1. Causes and numbers of deaths of radio-collared fishers (*Martes pennanti*) in south-central Maine, 1984–1989

	Trapped	Shot[a]	Vehicle	Other human-related[b]	Natural[c]
1984	2	0	1	0	0
1985	11	0	1	0	1
1986	6	0	0	0	0
1987	11	1	0	1	1
1988	5	2	0	0	0
1989	5	0	0	1	1
TOTAL	40	3	2	2	3

[a]Illegal.

[b]One unknown, one choked when its collar caught on a branch.

[c]One apparently killed by a coyote, one died from an infection, and one choked on deer cartilage.

fur trapping accounted for 40 deaths (80%), shooting (which was illegal) for three (6%), two (4%) fishers were hit by vehicles, one caught its radio collar on a branch, and the exact cause of death was undetermined in one case, although it appeared human-related. Of the three animals that died from natural causes, one juvenile male choked on deer cartilage, one adult male died from an undetermined infection, and an 8-year-old female was apparently caught on the ice and killed by coyotes (*Canis latrans*) (Table 8.1).

Seventy-six individuals, including 14 monitored both as juveniles and adults, were monitored for a total of 18,946 radio-days over six years (trapping = 2655 days, nontrapping = 16,291) (Table 8.2). Nine of the 25 adult males were monitored for more than one year (maximum three years for three fishers), and 7 of the 19 adult females were monitored for more than one year (in one case, for five years). Estimates for juvenile females were based on smaller samples than the other three classes (Table 8.2).

We lost contact with 21 fishers (12 adult females, 6 adult males, 2 juvenile females, 1 juvenile male). Three of these were later killed by fur trappers, and 10 were recaptured and recollared. Two recollared animals were lost a second time, and the fates of 10 fishers were undetermined.

During the nontrapping interval, we found no differences in the mean survival rates of juvenile males and juvenile females ($Z = 0.16$, $P = 0.87$, pooled rate = 0.72, CL = 0.53–0.99) or for adult males and adult females ($Z = 0.44$, $P = 0.66$, pooled rate = 0.89, CL = 0.81–0.99). During the trapping interval, multiple comparisons at the 95% level of significance showed no differences between the mean survival of juvenile males and juvenile females (pooled rate = 0.38; 0.25, 0.57), but all other comparisons

Table 8.2. Number radio-collared fishers (*Martes pennanti*) by age[a] and sex class monitored in south-central Maine and number of radio-days, 1984–1989

Year[b]	Juvenile males		Juvenile females		Adult males		Adult females	
	RD[c]	IF[c]	RD	IF	RD	IF	RD	IF
1984	296	4	104	1	301	6	335	2
1985	398	7	48	3	1812	10	1711	10
1986	464	6	0	0	2498	10	1733	7
1987	382	5	61	3	942	8	1578	6
1988	363	5	476	7	1167	5	1383	6
1989	113	3	119	2	1591	6	1071	7
TOTAL RD	2016		808		8311		7811	
TOTAL IF[d]		30		16		25		19

[a]Juveniles < 1 year old, adults ≥ 1 year old.

[b]From mean birthdate of one year (set at 15 March) to the same date next year.

[c]RD = radio-days, IF = individual radio-collared fishers.

[d]Some adults were monitored for more than one year and thus the adult IF columns are not additive (i.e., yearly totals in IF columns > number at bottom). Fourteen juveniles were also monitored after they became adults, and thus 76 individuals were radio-collared (i.e., 30 + 16 + 25 + 19 = 90; 90 − 14 = 76).

were significant (adult males = 0.57; 0.42, 0.78; adult females = 0.79; 0.64, 0.97).

Forty of the 41 deaths during the trapping interval were from fur trapping. The juvenile female that was shot (illegally) was excluded from the mortality estimate for the trapping interval in order to test for differential vulnerability of age and sex classes to trapping (i.e., we could not mix causes of death). Comparisons of mean trapping mortality rates indicate that adult females were 49% as likely to be trapped as adult males and 34% as likely as juveniles (Table 8.3). Inclusion of the juvenile that was shot had no effect on the mean estimate of juvenile mortality for the trapping interval.

Discussion and Management Implications

Because mortality can vary between years, rates should be calculated during intervals of less than one year (Heisey and Fuller 1985). Given our small annual samples, however, we combined data from six years into the two intervals, and probably violated the assumption of constant mortality (see also Powell, this volume). Pooling would cause the mortality estimates to be biased if our data were unequally distributed among years of differing mortality rates. Fur prices, and apparently trapping effort, increased gradu-

Table 8.3. Interval mortality rates of radio-collared fishers (*Martes pennanti*) in south-central Maine, 1984–1989[a]

Age and sex class[b]	Nontrapping[c]			Trapping[d]		
	Deaths	Radio-days	Mortality rates (95% CL)	Deaths	Radio-days	Mortality rates (95% CL)
Juv. ♂	3	1380	0.29 (0.00,0.52)	15	636	0.61 (0.37,0.76)
Juv. ♀	1	550	0.25 (0.00,0.57)	7	258	0.66 (0.25,0.85)
Juv. ♂ + juv. ♀	4	1930	0.28 (0.01,0.47)	22	894	0.62 (0.43,0.75)
Adult ♂	2	7388	0.09 (0.00,0.19)	13	923	0.43 (0.22,0.58)
Adult ♀	3	6973	0.13 (0.00,0.26)	5	838	0.21 (0.03,0.36)
Adult ♂ + adult ♀	5	14361	0.11 (0.01,0.19)		Not applicable[e]	

[a]Mortality rate = 1 − survival rate. One juvenile female was shot during trapping season (excluded from this analysis); all other deaths during trapping interval were due to trapping.

[b]Juveniles (<1 year old) became adults (>1 year old) on 15 March.

[c]Nontrapping interval for juveniles from 1 September to start of the trapping season plus from end of trapping season to the 15 March birthdate (156 days). Nontrapping interval for adults from 15 March of one year to 15 March of the next year, excluding the trapping season (326 days).

[d]Trapping season = 39 days.

[e]Because of the significant difference in trapping mortality, group data for adult males and females were not pooled.

ally from 1984 to 1987, but declined sharply in 1988 and 1989 (the average price of female pelts dropped from $180 in 1987 to $50 in 1989). Sample sizes were relatively evenly distributed across years (Table 8.2), however, so we believe the mortality estimates represent unbiased mean rates during the six years of study.

Our results support the suggestion of Strickland and Douglas (1981) and Strickland (this volume) that juvenile fishers are more vulnerable to trapping than are adult females and are overrepresented in the harvest compared with their actual occurrence in the population. Thus, variation in harvest intensity might change the ratio of juveniles to females over 2 years old in the harvest. This ratio, however, can be affected by reproductive success, preharvest survival, and trapper behavior (see Buskirk and Lindstedt 1989); such factors should be assessed before the ratio is used for management.

Our data also support Strickland and Douglas's (1981) suggestion that adult males suffer greater total mortality than adult females, and that the adult population consists of more adult females than adult males. Strickland and Douglas (1981) thought that the greater abundance of adult females was due to higher nontrapping mortality of adult males. We found no difference by sex in nontrapping mortality of adults, however; the higher trapping mortality of adult males relative to adult females most likely caused the unequal sex ratio among adults.

We do not know why adult females were less susceptible to trapping than the other age and sex classes. Inexperience of juveniles (Strickland and Douglas 1981) may cause them to be more susceptible to baiting, and their dispersal during fall and winter (Arthur et al. 1993) may cause them to encounter traps more frequently. Although mean home range sizes of adult fishers in Maine did not differ by sex ($P = 0.20$), most ranges of adult males were larger than those of adult females (Arthur et al. 1989a). Furthermore, following the spring breeding season, males often established home ranges in areas they did not previously use, whereas females occupied essentially the same range throughout their adult lives (Arthur et al. 1989a). Mean distances between consecutive independent locations during autumn were greater for adult males than adult females (2.13 vs. 1.80 km), but the difference was not significant at the $P = 0.05$ level (Arthur and Krohn 1991: table 1). Adult males would encounter more traps than adult females if, in fact, adult males had larger home ranges and were more mobile than adult females, given uniform trapping effort across the study area. By not spending their entire adult life in the same home range, as do adult females, adult males may be less familiar with their home range and less wary of traps. Also, males might be more susceptible than females to bait because of their larger size and presumably greater food requirements.

Previous models of fisher populations did not consider differential vulnerability among age and sex classes (e.g., Powell 1979c). Population models are sensitive to small changes in mortality, particularly among females. For example, a simple population model that we constructed suggested that a stable population with the mortality rates we observed in south-central Maine requires an annual fecundity rate of approximately 1.5 offspring per adult female (≥ 2 years old). This rate is similar to the actual fecundity rate of about 1.3 found by a concurrent study in our area (Paragi 1990). Using the observed fecundity rate, our model predicted that the population would decline by about 2% per year, which agrees with the perceptions of biologists and fur trappers in the area regarding actual population change. However, when we calculated mortality irrespective of differences among age and sex classes by pooling all animals into one class, the annual mortality rate was 0.53, and the fecundity rate necessary for a stable population was 4.8 per adult female. With the observed fecundity rate and the single mortality rate, the model predicted that the population would decline by about 31% per year and would be extinct after 20 years. Management actions based on the results of the second model would result in restricting harvest far below levels that the population can actually sustain.

Buskirk and Lindstedt (1989) noted that in trap captures of mustelids, sex ratios are generally biased toward males. This bias can result from the patterns of trapping vulnerability measured by our study, in which an adult female was about half as likely as an adult male and only one-third as likely as a juvenile of either sex to be captured. Thus, bias in trapped samples, at least in the case of the fishers we studied, was more a function of age than sex, and age ratios must be studied in concert with sex data.

We can only speculate as to what mortality patterns would be like in an unharvested fisher population. Trapping mortality was obviously high for juveniles in our study area, but we believe that in areas without trapping, natural mortality would replace much of the trapping mortality we observed. In an untrapped population, adults would occupy essentially all habitats and turnover would be low, leaving few areas for juvenile dispersal. It seems likely that juveniles would have difficulty finding unoccupied areas and would have to travel greater distances than observed in this study (Arthur et al. 1993). We hypothesize that properly regulated fur trapping removes juveniles that for the most part would have died of natural causes, although in the case of our study, clearly sooner than would occur naturally (i.e., trapping mortality in October and November instead of stress-starvation mortality in January through March). Empirical data are needed on mortality patterns in lightly harvested and unharvested fisher populations to understand more fully the biological significance of our observations.

Differences in vulnerability to trapping among age and sex classes probably vary among areas with different harvest intensities. In an area that is less intensively harvested than ours for example, survival of adult males could be higher, unoccupied areas would be less common, and home ranges of resident males might be smaller and more stable from year to year. These conditions might reduce the difference in movement patterns between adult males and adult females and thus reduce differential vulnerability. Conversely, Strickland and Douglas (1981) suggested that greater harvest intensity would also result in less differential vulnerability because, if a greater proportion of a population was harvested, the harvests would more closely represent the age and sex composition of the population. We recommend that studies of fisher survival also measure trapping vulnerability and that future models of fisher populations include age- and sex-specific survival rates. Finally, we concur with Powell (this volume) that the population characteristics of unharvested fisher populations also need to be documented if the population ecology of this species is to be more fully understood.

Acknowledgments

The Maine Department of Inland Fisheries and Wildlife (MDIFW) funded this research with Federal Aid in Wildlife Restoration Project W-82-R. The Maine Trappers' Association and the Maine Cooperative Fish and Wildlife Research Unit (MDIFW, U.S. Fish and Wildlife Service, University of Maine, and Wildlife Management Institute, cooperating) provided additional support. R. A. Cross prepared and aged the fisher teeth, and H. C. Gibbs performed the necropsies. Many students from the University of Maine, Orono, and Unity College helped with the fieldwork. D. J. Harrison provided helpful reviews of the manuscript. This chapter represents publication no. 1581 of the Maine Agricultural and Forestry Experiment Station.

PART III

Management of Populations

Introduction

Population management for martens and their allies takes various forms, but in each case depends on scientifically valid techniques, including proper estimates of population structures and trends. This section examines well-tested techniques for estimating harvest rate, for estimating age class and sex of harvested individuals, for translocating, for humane trapping, and for monitoring population size or trends. The chapters also reveal, however, that we still have much to learn on all these subjects.

Strickland discusses several general methods for estimating harvest rate from trapping returns: the proportion of the pretrapping population harvested, the proportion of tagged animals killed by trapping, and ratios of juveniles per mother (or adult female) or of females to males. All three methods provide valid, if inferential, indices of harvesting rate. But Strickland also notes that reliable harvest-independent methods of population estimation for fishers and American martens are not yet developed. She cautions that all methods can be misleading when harvests are small (less than 100).

Poole et al. describe three techniques for determining the age of martens and fishers, all based on samples from carcasses. These include counts of cementum annuli in teeth, the ratio of pulp width to total width of canine teeth from radiographs, and measurements of temporal muscle coalescence over the top of the skull. All are precise in separating juveniles from adults, in the sense of providing consistent estimates among observers or techniques. But these authors caution that because the true age of animals is seldom determined, accuracy or agreement between true and estimated age is largely unknown. They conclude that we require studies of wild animals with known ages and the development of methods for use with live animals.

Few studies have experimentally evaluated the effects of trapping on populations. Fortin and Cantin present results of their unprecedented evaluation of harvesting of American martens in a large area that had been exposed to no legal trapping for more than 100 years. They found a rapid decline in trapping success and in the population's sex ratio over the four years of the study. They conclude that their study population was subjected to excessive harvest within four years and that a particularly heavy harvest in year 2 had repercussions for subsequent years.

Much trapping for marten, sable, and fisher fur is done with methods and technology that are over 150 years old. To address modern concerns about humaneness and crippling losses, which generally are not considered in furbearer management, Proulx and Barrett report on the development of more humane kill-trapping devices. They found only one commercially available killing trap for American martens that meets humane standards and no such trap for fishers. Reinforcing Strickland's recommendations, these authors emphasize the potential utility of live trapping.

Translocation has been widely practiced to create or restore *Martes* populations. Slough reviews the attributes of 47 translocations of American martens and evaluates factors in their success. He concludes that the number of translocated individuals is the critical variable; nearly all translocations of 30 or more animals resulted in establishment of self-sustaining populations, whereas releases of fewer than 12 nearly always failed. Slough suggests that season of release, amount of high-quality habitat in the release area, use of gentle release techniques, and attention to the age and sex of the released animals all contribute to success. The specific issue of timing of release is evaluated for fishers by Proulx et al., who compare site fidelity of animals released in March or June. These authors contradict the results of Slough, finding that June releases were more successful than those in March.

Population management requires estimates of distribution and abundance, and I review techniques of monitoring these attributes in marten and fisher populations. For large-scale studies, where the objective is to compare abundance among large areas or on a single area over time, baited tracking plates or cameras offer great promise because both can be deployed in controlled designs to yield standardized indices. As with studies of harvest rate, however, we lack examples comparing such indices to true population size.

Martin G. Raphael

9   Harvest Management of Fishers and American Martens

Marjorie A. Strickland

Caughley (1977:168) identified three possible goals for population management: conservation (to increase the density of small or declining populations), sustained yield (to harvest at a rate that does not cause the population to decline), and control (to stabilize or reduce a population's density). The goal chosen for a particular population depends on social and economic considerations and on knowledge of the population and its environment. Management to produce a sustained yield is the usual goal in jurisdictions with adequate populations of fishers and American martens, but conservation may be the goal where densities are low and protection is required. Furbearer management may involve both indirect management through habitat manipulation and direct management by translocating animals or controlling harvests. The emphasis in this chapter is on sustained yield management through harvest controls.

I discuss only the North American *Martes*: fishers (*M. pennanti*) and American martens (*M. americana*). Both are easily trapped, have high pelt values, show relatively low recruitment rates, and are vulnerable to overharvesting. Neither is endangered or threatened throughout North America, although both are classed as such in some parts of their peripheral range.

Historical Harvests

Because fishers and American martens have always been highly valued in the fur trade, and because trapping was virtually unregulated before the 1920s, populations of both were severely reduced by the early 1900s. Local extirpation occurred, especially in the southern parts of their ranges, as a

149

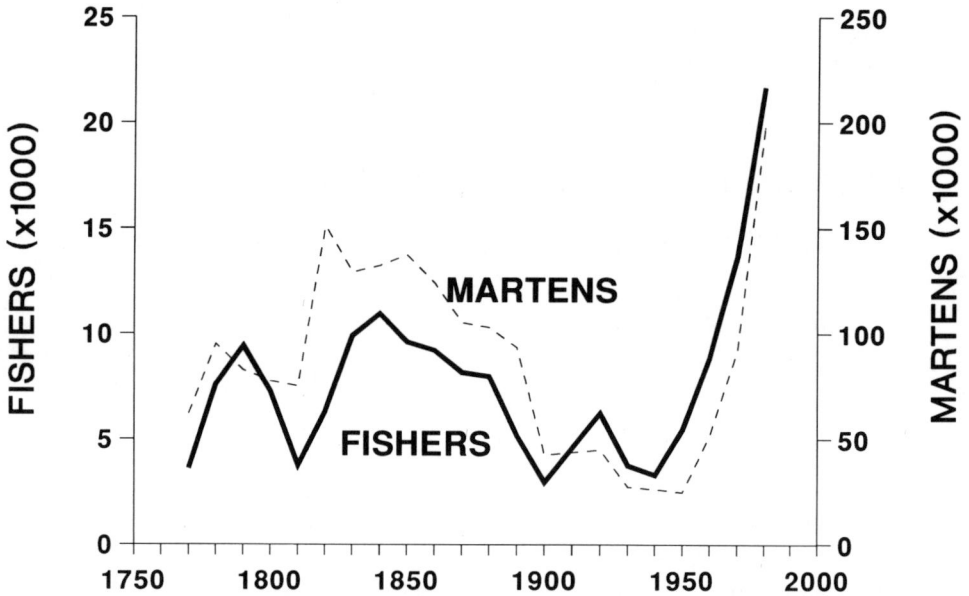

Figure 9.1. Mean annual harvest, averaged by decade, of fishers (*Martes pennanti*) and American martens (*M. americana*) in North America, 1770–1990. Adapted from Novak et al. 1987.

result of the excessive harvesting and extensive habitat loss following human settlement (Douglas and Strickland 1987, Strickland and Douglas 1987).

About 11,000 fisher pelts were harvested in North America annually in the 1840s (Fig. 9.1). Harvests gradually declined, and by the 1940s the average annual catch was 3300. As protective laws, reintroductions, and habitat recovery allowed fisher populations to grow, harvests across North America increased and averaged 17,000 pelts annually by the 1980s (Obbard et al. 1987). Annual American marten harvests in North America exceeded 140,000 in the 1810s and 1820s but declined to about 25,000 in the 1950s (Fig. 9.1). Subsequently, increased populations, improved access, and high demand increased harvests to a yearly average of 192,000 in the 1980s (Obbard et al. 1987).

Record harvests in the 1980s indicated that populations of fishers and American martens were high, but also that they were being heavily trapped. Canada produced more than 75% of the North American harvests of both species in the 1980s. Reduced popularity of furs in the 1990s has led to a decline in fur prices, and trapping pressure has eased.

Harvest Management

Populations of fishers and American martens that have been reduced to low levels by excessive harvest may take years to recover, and long-term loss of genetic variation may result. Fecundity may be altered; Strickland and Douglas (1981) believed that an increase in unbred females among fishers in southern Ontario resulted from excessive losses of mature males by intensive trapping. Low-density populations are the least resilient and the most difficult to assess, and only well-established populations should be harvested.

Unharvested populations of fishers and American martens are seldom stable (Powell, this volume). Their size varies from year to year as reproduction and mortality respond to the fluctuating food base. American marten populations rise and fall with changes in the abundance of small mammals, especially *Clethrionomys* spp. (Hawley and Newby 1957a, Weckwerth and Hawley 1962, Strickland and Douglas 1987, Thompson and Colgan 1987a), and fisher populations exhibit regular 10-year cycles that correlate with, and lag about three years behind, the snowshoe hare (*Lepus americanus*) cycle (Powell, this volume).

Trapping mortality may be additive to natural mortality. Thompson and Colgan (1987a) found that trapping mortality of American martens in Ontario was additive when food was scarce, reproduction was low, and adult animals had been forced to disperse in search of food. Population growth in the Yukon was highest among unharvested populations of American martens (Archibald and Jessup 1984), suggesting that harvest mortality was an additive factor. For fishers, trapping is often the main mortality factor. Krohn et al. (this volume) report that 80% of mortality of radio-tagged fishers ($n = 50$) in a heavily trapped area of Maine was due to fur trapping. The mortality resulting from trapping directed mainly at juveniles is most likely to be compensatory, inasmuch as juveniles have higher natural mortality than adults (Krohn et al., this volume). Therefore, a good harvest strategy is to exploit juvenile animals and preserve breeding stock.

One aim of sustained yield management is to reduce variability in population size so that harvests are predictable and consistent. Although this may be best for the trapper and the manager, it may have ecological consequences (Powell, this volume).

Management Plans and Targets

Because a sustained yield can be produced at many population levels as long as total mortality balances recruitment, targets for both the population size and harvest must be decided. Managers should consider, first, whether the population can sustain any harvest, and second, how large the annual harvest can be. Ideally, the answers would be based on estimates of population size, density, and distribution; on habitat capability, fecundity, and natural mortality by age class; and on an understanding of public demands and socioeconomic and political issues. From these data, options for population and harvest levels can be determined and a management plan produced (Gilbert and Dodds 1987).

Maine's strategic management plans for fishers (Clark 1986) and American martens (Ritter 1986) set population and harvest targets using field research, habitat models, and harvest data. Carrying capacity of each area was determined based on habitat suitability index (HSI) models, and maximum supportable winter densities (0.6 American martens/km^2; 0.26 fishers/km^2). Current population size was estimated from telemetry studies in high-quality habitat. Track counts and harvests were used to estimate abundance in other areas relative to the telemetered area. Harvest targets were set at 25–30% of the fall population for fishers, a safe level predicted from population modeling. Projected changes in habitat and user demands were employed to estimate future targets. These comprehensive plans, although costly, established a model for other jurisdictions.

Often management plans have not been developed, and harvest targets are based on data from past harvests and perhaps on carcass collections, track counts, or questionnaires (Anderson 1987, DiStefano 1987, Hamilton and Fox 1987, Melchior et al. 1987, Slough et al. 1987). The decision to initiate a harvest in a formerly closed area is often based on "suggestive" data such as increases in sightings, accidental catches, and road kills, and on the opinions of trappers and managers.

Reliable harvest-independent techniques of population estimation and evaluation for fishers and American martens have not yet been developed. Tagged or radio-collared animals can provide information on population density and mortality in a small area (Clark 1986, Ritter 1986, Thompson and Colgan 1987a, Arthur et al. 1989a, Krohn et al. 1989, Paragi 1990), but they are too costly for large-scale use. Computer models that incorporate data on fecundity and mortality, collected for an area of interest or from the literature, are sometimes used to aid in target setting. Douglas and Strickland (1987) described those used for fishers in Ontario and Minnesota. Models permit the trial of alternative scenarios and future harvests, but they should be used cautiously because future variation in recruitment and mortality cannot be predicted.

Managers must consider the demands of all users when setting management targets, especially for public lands. Although many trappers claim to trap mostly for recreation (Todd and Boggess 1987), they prefer species that have high pelt values and their expectations for fisher and American marten harvests often exceed allowable levels (Clark 1986). On the other hand, small-game hunters and farmers may want numbers of fishers and American martens reduced because they view them as competitors. Naturalists, in turn, may wish to preserve populations, and, foresters and timber companies may want increased fisher populations in order to control porcupines (Coulter 1966). Lastly, trapping is criticized because it uses traps that cause pain (Proulx and Barrett, this volume). Animal rights activists aim to stop all trapping by eliminating the market for fur products. This group has strong financial and media backing (Herscovici 1985) and has become more vocal and militant since the 1980s.

Harvest Dynamics

Because harvests are not random samples of populations, the sex and age distribution in a harvest will represent that of the population only if the vulnerability of different sex and age classes is equal and total effort is constant over space and time (Gilbert 1979). Juveniles and males of both fishers and American martens are more vulnerable to trapping than adult females. Krohn et al. (this volume), studying telemetered fishers in Maine, found that yearly trapping mortality was 62% among juveniles of both sexes, 43% for adult males, but only 21% for adult females. Soukkala (1983) reported that annual harvest rates for American martens in Maine were 34% for juvenile males, 46% for juvenile females, and 43% for adult males, but only 17% for adult females. He believed that the harvest rate for juvenile males was likely higher but that their dispersal from the study area had hidden these deaths.

This differential vulnerability to trapping has been attributed to age and sex differences in home range size, to distance traveled, and to behavior (Krohn et al., this volume). In both fishers and American martens, home ranges of adult males are larger than those of adult females, and males travel greater distances (Powell, this volume). Males therefore are exposed to more traps and have an increased chance of capture. The smaller home ranges of females reduce their exposure to traps (Buskirk and Lindstedt 1989). If trap density is high, however, the advantage of a smaller home range is negated (Gilbert 1979). Juveniles move considerable distances, especially when they disperse in autumn, and may be less wary of traps because of inexperience. Because juveniles are more vulnerable than adults to capture, the ratio of juveniles to adult females ($\geq$1.5 years) in the harvest is greater than in the population. This ratio varies with recruitment, with spacing (Powell, this volume), and with trapping intensity.

Harvest Concepts

Harvest, viewed simply, is a function of the abundance of animals in the population, their vulnerability to trapping, and trapping effort expended (Erickson 1982, Banci 1989). Thus harvests are modified by a number of variables. First, there are factors that affect population size, such as fecundity, recruitment, mortality, immigration, and emigration. Second, such variables as behavior, weather, food availability, and season of the year affect vulnerability of an animal to trapping. Third, trapper effort will vary with pelt prices, trapper numbers, regulations and quotas, access, weather, and cultural values (Gilbert 1979, Todd and Boggess 1987, Thompson et al. 1989).

Harvest depends on population size and harvest rate:

$$\text{Harvest} = NH$$

where N is the size of the pretrapping population, and H is the harvest rate (i.e., the proportion of the pretrapping population harvested). H depends on the vulnerability of the animals to trapping and on trapper effort (Gilbert 1979). Knowing the harvest rate and the number of animals harvested, one can calculate the size of the population from which the harvest was taken:

$$N = \frac{\text{harvest}}{H}$$

Good estimates of total harvest are usually available. Estimates of harvest rate, however, require a sound estimate of population size, and population size is usually what one is trying to determine! The following are three ways to estimate harvest rate, or correlates of harvest rate:

Harvest rate can be estimated from the proportion of tagged or radio-collared animals killed by trapping (Krohn et al. 1989). One must assume that this rate is valid for the whole population at that density and trapping effort. This assumption does not allow for behavioral differences in trapability; trapped and released animals may become trap-shy or trap-happy, depending on the reinforcement they receive. Krohn et al. (1989) reconstructed their preharvest fisher population using harvest rates and total harvest and found that it compared favorably to an estimate derived independently from home range data. This method is too costly for most agencies. It does, however, provide support for other estimates of harvest rates.

Harvest per unit of trapping effort (HUE) can be used as an indirect indicator of population trends (Dixon 1981):

$$\text{Population trend} = \frac{\text{harvest}}{\text{unit of effort}}$$

Using HUE for this purpose assumes population closure (i.e., that changes from immigration, emigration, births, and deaths are negligible) (Davis and Winstead 1980) and that vulnerability to trapping does not vary between trapping periods. It also requires that the unit of effort be consistent from year to year. Unit of effort can be affected by type of trap, trap set, weather, and trap density (Caughley 1977). Using HUE further requires that trappers keep accurate trapping records (Rollins 1989). Fortin and Cantin (1990), studying American martens in Quebec, found that HUE correlated strongly ($r = 0.99$) with marten density during a four-year period when marten densities declined from $1.5/km^2$ to $0.4/km^2$.

Indices such as harvest per trapper or harvest per successful trapper are less useful because they measure only success rate. Total harvest and success rates can remain high even when the population declines, because trappers expend greater effort (Thompson and Colgan 1987a). Conversely, success rates may decline because of an increase in the number of trappers, which means that more individuals are taking the same number of animals, with lower catches per trapper (Erickson 1982).

A third method uses the animals themselves as integrators of the factors acting on them. In Ontario, I compared the sex and age ratios in fisher harvests for each of 17 years (1973–1989) with harvest rates estimated from a simulation model (Douglas and Strickland 1987). The ratios and rates were significantly correlated (Figs. 9.2–9.4) and appear to integrate the effects of recruitment and vulnerability of fishers with trapper effort. Three ratios can be used:

1. The number of juveniles per mother (females ≥ 2.5 years) in the harvest correlates strongly with harvest rate ($Y = 156.35X^{-1.005}$, $r = 0.84$, df = 15, $P < 0.001$; Fig. 9.2), but requires that adult females be aged by counting cementum annuli to separate yearlings from those more than 2.5 years old (Strickland et al. 1982c).
2. The number of juveniles per adult female (≥ 1.5 years) in the harvest correlates strongly with harvest rate ($Y = 25.37X^{-0.63}$, $r = 0.82$, df = 15, $P < 0.001$; Fig. 9.3). This means that fishers need only be classed as juveniles or adults (≥ 1.5 years), a procedure that can be done economically using radiographs of the canine teeth (Poole et al., this volume).
3. The ratio of females to males (all ages) is correlated with harvest rate ($Y =$

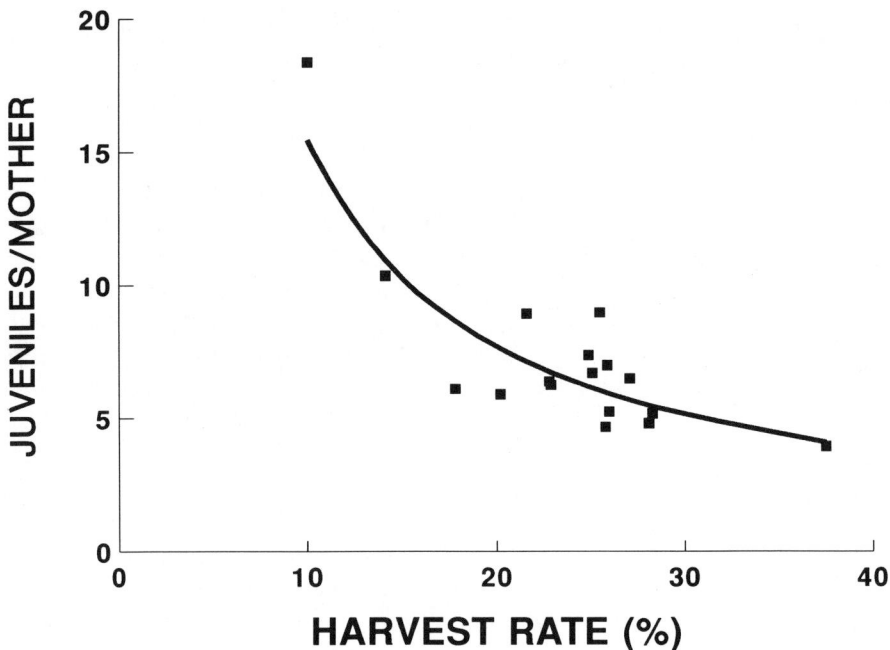

Figure 9.2. Ratios of juvenile fishers (*Martes pennanti*) to mothers (females ≥2.5 years old) in actual harvests versus harvest rates, calculated by a computer model (Douglas and Strickland 1987), in Algonquin Region, Ontario, 1973–1989. Data fitted to power curve: $Y = 156.35X^{-1.005}$, df $= 15$, $r = -0.84$, $P < 0.001$.

$0.698 + 0.0196X$, $r = 0.70$, df $= 15$, $P < 0.01$; Fig. 9.4). But it is less reliable than age ratios as an indicator of harvest rate, perhaps because trappers may select female fishers for their higher pelt price by setting traps where signs indicate females may be caught or by disposing of unwanted males illegally. This is most likely to occur where harvest limits are strictly enforced.

In the Algonquin Region of south-central Ontario, where sex and age ratios have been used for management decisions, fisher populations have been stable during the period 1979–1989 (Fig. 9.5). A sustained annual harvest of about 1 fisher/25 km² has been produced at an estimated harvest rate of 20–25% and juvenile per mother ratios of six to eight (Fig. 9.2). Using maximum supportable densities of 0.26 fishers/km² (Clark 1986), we can calculate the carrying capacity for the Algonquin Region to be about 8000 fishers, and the pretrap population to be 5000 fishers (1/6.2 km²), or 60% of carrying capacity, which is similar to that for Maine. These estimates

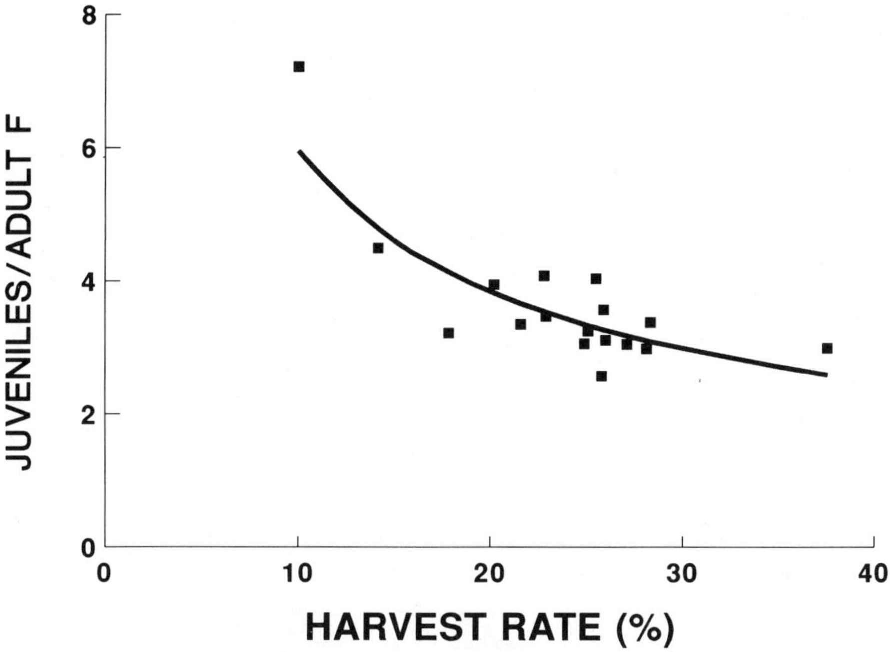

Figure 9.3. Ratios of juvenile fishers to adult females (≥1.5 years old) in actual harvests versus harvest rates, calculated using a computer model (Douglas and Strickland 1987), in Algonquin Region, Ontario, 1973–1989. Data fitted to power curve: $Y = 25.37X^{-0.63}$, df = 15, $r = -0.82$, $P < 0.001$.

have not been confirmed by independent evaluations of population size, but population trends agree with those reported by trappers in questionnaires.

Powell (this volume) notes that trapping generally leads to a younger age structure in the population. Thus the criteria used for the heavily trapped Algonquin Region may not be applicable to jurisdictions where harvest is less intensive or has recently been initiated. When Wisconsin began a limited fisher season in 1985, the small harvests (<300) in the first three years had ratios of less than four juveniles per mother (females ≥2.5 years), even though independent estimates indicated the population was increasing and that the harvest rate was <10% (Kohn et al. 1991). This previously untrapped population had low recruitment (few corpora lutea per pregnant female and low percentage of pregnant females, compared with Ontario fishers), and Ontario's ratios for a heavily harvested population were not applicable. Minnesota, which traps more than 1000 fishers annually at harvest rates of 15–20%, has juvenile per mother ratios of more than five, only slightly lower

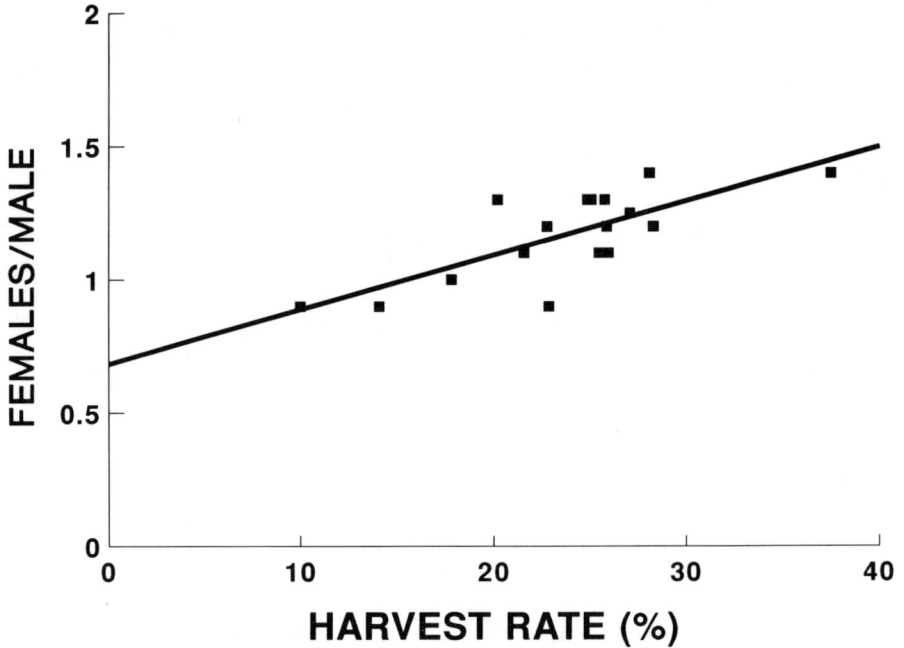

Figure 9.4. Sex ratios in actual fisher harvests (females/male, all ages) versus harvest rates, calculated by a computer model (Douglas and Strickland 1987), in Algonquin Region, Ontario, 1973–1989. Data fitted to linear regression: $Y = 0.698 + 0.0196X$, df $= 15$, $r = 0.70$, $P < 0.01$.

than would be expected in Ontario (W. Berg, Dep. Nat. Resour., Minn., pers. commun.).

A similar relationship between harvest rates and sex and age ratios in the harvest has been reported for American martens (Strickland and Douglas 1987; Thompson and Colgan 1987a; Fortin and Cantin 1990; W. Berg, Dep. Nat. Resour., Minn., pers. commun.). Age ratios in marten harvests are similar to those of fishers (Figs. 9.2, 9.3), but sex ratios differ. Light trapping results in two or three males per female; heavy trapping yields a ratio approximating one to one (Yeager 1950, Quick 1953, Soukkala 1983, Archibald and Jessup 1984, Fortin and Cantin 1990). A suitable population model for American martens is still required to estimate harvest rates.

The three methods of estimating harvest rates (recapture of marked animals, HUE, and sex and age ratios) can be used with any type of management (seasons, quotas, or refuges) to assess the impact of harvesting. Although these methods may not be reliable individually, they are valuable in

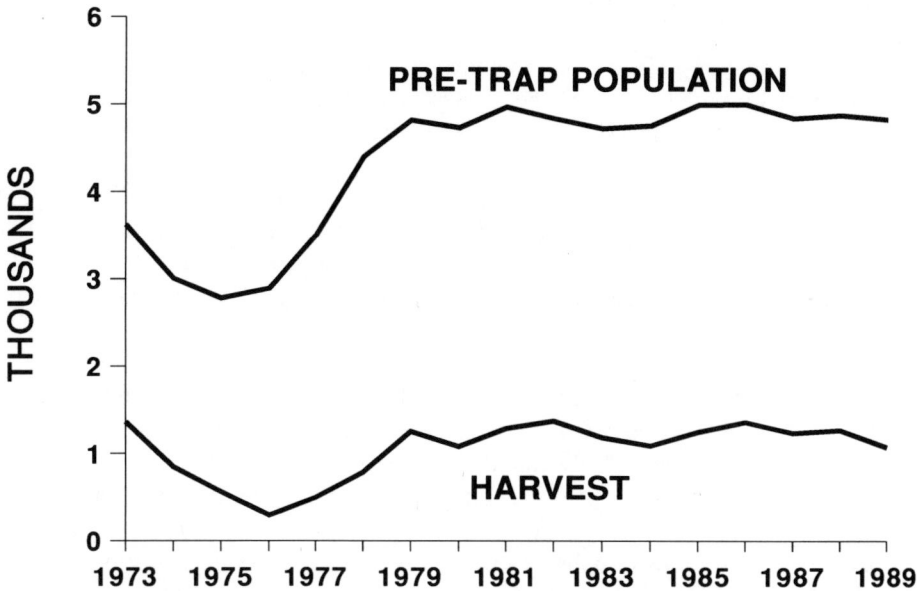

Figure 9.5. Actual fisher (*Martes pennanti*) harvests and estimates of fisher population from computerized model (Douglas and Strickland 1987) in Algonquin Region, Ontario, 1973–1989.

combination with each other or other sources of data. When harvests are small (<100) the potential for error is high, and all three methods can be misleading, making it particularly difficult to assess a restricted harvest in a newly opened area.

Regulating the Harvest

Controls must be implemented to prevent harvest targets from being exceeded. To be effective, controls require the cooperation of the trappers and strict enforcement.

Licensing

Most jurisdictions license trappers as a means to control trapper numbers and to communicate with trappers. Licenses are also a source of revenue for agencies.

Seasons

Specifying trapping seasons ensures that harvests occur when pelts are prime and most valuable (Novak 1987). The quality of fisher and marten pelts is usually highest in late November and declines by mid-February (Coulter 1966, Soutiere and Steventon 1981, Powell 1985, Fortin 1988). Time of best pelt condition varies with latitude (Slough et al. 1987), however, so seasons should coincide with the local peak. Where both fishers and American martens are trapped, concurrent seasons are preferable.

Some jurisdictions use short seasons to limit trapping opportunities and thereby restrict the harvest (Anderson 1987). Shortened seasons without quotas, however, may concentrate effort without decreasing the harvest. Clark (1986) determined that 79% of the fisher harvest in Maine occurred in the first 15 days of a 40-day season. To decrease the harvest effectively, season length must therefore be substantially reduced.

If populations are harvested excessively, closure for a year or more may be necessary to allow recovery. Early in this century most jurisdictions had long-term closures to allow populations of fishers and American martens to recover (de Vos 1952, Powell 1982). Open and closed seasons can be alternated, but if seasons are not combined with other controls, this approach can reduce populations below viable levels. Even when seasons are closed, fishers and American martens often are captured accidentally. Trappers in Ontario reported that 20–30% of fisher harvests and 9–14% of those of American marten were taken in traps set for other species (Strickland and Douglas 1984).

Length and timing of seasons also affects the sex and age composition of the harvest. Because juveniles are more vulnerable than adults, they predominate in the catch early in the trapping season. In Ontario, juveniles accounted for 71% of the fishers and 67% of the American martens taken in November. In comparison, only 54% of the fishers and 38% of the American martens harvested in February were juveniles (Douglas and Strickland 1987, Strickland and Douglas 1987). The percentage of adult females in the harvest increased in late winter. Likewise, the proportion of adult male fishers in the harvest increased in February (Douglas and Strickland 1987), perhaps because of behavioral changes associated with the approach of mating season (Coulter 1966, Powell 1982, Arthur et al. 1989a). Male American martens are less affected by seasonal timing because mating is in summer (Mead, this volume).

Harvest Limits

A bag limit (e.g., one animal per trapper) is a simple means of distributing allowable harvest among trappers, provided that the number of trappers can be limited. Another option is the use of a quota for the management unit, to be filled on a first-come, first-served basis. This technique requires immediate reporting of harvest to know when the quota is filled, however, and quick communication with trappers in remote areas.

Area-specific quotas, combined with registered traplines such as those used in Canadian provinces (DiStefano 1987, Gilbert and Dodds 1987, Novak 1987), best control the harvest. Registered traplines, which allocate exclusive trapping rights for a defined area to one trapper, will help distribute the harvest geographically, minimize competitive trapping, encourage stewardship of trapline resources, and reduce poaching. Most U.S. jurisdictions do not use this type of management; instead, many trappers in a single area may compete for the same animals. This pattern, which is aggravated by easy access, forces a more conservative harvest target. Because fishers travel over areas that may encompass more than one trapline, some harvest competition exists even with registered traplines. American martens, because of their smaller home ranges (Powell, this volume), may be managed on a trapline basis if the trapline is large.

Refugia

Several relict populations of fishers and American martens across North America owe their existence to refuges where all land furbearers are protected (Strickland et al. 1982*a,b*). "Natural" refuges also exist where access for humans is difficult. Refuges provide population reservoirs for dispersal to surrounding areas in many parts of North America (de Vos 1951*a*, Coulter 1966, Strickland and Douglas 1987, Banci 1989). Quick (1956) believed that local populations of American martens in British Columbia, eliminated by heavy trapping, would be repopulated within one year through immigration from adjacent untrapped areas. Untrapped areas were important American marten refuges in the Yukon Territory (Archibald and Jessup 1984) and British Columbia (Blood 1989). In these areas, dispersing American martens, predominately juveniles, could be harvested exhaustively, provided that the refuge was left untrapped. Untrapped areas at least 10 km wide were left between adjacent traplines, a size chosen because female American martens' nonoverlapping home ranges average 3 km in diameter. Harvests may vary annually, regulated by the reproductive success of the protected animals (Archibald and Jessup 1984). Management by refuges may need to be combined with quotas in years of food scarcity because home ranges may expand or adult martens may become transient (Powell, this volume) and recruitment may decrease (Thompson and Colgan 1987*a*). Also, managers using the refuge concept should recognize that dispersal tends to follow habitat quality gradients. Trapline refuges are less useful for managing fishers owing to their larger home range size (Powell, this volume); refuges to protect fisher populations would need to be very large.

Selective Trapping

Long-lived species such as fishers and American martens are best managed by harvesting juveniles and manipulating sex ratios in the harvest to preserve breeding females (Novak 1987). This strategy can be facilitated with live-holding traps that provide an opportunity for the trapper to determine the animal's sex, age, and pelt primeness before deciding whether to harvest it. Livetraps must be checked at least daily (Proulx and Barrett, this volume), which may be difficult where traplines are large or access is poor. Reliable field determinants of sex and age are needed. Sex can usually be determined by size and morphology (Strickland and Douglas 1987, Douglas and Strickland 1987), but none of the more reliable aging methods are suitable for animals held in livetraps.

Assessing the Harvest

Total Harvest

Selection of methods for estimating total harvests involves trade-offs between precision and accuracy on the one hand and cost on the other. Inasmuch as harvest data form the basis of most management assessments, their accuracy is important. Mandatory registration, stamping pelts, or turning in carcasses provides a count of the animals taken legally by each trapper (Novak 1987). When combined with a computerized inventory system, it provides an efficient and reliable method for collecting and handling harvest data. Other sources of data on total harvest include fur dealer returns, export permits, trapper license returns, and telephone or mail surveys; the latter can also provide data on effort (Anderson 1987, Rollins 1989, Fortin and Cantin 1990). Some jurisdictions assess precision by comparing data from more than one source (Anderson 1987, DiStefano 1987). But none of these methods provides information on wounding losses, estimated to be 12% of the legal harvest in Ontario (Douglas and Strickland 1987), or illegal harvest. In Wisconsin, where trapping for fishers is limited, registered and unregistered harvests were believed to be about equal; in Minnesota, illegal harvest was estimated to be 22% of the registered take (Kohn et al. 1991).

Biological Data from the Harvest

Carcasses can provide data on sex, age, reproductive status, body condition, diets, parasites, and diseases. Carcass submissions may be mandatory or voluntary; partial collections should be representative of the total harvest. If only sex and age data are required, whole fisher and marten carcasses are not necessary. Both sex and age can be determined from teeth (Poole et al., this volume). Radiographs are a quick and reliable method of distinguishing juveniles from animals 1.5 years old or older (Poole et al., this volume).

Ovarian sections and blastocyst counts provide data on rates of fecundity and pregnancy. Placental scars, when visible in fresh specimens, indicate advanced pregnancies and suggest numbers of animals that are whelped (Paragi 1990). These data help predict recruitment; if fecundity is low, reduced recruitment can be expected, and harvest targets should be adjusted accordingly. But high or normal fecundity does not guarantee high recruitment; other factors may reduce recruitment before the trapping season.

Track Counts and Questionnaires

Independent estimates of population trends can be derived from tracks on snow transects. This method requires good tracking conditions and standardization of variables, including transect selection, time following snowfall, and weather conditions (Johnson 1984, Kohn et al. 1991). Trappers observe animal signs in their regular trapping activities and develop subjective judgments on the status of populations of fishers and American martens and their prey. This information can be collected in a standardized format by means of questionnaires or telephone interviews (Strickland and Douglas 1984, Slough et al. 1987, Banci 1989, Rollins 1989). Although cost-effective, these data may be biased and should be combined with other information.

Small Mammal Populations

American martens feed primarily on microtine and cricetine rodents (Martin, this volume), and changes in the size of small mammal populations may be used to predict changes in marten populations. Data on relative abundance of small mammals can be obtained by using a snap-trap index or track counts (Thompson and Colgan 1987a). Small mammal populations in turn fluctuate with food availability (Obbard 1987a).

The Future

Management of furbearer harvests may be of only academic interest in the future. Declines in fur prices since the late 1980s may mean the death of the trapping industry as we know it. Increased production of ranched furs and pressure from antitrapping movements may reduce demand for wild fur products, and without a market there can be no industry. Even recreational trapping requires some commercial outlet for the pelts. On the other hand, low prices may be just another dip in a cyclic trend. Fur prices are controlled mostly by fashion, and they have fluctuated unpredictably in the past. They may continue to do so in the future. But fishers and American martens have intrinsic value, apart from their commercial value as furbearers. The biggest threat to future fisher and marten populations, whether trapped or not, will be the loss of suitable habitat as humans continue to alter the landscape through logging and development.

Acknowledgments

This chapter is dedicated to the memory of Carman W. Douglas (1924–1987). His vision was to improve fisher and American marten management in Ontario by developing a rational, biologically based method for setting quotas. He lived just long enough to realize that dream. I also thank Roger Powell and three anonymous reviewers for their constructive criticisms and excellent editing of this chapter.

10 Translocations of American Martens: An Evaluation of Factors in Success

Brian G. Slough

American martens (*Martes americana*) have been extirpated in southern portions of their range since Europeans first settled North America (Strickland and Douglas 1987; Gibilisco, this volume). Reintroductions to former habitats and introductions to new range have been practiced extensively since 1934, typically to restore a component of wilderness systems, although the economic value of American martens as a furbearer is also a factor. The International Union for the Conservation of Nature (IUCN) has made a position statement on translocation of living organisms that proposed conditions under which translocations should be undertaken and outlined the components of a well-designed translocation program (IUCN 1987). Reintroductions should take place only after the original causes of extinction have been removed and if the habitat is suitable. Restocking has been recommended if genetic problems reduce the viability of the existing population or if a population has dropped below critical levels for population maintenance. IUCN discouraged introductions to natural habitats, or to areas where they may spread to natural habitats, because of potential damaging effects on native species. Islands, including isolated biological systems, are considered ecologically vulnerable to introductions because of their typically high endemism. A reintroduction program should include four phases: a feasibility study, preparation, release, and assessment.

The feasibility of translocations of American martens depends on several factors. Sufficient funding must be available to complete the project, including a follow-up phase. Genetically suitable source stock, preferably from the wild, must be available. Public attitudes must favor the protection of the translocated martens and the management of the founded population, and public support must be expressed politically. The critical feasibility factor is

165

the presence of high-quality habitat in an area large enough to guarantee the survival of a minimum viable population over the long term. Biological questions include the number of martens to be released, their sex and age, capture and release techniques, and scheduling. All logistics and equipment, including livetraps, transport containers, holding pens, and food, must be in place. The monitoring of released martens should be an integral part of any reintroduction program. Reasons for success or failure should be communicated to other wildlife managers through the relevant media, including publications and symposia.

Because of economic and logistical constraints, efforts made to translocate American martens are often concentrated on the preparation and release phases, with little effort expended on feasibility studies or assessment. The Canadian Parks Service has studied feasibility before reintroductions to three Atlantic Region national parks (Bateman 1980, 1982a; Evans 1986). Some American states also produced recovery plans for American martens (Gieck 1986, DiStefano et al. 1989). Handling and release strategies are well documented (Churchill et al. 1981, Evans 1986), and post-release monitoring methods have been described by Schupbach (1977), Davis (1983), Kohn and Eckstein (1987), and Slough (1989). Berg (1982) reviewed some translocations of American martens, fishers (*M. pennanti*), and river otters (*Lutra canadensis*). Marten ecology has been considered in transplant feasibility studies (Evans 1986), but the level of success of the translocation, relative to the initial biological considerations, is rarely evaluated. Furthermore, success of translocations is seldom evaluated beyond the short term. Consideration of marten ecology in the feasibility and preparation phases of translocations will increase the likelihood of success.

Methods

To assess the status of several completed and ongoing American marten transplants and to evaluate common elements of success, I read published reports and sent questionnaires to all states and provinces where American marten translocations had takenn place. (See Acknowledgments, below.) Questions covered four areas: feasibility, translocation details, trapping regulations, and assessment of success. Respondents provided unpublished reports, file notes, and personal recollections. Total numbers of American martens released and numbers of females released were compared between successful and unsuccessful translocations with Mann-Whitney U tests.

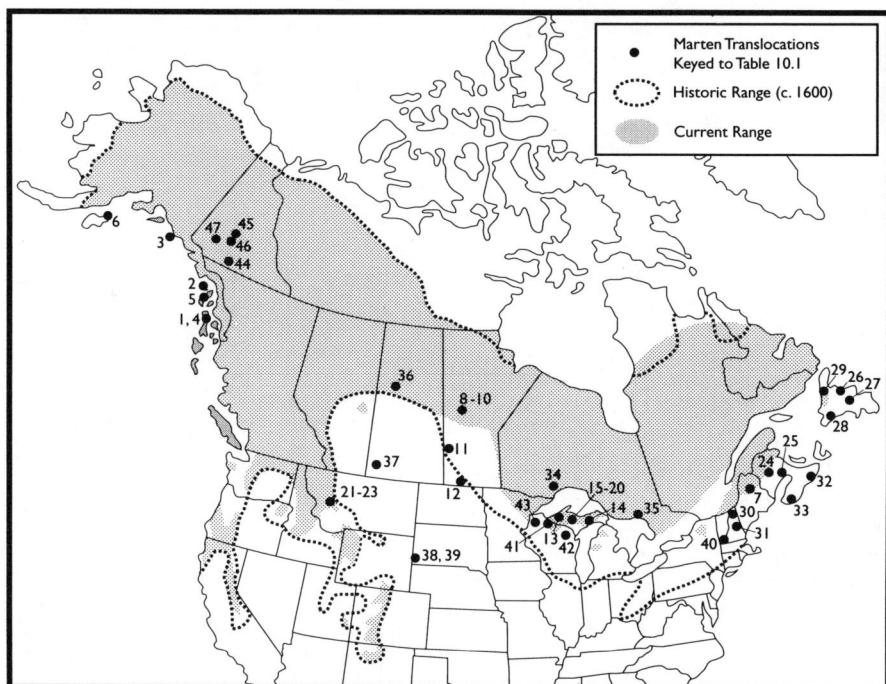

Figure 10.1. American marten (*Martes americana*) translocation sites, 1934–1991.

Results

I obtained data on 38 reintroductions and 9 introductions of American martens in eight American states and seven Canadian provinces and territories (Fig. 10.1, Table 10.1). Some of these translocations involved releases into several contiguous target areas in Michigan, Yukon, and South Dakota (nos. 15, 16, 17, 38, 39, 44, 45, 46, 47, Table 10.1). These were treated as separate translocations if investigators believed that separate populations would initially be established, even though they may join later. Some of the translocations may be more correctly termed restockings when some native American martens were present, as some managers believed for the Prince of Wales Island, Maine River, Cypress Hills, and Yukon translocations (nos. 1, 29, 37, 44–47, Table 10.1). Also, many small-scale undocumented American marten translocations were conducted by wildlife managers in Ontario in the 1950s and 1960s (M. Novak, Ont. Minist. Nat. Resour., pers. commun.) and by private individuals in Alaska (R. Flynn, Alas. Dep. Fish and Game,

Table 10.1. American marten (*Martes americana*) translocations, 1934–1991

Translocation	Date	Translocation type[a]	n(♀)	Source[b]	Release method[c]	Protection[d]	Status[e] (1991)	Source
1. Prince of Wales I., Alas.	1934	I	10(4)	Alas. (H)	Q	N	S/T	1–3
2. Baranof I., Alas.	1934	I	7(3)	Alas. (H)	Q	N	S/T	1–3
3. Kayak I., Alas.	1940s	I		Alas. (H)	Q	N	S	1,4
4. Patterson I., Alas.	1940	I		Alas. (H)	Q	N	S	1,5
5. Chichagof I., Alas.	1949–1952	I	21(≥4)	Alas. (H)	Q	N	S/T	1–3,6
6. Afgonak I., Alas.	1952	I	20(12)	Alas. (H)	Q	N	S	1,3,6,7
7. Northern Maine	1983	R	76	Me. (H)	Q	N	S/T	8
8. Barb Lake, Man.	1960–1961	R	11(≥3)	Man. (U)		M	F	9–11
9. Tramping Lake, Man.	1961	R	2(1)	Man. (U)		M	F	9–11
10. Minago River / William Lake, Man.	1967–1968	R	99	B.C. (H)		M	S/UT	10,11
11. Duck Mt., Man.	1969	R	42	Ont. (H)		M	S/T	10,11
12. Turtle Mts., Man.	1990	I	10	Man., Ont.	Q	M	Unk.	12
13. Porcupine Mt. State Park, Mich.	1955–1957	R	29(11)	Ont. (U)	Q	P	F	13,14
14. Hiawatha Natl. For., Mich.	1968–1970	R	99(37)	Ont.	Q	M	S	14,15
15. Huron Mt., Mich.	1979–1980	R	78(31)	Ont. (U)	G/Q	M	S	14,16
16. McCormick Exp. For., Mich.	1980	R	22(13)	Ont. (U)	G/Q	M	S	14,16
17. Iron Co., Mich.	1980–1981	R	48(27)	Ont. (U)	G/Q	M	S	14,16
18. N. Otsego & S. Cheboygan Cos., Mich.	1985	R	49(24)	Ont. (U)	Q	M	S	14
19. Northeast Lake Co., Mich.	1986	R	36(17)	Ont. (U)	Q	M	S	14
20. Chippewa Co., Mich.	1989	R	20(9)	Ont. (U)	Q	M	Unk.	17

	Location	Years		Count	Region				Ref.
21.	Silver Bow Co., Mont.	1944	R	12	Mont.				18
22.	Lincoln Co., Mont.	1955	R	21	Mont.				18
23.	Meager Co., Mont.	1956–1957	R	9(5)	Mont.			F	18,19
24.	Acadia For. Res. Stn., N.B.	1967–1968	R	<10	N.B. (H)		P	F	20
25.	Fundy Natl. Park, N.B.	1984–1991	R	44(≥17)	N.B. (H)	G/Q	P	S	21–26
26.	Siviers I., Nfld.	1976	I	3(2)	Nfld. (U)	Q	M	F	27–29
27.	Terra Nova Natl. Park, Nfld.	1982–1983	R	8+2 litters	Nfld. (U)	G	P	S	25,30–32
28.	LaPoile River, Nfld.	1975	R	3(2)	Nfld. (U)	Q	M	S	28,29,33
29.	Maine River, Nfld.	1976–1978	R	11(4)	Nfld. (U)	Q	M		28,29,33
30.	Second College Grant, N.H.	1953	R	2(1)	Ont.		P	Unk.	34
31.	White Mt. Natl. For., N.H.	1975	R	29(9)	Me.		P	Unk.	35,36
32.	Liscomb Game Sanctuary, N.S.	1956	R	12(7)	Ont. (H)		P	F	37,38
33.	Kejimkujik Natl. Park, N.S.	1987–1990	R	80(35)	N.B. (H)	G/Q	P	Unk.	23,39,40
34.	Sibley Prov. Park, Ont.	1950–1951	R	47(16)	Ont.		P	Unk.	41,42
35.	Parry Sound Dist., Ont.	1956–1963	R	248(94)	Ont. (U)	Q	N	S/T	43,44
36.	Prince Albert Natl. Park, Sask.	1954	R	24(12)	Alta. (U)	Q	P	S	45,46
37.	Cypress Hills Prov. Park, Sask.	1986–1987	I	33(19)	Alta./Yukon (U)	G	P	S	47,48
38.	N. Black Hills, S.D.	1980–1981	R	42(17)	Id. (H)	G	P/M	S	49
39.	Central Black Hills, S.D.	1989–1990	R	39(18)	Id. (H) / Colo. (H)	G	P/M	S	50
40.	Green Mt. Natl. For., Vt.	1989–1990	R	69(17)	N.Y. / Me. (H)	Q	M	Unk.	8,51
41.	Stockton I., Wisc.	1953	R	5	Mont.		S/UT		52,53

(continued)

169

Table 10.1. *(Continued)*

Translocation	Date	Translocation type[a]	n(♀)	Source[b]	Release method[c]	Protection[d]	Status[e] (1991)	Source
42. Nicolet Natl. For., Wisc.	1975–1983	R	172(51)	Ont. / Colo.	G/Q	P	S	52,54–56
43. Chequamegon Natl. For., Wisc.	1987–1990	R	139(45)	(H,F[n = 4])		P		57
44. Takhini Lake / Wheaton River, Yukon	1984–1986	R	31(17)	Yukon (H)	Q	M	Unk.	58
45. Braeburn, Yukon	1984–1986	R	26(12)	Yukon (H)	Q	M	Unk.	58
46. Takhini River, Yukon	1985–1986	R	63(21)	Yukon (H)	Q	M	S/UT	58
47. Haines Junc., Yukon	1984–1987	R	51(17)	Yukon (H)	Q	M	S/UT	58

Note: Blank cells = not reported.
[a] I = introduction, R = reintroduction.
[b] H = harvested population, U = unharvested population, F = fur farm.
[c] Q = quick release, G = gentle release.
[d] P = protected (no trapping), N = not protected, M = no marten-specific trapping.
[e] S = self-sustaining, T = harvested, UT = low-density unharvested, F = failure, Unk. = unknown.

Sources:
1. Burris and McKnight 1973
2. R. Flynn, Alas. Dep. Fish & Game
3. H. R. Melchior, Alas. Dep. Fish & Game
30. Bateman 1982b
31. Bateman 1984
32. Hoffman 1983

4. H. J. Griese, Alas. Dep. Fish & Game
5. D. N. Larsen, Alas. Dep. Fish & Game
6. Nelson 1952b
7. Nelson 1952a
8. K. D. Elowe, Me. Dep. Inland Fish & Wildl.
9. Miller 1961
10. C. S. Johnson, Man. Dep. Nat. Resour.
11. R. Stardom, Man. Dep. Nat. Resour.
12. G. Armstrong, Man. Trappers' Assoc.
13. Harger and Switzerburg 1958
14. R. D. Earle, Mich. Dep. Nat. Resour.
15. Schupbach 1977
16. Churchill et al. 1981
17. D. M. Elsing, USDA For. Serv.
18. Rognrud 1983
19. Thompson 1949
20. D. Cartwright, N.B. Dep. Nat. Resour. & Energy
21. Sullivan 1984
22. Quann 1985
23. Sinclair 1986
24. Sinclair 1987
25. G. Corbett, Environ. Can., Can. Parks Serv.
26. G. Sinclair, Environ. Can., Can. Parks Serv.
27. O. Forsey, Nfld. & Lab. Dep. Environ. & Lands
28. L. Mayo, Nfld. & Lab. Dep. Environ. & Lands
29. Bissonette et al. 1988

33. Evans 1986
34. Silver 1957
35. Soutiere and Coulter 1975
36. E. Orff, N.H. Dep. Fish & Game
37. B. Sabean, N.S. Dep. Lands & For.
38. Dodds and Martell 1971a
39. Boss et al. 1987
40. Drysdale and Charlton 1988
41. de Vos 1952
42. M. Novak, Ont. Minist. Nat. Resour.
43. Rettie 1971
44. M. A. Strickland, Ont. Minist. Nat. Resour.
45. P. Galbraith, Sask. Dep. Parks & Renewable Resour.
46. R. Leonard, Environ. Can., Can. Parks Serv.
47. Hobson et al. 1989
48. W. Runge, Sask. Dep. Parks & Renewable Resour.
49. Frederickson 1983
50. L. Frederickson, S.D. Dep. Game, Fish & Parks
51. J. J. DiStefano, Vt. Dep. Fish & Wildl.
52. Davis 1978
53. Gieck 1986
54. Kohn and Eckstein 1987
55. Davis 1983
56. T. Rinaldi, USDA For. Serv.
57. B. E. Kohn, Wisc. Dep. Nat. Resour.
58. B. G. Slough, Yukon Dep. Renewable Resour.

pers. commun.). Such translocations may have occurred in other jurisdictions as well. Richardson et al. (1986:171) incorrectly credited Arkansas (instead of Alaska) and Clark et al. (1989:587) incorrectly credited British Columbia, Arkansas, Oregon, Washington, and Colorado with translocations of American marten.

Releases generally occurred in nonsummer months, and involved 2–249 American martens ($\bar{x} = 43.1 \pm 48.2$ [SD]) including 1–94 females ($\bar{x} = 18.2 \pm 18.3$). In 21 cases, 30 or more martens were translocated ($\bar{x} = 75.5 \pm 53.5$) and in 11 translocations, 50 martens or more were released ($\bar{x} = 106.7 \pm 58.6$). There were significant differences between both total numbers of American martens released ($P = 0.012$, $U = 130.0$, $N_1 = 26$, $N_2 = 6$) and numbers of females released ($P = 0.015$, $U = 77.5$, $N_1 = 18$, $N_2 = 5$) in successful and unsuccessful translocations, with higher numbers resulting in more success. In most cases (23 out of 27) American martens were released using the quick method (i.e., not held at release site for acclimatization). Only 4 translocations used the gentle release method, in which martens were held on site for more than 2 days. Six other translocations employed both methods. Martens were protected by trapping prohibitions on all furbearers in 15 cases; marten-specific trapping was prohibited in 22 cases, and no protection from trapping was provided in 8 cases. Many translocations entailed several annual releases. Sources of stock were generally as near the release area as possible, except in the Minago River/William Lake release (no. 10, Table 10.1), in which American martens from a darker race from British Columbia were preferred. Captive-bred American martens may have been used in the Patterson Island and Chequamegon National Forest releases (nos. 4 and 43, Table 10.1), when fur farm stocks were released.

Success of Translocations

In some cases, management objectives included establishing a harvestable population as well as maintaining the genetic variation of the founding population. An American marten translocation usually was considered successful by managers if it resulted in a self-sustaining population, as indicated by survival and reproduction, through the assessment period. Specific population goals were rarely defined, and success in reaching those goals was not systematically evaluated. Populations of 300 American martens were the targets of the Vermont and Wisconsin marten recovery projects (Gieck 1986, DiStefano et al. 1989)

At least 27 American marten populations considered self-sustaining have reportedly been established, and 6 of these have been harvested. At least 8 translocated populations were not meant to be harvested. Managers believed that natural population expansion, as opposed to translocation, was responsible for repopulating the Parry Sound and Prince Albert National Park release sites (nos. 35 and 36, Table 10.1). Unofficial translocations and the presence of native American martens may have been factors in other repopulations as well.

Factors Contributing to Success

Factors in the success of translocations are difficult to identify because they are often interrelated and unknown for specific translocations. The factors most clearly associated with the success of American marten translocations were habitat quality of target areas and the number of martens released (Table 10.1). None of the translocations of 30 martens or more was considered a failure, and 16 of 27 successful translocations were from this group. Seven of 8 known unsuccessful American marten translocations involved 12 martens or less, and 1, in Porcupine Mountain State Park (no. 13, Table 10.1), involved the release of 29 animals. Some factors had no apparent association with success or failure. The sex ratio of 1320 American martens used in 31 translocations was 1.4 males to each female. Both male- and female-biased ratios produced at least some successes. Similarly, source of translocation stock, release method, and protection from trapping produced both successes and failures.

Assessment

Before 1970 little or no assessment was made of the success of American marten translocations, nor was the need for further translocations determined. Early assessment programs used a variety of techniques: winter track-counts, snow tracking, live-trapping, fur harvest returns, and anecdotes (van Zyll de Jong 1969, Schupbach 1977). Later programs used radiotelemetry to collect data on post-release survival, habitat use, and movements by martens (Churchill et al. 1981, Davis 1983, Bateman 1984, Quann 1985, Drysdale and Charlton 1988). These programs were typically conducted within one year of the translocation, and few were conducted three or more years after release (van Zyll de Jong 1969, Schupbach 1977). By contrast, most American marten translocations initiated since 1980 have included short- and long-term monitoring (Evans 1986, Gieck 1986, DiStefano et al. 1989), although few results have been reported so far.

Discussion

Habitat

Griffith et al. (1989) found that releases in good to excellent habitat were clearly associated with success for 416 attempted translocations of native birds and mammals. Habitat assessment is frequently a component of feasibility studies for American marten translocations (Evans 1986). More detailed habitat assessments may involve surveys of microtine rodents, forest structure (especially of coarse woody debris), proximity to water, and availability of denning and resting sites that provide protection from cold. Habitat quantity and quality within target areas will ultimately determine their carrying capacity and strongly influence success of the translocation.

Spacing Patterns and Movements

Home range shifting and dispersal may occur between February and September (Slough 1989). Release between October and January would therefore avoid these peak periods, encouraging the establishment of home ranges near release sites. Stress and disorientation can be minimized with brief handling and transport periods, and with gentle release procedures, in which the animals are held on site for at least five days (Davis 1983). A social unit including a female and her kits also may be more sedentary than individual martens (Hobson et al. 1989). Because transients are always present in American marten populations (Powell, this volume), releasing large numbers of martens throughout a target area will allow for some movement out of the area. Prior scent marking by managers with a collection of feces, urine, or anal scent has not been attempted with carnivores but, through orientation, may help to establish normal spacing patterns, as Pulliainen (1982) suggested for Eurasian pine martens (*M. martes*). Several releases over time, rather than a single release, may also help the development of a social system, inasmuch as released animals may read scent marks of animals previously released and be induced to assume residency (Pulliainen 1982).

Reproduction

American martens become sexually mature at a late age and have small litters for their body size (Mead, this volume), thereby limiting the rate of increase of a translocated population. Adult female American martens have higher fecundity than yearlings (Mead, this volume). Short of holding animals in captivity until they mature and selecting adults for the release stock, trapping from an unharvested or lightly harvested source population will increase the likelihood of a higher proportion of adults than is found in an annually harvested population (Lensink 1953). Archibald and Jessup (1984), for example, reported 87% adults and 13% yearlings in an area that had not been trapped for 10 or more years, whereas annually harvested traplines yielded 10% adults, 23% yearlings, and 67% young-of-the-year. Vacant areas created by trapping were colonized largely by young-of-the year. Obtaining and releasing the source stock during the period of delayed implantation (September through March) will further increase the likelihood of a near-optimal rate of increase the first year of the translocation.

Martens should not be captured between April and August to avoid taking lactating females, juveniles dependent on maternal care, or breeding adults. Releasing pregnant females just before parturition or releasing females with their litters may, however, encourage the prompt establishment of a home range owing to a need to locate a den site. Social units require gentle release to help maintain their integrity.

Population Structure

Trapping American martens usually produces a male-biased sex ratio (Buskirk and Lindstedt 1989). Many researchers translocated all animals caught, regardless of their sex. Others used an even or female-biased sex ratio by releasing some males back to the source population. They assumed that females were more valuable to the growth capability of the founders and that because male home ranges are larger and may overlap several female ranges, a female-biased founder population would not reduce breeding success. This assumption depends on the ability of founders to establish contiguous intrasexual home ranges and overlapping intersexual home ranges. Male American martens, especially juveniles, are known to make more extensive movements than females in both natural and translocated populations (Strickland and Douglas 1987). A male-biased sex ratio may therefore increase the likelihood of establishing a normal spacing pattern and maximize reproductive success.

Metabolic Demands

American martens have high metabolic demands when not at rest because of their limited insulation (Buskirk et al. 1988). Adverse weather conditions at the time of release could increase mortality through sustained negative energy budgets. Wetness and ambient temperatures of 16°C or less increase heat loss. Insufficient insulating snow cover and a lack of subnivean resting sites further increase energy demands. Release during favorable weather and snow conditions is recommended. Catastrophes resulting from a single hard winter can be avoided by releasing martens over two to three years.

Stress can be minimized by brief handling periods, gentle release, and proper husbandry techniques. Animals in which radios are implanted require a postoperative recuperation period before release (Slough 1989). Supplemental feeding, such as the placement of beaver (*Castor canadensis*) or ungulate carcasses at the release site, may help reduce the stress of adapting to unfamiliar habitats and food species.

Genetics

Translocated American martens must be adequately adapted to their new environment and be able to maintain a level of genetic diversity characteristic of wild populations. For these reasons, wild-caught American martens may be preferable to captive-bred stock, which may be inbred. The source population should come from a nearby area and a similar habitat. Breeding in captivity is recommended primarily to ensure against extinction in the wild, as well as to produce stock for release into the wild, to collect life history and behavioral data, and to develop public education programs that will aid in enlisting support for conservation programs for rare species (Richardson et al. 1986). Captive breeding has been proposed for reintroducing the Newfoundland subspecies of American marten (*M. americana atrata*) (L. Mayo, Nfld. and Labrador Dep. Environ. and Lands, pers. commun.).

Shaffer (1981:132) tentatively defined a minimum viable population as the "smallest isolated population having a 99% chance of remaining extant for 1,000 years despite the foreseeable effects of demographic, environmental and genetic stochasticity, and natural catastrophes." The probability of survival and time period may in fact be much smaller for some practical applications. Genetic diversity is rarely measured in founded populations, but it is likely that populations founded by fewer than 10 individuals might suffer from immediate negative effects of inbreeding and loss of genetic variability (Leberg 1990). Fifty or more individuals may be required to preserve allelic variation, a number that might be lower with the migration or addition of individuals from a source population that has higher genetic variation than the sink population (Weishampel 1990). A population may need to have 500 or more individuals to guarantee the long-term fitness and genetic adaptability of a species; over the short term, however, a population size of 50 should be sufficient to prevent immediate loss of fitness (Franklin 1980, Lehmkuhl 1984). Clark (1989) calculated a minimum viable population size for black-footed ferrets (*Mustela nigripes*) to be 214 based on modeling estimates, and 120 based on modeling of demographic and environmental stochasticity. Minimum effective population size depends on the degree of genetic variation and the breeding structure of the population. The genetic variation of a sample of American martens from Wyoming was higher than average for terrestrial Carnivora (Mitton and Raphael 1990).

Translocations of at least 50 American martens of an even sex ratio are recommended to preserve allelic diversity and to prevent the loss of fitness associated with inbreeding. If fewer than 50 American martens are translocated, then natural immigration or the regular addition of individuals from a source population of higher genetic diversity, as practiced in Fundy National Park (no. 25, Table 10.1), will help prevent inbreeding. Based on the range of stable American marten densities of 0.4–1.2 animals/km² observed across its range (Strickland et al. 1982*a*), the area required to sustain 50 individuals is 42–125 km². At least 420–1250 km² of habitat capable of supporting 500 American martens is recommended for long-term translocation success.

Interspecific Interactions

Fishers may compete with American martens for food, but the fisher specializes less on microtine rodents, allowing the two species to coexist (Rosenzwieg 1966). The American marten, more specialized with regard to prey, may show better adaptation to some sites if prey larger than microtines are not available. Observations that the two species are often inversely proportional in abundance (de Vos 1952, Douglas and Strickland 1987) may reflect the dynamics of food competition and fisher predation on American martens. Alternatively, they may indicate only spatial variation in habitat conditions (Strickland and Douglas 1987).

American martens are occasionally preyed on by larger carnivores such as canids and birds of prey (Clark et al. 1987), although such predation is likely insignificant in limiting their numbers. Trapping by humans, however, can reduce American marten numbers to the point of local extinction (de Vos 1952, Soper 1970). Protection from marten-specific trapping or, if possible, all land-based trapping using nondiscriminating sets, would enhance the likelihood of success of American marten translocations.

Diseases and Parasites

Diseases and parasites are not known to limit American marten populations. On the other hand, not all marten populations have encountered the same pathogens; therefore, immunity to diseases could vary among geographic regions, and native martens and other species could contract diseases such as canine distemper or rabies from translocated American martens.

Bissonette et al. (1988) suggested that 25% of 40 American martens studied in Newfoundland had died from encephalitis caused by the canine distemper virus. The virus caused 59% of the known deaths and may be the reason for the marten population decline on the island in the early 1980s. A quarantine period may therefore be advisable for translocated animals. Obtaining stock from nearby geographic areas should remove this concern.

Acknowledgments

The following individuals graciously assisted by contributing unpublished reports and information: H. J. Griese, D. N. Larsen, R. Flynn, and H. R. Melchior of the Alaska Department of Fish and Game; K. D. Elowe of the Maine Department of Inland Fish and Wildlife; C. S. Johnson and R. R. P. Stardom of the Manitoba Department of Natural Resources; G. Armstrong of the Manitoba Trappers' Association; R. D. Earle of the Michigan Department of Natural Resources and Energy; C. J. Gibilisco of Montana State University; D. J. Cartwright of the New Brunswick Department of Natural Resources and Energy; O. Forsey and L. Mayo of the Government of Newfoundland and Labrador Department of the Environment and Lands; E. P. Orff of the New Hampshire Department of Fish and Game; B. Sabean of the Nova Scotia Department of Lands and Forests; M. A. Strickland and M. Novak of the Ontario Ministry of Natural Resources; W. Runge and P. Galbraith of the Saskatchewan Department of Parks and Renewable Resources; L. F. Fredrickson of the South Dakota Department of Game, Fish, and Parks; J. J. DiStefano of the Vermont Department of Fish and Wildlife; B. E. Kohn of the Wisconsin Department of Natural Resources; T. Rinaldi and D. M. Elsing of the USDA Forest Service; M. H. Davis of the U.S. Bureau of Land Management; J. A. Bissonette of the Utah Cooperative Wildlife Research Unit; and G. Corbett, G. Sinclair, and R. Leonard of the Canadian Parks Service, Environment Canada.

11 The Effects of Trapping on a Newly Exploited American Marten Population

Clément Fortin and Michel Cantin

Without proper management populations of American martens (*Martes americana*) can be severely depleted. Because they are easy to trap and prices for their fur have remained high in recent years, despite a general decline in the fur market, management must be improved to prevent overharvesting. Such refinements must be based on sound biological data. Few studies have been designed to specifically measure the effects of trapping on marten populations.

Since the work of Yeager (1950) and Quick (1956), it has been known that the male and juvenile cohorts of a population are most vulnerable to trapping. A sex ratio equal or favoring females likely indicates overharvesting (Soukkala 1983, Archibald and Jessup 1984), as does a low proportion of juveniles and a high proportion of adult females in the harvest (Strickland and Douglas 1987). Assuming that a difference in vulnerability between sex and age classes explains the preponderance of males and juveniles in the harvest, Strickland and Douglas (1987) used age and sex ratios as indices of harvest intensity. However, these indices are often post-facto indicators and cannot prevent overharvesting.

The goals of our study were to assess harvest intensity through the use of sex and age ratios in the harvest, to estimate natural and trap-induced mortality, and to propose a model to predict marten harvest. We used the unprecedented opportunity to study an area that had undergone no legal trapping for over 100 years.

Study Area and Methods

The Laurentides Wildlife Reserve is located about 40 km north of Quebec City and encompasses 7934 km². This predominantly wilderness area is part

179

of the Laurentides Highlands, the largest and highest section of the Laurentian Precambrian shield. Elevations range between 350 and 1090 m; mean temperatures in the central part of the study area were 15.5°C in July, and −17°C in January. Annual precipitation as snow averages 4.3 m. Typically, the boreal forests are dominated by black spruce (*Picea mariana*) and balsam fir (*Abies balsamea*).

A policy on furbearers, implemented in the summer of 1984 by the Quebec government, opened most Quebec wildlife reserves to trapping. Because the Laurentides Wildlife Reserve had not been trapped for over a century and had acted as a reservoir for adjacent areas, we chose it as an experimental area. Legal trapping on the study area began in the fall of 1984, with no restriction on the number of martens each trapper could take. Pelt prices dropped in 1985–1986, then nearly doubled in 1986–1987 and 1987–1988. Harvest was monitored from the fall 1984 until the spring of 1988, for four trapping seasons, during which the trapping season extended from 18 October to 1 March. Trapping territories ($n − 161$) were allocated to individual trappers through a computerized lottery system. All data were collected on a trapline basis. Meetings were held with trappers before each trapping season to explain the purpose and methods of the study, and participating trappers were given an information package. Just before the beginning of the season, all trappers were contacted by mail to encourage their participation.

Two measures of total harvest were used: the trapper's mandatory report (required by the government) and the notes kept in the trapper's daily logbook. Trapping effort (no. of traps × no. of nights [tn]) was determined from the daily logbook, in which trappers indicated how many traps were set. It is illegal to use foothold traps to capture martens in Quebec; all trappers used Conibear traps. Carcass deposit bins were placed on the five main access highways to the reserve. Trappers were asked to place each carcass in a plastic bag and identify it with the trapper's name, trapline number, and date of capture.

Carcasses were collected weekly, and sex and age were determined with radiographs of root canals (Dix and Strickland 1986a, Fortin et al. 1988). A random subsample of the adult segment of two harvests, the 1984–1985 ($n = 313$) and 1985–1986 ($n = 141$), was aged using cementum annuli (Stone et al. 1975). Microtome sections were taken from the ovaries of adult females in the random subsample and stained with a mixture of hematoxyline, phloxine, and saffron. Corpora lutea were counted to estimate ovulation and fecundity rates. Data were analyzed using SAS/STAT (1988). Natural mortality was calculated for martens four years old or older, with sexes combined. Because these individuals were all adults before trapping began in the reserve, we assumed that all mortality before trapping was natural.

For our analyses, all values obtained from a ratio were weighted for the denominator. Yield (captures[c]/10 km²) and trapping pressure (tn/10 km²) were weighted for area, and trapping success (c/100 tn) was weighted for trapping effort. We used chi-square analysis to examine age and sex composition within and among seasons and to compare ovulation rate among age classes. When comparing ovulation rates between females aged 1.5 years or more, the chi-square was corrected for continuity. Differences in sex ratio between periods (18 October through 26 December, vs. 27 December through 1 March) were tested with the Friedman's test (Sokal and Rohlf 1981). Analysis of variance (ANOVA) with the SAS-GLM procedure was used to account for variables explaining yield and trapping success and to test the differences betweeen fecundity rate by season and age of the adult females. Because of small sample size, individuals more than 5 years old were grouped. The multiple regression calculated to explain the harvest was taken from the SAS-REG stepwise procedure, which also was used to test relationships among trapping success and sex and age ratios. Natural mortality rate was estimated with two methods from the fisheries literature: the coded ages method (Robson and Chapman 1961) provides an estimate of the survival rate using a segment of the catch curve, assuming year class strength and annual survival rates are constant; and the capture regression method (Ricker 1975) allows an estimation of the survival rate using the natural logarithm of the number of individuals remaining in each year class. The harvest rate was measured by age composition of the harvest, as in Fraser 1984.

Results

Harvest

Over four years, trappers supplied us with 4611 American marten carcasses, 74% of the total harvest. Except for the first year, when the harvest was relatively stable over the season (Fig. 11.1), most of the harvest (81%) was taken by the fifth two-week period, around Christmas. Trapping effort doubled between years 1 and 2 (Table 11.1), although pelt prices dropped from $57 (Canadian) to $43. That value nearly doubled to $82 in 1986–1987 and remained high at $85 in 1987–1988, while effort remained high and success dropped sharply for the last two seasons. Harvest was highest in year 2 of the study (Table 11.1). The harvests in the first (1984–1985) and fourth (1987–1988) seasons were similar; however, trapping effort had doubled by the fourth season, reflecting a sharp decline in trapping success.

Trapping success varied little within year 1, from 1.4 c/100 tn early in the season to 1.3 c/100 tn in the ninth period ($\bar{x} = 1.3$ c/100 tn for year 1) (Fig.

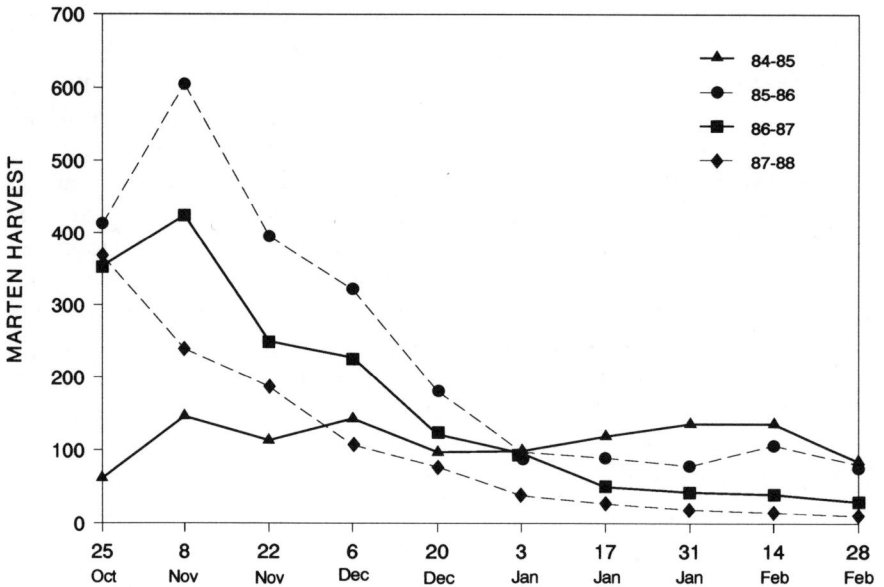

Figure 11.1. Harvest of American martens (*Martes americana*) in relation to date, in Laurentides Wildlife Reserve, Quebec, 1984–1988.

11.2). This pattern changed in the tenth trapping period, when 2.6 c/100 tn were recorded. Trapping success also increased at the end of each of the other three trapping seasons of the study (Fig. 11.2). Trapping success varied more within year 2, although the mean success, 1.3 c/100 tn, was similar to that of year 1. The main difference between years 1 and 2 was trapping effort and harvest, which increased twofold. In the last two years of the study, the success rates followed similar seasonal trends, but with lower mean success

Table 11.1. Characteristics of American marten (*Martes americana*) harvest in the Laurentides Wildlife Reserve, Quebec, 1984–1988

	Harvest	Effort (tn × 1000)	Success (captures/ 100 tn)	Yield (captures/ 10 km²)	Pressure (effort/ 10 km²)
1984–1985	1134	86.6	1.3	2.1	163.6
1985–1986	2368	174.1	1.4	3.8	276.5
1986–1987	1635	212.6	0.8	2.5	319.7
1987–1988	1091	185.1	0.6	1.6	270.1
TOTAL	6228	658.4	0.9	2.5	262.3

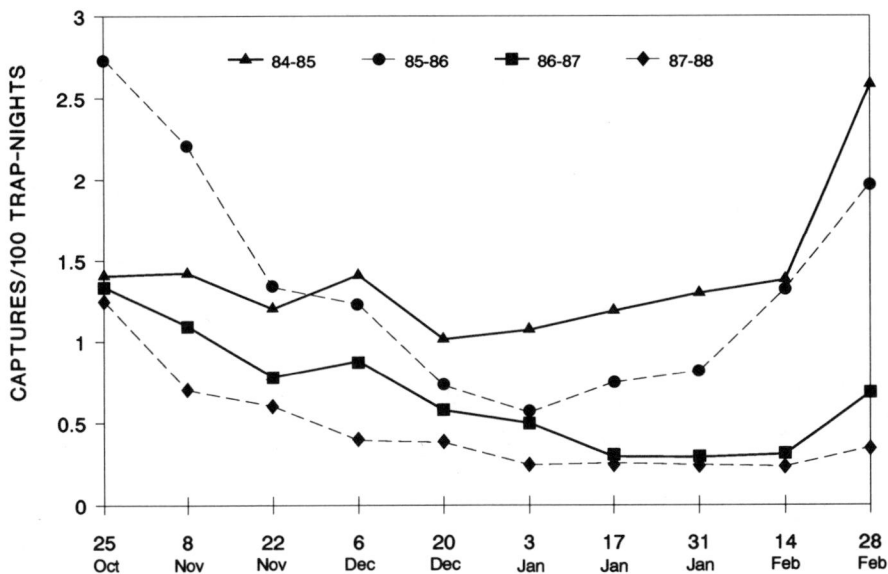

Figure 11.2. Trapping success for American martens (*Martes americana*) in relation to date, in Laurentides Wildlife Reserve, Quebec, 1984–1988.

rates (0.7 c/100 tn in 1986–1987, 0.5 c/100 tn in 1987–1988). Thus, the combined average success of the last two years (0.6 c/100 tn) was half that of the first two years (1.3 c/100 tn).

Reproduction

Seventy-eight of 110 adult females harvested in year 1 and 59 of 195 adult females harvested in year 2 were examined for corpora lutea. The ovulation rate for adult females was similar: 79% in year 1 (1984) and 78% in year 2 (1985) had corpora lutea. Because ovulation rate did not vary during the study, data were combined to examine age differences (Table 11.2). Ovulation rates varied significantly among age classes (chi-square = 31.3, df = 5, $P < 0.001$). Forty-six percent of females aged 1.5 years had ovulated, whereas 92% of older females had done so. The mean number of corpora lutea per ovulating female (4.25) did not vary significantly among years ($F = 1.08$, $df_1 = 1$, $df_2 = 135$, $P = 0.30$). Inasmuch as there was no significant difference in the number of corpora lutea produced per adult female between years 1 and 2, we pooled these data ($\bar{x} = 3.35$ corpora lutea, $n = 137$) (Table 11.2). Fecundity did not differ among trapping seasons ($F = 0.04$, $df_1 = 1$,

Table 11.2. Reproductive characteristics of American martens (*Martes americana*) in the Laurentides Wildlife Reserve, Quebec, by age class, 1984–1985 and 1985–1986

	Age						Mean or total
	1	2	3	4	5	>5	
Ovulation rate							
% females ovulating	46	88	100	100	91	79	79
n	(41)	(17)	(12)	(18)	(11)	(14)	(137)[a]
Production of corpora lutea							
Ovulating females							
Mean	3.43	4.13	4.42	4.61	4.60	4.45	4.25
(± SE)	(±0.22)	(±0.27)	(±0.15)	(±0.20)	(±0.27)	(±0.39)	(±0.09)
n	19	15	12	18	10	11	108[b]
Adult females							
Mean	1.63	3.65	4.42	4.61	4.00	3.50	3.35
(± SE)	(±0.30)	(±0.41)	(±0.15)	(±0.20)	(±0.47)	(±0.59)	(±0.17)
n	(41)	(17)	(12)	(18)	(11)	(14)	(137)

[a]Twenty-four females whose age was unknown were classified as adults using radiographs.
[b]Twenty-three females whose age was unknown were ovulating.

$df_2 = 101$, $P = 0.84$), but did among age classes ($F = 11.6$, $df_1 = 5$, $df_2 = 101$, $P < 0.0001$).

Demography

Based on examination of 4411 carcasses submitted by trappers (Table 11.3), the ratio of juveniles to adults in the harvest differed significantly among years (chi-square $= 270$, df $= 3$, $P < 0.001$). Forty-eight percent of the martens harvested in year 1, 35% in year 2, and 45% harvested in year 4 were adults. Many authors have reported that early season harvests include mostly juveniles (Quick 1956, Francis and Stephenson 1972). This situation did not occur in year 1 of our study (Fig. 11.3), presumably because the initial population was composed mostly of adults. In year 2, a greater proportion of juveniles were trapped early in the season, but the number decreased after the fourth two-week period. Although adults outnumbered juveniles in each period of the harvest in year 3, juveniles were more numerous early in the season. In year 4, a larger proportion of juveniles was observed in the early harvest, but a rapid inversion in favor of adults took place by the fifth period. For the last three years of the study, the proportion of juveniles in the harvest was significantly higher early in the season than late in the season (Friedman's test, chi-square $= 24.4$, df $= 9$, $P < 0.005$). The male-to-female ratio in the harvest declined significantly (chi-square $= 67.7$, df $= 3$, $P < 0.001$) during the study (Table 11.3), from 2.4 males per female in year 1 to 1.1 males per female in year 4.

Although the sex ratio in the total harvest of year 4 was 1.1:1, it was 0.9:1 among adults (Table 11.3). Also, whereas the sex ratio of the total harvest dropped by half between years 1 and 4, it dropped by two-thirds among adults. By year 4, more females were caught than males (chi-square $= 78.7$, df $= 3$, $P < 0.001$). Males predominated in the harvest early in the season in each year (Fig. 11.4), but for all years combined the period of capture had no

Table 11.3. Age and sex distribution of American martens (*Martes americana*) harvested in the Laurentides Wildlife Reserve, Quebec, 1984–1987

	Age		Sex ($\male$:$\female$)				Juv. and ad. $\female$	
	n	Juv.:ad.	n	All	Juv.	Ad.	n	Juv.:ad. $\female$
1984–1985	848	1.1	871	2.4	2.0	2.7	548	4.0:1
1985–1986	1756	1.8	1868	1.5	1.3	2.2	1334	5.8:1
1986–1987	1010	0.5	1063	1.2	1.3	1.2	634	1.1:1
1987–1988	797	1.2	803	1.1	1.3	0.9	628	2.3:1
TOTAL	4411	1.1	4605	1.5	1.4	1.6	3144	2.9:1

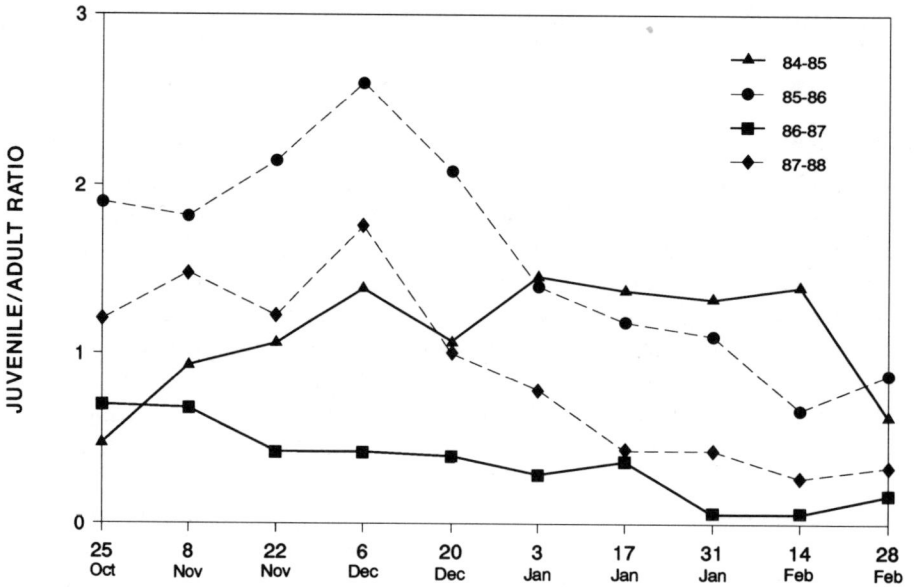

Figure 11.3. Ratio of juvenile to adult American martens (*Martes americana*) harvested during the season in Laurentides Wildlife Reserve, Quebec, 1984–1988.

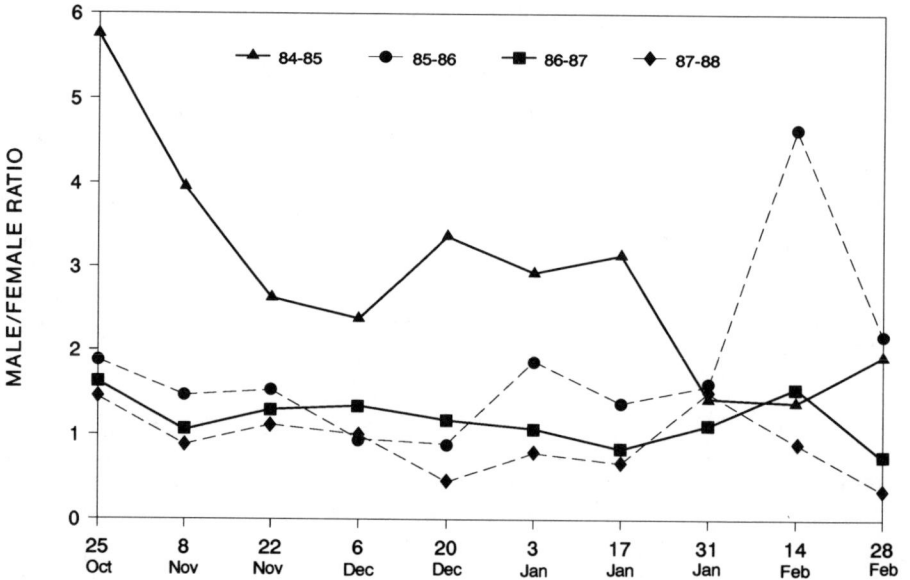

Figure 11.4. Ratio of male to female American martens (*Martes americana*) harvested during the season in Laurentides Wildlife Reserve, Quebec, 1984–1988.

significant effect on the proportion of males in the harvest (Friedman's test, chi-square = 9.65, df = 9, P = 0.36). Overall, the proportion of males was slightly higher late in the season (62%) than early (59%).

Mortality Factors

Using the coded ages method (Robson and Chapman 1961), we calculated a mean mortality rate of 40% (SE = 2%, n = 9), whereas the capture regression technique yielded a mortality rate of 35% (SE = 3%, n = 9) (Ricker 1975). With the method of Fraser (1984) for a population in which one sex is more vulnerable to trapping than the other, the proportion of specimens harvested each year is inversely proportional to the number of years required to reach an even sex ratio. Because in our study parity was reached in year 4, the average proportion harvested annually represents about 25% of the total population.

Index of Harvest Level

Sex Ratio. Male martens' vulnerability to trapping can be two to three times that of females (Strickland and Douglas 1987). When the sex ratio of the harvest is equal or favors females, overharvesting is likely (Yeager 1950, Quick 1953, Soukkala 1983), whereas a ratio of two or more males per female is considered safe (Strickland 1989). Based on these criteria, the harvest in our study was at a safe level in year 1 only (Table 11.3). Although the ratio at the beginning of year 2 (1985–1986) favored males, females were favored by the fourth two-week period (Fig. 11.4). The results for periods 4 and 5 combined show that almost half the adults captured were females (44.5%), indicating a high harvest level. In year 2, yield and success rate were the highest for the four years. The proportion of juveniles harvested in year 2 was 65%. During year 3, the sex ratio was about even for adults and all animals (Table 11.3), but only 33% of the harvest was juveniles. Trapping success that year dropped to 0.8 c/100 tn.

Ratio of Juveniles to Adult Females (j:af). Strickland and Douglas (1987) reported that the ratio of juveniles to adult females (≥1.5 year) is a better indicator of harvest intensity than the sex ratio and that harvest level is acceptable when j:af is 3 or higher. This condition was met only in the first two years of our study, when j:af was 4.0:1 and 5.8:1, respectively (Table 11.3). Early in year 1, j:af was 3.0:1 (Fig. 11.5), but this ratio increased in the sixth two-week period, reaching 9:1, then rapidly declined to 1.8:1 at the end of the season. In year 2, j:af was 8.8:1 early in the season but fell gradually to 2.3:1 by the eighth period, and then increased to 3.5:1 by the end

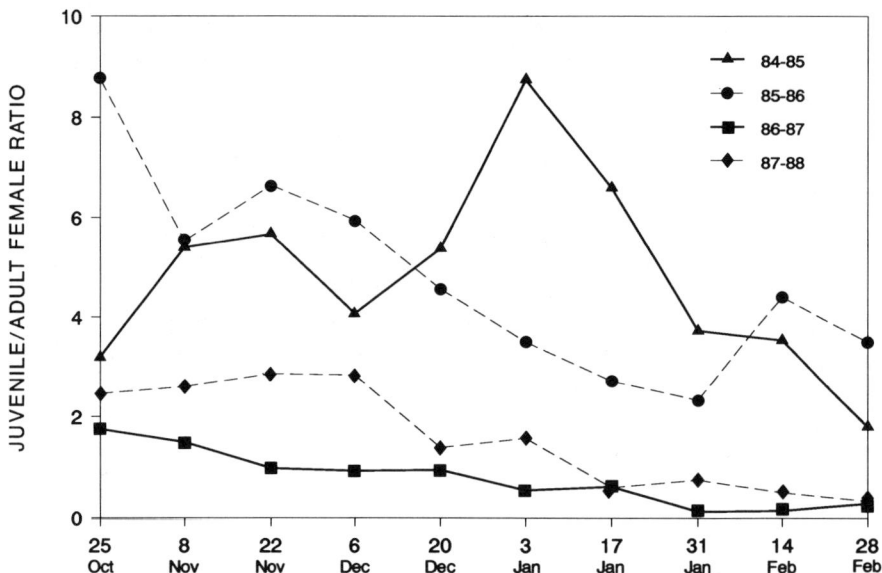

Figure 11.5. Ratio of juvenile to adult female American martens (*Martes americana*) harvested during the season in Laurentides Wildlife Reserve, Quebec, 1984–1988.

of the season. During the last two years, j:af was less than 3:1 during the entire season.

Harvest Prediction. The relationship between total harvest, trapping effort, and area trapped was highly significant ($F = 55.0$, $df_1 = 3$, $df_2 = 434$, $P = 0.0001$), and this regression could be used to predict the harvest for a given year (year i):

$$H_i = 0.004E_i - 0.003E_{i-1} + 0.12A_i + 0.40 H_{i-1}$$

where		SE	t	P
H_i = harvest at year i				
H_{i-1} = harvest at year $i - 1$		0.0498	7.97	0.0001
E_i = effort (tn) at year i		0.0004	9.76	0.0001
E_{i-1} = effort (tn) at year $i - 1$		0.0006	−5.54	0.0001
A_i = area trapped (km²) at year i		0.0169	7.30	0.0001

Discussion and Management Implications

In our view the positive correlation between harvest in year i and that of the preceding year reflects the quality of the habitat: given the same effort, an area that yields a high harvest one year could be equally productive the following year, provided the food base is sufficient and reproduction normal. The positive correlation between harvest and effort at year i might be an incentive to increase trapping activity, but the negative relation between harvest at year i and effort of the preceding year should instill some caution. The effects of heavy trapping pressure are not only visible during a given year but also the year after as well.

Except for the first year, most of the harvest was taken before Christmas. This pattern is due to the high trapping effort and availability of martens early in the season, combined with the harsh weather that makes travel increasingly difficult and costly after Christmas. Also, martens are less active in late winter than in early winter (Buskirk 1983, Thompson 1986), reducing their probability of capture.

A sharp decline was registered between year 1 and year 4 in trapping success and presumably in the number of martens available. Trapping success increased at the end of each season. This increase, which began between 7 February and 20 February, could be attributed to the beginning of postwinter dispersal, as identified by Archibald and Jessup (1984), and possibly to more compact snow, which makes movement easier.

Reproductive performance was among the highest reported in North America. Although the ovulation rate was lower than in southern Ontario (87%) (Strickland and Douglas 1987), it was higher than in the Yukon (74%) (Archibald and Jessup 1984) and well above the rate reported in northern Ontario (40%) when the population was subjected to a food shortage (Thompson and Colgan 1987a). This ovulation rate among females aged 1.5 years was lower than observed in southern Ontario, but the rate was similar to that region for older females. In southern Ontario, the ovulation rate for females aged 1.5 years was 80%, whereas it was 93% for older females (Strickland and Douglas 1987). In northern Ontario, Thompson and Colgan (1987a) observed an ovulation rate when food was scarce of only 7% among females aged 1.5 years, and 59% among older females. In the Yukon, Archibald and Jessup (1984) reported no significant difference in ovulation rate between females aged 1.5 years and older females.

The mean number of corpora lutea per ovulating female in our study was higher than reported in other North American marten populations. In Alaska, Lensink (1953) found a mean of 1.8 corpora lutea per ovulating female, whereas in Montana it was 3.0 (Wright 1963). Strickland and Douglas (1987) observed 3.5 corpora lutea per ovulating female in southern Ontario, and Thompson and Colgan (1987a) reported 3.2 in northern Ontario. The mean number of corpora lutea per adult female in our study was also the highest recorded in North America. In southern Ontario, Strickland and Douglas (1987) observed a mean of 3.0 corpora lutea per adult female, whereas in northern Ontario Thompson and Colgan (1987*a*) observed an average of 1.3 corpora lutea per adult female in a population with food limitations. Archibald and Jessup (1984) observed a mean fecundity of 3.3 corpora lutea in females aged 1.5 years and a mean of 3.8 in older females in the Yukon. As reported by Archibald and Jessup (1984), we observed a significant difference in fecundity between females 1.5 years and older, whereas Strickland and Douglas (1987) found differences in fecundity between trapping seasons and ages.

Nearly half the martens harvested in year 1 of our study were adults. Lensink (1953) asserted that territories that had been harvested lightly or not at all showed a higher percentage of adults than did heavily harvested areas. In New York State, 70% of marten carcasses collected after a 40-year ban on trapping consisted of adults (Brown 1980).

In year 2 of the study, juveniles made up 65% of the harvest. Similar results were obtained in the Yukon, where 67% of the three-year harvest consisted of juveniles (Archibald and Jessup 1984), perhaps indicating heavy exploitation. A significant difference was evident between the age distribution of the second and third seasons (chi-square = 267, df = 1, $P < 0.01$), which may be due in part to heavy exploitation and to bad weather in the spring and summer of 1986 that reduced food abundance. Scarce food could have reduced survival of juveniles and females, both of which are more prone to starvation than males because of high energy requirements (Hawley and Newby 1957*a*). Also, when food is scarce, adults enlarge their home ranges, increasing their chances of capture (Thompson and Colgan 1987*a*).

Our data are unique in showing a rapid decline in the sex ratio of a newly harvested population over a short period: the sex ratio of the harvest declined to parity by year 4. Several authors have reported that male mustelids are generally more vulnerable to trapping than females (Soukkala 1983, Archibald and Jessup 1984). Strickland and Douglas (1987) observed that two to three males were caught for each female in a population not subjected to adverse food conditions or intense exploitation. In the Yukon, Archibald and Jessup (1984) obtained a sex ratio of 1.4 males per female over a three-year period in a heavily harvested population. Quick (1953) reported an even sex ratio in an area of intensive trapping, whereas de Vos (1952) obtained a ratio of 1.2:1 in Ontario.

Inasmuch as both juveniles and males are more vulnerable to trapping than adult females, a high proportion of juvenile males in the total harvest could conceal an imbalanced sex ratio in the adult segment of the population, which could affect reproduction. Although Yeager (1950) reported that trapping in the fall (Oct–Dec) should yield a higher percentage of males in the harvest than winter trapping, that situation did not occur in all years of our study. The high proportion of juveniles recorded for year 2 suggests that heavy trapping had created vacant areas that were later occupied by juveniles dispersing from untrapped portions of the reserve. The same situation was observed in the Yukon (Archibald and Jessup 1984). In our view the harvest level in the second year of the study was excessive and had repercussions for subsequent years.

The Laurentides Reserve marten population, despite its newly exploited status and high reproductive rate, was subjected to excessive harvest within four years of the start of trapping. Scarce food that could have affected the survival of the young and a long trapping season with no restrictions on the number of martens harvested, especially when fur prices were high, contributed to this situation. Because the harvest level indices actually in use are post-facto management tools that are not always consistent, new density-related indicators should be evaluated for possible use in conjunction with currently employed indicators. We suggest field testing of this harvest prediction model and investigation of the relation between trapping success and density for different habitat types throughout the American marten's range. If a strong relationship exists between those two variables, trapping success could then be used as a measure of marten abundance for large territories. Where harvest numbers are available it could allow the estimation of harvest rates. Sex and age ratios should also be related to density to see how they fluctuate with marten abundance.

Acknowledgments

We thank M. A. Strickland of the Ontario Ministry of Natural Resources for her invaluable cooperation. We also thank J. Beauchemin, C. Caron, J.-G. Frenette, J.-L. Brisebois, and S. St-Onge for their technical assistance. We thank H. Jessup of the Yukon Fish and Wildlife Division for arranging for translation of the original French version. We are particularly grateful to the trappers of the Laurentides Wildlife Reserve and to J.-B. Gagnon and C. Thibault, former presidents of the Association Provinciale des Trappeurs Indépendants Inc. (APTII) (Regions 02 and 03), without whom this study could not have been done.

12 Ethical Considerations in the Selection of Traps to Harvest American Martens and Fishers

Gilbert Proulx and Morley W. Barrett

American martens (*M. americana*) and fishers (*Martes pennanti*) are high-ly valued in the fur industry because of their long, luxurious fur (Obbard et al. 1987). Trapping is criticized, however, because it involves capture de-vices that cause pain (Proulx and Barrett 1989*a*, 1991*a*). Such criticism led to the creation of major humane trapping research programs (Barrett et al. 1988), and, in Canada, resulted in the recommendation that American mar-tens and fishers be harvested with quick-kill or live-holding traps (Fur Insti-tute of Canada 1988). But traps that do not kill the animals quickly are still being used in the field (Proulx and Barrett 1991*a*), and foothold traps still appear to be the most common type used by North American trappers (Bog-gess et al. 1990).

In this chapter we review traditional and new technologies as they relate to American marten and fisher trapping, and identify the ethical decisions that must be taken to improve harvest programs.

The Humaneness Criteria

The issue of humaneness in wildlife trapping is a major societal concern (Barrett et al. 1988). Obviously, depending on the criteria used, many traps could be classified as humane. We define a humane livetrap as a device that holds an animal with minimal distress and trauma—in other words, without serious damage such as dislocated joints, fractures, and amputations (Olsen et al. 1988, Proulx 1990). We define a humane killing trap as a device that has the potential, at a 95% confidence level, to render 70% or more of target animals irreversibly unconscious within three minutes (Barrett et al. 1989; Proulx et al. 1989*a*, 1990, 1993).

Traditional Mechanical Traps

Foothold Traps as Live-Holding Devices. Although Krause (1989) recommended the nos. 0 and 1 foothold traps (Woodstream Corp., Lititz, Pa.) for American martens and the no. 1½ for fishers, these traps are known to cause distress in both species. De Vos and Guenther (1952) used steel traps with padded jaws to capture martens alive. They concluded that the foothold traps were not appropriate because martens struggled violently and broke their legs. In the Northwest Territories, Proulx (unpubl. data) found that 15 of 31 foot-captured American martens had severed phalanges or broken legs. Fishers caught in foothold traps also sustain severe injuries such as broken bones and amputations (Quick 1953; Coulter 1960; A. W. Todd, Alta. Fish and Wildl., pers. commun.).

Foothold Traps as Killing Devices. Foothold traps are also used in killing sets for American martens (Barnes 1988) but do not appear to be humane. Mechanical evaluations of the nos. 1- 1/2 and 3 foothold traps (R. Drescher, Alta Res. Counc., unpubl. rep.) and the no. 4 (S. R. Cook, Alta Environ. Cent., unpubl. rep.) foothold trap revealed that these traps did not have the potential to render martens struck in the head-neck region irreversibly unconscious in three minutes or less. Barrett et al. (1989) also found that nos. 3 and 4 foothold traps often struck martens in the thorax but did not cause major traumatic lesions that could lead to quick kills.

Conibear traps. The Conibear nos. 120 and 160 traps (Woodstream Corp., Lititz, Pa.) are considered state-of-the-art trapping devices that kill American martens quickly (Standing Committee on Aboriginal Affairs and Northern Development 1986), and their use is taught in trapping and conservation manuals (Canadian Trappers Federation 1984, Alberta Vocational Centre 1987). Cook and Proulx (1989), however, reported that the impact and clamping energies of the Conibear no. 120 trap were lower than the kill threshold standards for American martens of the Canadian General Standards Board (1984). Proulx et al. (1989*b*) also found in a series of tests in simulated natural environments that this trap did not consistently render American martens unconscious in three minutes or less, even when the animals were struck in the head-neck region. Finally, Barrett et al. (1989) found that at least 15 of 30 American martens were improperly struck in Conibear nos.

120 and 160 traps equipped with a two-prong trigger. By comparing strike locations and lesions induced by the traps in martens captured in simulated environments and on traplines, Barrett et al. (1989) concluded that these standard Conibear traps did not have the potential to consistently render American martens quickly unconscious.

The Conibear nos. 160 and 220 traps (Woodstream Corp., Lititz, Pa.) are recommended for the capture of fishers (Alberta Vocational Centre 1987, Krause 1989). However, a mechanical evaluation of the Conibear no. 220 trap, the more powerful of the two, showed that it did not have the potential to kill fishers quickly (Proulx 1990). In simulated natural environments, Proulx and Barrett (1993a) also found that a Conibear no. 220 trap modified with stronger springs and clamping bars failed to consistently render fishers irreversibly unconscious in no more than three minutes. Many animals struggled violently in this trap and sustained pelt damage in the neck area (see also Obbard 1987b). Despite their reputation as quick-killing devices, the Conibear nos. 160 and 220 traps did not meet the criteria of a humane trap.

Trapping Alternatives

Livetraps. Hamilton and Cook (1955) suggested that fishers could be taken in padded-jaw steel traps. It has now been well demonstrated that limb injuries to a broad range of terrestrial furbearers can be markedly reduced with these traps (Olsen et al. 1988, Onderka et al. 1990). Padded-jaw foothold traps may be less injurious than steel ones, although this advantage may disappear when animals are held for long periods (Proulx 1990). Padded-jaw foothold traps should be tested in controlled conditions before being recommended as humane livetraps for fishers.

Box traps have been successfully used for American martens (trap size 23 × 23 × 43 cm and larger) (de Vos and Guenther 1952; Newby and Hawley 1954; Slough and Smits 1985) and for fishers (trap size 25 × 31 × 81 cm and larger) (Irvine et al. 1964, Kelly 1977). The animals may damage or lose teeth by chewing on traps (Arthur 1988; P. Cole, Alta. Res. Counc., pers. commun.), but these injuries can be minimized by constructing traps with no openings larger than 2.5 cm square (Arthur 1988) or by building them out of wood. The use of box traps is not recommended during the coldest months of the trapping season (de Vos and Guenther 1952; Slough and Smits, B. G. Slough and C. M. Smits, Yukon Dep. Renew. Resourc., unpubl. rep.) or if the size of the trapline does not allow it to be checked daily. Trappers should not dismiss the use of box traps just because they are cumbersome (and therefore impractical); such an argument does not hold up for trappers who use motorized vehicles for travel (Todd and Boggess 1987).

Killing Traps. The C120 Magnum trap (Les Pièges du Québec Inc., St.-Hyacinthe, Que.), a mechanically improved rotating-jaw trap, has more than double the striking and clamping forces of the Conibear no. 120 trap, and rates above the Canadian General Standards Board's 1984 kill threshold for American martens (Proulx et al. 1989*a*). When equipped with a pitchfork trigger, this trap consistently strikes martens in vital regions and renders them irreversibly unconscious in three minutes or less (Proulx et al. 1989*a*). On traplines, this trap is as efficient as the commercial devices commonly used by trappers; it can be safely handled (Proulx et al. 1989*a*, Proulx 1991), and it does not damage the pelts (Barrett et al. 1989). Also, this trap is manufactured commercially (Proulx 1990).

To our knowledge, no commercial trap has the potential to consistently render fishers unconscious in three minutes or less. Proulx and Barrett (1993*b*), however, reported that the Bionic trap (W. Gabry, Vallenby, B.C.), a large mousetrap-type device (see Proulx and Barrett 1991*b* for design) that consistently strikes fishers in the head and neck region, rendered them unconscious in less than three minutes in simulated natural environments. The trap was redesigned to simplify its manufacture and improve its longevity (Proulx 1991) before being tested on traplines.

Ethical Decisions

Public concerns regarding the welfare of trapped animals are valid, and wildlife professionals must incorporate new trapping technology in their programs. The use of humane traps is part of sound stewardship of wildlife resources and is necessary to ensure the future of wildlife management programs that use trapping (Proulx and Barrett 1989, 1991*a*). It is therefore vital that the use of steel foothold traps for trapping American martens and fishers be abolished as soon as possible. Because these mustelids are often taken in baited sets intended for other furbearers (Strickland, this volume), research is also needed on other species to develop humane trapping systems that are more species-selective.

The harvest of American martens and fishers should be conducted with box traps or humane killing devices that have been thoroughly tested. In the case of martens, the C120 Magnum is the only fully tested trap that has met the highest standards for trap performance (Barrett et al. 1989, Proulx et al.

1989*a*). Killing devices that have not met performance criteria, such as the Conibear no. 120, should be replaced by the C120 Magnum trap. Such replacement has already been initiated by trappers and government agencies (Blackwell 1992, Northwest Territories Renewable Resources 1992). In the case of fishers, wildlife agencies should ban the use of the Conibear 160 trap because it is weaker than the Conibear no. 220 trap, which failed the criteria for humane killing. If the Bionic trap meets the criteria for humaneness in field tests, it should be phased in to replace the Conibear no. 220 trap.

If trapping of American martens and fishers is to continue in North America, both furbearer management and trap technology must evolve with public sentiments and meet the challenges of humane trapping and conservation.

Acknowledgments

We thank A. W. Todd and F. Neumann, Alberta Fish and Wildlife Division, for reviewing an earlier version of the chapter.

13 Post-Release Movements of Translocated Fishers

Gilbert Proulx, Alfred J. Kolenosky, Micheal J. Badry,
Randy K. Drescher, Ken Seidel, and Pam J. Cole

The original range of fishers (*Martes pennanti*) in Alberta extended from south of the North Saskatchewan River to the 15°C July isotherm (Hagmeier 1956*a*, Hall and Kelson 1959, Douglas and Strickland 1987). By the early 1900s, particularly in the southern part of the original range, the loss of forests through logging, fire, and settlement reduced the number of fishers. This habitat loss, along with unregulated trapping and the use of strychnine for harvesting and predator control, severely reduced or eliminated fishers from much of their readily accessible range (Douglas and Strickland 1987). Although fishers are still present in Alberta, a major population decrease was observed between 1982 and 1988 (Alberta Forestry, Lands and Wildlife 1989). Reintroduction programs are one management option that can reestablish fisher populations (Banci 1989), and they were judged necessary in central Alberta (Proulx 1990).

The fisher is a frequently reintroduced furbearer, and Berg (1982) reported 19 fisher translocations between 1955 and 1981 (6 in Canada and 13 in the United States). Several more occurred in the 1980s (Davie 1984, Banci 1989, Elowe 1989, Rego 1989, Roy 1991). The results of only a few introductions have been published; many exist as either unpublished file reports or undocumented recollections (Berg 1982). Successful reintroductions have depended on the presence of suitable habitats (Banci 1989), protection from trapping (Irvine et al. 1964, Weckwerth and Wright 1968), and sex ratios that were equal or that favored females (Petersen et al. 1977, Berg 1982). Also, in successful translocations, the release sites were stocked with food such as deer (*Odocoileus virginianus*) and beaver (*Castor canadensis*) carcasses (Irvine et al. 1964, Churchill et al. 1981); fishers were quickly transported to the release sites, and, after an acclimatization period (Davis 1983), were released over a limited and intensive period (Petersen et al. 1977). Berg (1982) also recommended that release sites be spaced to minimize crowding.

Despite this extensive experience, the best time to release fishers is not known. The times of many releases were not reported (Dodds and Martell 1971, Brander and Books 1973, Cottrell 1978, Wallace and Henry 1985). Most reintroductions occurred in winter (Berg 1982) for no apparent reason other than seasonal availability of animals (Weckwerth and Wright 1968, Pack and Cromer 1981, Strickland and Douglas 1981). Winter releases may not be advantageous because fisher activities may be impeded in snowy areas (Cahalane 1961, Leonard 1980, Raine 1983, Banci 1989) and the animals may be more vulnerable to predation (Roy 1991). Roy (1991) also reported that fishers released in January and February abandoned his study area in spring. He suggested that winter releases did not allow sufficient time for the development of home ranges prior to the long-distance movements of the March–April breeding season, and he recommended fall release programs. A fall reintroduction in Alberta, however, was also accompanied by long-distance movements (15–100 km) and abandonment of the release sites (Davie 1984).

Irvine et al. (1964) suggested live-trapping fishers in March because of the mating period and milder weather. Douglas and Strickland (1987) also recommended late winter releases because females are nearing term and may be more likely to den close to the release site, thus establishing site recognition for their young. Fishers translocated in March, however, are also known to wander away from their release sites (Weckwerth and Wright 1968, Roy 1991). No release program has been conducted in summer (Berg 1982, Banci 1989).

The objective of our study was to assess and compare the post-release movements of fishers reintroduced at breeding time in March, when there is reduced vegetative cover, with those of fishers reintroduced in June, after the breeding season, when deciduous trees are in leaf.

Study Area

The release program was conducted in the Elk Island National Park–Blackfoot Recreation Area–Ministik Lake Bird Sanctuary complex, a 400-km^2 area near the City of Edmonton (Fig. 13.1). The relief is gentle, with a few groups of widely separated low hills. This region is part of the central Parklands, which consist of open grassland alternating with groves. The groves comprise predominantly quaking aspen (*Populus tremuloides*), together with balsam poplar (*Populus balsamifera*), willows (*Salix* spp.), white

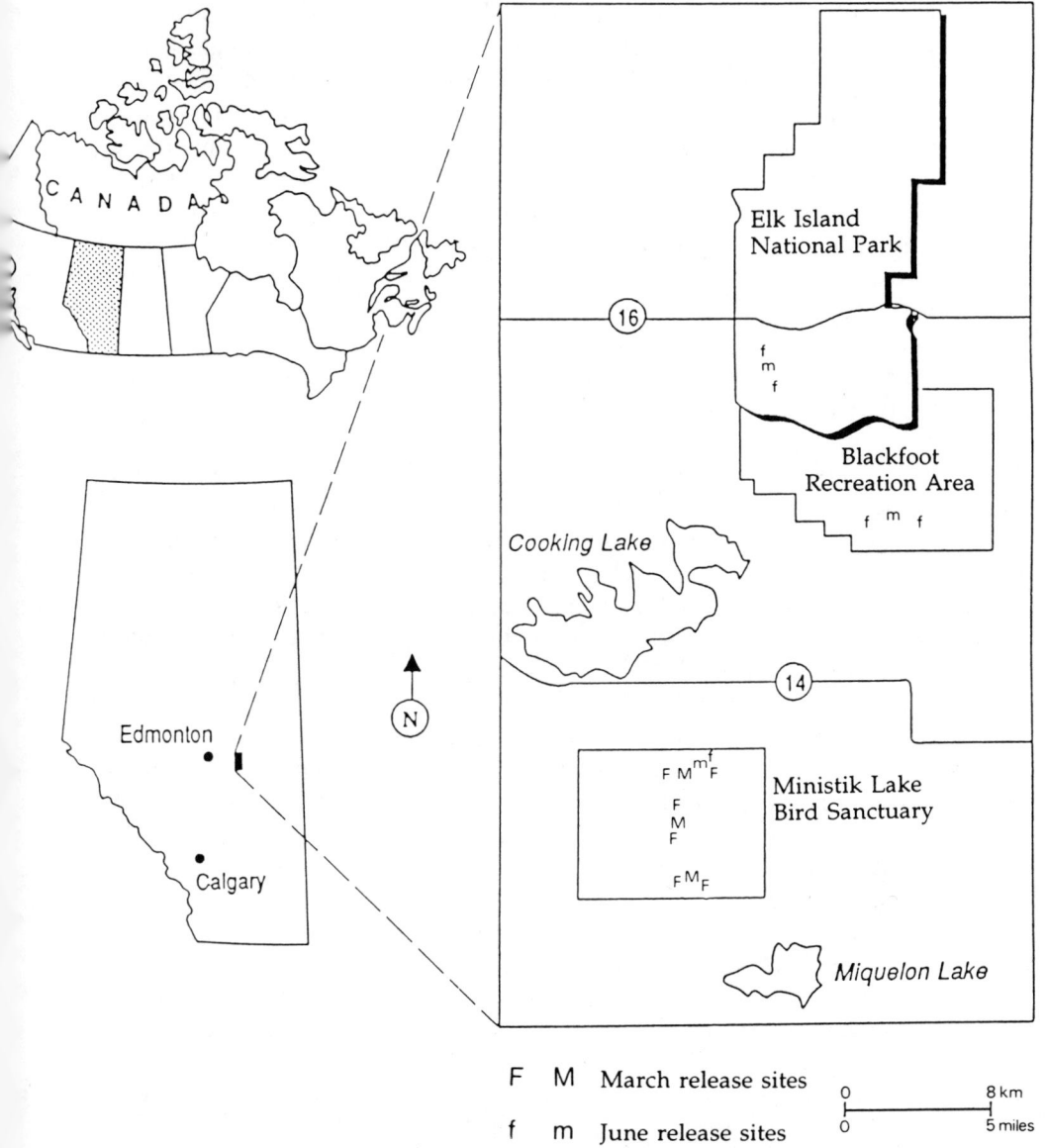

F   M   March release sites

f   m   June release sites

Figure 13.1. Sites where 10 fishers (*Martes pennanti*) were released in the parklands of Alberta in March and June 1990.

birch (*Betula papyrifera*), and white spruce (*Picea glauca*) (Hardy Assoc. Ltd. 1986, Looman and Best 1987). The study area is dominated by deciduous forest stands. The Ministik Lake Bird Sanctuary also has large mixed coniferous-deciduous forest stands characteristic of the boreal forest (Rowe 1972). The area has dense populations of cervids (an important source of carrion), small mammals, porcupines (*Erethizon dorsatum*), and waterfowl. The whole complex is partially or fully protected from human use. No commercial trapping is allowed in the park. In the Blackfoot and Ministik areas, beavers (*Castor canadensis*) and a few coyotes (*Canis latrans*) are harvested annually.

Methods

The release program involved 17 fishers captured in Ontario and Manitoba (3 were born in captivity). The animals were held in captivity in Vegreville, Alberta, for 18–24 months for behavioral studies. Observations in enclosures before release did not reveal any anomaly in the animals' behavior. Regardless of their origin, all the animals avoided human beings.

Nine animals (three groups, each with two nonpregnant females and one male) were released in March, in mixed forest stands of the Ministik Sanctuary. In June, one group of two females and one male was released in the park, and another such group was released in the Blackfoot area. The release sites consisted of deciduous forest stands. One male and one female were also released in Ministik mixed stands. All the March and June release sites were in continuous forest stands. The fishers were kept at the sites for three to five days in their respective nest box and connecting wire-mesh holding pen. Within a group, adult females were released one day before the males in the hopes that they would scent-mark the grounds and draw the males toward their vicinity. Males were released between female release sites (Hobson et al. 1989). Beaver meat was kept in the nest box during the week following the release of the animals. In March, four more beaver carcasses were left near each nest box in an attempt to keep the fishers in the vicinity.

All the fishers were ear-tagged and equipped with SMRX-3 radio transmitters with STO-1 mortality sensors (Lotek Engineering Inc., Aurora, Ont.). Because of mortalities, losses of collars, and time constraints, only the movements of two males (one born in captivity) and three females released in March were compared with those of two males (one born in captivity) and three females released in June (Fig. 13.1). The animals were monitored at least once a week, with two radiolocations taken within 10 minutes of each other, on the ground. Most fixes were taken 2 km or less from the animals. Telemetry error was assessed by comparing known locations of transmitters with those determined by triangulation. We were able to determine the position of a fisher within 25 m of its true position when the fixes were obtained 1 km or less from the animal. At 2 km, error increased to 185 m. We discarded incorrect readings resulting from signal bounce (Lee et al. 1985).

The distances between the animals and their release sites were determined 24 hours and four to seven days after release in March and June. From the beginning of the second week to the end of the fourth week, the animal movements varied greatly. During this period, several fixes were taken and all were used in a comparison of the release groups and the sex classes. Because of the small sample size, the Mann-Whitney U test (an excellent alternative to the parametric *t* test; see Siegel 1956) was used to test whether the distance between the animals and their sites differed between March and June, and between sexes.

We sampled habitat use by determining forest types at fisher radiolocations. The forest (total canopy $\geq$ 50%) surrounding each location was classified during ground surveys as deciduous (>75%), coniferous (>75%), or mixed (neither type > 75%). Locations could also be classified as scrub (total canopy < 50%), grassland (open field and pasture), and wetland (bog, meadow, or shallow marsh with emergent vegetation) (Arthur et al. 1989*b*). Forested stands of 1 km² or more were considered continuous areas. Smaller stands were classified as woodlots.

Results

Distances between Radiolocations and Release Sites. Twenty-four hours after their release, fishers were significantly ($P < 0.05$) further away from their release sites in March than in June (Table 13.1). Radiolocations taken four to seven days after the release indicated that March-released fishers continued to move away from their release sites and were found at distances greater ($P < 0.05$) than those of the June-released animals (Table 13.1).

From the beginning of the second week to the end of the fourth week following the animals' release, 18 and 23 radiolocations were recorded for the fishers released in March and June, respectively. Again, the distance between the animals and their release sites was greater ($P < 0.05$) for the March-released fishers than the June-released ones (Table 13.1). All the males released in March were found more than 30 km away from release sites. In contrast, the June-released males moved less (i.e., 0.9–16 km, $P < 0.05$). All the females released in March were more than 9 km away from release sites. The June-released females remained closer to release sites, 8 of 12 radiolocations being less than 9 km away. All the locations of the males released in March were farther away than those of the females. Males and

Table 13.1. Distance between fishers (*Martes pennanti*) and their release sites at three periods after their release

		Distance from release site (km)					
		24 hours		4–7 days		2–4 weeks	
Release date	*n*	$\bar{x}$	Range	$\bar{x}$	Range	$\bar{x}$	Range
March 1990							
Males	2	4.8	1.1–8.5	24.3	22.5–26.0	39.1	30.2–72.5
Females	3	5.4	3.8–6.0	22.3	14.9–29.6	15.1	9.8–17.7
All	5	5.1	1.1–8.5	23.3	14.9–29.6	23.1	9.8–72.5
June 1990							
Males	2	0.6	0.1–1.1	3.7	2.3–5.0	9.6	0.9–16.0
Females	3	2.2	0.8–3.6	5.1	4.1–6.3	6.2	2.3–12.1
All	5	1.6	0.1–3.6	4.5	2.3–6.3	7.8	0.9–16.0

females released in June, however, were found at similar ($P > 0.05$) distances from their release sites.

Habitats. We visited the sites of 16 radiolocations of March-released fishers. Twelve sites were 0.4 km^2 or less: 7 deciduous woodlots, 1 mixed woodlot, 1 patch of scrub, 1 marsh, and 2 grasslands. The other 4 sites were part of the continuous deciduous forest. In contrast, 13 of 16 June-released locations (15 in deciduous forest and 1 in wetland) were in continuous forest.

Discussion

Fisher reintroduction programs are expensive, and one needs to carefully determine where and when they must be carried out. Irvine et al. (1964) suggested March as a time to live-trap and release fishers because of the mating period and milder weather. Others suggested that the animals be "slow" released—that is, after approximately five days at the release site—to acclimatize the animals to their new surroundings, and that food such as beaver carcasses be available at the release sites (Davis 1978, Churchill et al. 1981, Berg 1982). In March, we followed all these recommendations. We also released the animals in mixed forest stands that are known to be used by fishers (Strickland et al. 1982*b*) and are among the most untouched environments of the region (Anonymous 1989). Despite all these precautions, our fishers traveled long distances and left the study area less than a month after their release.

The long-distance movements of March-released fishers may well have been related to breeding (Coulter 1966, Kelly 1977). Roy (1990, unpubl. rep.) also reported long movements by fishers released in January and March; his animals never returned to the study area and he concluded that the release dates did not allow sufficient time for the development of home ranges before the long-distance movements associated with breeding. These long movements may also have been related to exploration of unfamiliar environment, at a time of the year when the vegetation cover is reduced and food sources are limited. Our fishers left the deciduous forest, crossed roads, grasslands, and wetlands, and often relocated in woodlots. With reduced canopy, the animals may have been more vulnerable to predation and humans. Four March-released fishers, other than those studied here, died during the first month of their release of injuries received in intra- and interspecific fights (as determined by the wounds, the proximity of two male fishers, and the presence of predators) and in a car collision.

In June, there was a dense canopy and abundant food sources, such as fruits, small mammals, birds, eggs, and frogs. Furthermore, the animals were not searching for mates. Perhaps for these reasons, the fishers remained in the release area. Because of the increased vegetation cover, the animals could more safely move around, and no deaths were recorded. Our regular follow-up on the animals' movements indicated that six months after their release, the fishers were still in the vicinity of their release sites. Of course, in future breeding seasons, these animals will move about in search of mates. By then, however, the animals will have established home ranges. Several of our fishers already use areas that are close to each other, and we expect that the movements associated with the breeding season may be limited to our study area.

Our data suggest that fishers introduced in June are more likely to settle in the vicinity of their release sites than those released in March. Females captured in March, before parturition, however, could be kept in captivity until June and be released with their young. Hobson et al. (1989) showed that under these circumstances, female American martens (*M. americana*) remained near their release sites. We hope our conclusions will be tested further through comparison of release times in different ecosystems.

Acknowledgments

We thank the staff of the Elk Island National Park and Alberta Fish and Wildlife Division for their cooperation and D. Berndt and T. McKenzie for their technical help. This study was supported by the Fur Institute of Canada, the Alberta Fish and Wildlife Division, the Alberta Research Council, and the Alberta Agriculture Department.

14 Age and Sex Determination for American Martens and Fishers

Kim G. Poole, Gary M. Matson, Marjorie A. Strickland,
Audrey J. Magoun, Ron P. Graf, and Linda M. Dix

Knowledge of the age of an individual mammal or the age structure of a population is an essential component of population and harvest management (Johnston et al. 1987). This information is especially important when dealing with American marten (*Martes americana*) and fisher (*M. pennanti*) populations. Age and sex ratios in harvests of these species provide an indication of harvest intensity (Strickland and Douglas 1987; Douglas and Strickland 1987; Strickland, this volume). Aging is also used to determine age-specific fecundity and mortality in order to understand population dynamics (Krohn et al., this volume; Powell, this volume; Strickland, this volume). In areas where large marten harvests are examined each year, the ability to reliably determine sex and age of each specimen by examination of heads alone can save considerable time and expense. Aging and sexing techniques have been reviewed briefly for American martens (Strickland et al. 1982a:607–609, Strickland and Douglas 1987:539–540) and for fishers (Strickland et al. 1982b:594–595, Douglas and Strickland 1987:521–522). Johnston et al. (1987) discussed aging of furbearers using tooth structure.

In this chapter we review techniques currently used to age and sex American martens and fishers, and we report on the methodology, time- and cost-effectiveness, and reliability of these techniques. The chapter combines a review of techniques and results of original research on three methods of age determination (counts of cementum annuli, percent pulp cavity in canine teeth, and degree of temporal muscle coalescence) and three methods used to determine sex from skull or tooth measurements. We conclude with guidelines for using the various techniques and suggestions for future research. The techniques reviewed here may be applicable, with appropriate testing, to Old World *Martes*.

204

We define *accuracy* as the agreement between the true and the estimated age or sex of the animal. Because the chronological age is seldom known, we use *reliability* to mean agreement between the age estimated by the technique in question and the age derived from cementum analysis. *Precision* is the variability or repeatability of one or more estimates—for example, between two teeth from the same animal or between two observers examining the same section. Because of problems with accuracy, we have reported statistics in a conservative manner; see the discussion of statistical analysis by Johnston et al. (1987:237–238).

Age Determination

Young American martens often attain adult length and weight by early fall (Brassard and Bernard 1939, Lensink 1953) and fishers reach adult size by midwinter (Rego 1984, Douglas and Strickland 1987); thus techniques other than gross examination are necessary to categorize both species into age classes: juvenile (<1 year) and adult (≥1 year) (Tables 14.1, 14.2). Because up to 70% of most harvests are juveniles, using these techniques to separate age classes can save considerable time and expense. Cementum analysis is the only technique that provides an estimate of age of adult animals. It can also be used to age live-captured animals if the first premolar (pM1) is removed from anesthetized individuals.

Counts of Cementum Annuli

Although cementum analysis has been used for many species (Larson and Taber 1980, Johnston et al. 1987), no published studies have verified the technique in American martens and only one has studied its accuracy for fishers (Strickland et al. 1982c). Animals that had teeth removed at known intervals of one to three years provide material for determining characteristics of annual layers, but not for evaluating the accuracy of the technique. Although we had no martens of known age to assess accuracy, our study determined the precision attained when using two observers and two or more tooth types to age the same marten.

Methods. Teeth were obtained from 86 American martens harvested in five geographic locations: Oregon (*n* = 6), Alaska (*n* = 10), Northwest Territories (*n* = 6), Ontario (*n* = 59), and New Brunswick (*n* = 5). Animals of both sexes, primarily adults, were selected for the study. Duplicate sets of canine and premolar (pM3 and pM4) teeth from each marten were processed

Table 14.1. Techniques for age estimation of American martens (*Martes americana*)

Method	Distinguishes:	Reliability	Cost and time	Reference
Pelage characters	Juveniles from adults	Useful until late summer	Inexpensive, rapid	Newby & Hawley 1954
Tooth replacement	Juveniles from adults	Useful until 18 weeks	Inexpensive, rapid	Brassard & Bernard 1939
Fusion of epiphyses	Juveniles from adults	Useful to end of October	Inexpensive, moderate	Dagg et al. 1975
Suprafabellar tubercle	Juveniles from adults	Useful year-round	Inexpensive, moderate	Leach et al. 1982
Cranial sutures	Juveniles from adults	Most fused by 12 months	Inexpensive, slow	Brown 1983
Sagittal crest (cleaned skulls)	Juveniles from adults	Useful for males, not useful for females, too variable	Inexpensive, slow	Marshall 1951*b*, Brown 1983
Tooth wear	Juveniles from adults	Not useful; wear variable	NA	Strickland & Douglas 1987
Nipple development	Nonparous from parous females	6 of 44 correct	NA	Strickland & Douglas 1987
Baculum weight	Male juveniles from adults	Useful for males during trapping season	Inexpensive, slow	Marshall 1951*b*, Brown 1983, Strickland & Douglas 1987
Percent pulp cavity in lower canines	Juveniles from adults	Good	Moderate, moderate	Dix & Strickland 1986*a*, Nagorsen et al. 1988
Temporal muscle coalescence	Juveniles from adults	Good	Inexpensive, rapid	A. J. Magoun et al., unpubl. data
Counts of cementum annuli	Ages of adults	Good	Expensive, slow	Strickland & Douglas 1987

Note: NA = not applicable.

206

Table 14.2. Techniques for age estimation of fishers (*Martes pennanti*)

Method	Distinguishes:	Reliability	Cost and time	Reference
Pelage characters	Juveniles from adults	Useful until late summer	Inexpensive, rapid	Douglas & Strickland 1987
Tooth replacement	Juveniles from adults	Useful until about 7 months	Inexpensive, rapid	Douglas & Strickland 1987
Fusion of epiphyses	Juveniles from adults	Useful until about 7 months	Inexpensive, slow	Wright & Coulter 1967, Dagg et al. 1975
Suprafabellar tubercle	Juveniles from adults	Useful year-round	Inexpensive, moderate	Leach et al. 1982
Cranial sutures	Juveniles from adults	Some fused before trapping season	Inexpensive, slow	Coulter 1966, Dagg et al. 1975
Sagittal crest (cleaned skulls)	Juveniles from adults	Useful for males, variable for females	Inexpensive, slow	Coulter 1966, Douglas & Strickland 1987
Baculum weight	Male juveniles from adults	Useful until at least 1 January	Inexpensive, slow	Wright & Coulter 1967, Douglas & Strickland 1987
Percent pulp cavity in lower canines	Juveniles from adults	Good	Moderate, moderate	Kuehn & Berg 1981; Jenks et al. 1984, 1986; Dix & Strickland 1986b
Counts of cementum annuli	Ages of adults	Good	Expensive, slow	Strickland et al. 1982c, Douglas & Strickland 1987

by two laboratories, Matson's Laboratory (Milltown, Mont.) and the Algonquin Region Laboratory, Ontario Ministry of Natural Resources (OMNR). One laboratory processed teeth from the left mandible and the other from the right mandible of the same animal. The OMNR laboratory also processed the first (pM1) and second (pM2) premolar from some of the martens. The teeth were decalcified, embedded, and sectioned longitudinally on the midsagittal plane of the root (Humason 1972, Johnston et al. 1987). Sections were examined with a compound microscope, using transmitted white light and magnifications of 50–400×. The animals' ages were estimated from tooth sections by two technicians working independently: L. Dix (OMNR) was more experienced with pM3 and pM4 (>6000 examined), whereas G. Matson (Matson's Laboratory) was more experienced with canines (>10,000 examined). Dix preferred the pM4 because of the clarity of the cementum annuli in the arch between the two roots. Matson preferred the canine tooth because of its thicker cementum. These observers also subjectively ranked their confidence in each age estimate ("A": reliable, "B": some doubt, "C": unreliable) based primarily on the distinctness of the cementum annuli. Ages with "C" rating were not analyzed further.

Results: Agreement between Technicians. When the two observers estimated the age of the same tooth section, they agreed in almost 80% of the pM2, pM4, and canine samples but in only 53% and 67% of the pM1 (n = 15) and pM3 (n = 67) samples, respectively. Agreement between observers for pM2 was 78% (n = 41); for pM4, 79% (n = 146); and for canines, 78% (n = 101). When the technicians examined the tooth type with which they had the greatest experience, their age estimates for the same marten agreed in 73% of the samples (Fig. 14.1). Greatest agreement occurred in the juvenile and yearling age classes (84%). About 75% of the samples were estimated as 2 years or less and about 50% were aged as 1 year old (Fig. 14.1). Most frequently, the difference between the ages determined by the two technicians was within one year. Where differences occurred, age estimates by Dix generally were older than those of Matson (Fig. 14.1).

Results: Agreement among Tooth Types. Ages agreed in 69–80% of the paired comparisons of three tooth types; 88–100% of the ages agreed within one year. Agreement between pM3 and pM4 was 80% (n = 147); between canine and pM3, 69% (n = 137); and between canine and pM4, 69% (n = 139). When three teeth from the same animal were compared, only 62% of the ages agreed in all three teeth (n = 128). However, 95% of the age estimates agreed in two of the three teeth (n = 130).

AGE IN YEARS (DIX USING FOURTH PREMOLAR)

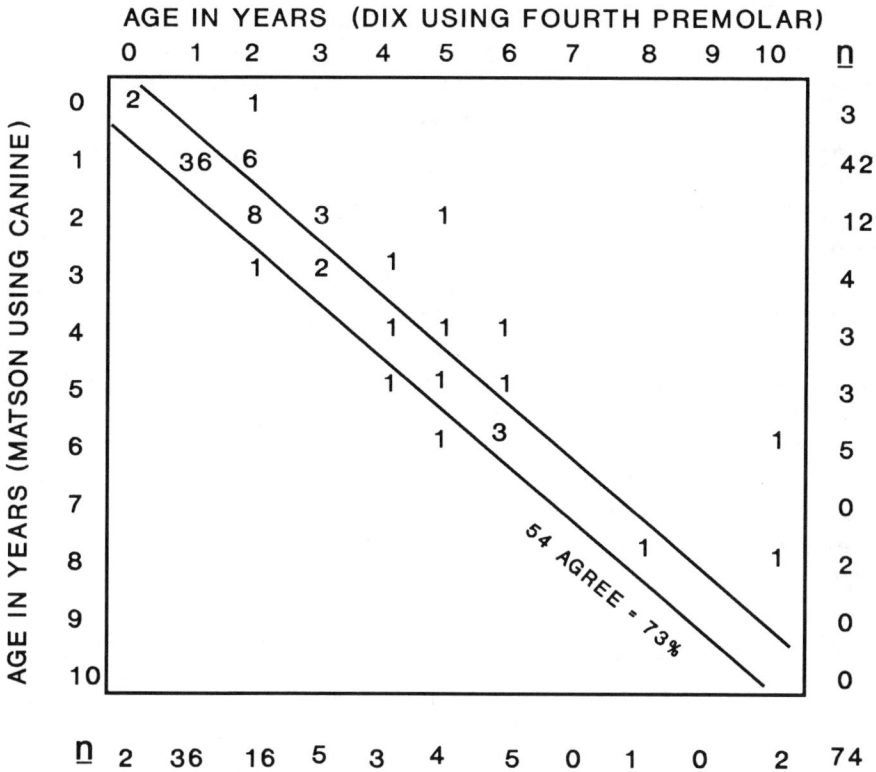

Figure 14.1. Agreement in ages when two technicians aged the same American marten (*Martes americana*) using the tooth types with which they were most experienced. Both technicians aged teeth prepared by their own laboratory. All sections with low reliability ("C") ratings were deleted.

Results: Other Variability. Our sample sizes were too small to examine variation in tooth section characteristics related to geographical area. American martens from Oregon, however, appeared to have the least distinct annuli, and martens from the Northwest Territories had the most distinct annuli. More extreme climates in the north have been thought to contribute to more distinct annuli, but the relationship between environmental factors and cementum annuli is not well understood (Johnston et al. 1987). No differences were observed in the characteristics of tooth sections of males and females. In confidence rankings, Dix assigned more "A" rankings to females, but Matson gave more to males.

Discussion. Disagreement in age estimates is evidence that the estimates of one or both technicians were in error. Consistency in the direction of the disagreement suggests that they may have used different criteria to determine age. Precision can be improved by standardizing methods and creating aging models—detailed written descriptions of the characteristics of annuli and how to interpret them—to ensure consistency in interpreting annuli.

This study shows that two technicians examining the same tooth section, or one technician examining any two teeth from the same animal, or two technicians working with the teeth with which they have the most experience, produced the same age estimate 80% of the time at best. Because precision was higher in young martens and 50% of the study sample was aged as yearling, precision likely would be lower if the sample contained more older animals. At the conclusion of this study, Matson agreed with Dix that pM4 had the most distinct annuli and is preferable over other teeth for aging. Use of pM1 teeth provided the least agreement between technicians and should be restricted to age estimates of live animals (cf. Arthur et al. 1992).

Few specimens of known age are available to test the accuracy of cementum aging. Also, there are several problems associated with using known-age material for validation. First, the samples often lack technical uniformity, having been obtained over more than one season and with different laboratory techniques. Second, animals of known age are often raised in captivity, and so cementum development may differ from that in wild animals. Third, knowledge of true age of a specimen could bias an observer who is interpreting sections (Johnston et al. 1987:236). We recommend that each laboratory using cementum annuli for age estimation test its own precision and accuracy. Tests might include examining teeth from animals of known age, working with "blind" duplicates (two teeth from the same animal), and reexamining the same tooth section several times.

Although cementum analysis is the only technique that provides age estimates of adult martens and fishers, the cost and time required is high relative to techniques that determine only age class. Initial equipment costs, complexity of preparation, and the experience required to produce precise results will discourage some researchers from setting up their own facilities. Commercial facilities are available, however, that will section and age teeth for US$2.50–3.25 per tooth (1993 costs), which is reasonable for many studies.

Pulp Cavity in Canine Teeth

Age classification using radiographs of canine teeth to measure the ratio of pulp cavity width to tooth width has been described for many species (Dix and Strickland 1986a, Johnston et al. 1987). The pulp cavity narrows with age as a result of deposition of dentine (Johnston et al. 1987); this deposition occurs in martens to at least 3 years of age (Dix and Strickland 1986*a*, Nagorsen et al. 1988). Percent pulp has been used successfully to separate juvenile and adult age classes in American martens (Dix and Strickland 1986*a*, Nagorsen et al. 1988) and fishers (Kuehn and Berg 1981; Jenks et al. 1984, 1986; Dix and Strickland 1986*b*), although overlap in tolerance limits does not allow separation of adult age classes (Dix and Strickland 1986*a*,*b*; Jenks et al. 1986; Nagorsen et al. 1988). Here we summarize further testing on the technique as applied to martens from northwestern Canada.

Methods. Marten carcasses ($n = 2132$) were collected from trappers in five areas of the Northwest Territories during the 1988–1989 and 1989–1990 trapping seasons (25 October to 15 March). Trapping areas sampled were up to 1200 km apart, from the Colville area (67°N, 127°W) to east of Fort Smith (60°N, 110°W). Lower canines were extracted by simmering the mandible in hot water (90–95°C) for one hour or less. A canine from each specimen was glued buccal side up on 190 × 270 mm file folders, up to 140 per sheet, and radiographed with a Senograph 600T Mammo Unit and Kodak Mammography film exposed at 30 Kv and 7 Mas. Following procedures outlined by Dix and Strickland (1986*a*), we measured maximum pulp cavity width and tooth width on the radiographs (Fig. 14.2) using a Canon microfiche reader. The radiographs were magnified 23.5×, and the images were measured to the nearest 0.1 mm using dial calipers. Percent pulp was calculated as pulp cavity width/tooth width × 100. Approximately 2% of the images were unreadable. To examine the precision of the technique and compare measurement methods, two experienced technicians independently measured a sample of 240 canine teeth collected from the Colville area in 1990–1991; they used both the microfiche reader and a 12× dissecting microscope equipped with a micrometer eyepiece.

The ages of marten canines were estimated by cementum analysis at a commercial laboratory (Matson's Laboratory, Milltown, Mont.) after pulp cavity measurement. To save money, we classified approximately 25% of the animals, those with no temporal muscle coalescence (TMC) and more than 52% pulp cavity, as "definite" juveniles and were not aged by cementum analysis. These divisions were based on previous results (K. Poole and R. Graf, unpubl. data) and published literature (Dix and Strickland 1986*a*, Nagorsen et al. 1988). Only 2 of 170 teeth with more than 52% pulp cavity that were aged by cementum analysis were not from juveniles, and 1 of these 2 had no TMC, suggesting that it actually was a juvenile tooth. This finding provides support for use of more than 52% pulp cavity measurement and no TMC for classifiying "definite" juveniles, and it suggests that use of these

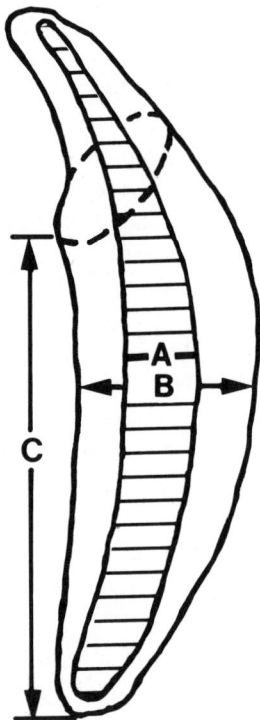

Figure 14.2. Measurements taken from radiographs of lower canine teeth of American martens (*Martes americana*). A = maximum pulp width, B = tooth width, C = root length. For sex determination, tooth width and root length can also be measured from extracted teeth.

animals in calculation of the dividing points likely contributed a marginal degree of error to the study. All borderline or questionable specimens were aged by cementum analysis, and some specimens had their age estimated twice if the first cementum age did not correspond with the other age class determination techniques.

For each trapping area, dividing points to separate juveniles from adults were calculated for males and females using the formula of Dix and Strickland (1986a):

$$D = \bar{x}_a + [(\bar{x}_b - \bar{x}_a)(\text{SD}_a)/(\text{SD}_a + \text{SD}_b)]$$

where SD is the standard deviation of each group, $\bar{x}$ is the mean of each group, and $\bar{x}_a < \bar{x}_b$.

Results. A brief visual inspection of the radiographs was often sufficient to identify juveniles (Fig. 14.3). On the basis of measurements of percent pulp cavity, an average of more than 90% of juveniles and more than 97% of adults were correctly assigned to age class (Table 14.3). Since none of the animals were of known age, it is possible that some of the detected "errors"

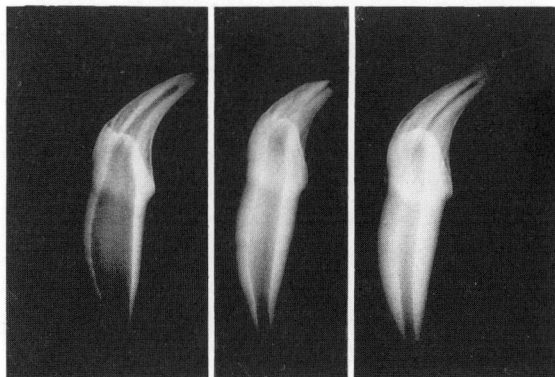

Figure 14.3. Radiograph showing the pulp cavity of (*left to right*) juvenile, yearling, and adult (2-year-old) male American martens (*Martes americana*).

may have resulted from inaccurate cementum aging. Dividing points for separating juveniles from adults were higher in males (range: 46.5–52.2%) than in females (43.3–46.6%). In juveniles, percent pulp decreased significantly as the season progressed ($P < 0.001$), decreasing on average 1% per 15 days in males and 1% per 21 days in females. During the Northwest Territories' 140-day trapping season, this could result in a decrease in average percent pulp in juveniles of 9% in males and 6.5% in females.

Differences in percent pulp cavity estimated by the two technicians were less with the microfiche reader ($\bar{x} = 1.0\% \pm 1.04\%$ [SD]) than with the dissecting microscope ($\bar{x} = 2.4\% \pm 1.94\%$). Within each method, measure-

Table 14.3. Lower canine pulp cavity criteria for separating juvenile and adult American martens (*Martes americana*), Northwest Territories, 1988–1989 and 1989–1990

Area	Sex	Dividing point (% pulp cavity/ tooth width)	Juveniles aged correctly %	Juveniles aged correctly n	Adults aged correctly %	Adults aged correctly n
Colville	M	52.2	84.6	13	100	232
	F	46.2	96.7	30	98.7	156
Fort Rae	M	48.3	88.2	17	95.6	45
	F	43.5	95.5	22	97.6	42
Fort Simpson	M	50.4	100	11	100	80
	F	46.3	95.7	47	100	44
Trout Lake	M	48.1	88.9	18	97.0	66
	F	46.6	98.2	56	95.5	44
Fort Smith	M	46.5	92.3	13	97.6	82
	F	43.3	94.4	36	92.9	42
TOTAL	M		90.3	72	98.8	505
	F		96.3	191	97.6	328

Note: Juveniles ≥ dividing point; adults < dividing point.

ments obtained by the technicians resulted in disagreement in assigning age class in three (microscope) and one (microfiche) of the 240 teeth examined, or less than 1.3% of the sample.

Discussion. Measurement of percent pulp cavity often allowed more than 95% of the marten samples to be correctly identified as juveniles or adults. The percentage of juveniles aged correctly was almost certainly underestimated because the majority of juvenile teeth with wide pulp cavities were not subjected to cementum annuli counts. The degree of reliability we obtained is similar to that of Nagorsen et al. (1988), although it is generally not as high as the more than 98% accuracy ($n = 737$) reported for an Ontario study (Dix and Strickland 1986a). As noted by Nagorsen et al. (1988), some of this variation may have resulted from different trapping season lengths and the corresponding decrease in percent pulp during this period. The trapping season in Ontario is November and December only, whereas that in the Northwest Territories runs from late October to mid-March. Where collection date is known, grouping martens into monthly samples may improve the effectiveness of the technique (Nagorsen et al. 1988). The precision of the technique was high but was influenced by the measurement method used. Using a microfiche reader appears to be the better of the two measurement methods examined.

The pulp cavity criterion of 46–52% for males and 43–46% for females that we used to distinguish juveniles from adults in martens from the Northwest Territories are considerably higher than the cut-off levels used elsewhere (males 30–38%, females 27–34%; Dix and Strickland 1986a, Nagorsen et al. 1988). This difference may be a result of slower development by martens from the Northwest Territories or of a later birth date, which would give juvenile martens less time to develop before the start of the trapping season.

The only published account that provides quantitative data using this technique for fishers comes from a sample of 440 fisher carcasses collected in Maine (Jenks et al. 1984). Application of this method gave juvenile-adult dividing points of 46.5% pulp cavity for males and 39.5% for females, which correctly aged more than 95% of the specimens as juveniles or adults (Dix and Strickland 1986b). Unpublished data on 810 Ontario fishers gave dividing points of 48.8% for males and 35.0% for females, which correctly classified 99% of the males and 100% of the females (M. Strickland, unpubl. data).

For both martens and fishers, the pulp cavity technique produces results that are acceptable for most management purposes. If a more detailed breakdown of the age structure of a group of carcasses is necessary, using radiographs to select out most of the juveniles will cut down on the number of samples needing cementum analysis, thereby decreasing costs. With cooperation of a local hospital, few costs are associated with this technique. We found that teeth from up to 200 martens per day could be extracted and mounted on file folders, and up to 400 teeth per day could be measured carefully. More teeth could be examined in a day if precise measurements were not required.

Because of regional differences, to obtain the greatest degree of reliability, one must calculate dividing points for percent pulp for each sex from a large sample of cementum-aged specimens for each region. Dividing points may also need to be recalculated if major changes in length or timing of trapping seasons occur. Radiograph quality must render images sharp and readable. We found that some teeth were unreadable because of poor image quality; these were generally on the edges of the plates. A small proportion of the teeth had abnormal histology, making measurement impossible and perhaps necessitating cementum analysis. These problems generally affected less than 2% of the samples.

Temporal Muscle Coalescence

Estimating *Martes* ages by using cementum annuli or measurement of pulp cavity requires that teeth be extracted and processed. To develop an inexpensive, simple, and relatively reliable technique to distinguish juvenile, yearling, and adult (≥ 2 years) age classes, Magoun et al. (unpubl. data) examined temporal muscle development. Temporal muscles, the major adductors of the jaw, originate from the dorsal cranium along the temporal ridges. In young animals of both sexes, the temporal ridges are widely separated; but as the animals mature, the temporal muscles and ridges coalesce and the latter form a sagittal crest (Marshall 1951*b*, Brown 1983). Length of TMC closely approximates the length of the sagittal crest. Magoun et al. (unpubl. data), in their preliminary study in Alaska, found that TMC correctly classified most juveniles and a large proportion of adults, but the yearling category could not be readily identified. They recommended further testing of the technique. Here we report on results of our examination of the technique, using martens from Alaska and the Northwest Territories.

Methods. A total of 670 marten carcasses collected from trappers in four areas in interior Alaska from 1983 to 1989 were examined. The four areas were separated by 600 km or less and were concentrated in the area between McGrath and the Brooks Range (63°–67°N, 154°–156°W). Marten carcasses from the Northwest Territories are described in the section on pulp cavity in canine teeth. Aging was performed by cementum analysis (Alaska sample

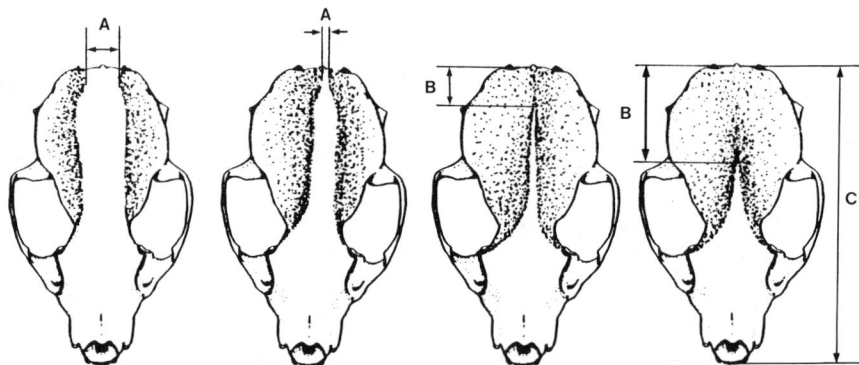

Figure 14.4. Skull measurements. A = width between temporal muscles (WBTM), B = length of temporal muscle coalescence (LTMC), and C = total skull length. Both LTMC and total skull length are taken from below the lambdoidal crest at the back of the skull. Drawing of American marten skull by Denise Casey, courtesy of Tim W. Clark and Denise Casey.

and most of the Northwest Territories sample) and percent pulp cavity. All carcasses were kept frozen until examination.

The fascia covering the skull was removed from martens in the Alaska study but not from those in the Northwest Territories sample. On skulls where TMC had occurred, the length of TMC (LTMC) was measured using dial calipers from the notch below the lambdoidal crest to the point where the temporal muscles begin to separate on the anterior cranium (Fig. 14.4). On skulls of females where TMC had not occurred, the minimum width between temporal muscles (WBTM) was measured with dial calipers (Fig. 14.4). Two technicians separately examined 87 skulls from the Northwest Territories. Data were plotted to examine the correlation between LTMC or WBTM and the age as derived from tooth examination. Five percent (107/2132) of the Northwest Territories martens and 1.6% (9/552) of a sample of the Alaska martens were unreadable because of damaged skulls.

Results. Measurements of LTMC were more reliable estimators of age class for males than for females (Tables 14.4, 14.5). Among males, results were similar between samples from the Northwest Territories and Alaska; more than 97% of males with LTMC less than 10 mm (Alaska) and less than 20 mm (Northwest Territories) were juveniles, whereas less than 7% of males with LTMC greater than 30 mm were juveniles (Table 14.4). Juvenile and adult martens were found in roughly equal proportions in the 10- to 30-mm LTMC classes in the Alaska sample and in the 20- to 30-mm class in the

Table 14.4.   Correct age classification of juvenile male American martens (*Martes americana*), by length classes of temporal muscle coalescence (TMC) in Alaska and the Northwest Territories (NWT), 1988–1990

Length TMC (mm)	Alaska		NWT	
	%	n	%	n
≤0	98.2	110	98.8	333
0–9	94.1	17	97.3	111
10–19	50.0	12	96.8	93
20–29	26.7	15	62.9	35
≥30	0	130	6.4	625

Northwest Territories sample. Males with LTMC of 10–30 mm comprised only 2% of 1451 harvested male martens in Alaska.

Whereas less than 3% of females with some measurable TMC were juveniles in the Alaska sample, 11% were juveniles in the Northwest Territories sample (Table 14.5). No obvious explanation of this difference is available; unfamiliarity with the technique by some technicians may have contributed some error. Over 93% of females with WBTM greater than 3 mm were classified correctly as juveniles. Females with WBTM of 0–2.5 mm were about evenly divided between juveniles and adults. Females in this range were about 13% of 1065 female martens collected from Alaskan trappers. Some other aging technique (percent pulp cavity or cementum analysis) would be required to properly classify this relatively large and important segment of the harvested population.

Table 14.5.   Correct age classification of juvenile female American martens (*Martes americana*), by length classes of temporal muscle coalescence (TMC) and width between the temporal muscles (WBTM) in Alaska and the Northwest Territories (NWT), 1988–1990

Class (mm)	Alaska		NWT	
	%	n	%	n
WBTM ≥ 4	100	109		
WBTM = 3	92.0	25	93.9	247[a]
WBTM = 2	71.4	35	56.0	25
WBTM = 1	33.3	36	50.0	4
Length TMC > 0	2.8	181	11.3	230

Note: WBTM narrows and becomes 0 when the temporal muscles coalesce and TMC is measurable.

[a]This cell represents WBTM ≥ 3 mm.

All but one of the Northwest Territories martens measured independently by two technicians were assigned to the same age class, although absolute measurements of WBTM varied (t-test, $P = 0.006$). These differences were attributed to difficulty in identifying the point of muscle separation or in determining whether TMC had actually taken place.

Regional differences in skull sizes or in growth rates may influence this technique. In a 1989–1990 study in interior northern British Columbia (55°N, 127°W) that used a dividing point of greater than 0 mm LTMC to identify adult females (no measurement of WBTM) and greater than 10 mm LTMC to identify adult males, 93–95% of age classes were correctly classified in a sample of 384 martens (D. Steventon, Wildl. Branch, Minist. of Environ., B.C., pers. commun.). A higher error rate was obtained in examination of a sample of Ontario marten carcasses using the British Columbia criteria: 55% of 81 adult (≥ 1 year) females had LTMC = 0, and 14% of 191 adult (≥ 1 year) males had LTMC less than 10 mm. However, 99% of 471 juveniles of both sexes were aged correctly by these criteria (M. Strickland, unpubl. data).

Discussion. Our results suggest that TMC can be used to distinguish juvenile from adult American martens, at least in northwestern North America. The method correctly identified juveniles from Ontario, but many adults, especially females, were not correctly identified. The reasons for these regional differences in the reliability of the technique are unclear. For any area where the technique has not been used, the technique should be compared with some other reliable aging technique and the appropriate classification criteria established.

The main advantages of this technique are its low cost and speed. Also, it is the only reliable technique that can be used by trappers in the field to monitor their harvest. Age classes showing the highest variability in TMC can be determined by other methods. Measurement of TMC, however, is difficult for the inexperienced technician or trapper, especially in female specimens where WBTM is a critical factor in reliable age determination, and training should be required. For female specimens in which the temporal muscles are almost touching, it is often difficult to determine whether TMC has actually occurred. The two temporal ridges must be fused and form a sharp edge for TMC to have occurred and be measured. This distinction is important because females with *any* TMC are classified as adults. Each group of researchers should standardize and test techniques for measuring TMC; otherwise, variation in observers' interpretations may confound their results. This technique requires great diligence, especially when measuring female martens. Dried-out skulls are often difficult to measure, and trappers should be encouraged to store collected skulls or carcasses in plastic bags.

We are not aware of any evaluation of this technique for fishers, but similar success might be expected. Males may be more difficult to age than females, because the temporal ridges coalesce in male fishers any time between 8 and 12 months of age.

We found that up to 5% of the heads examined showed hemorrhaging or fractures as a result of trappers hitting the animal on the head to kill it or because of contusion from the bars of some of the more powerful quick-kill traps. Such damage made measurement of TMC unreliable or impossible. Increased use of quick-kill traps may decrease the utility of this technique.

Sex Determination

Management agencies often must determine sex and age ratios of hundreds or thousands of American marten and fisher carcasses annually (Powell, this volume; Strickland, this volume). To reduce costs of collection, transportation, and processing, researchers have tried to determine sex from skull or tooth characteristics alone (Table 14.6). Because fishers have more sexual dimorphism than martens (Holmes and Powell, this volume), techniques involving skull or canine measurements generally have proven to be 100% accurate in determining sex of fishers but are slightly less reliable for martens. Here we examine the accuracy of three methods to determine sex from skull or tooth measurements in martens from the Northwest Territories.

Methods. Sex was determined by examining the genitalia of 2132 marten carcasses collected from trappers in the Northwest Territories. Total skull length on uncleaned skulls was measured from below the lambdoidal crest to the point where the incisors insert into the skull (Fig. 14.4). The width and length of the root of extracted lower canines (Fig. 14.2) were measured to the nearest 0.1 mm using dial calipers. Samples were separated by geographical area to account for population variation, and discriminant function analysis was used to identify the best determinant of sex. The "distance" between the sexes was calculated by dividing the difference between the sample means by the mean standard deviation (Snedecor and Cochran 1967:415) and was used to judge which variable had the highest degree of accuracy. This value must exceed 3.0 for a high degree of accuracy. The SAS-DISCRIM procedure (SAS Institute, Inc. 1988) produced a generalized squared distance between groups that also approximated this distance (Snedecor and Cochran

Table 14.6.　Techniques used to sex American martens (*Martes americana*) and fishers (*M. pennanti*) from skull and tooth characteristics

Method	% Accuracy	Cost and time	Reference
MARTENS			
Upper or lower canine root width	92.2–100	Inexpensive, moderate	Brown 1983, Dix & Strickland 1986*a*, Nagorsen et al. 1988; see text
Lower canine root length	97.7–98.1	Inexpensive, moderate	Dix & Strickland 1986*a*; see text
Total skull length or condylobasal length	99.3–100	Inexpensive, rapid	Giannico 1986; see text
Zygomatic width and most other skull measurements	95–100	Inexpensive, rapid	Brown 1983
FISHERS			
Width of lower canine	100	Inexpensive, moderate	Parsons et al. 1978
Length of lower canine	100	Inexpensive, moderate	Jenks et al. 1984, Dix & Strickland 1986*b*
Skull length, zygomatic width	98–100	Inexpensive, rapid	M. J. Strickland, unpubl. data

1967:415) and provided error estimates. Dividing points to separate sexes were derived from the formula of Dix and Strickland (1986*a*). Total skull length of 87 martens was measured independently by two technicians.

Reliability of sex determination from skull length was tested using 1005 martens from Alaska and 423 martens from northwestern interior British Columbia (D. Steventon, Wildl. Branch, Minist. of Environ., B.C., pers. commun.). In the Alaska sample, skull length was measured to the nearest millimeter, and the dividing point was based on frequency distribution to give the overall lowest error.

Results. In all samples from the Northwest Territories examined, total skull length was most accurate for determining sex (mean distance between sexes = 3.54, range: 2.91–4.28), followed by width of canine tooth (mean distance = 3.20, range: 2.63–3.90) and length of canine root (mean distance = 2.63, range: 1.58–3.27). Although both skull length and canine width exceeded the minimum standard required to determine sex accurately, total skull length was chosen as the preferred measurement because of its higher

Table 14.7. Total skull lengths for distinguishing male from female American martens (*Martes americana*) harvested in northwestern North America

Location	Dividing point (mm)	Males classified correctly		Females classified correctly	
		n	%	*n*	%
Colville	82.7	411	96.9	302	98.4
Fort Rae	82.5	91	100	77	100
Trout Lake	81.5	270	93.8	200	93.0
Fort Simpson	81.2	158	92.4	95	96.0
Fort Smith	80.5	255	93.1	145	94.8
TOTAL NWT		1185	95.0	819	96.2
Alaska	81.5	524	98.9	472	99.4
British Columbia[a]	79.6	265	98.5	150	97.4

Notes: Males ≥ dividing point; females < dividing point. NWT = Northwest Territories.
[a]D. Stevenson, Wildl. Branch, Minist. of Environ., B.C. pers. commun.

accuracy and ease of measurement; no teeth need to be extracted. The dividing points and accuracy of classification varied slightly among areas (Table 14.7). Skulls measured independently by two technicians were assigned the same sex in 100% of cases; the mean absolute difference between measurements was 0.51 ± 0.44 mm.

In a blind test, 97.9% of 243 martens from the Colville area examined in 1990–1991 were sexed correctly using a skull length of 82.7 mm or greater as the dividing point between for males and females.

Discussion. Total skull length is easy to measure and is highly accurate in determining sex for American martens. Total skull length can be measured with no preparation, saving considerable time. Because of between-area differences in skull size, region-specific dividing points for distinguishing sex by skull length should be determined.

Between-technician differences may have resulted from different pressure exerted on the calipers in the layer of muscle at the back of the skull. Standardization of all measurements and training of all those taking measurements should increase the precision of the technique. Condylobasal length may also be used to separate sexes (Giannico 1986). Although it may be a more precise morphometric measurement, condylobasal length is more time-consuming because each carcass must be cut into.

One advantage of determining sex from canine tooth width and age class from canine tooth pulp cavity is that both age and sex can be estimated from the lower jaw, negating the need to collect the entire skull.

Recommendations for Users

A clear understanding of the hypothesis being tested is required before deciding which techniques to use to determine the age and sex of *Martes* (Johnston et al. 1987). Techniques and efforts will vary, depending on whether researchers need to determine the sex and age classes of a large sample or the sex and age of an individual. The precision and accuracy needed for the hypothesis test being performed also must be considered. Because there has been little or no verification of any of the aging techniques on animals of known age, the accuracy of aging techniques for martens or fishers remains uncertain.

Management of the harvest of marten and fisher populations generally requires determination of adult and juvenile age classes by sex to monitor harvest intensity (Strickland, this volume). Thus, the sex and age class of a large number of carcasses must be determined rapidly and cheaply. Age class (juvenile or adult) can be estimated by percent pulp cavity in canine teeth or by measurement of TMC. The relatively small number of animals for which this method is not reliable can be aged using cementum analysis.

If increased reliability is required, a combination of techniques may be employed. In the Northwest Territories, for instance, both TMC and percent pulp cavity of all American martens are measured. Samples classed as juveniles by both methods are considered juveniles; ambiguous or borderline cases are aged by cementum analysis. Some analyses of age and sex ratios in the harvest require that females 2 years old or older be identified (Douglas and Strickland 1987, Strickland and Douglas 1987). To accomplish this, researchers conduct cementum analysis only for females that are not clearly juveniles according to TMC or percent pulp cavity determination.

Sex can be determined reliably from total skull or condylobasal length or from canine tooth width. Valuable information may be lost, however, if only heads are collected. Carcasses provide data on body condition (Buskirk and Harlow 1989), diet, parasite loads, and female reproduction. The last-named is important in the interpretation of sex and age class ratios and as a possible indicator of resource scarcity (Thompson and Colgan 1987a). The presence of corpora lutea can provide a partial check of aging. Although not all females at least 1 year old will have corpora lutea, any female with detectable corpora lutea should not be classed as a juvenile (Strickland et al. 1982a).

Given population differences in animal size, birth dates, and development, adequate specimens of known sex and cementum-aged samples must be examined in each new area to define dividing points between age classes and sexes. Snedecor and Cochran (1967:511–519) give some guidelines for sample size requirements. While measurement techniques for all methods should be standardized and consistently applied, special testing and care should be taken when dealing with TMC in females.

Animals of known age are needed to verify accuracy of aging techniques (Dapson 1980, Johnston et al. 1987). Young animals could be marked as kits in the den or captured early in the summer, when age can be positively identified using, for example, the presence of deciduous teeth. Carcasses recovered from animals of known age should be thoroughly examined by persons experienced in the various techniques to ensure maximum information is obtained from these important specimens.

Researchers should investigate other techniques of age classification. For example, using TMC would be more useful if a characteristic other than analysis of tooth structure could be identified to separate females that are difficult to classify. Aging methods that are reliable and simple enough for trappers to use in the management of their traplines are also needed. If selective trapping using livetraps is to be a useful management tool (Strickland, this volume), methods are needed to determine the sex and age of live animals in traps.

Acknowledgments

We acknowledge with thanks all the marten and fisher trappers and conservation officers across North America who provided and collected carcasses for these studies and the many people who worked on the carcasses in the laboratories. D. Steventon, Wildlife Branch, Ministry of Environment, British Columbia, contributed data from martens in his area that he aged and sexed. Marten teeth used in the cementum analysis study were contributed by the late D. Cartwright, New Brunswick Fish and Wildlife Branch, and by C. Trainer, Oregon Department of Fish and Wildlife. We thank T. Clark and D. Casey for permission to use their drawing of the marten skull. T. Holmes and D. Nagorsen provided comments on an earlier version of the chapter, and A. Harestad added much editorial polish to the finished product.

15 Techniques for Monitoring Populations of Fishers and American Martens

Martin G. Raphael

Fishers (*Martes pennanti*) and American martens (*M. americana*) are secretive forest-dwelling carnivores that are difficult to observe and study. Both are associated with late-successional coniferous forest (Buskirk and Powell, this volume), and martens are used as ecological management indicators of old-growth conditions in most national forests of the western United States that fall within the distribution of the species. Because martens are vulnerable to the loss of habitat through timber harvesting, reliable detection and monitoring techniques and protocols are needed so that agencies can watch for population declines and modify resource management practices to halt them when they occur. Monitoring techniques are also needed by landowners, researchers, and naturalists.

The current population status of American martens and fishers has been incompletely documented especially in the western United States (Gibilisco, this volume). Fisher populations are believed to be low and declining in the Sierra Nevada of California (Schempf and White 1977), but comprehensive surveys have not been completed there. Sighting and trapping records indicate that fisher populations are also small in Washington (Aubry and Houston 1993), Oregon, and the northern Rocky Mountains, leading to a recent petition to list the species as threatened (*Federal Register* 56[8]:1159–1161). Populations of American martens have declined on Newfoundland Island (Thompson 1991) but are believed to be relatively high and stable in suitable habitat elsewhere. Comprehensive field surveys have not been undertaken, however. The primary obstacle to such surveys has been the lack of reliable broadly applicable survey and monitoring techniques.

In this chapter I review available techniques for detecting American mar-

224

tens or fishers. I also discuss the relative advantages and disadvantages of each technique in relation to monitoring objectives.

Habitat Survey

The most cost-effective, but likely the least precise, assessment of population status of American martens and fishers is based on an inventory of land area covered by suitable habitat. This method, used by many national forests of the United States, rests on two assumptions: that habitat suitability is well enough known to describe stands that are likely to meet the life requisites of martens or fishers; and that the amount of habitat classified as suitable is directly correlated with population size. The first assumption can be partially addressed through wildlife-habitat relationship models such as the Habitat Suitability Index (HSI) (Allen 1982, 1983; Schamberger and Krohn 1982; Cole and Smith 1983), Habitat Capability Model (Nelson and Salwasser 1982, Hoover and Wills 1984), and PATREC (Williams et al. 1977, Grubb 1988). The second assumption should be addressed through statistical comparison of model-based habitat evaluations with empirical data on abundance of a species in an area (Marcot et al. 1983, Salwasser et al. 1983, Fagen 1988). Unfortunately, existing models have received only limited testing (Thomasma et al. 1991). These models, to be reliable, must be tested, revised, and tested again in more than one study area. I know of no models that have received such testing. In addition, the second assumption is valid only if habitat is limiting. If other factors, such as predation, limit the population, population size will not be correlated with amount of habitat, and results of field tests of model reliability will be misleading.

The primary advantage of these models is their potential for evaluation of large areas of land without costly fieldwork. Existing timber inventory data may be sufficient to classify habitat suitability. For example, the HSI model for the fisher depends on four stand attributes: percent canopy closure, average diameter of canopy trees, number of canopy layers, and percent deciduous species in overstory (Allen 1983). All these attributes are routinely collected by land management agencies or can be derived from data they collect. For the American marten, however, stand attributes should include percent ground surface covered by woody debris greater than 7.6 cm in diameter (Allen 1982), an attribute not routinely collected. Application of the marten HSI model may therefore require additional field sampling.

The disadvantages of model-based monitoring are the high risks inherent in using untested relationships and the low resolution at which existing models classify habitat. These two disadvantages are multiplicative: in combination they could result in a high probability of failing to detect real trends in population abundance.

Animal Survey

Harvest Records

A commonly used technique, especially among state agencies, is to infer population trend from commercial trapping results, either total catches or catches per unit effort (Douglas and Strickland 1987, Strickland and Douglas 1987). Such techniques are described in detail by Strickland (this volume). Harvest records can yield large sample sizes and statistically precise estimates of population parameters, especially when data are aggregated over broad regions. These samples can be used to infer relative productivity or abundance of *Martes* in broadly defined regions (e.g., fisher productivity in ecoregions of North America as summarized by Banci 1989). Comparisons of yearly catch, assuming that we know the harvest effort, may provide good estimates of regional population trends over time, in contrast with trends derived from site-specific studies, especially if trapping effort is well documented. Data from individual traplines can also be gathered and may yield estimates of population status in local areas.

On the negative side, harvest rates may not reflect true population size for a variety of reasons (Strickland, this volume). First, the size of the harvest depends on complex interrelationships among market condition (pelt prices), social forces (animal welfare, animal rights), general economic conditions (more effort by trappers when trapping income is more important), and environmental concerns. Second, unless regulations (e.g., in much of Canada) prevent him or her from doing so, a prudent trapper may move to a new area if capture rate declines. This results in a lag between changes in capture rate and true population size. Third, managers, especially in the United States, may not control trapping locations and intensity well enough to assure consistent effort across space and time. Without such control, use of harvest data for survey and monitoring is subject to biases of unknown magnitude. Thus, the value of harvest data depends on the specific regulations in effect for the geographic area of interest.

Live-Trapping

Estimates of occurrence or relative abundance may be obtained by live-traps deployed singly or in small clusters to determine the presence or absence of a species, or the traps may be arranged in larger transect lines or grids to estimate relative abundance. General guidelines for designing live-trap experiments and for analyzing results are available in other reports (Seber 1973, Otis et al. 1978, Pollock et al. 1990) but are not specific to *Martes*.

Live-trapping offers many advantages over other detection techniques. First, captured animals can be identified to species and marked and released for mark-recapture estimates of population size. Because animals are not removed from the population, intensive periodic trapping within a prescribed study area can lead to enumeration of nearly all individuals. If samples are large, sex ratio, age distribution, and reproductive status can be estimated, especially for studies lasting several years or more.

Disadvantages of live-trapping are primarily related to the costs of labor and materials. Capture rates are generally low, so large investments in time, traps, and personnel may yield only modest returns. Mean capture rates from 12 published studies of American martens varied from 0.25 to 3.36 individuals/100 trap-nights ($\bar{x} = 1.31$, SD = 1.09) (Table 15.1). Capture rates likely are lower for fishers, because of their lower density (Powell, this volume). Thus, many traps must be set to ensure adequate numbers of captures if the objective is to estimate population size. Unless the number of captures is high, abundance estimates will be biased and subject to large variances. In addition, individual animals may become trap-happy (repeatedly enter traps) or trap-shy (avoid traps after first capture), which further complicates analyses of capture data. For the studies summarized in Table 15.1, total captures, including repeated captures of the same individuals, averaged 5.8 captures/100 trap-nights (SD = 3.6), indicating an overall crude rate of more than four captures per individual.

Live-trapping requires great care to avoid accidentally killing or injuring captured animals. Traps must be checked daily under favorable conditions and two or more times per day during cold or wet periods. Traps can be checked less often if they are fitted with protective nesting boxes. S. W. Buskirk (Univ. of Wyo., pers. commun.) experienced no trapping mortalities among more than 120 captures of American martens in such traps. Lowered labor costs and fewer risks to animals result if traps are fitted with radio transmitters to signal when captures occur (Arthur 1988).

Snow Transects

Snow transects (i.e., counting animal tracks intercepting a transect) have long been used in studies of *Martes* and have been recommended as a technique to estimate relative abundance (Douglas and Strickland 1987, Strick-

Table 15.1. Livetrap capture rates for American martens (*Martes americana*)

Location	Season	Total captures	No. indiv.	Total tn	Indiv./ 100 tn	Total captures/ 100 tn	Reference
British Columbia	Fall, winter	44	21	625	3.36	7.04	Miller et al. 1955
Montana	Summer, fall	70	29	1,025	2.83	6.83	Burnett 1981
Montana	Year-round	223	53	1,912	2.77	11.66	Newby & Hawley 1954
Montana	Year-round	778	112	6,507	1.72	11.96	Weckwerth & Hawley 1962
California	Year-round	72	18	1,566	1.15	4.60	Simon 1980
Wyoming	Summer	97	17	1,987	0.86	4.88	Campbell 1979
Maine	Summer	609	123	16,065	0.77	3.80	Soutiere 1979
Wyoming	Summer, fall	291	22	3,224	0.68	9.03	Clark & Campbell 1976
Wyoming	Summer, fall	142	20	3,234	0.62	4.39	Hauptman 1979
Idaho	Fall, winter	80	13	2,896	0.45	2.76	Koehler & Hornocker 1977
California	Year-round	49	11	3,641	0.30	1.35	Martin 1987
Ontario	Summer	573	89	35,670	0.25	1.61	Francis & Stephenson 1972

Note: tn = trap-night.

land and Douglas 1987). Thompson (1949) counted fresh tracks crossed in a day's travel and reported an apparent increase in marten population over a four-year period based on observations of increased numbers of tracks. Similarly, de Vos (1952) recommended snow tracking to obtain indices of abundance of martens and fishers. De Vos described differences in behavior between martens and fishers that influence track counts, especially the tendency of martens to meander more than fishers.

As used by contemporary investigators, snow transect sampling entails following predefined transect lines one or two days after snowfall. Transects can be traversed by skis, snowshoes, or snowmobile, the latter obviously permitting more extensive coverage. Observers tally each track that intercepts the transect line. Some investigators, (e.g., Thompson 1949, de Vos 1952) recommend not counting repeated crossings by the same animal. Because such determination is seldom possible, others (e.g., Thompson et al. 1989, Raphael and Henry 1990) recommend tallying all crossings.

Adjustments are necessary to account for time since last snowfall. Thompson et al. (1989) divided total numbers of tracks by the number of 12-hour periods since snowfall; Raphael and Henry (1990) divided by the number of 24-hour periods. Temperature and wind influence the quality and persistence of identifiable tracks. Temperature variation could have a major impact on results, so tracks should be counted only when weather conditions are suitable. Results can be reported as numbers of intercepts per kilometer. Mean numbers of intercepts per kilometer varied from 0.05 to 3.40 among seven studies (Table 15.2).

The basic assumption in this method is that number of intercepts is directly related to animal density. Because no existing sampling method is known to

Table 15.2.   Results of American marten (*Martes americana*) track counts from snow transects in boreal forests, North America

Location	Total length (km)	Mean intercepts (*n*/km)	Reference
Ontario	21	3.40[a]	Thompson et al. 1989
Washington	7	1.69	Koehler et al. 1990
Wyoming	329	0.81	Raphael & Henry 1990
Oregon	82	0.60	Bull et al. 1992
Washington	142	0.59	Jones & Raphael 1991
Manitoba	3102	0.09	Raine 1983
California	787	0.05[b]	Nelson 1979

[a]Averaged over 5 years, uncut forest only.
[b]Includes tracks of both American martens and fishers.

yield unbiased, accurate estimates of density of *Martes*, the relationship between track intercepts and density has not been tested. One can, however, compare variation in track-intercept counts against independently derived estimates of relative abundance, perhaps from an intensive telemetry study. Thompson and Colgan (1987*a*) compared track counts with numbers of trapped and marked martens (expressed as total captures/100 trap-nights) and to total animals taken by trappers on traplines near the study area. All three population indices (track counts, capture rates, trapper records) suggested a declining population trend over the five-year study.

Other studies have documented correlations between track counts and relative abundance. Keith and Windberg (1978) found a direct relationship between track counts and snowshoe hare abundance. Van Dyke et al. (1986) found that variation in density of radio-collared mountain lions (*Felis concolor*) explained 61% of variation in track counts on roads. Kutilek et al. (1983) reported an increase in track counts with increasing estimates of mountain lion density in four study areas. They urged caution in comparing results from one area with another, however, because differences in topography and vegetation may affect carnivore behavior. Similarly, Van Sickle (1990) found a strong relationship ($R^2 = 0.73$) between numbers of mountain lion tracks and numbers of home ranges intercepted by roads. He reported that monthly road-track counts could be used as an index of lion numbers over time, but, like Kutilek et al., he cautioned against using them to infer differences across study areas. Fitzhugh and Gorenzel (1983) and Bekker (1991) offered several sampling design considerations that, although directed toward other species, may be applicable to martens and fishers.

Numbers of track intercepts vary among months, possibly in relation to behavior. Raphael and Henry (1990), for example, computed mean numbers of monthly intercepts per kilometer from November to March over a four-year study in southeastern Wyoming. The mean was greatest in November and declined in successive months (analysis of variance, test for linear trend, $F = 7.83$, $P = 0.010$) (Fig. 15.1).

The primary advantage of the snow transect method is its lack of bias associated with baits or other attractants. Baits or scents introduce variables that are difficult to control. Animals are usually attracted to baits by odor. Distance of detection varies with topography, vegetation, type of bait, wind speed, temperature, and humidity. Thus, one cannot calculate the actual area sampled by a baited station. For most studies, tracks in snow reflect natural movements of the animals, rather than movements influenced by the attractiveness of a bait. In addition, snow tracking does not require special equipment for attracting, detecting, or trapping animals. Setup time is also minimized; all that is required is laying out transect lines and perhaps marking preassigned distance intervals.

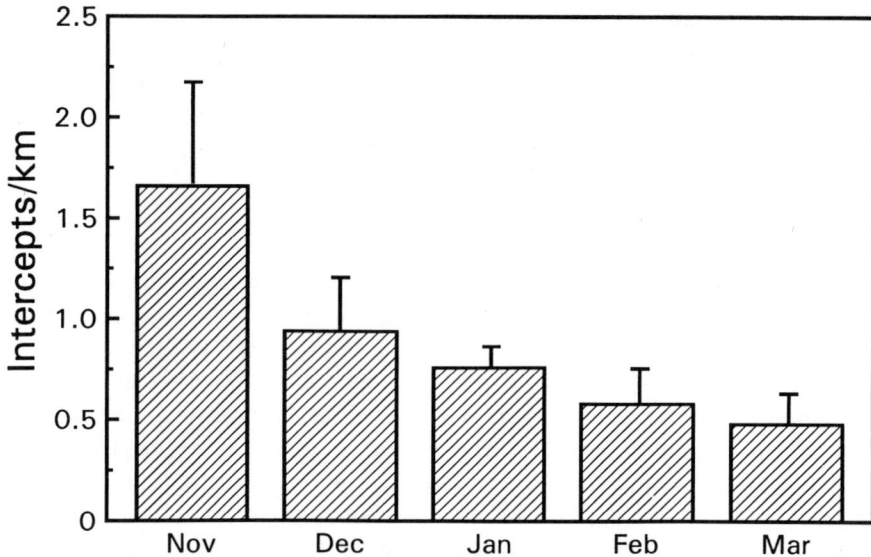

Figure 15.1. Mean numbers of intercepts of American marten (*Martes americana*) tracks in snow, 1986–1989, in the Sierra Madre, Wyoming. Vertical lines denote 1 standard error (Raphael and Henry 1990).

Snow transect's biggest disadvantage is its dependence on favorable snow conditions. Snow transects may never be feasible for western populations of fishers because they tend to occur at low elevations where snow is too infrequent for surveys. And in some areas, snow is too heavy or wet to provide adequate tracking conditions. In forests of western Washington, for example, I attempted snow tracking in the fall and winter of 1990 but was able to conduct surveys on only eight days despite repeated attempts; conditions were adequate for tracking on only one day. In contrast, researchers on the east side of the Cascade range, where snows are drier, were successful in conducting surveys (Bull et al. 1992). Similarly, I found conditions quite suitable in easterly ranges of the central Rocky Mountains. There, snowfalls were followed by long periods of snow-free weather, and snow was relatively dry and persisted from fall through spring. Adequate conditions for snow tracking are therefore highly site-specific.

Snow tracking can be expensive; snowmobiles are costly to purchase and operate. Without snowmobiles, researchers must check the transects on skis or snowshoes and, because little distance can be covered by an observer in one day, labor costs will be high.

Snow transect results also vary with the conditions of the transect, depending on placement, topography, habitat, home range spacing, and weather conditions. For ease of access and safety concerns, trackers on snowmobiles usually follow roads. But road placement is not random with respect to topography or habitat; roads are often associated with timber harvest and recreational areas. Also, wide roads may inhibit animals from crossing.

Sooted Track Plates

Track recording devices have been used extensively to provide indices of carnivore abundance. Typically, animals are lured to baited stations where tracks can be identified in fine soil (Cook 1949, Wood 1959, Linhart and Knowlton 1975, Lindzey et al. 1977, Roughton and Sweeny 1982). This technique has not been widely used to assess the abundance of martens and fishers because these species usually occur where soils are too rocky and where fine soils cannot be easily transported. A sooted surface was first used by Mayer (1957) to record tracks of small mammals. Barrett (1983) modified the technique for American martens in the Sierra Nevada of California. He used two aluminum plates measuring 814 × 407 × 0.6 mm for the tracking surface and applied soot by passing a torch of burning kerosene under the plates.

Plates must be protected from inclement weather, especially precipitation. A plastic tent could be placed over each plate. Barrett (1983) and Martin (1987) reported successful use of these box-enclosed plates in winter. They placed their plates inside rectangular wooden boxes attached to trees. Martin's plates measured 0.06 × 18 × 45 cm and were placed in 25 × 25 × 45 cm plywood boxes. Shephard and Greaves (1984) reported that a mixture of ethanol and lampblack, painted onto vinyl floor tiles, was resistant to light rain or running water, either of which render sooted plates unusable.

Track stations have been deployed in a variety of patterns. Barrett (1983) placed boxes at 200 × 200-m intervals in a large grid that covered a 2500-ha study area. Martin (1987) placed boxes at 400 × 400-m intervals within a grid over a 3900-ha area. Raphael (1988) placed uncovered track plates within each 10-ha study plot; adjacent plots were spaced at least 360 m apart.

Length of tracking session is an important consideration. Longer sessions increase the probability of detection at each station but decrease the total number of stations that can be deployed over several tracking sessions. Raphael and Barrett (1981) found that the cumulative richness of carnivore species increased up to day 11 of a 15-day survey and concluded that 8 days were sufficient to achieve high detection probabilities. Comparison of several studies (Table 15.3) shows that visitation rates of each species of *Martes*

Table 15.3. Results of surveys of American martens (*Martes americana*) and fishers (*M. pennanti*) conducted using sooted track stations

Location	No. stations	Avg. no. days per station	% Stations visited	Reference
		MARTENS		
Sierra Nevada, Calif.	42[a]	15	34	Barrett 1983
Sierra Nevada, Calif.	80[a]	28	38	Martin 1987
Western Cascades, Wash.	46	14	36	Jones & Raphael 1991
Southern Cascades, Wash.	57	30	33	Criss & Kerns 1990
		FISHERS		
Southern Cascades, Calif.	57	30	12	Criss & Kerns 1990
Klamath Mts., Calif.	466	8	12	Raphael 1988

[a]Track stations were placed in wooden boxes mounted on trees.

varied little among studies, even though the average sample period (number of days per station) varied markedly.

Track stations have many desirable characteristics. As long as bait is present, the plates will accumulate detections until they become smeared from too many tracks. Observers do not need to attend stations daily, unless weather conditions require plate replacement. As the number of days between observer visits increases, larger numbers of stations can be monitored by each observer. In addition, recordings of multiple visits by multiple species are possible (Raphael and Rosenburg 1983), making the technique highly efficient. Track impressions are easily saved for verification and voucher specimens by using transparent tape (Raphael and Barrett 1984, Raphael et al. 1986). Keys are available to aid identification (Taylor and Raphael 1988).

The technique also has disadvantages. Most important, the application of soot must be done under carefully controlled conditions to assure adequate sensitivity for clear prints. Such control is difficult to achieve under most field conditions. The plates are sensitive to temperature and moisture. Identifying tracks can be difficult, especially when tracks are only partial. Applying soot with a burning torch is hazardous and must usually be done at the shop or laboratory rather than at a field site to reduce risk. Finally, because animals must be attracted to the station with bait or other lures, the sampling radius is variable and difficult to control. For studies of trends over a wide geographic area, this variability may not be a problem as long as the same bait is used. Conversely, it is a considerable problem if one is attempting to assess microhabitat use.

Hair Snares

Hair characteristics such as shaft size, the amount and distribution of pigments, and the type of cuticular scale pattern can be used to distinguish mammalian species (Mayer 1957, Day 1966). Various snares have been developed to collect hair for subsequent identification. Suckling (1978), for example, lined small baited tubes of PVC pipe with double-sided adhesive tape to detect small mammals in Australia. Winnett and DeGabriele (1982) and Scotts and Craig (1988) described improvements to Suckling's design and reported successful detections of both small and medium-sized mammals.

Several investigators have used hair snares to survey *Martes*. In the Sierra Nevada of California, Barrett (1983) deployed 39 snares at 200-m intervals; the snares consisted of 610 × 254-mm cylinders of welded mesh wire containing coils of barbed wire and a bait box of hardware cloth, and were attached vertically to tree trunks. Snares were checked at about 15-day intervals for 22 weeks, resulting in an overall detection rate of 7%.

Nelson (1979) used three different designs to sample mustelid hairs in northwestern California. First, he wrapped barbed wire loosely 10–12 times around the trunk of a 20–50-cm dbh (diameter at breast height) tree and suspended a salmon carcass by rope from a higher branch. Second, he constructed a wire mesh cylinder with barbed wire spiraled within and bait near the top. Third, he used a 50-cm diameter barbed wire cage with salmon suspended in the center. All traps were attached to tree trunks 2–3 m above the ground, and all three types successfully snagged hairs. Traps were checked at approximately one- to two-week intervals over four months (Dec—Mar). From 210 traps, 62 hair samples were collected (at 35% of the traps), including 10 samples identified as fisher hair.

Jones et al. (1991) used a device similar to that described by Scotts and Craig (1988). They placed baited PVC tubes with sticky materials at 39 stations in the Washington Cascades for 410 days (averaging 10.5 days/station). Hairs were collected on only four occasions, including two detections of marten.

Hair snares allow repeated visits by animals without requiring an observer to reset the station; they are low in cost because materials are readily available; and they require almost no maintenance except to check bait. Most important, they can be used in any weather and are not affected by precipitation, provided that bait is protected. On the other hand, like other baited devices, biases caused by different bait types are difficult to control. Barbed wire is awkward to handle and transport, and hair snares require considerable setup time. Not all hairs are easy to identify; existing keys are usually based on dorsal guard hairs (Day 1966). So far, all investigators report low detection rates, implying that animals are either rare or reluctant to enter such devices, or that when they do, they fail to leave hairs.

Cameras

Use of remote camera systems to photograph *Martes* is a relatively recent and promising technique. This approach involves setting a flash-equipped camera system activated by a mechanical or electronic device. The systems range from simple and inexpensive plastic cameras (Joslin 1977, 1988; Jones and Raphael 1993) to complex and expensive motor-driven 35-mm cameras (Mace and Manley 1991).

Jones and Raphael (1991) tested low-cost camera systems in California and Washington. They reported results of two years of summer field tests with small plastic cameras that use 110 film cartridges, flashbars (plug-in flash units that do not require batteries), and mechanical linkages between bait and shutter release. Cameras were set out at about 1-km intervals and checked every 1–2 days over 10-day sessions. Numbers of stations varied from 23 to 1081 and detection rates (percentage of stations where *Martes* were detected) varied from 0 to 57%, averaging about 14%. Bull et al. (1992) tested low-cost cameras during winter and recorded martens at 30% of 47 stations, each run for five weeks.

More sophisticated systems have been used primarily to detect larger carnivores such as brown bears (*Ursus arctos*) but have incidentally recorded *Martes* (e.g., Mace and Manley 1991). These systems consist of motor-driven 35-mm cameras equipped with battery-powered flash units and date-time recorders. They are triggered by emitted infrared radiation and respond to the body heat of an animal within range. In tests conducted on 246 stations (14–18 nights per station) from 1989 to 1990 in Montana, 3 American martens and 40 fishers were photographed. Assuming each photograph represented a different animal, the detection rate was about 17%.

Camera systems offer ease of standardization for comparing results from different study areas, but they are subject to sampling biases becauses they use bait to attract animals to the station. Preliminary results suggest that detections recorded by photography are more often identified than tracks on sooted plates, but further research is needed to test the two techniques fully. Photographs certainly provide a valuable permanent record of occurrences. Coupled with techniques to mark individuals, photographs can be used in mark-recapture experiments. For example, I have used photographs to tally uncaptured American martens and previously captured animals equipped with conspicuous radio collars. Most camera systems are easily protected from weather, making them useful over a broad geographic range.

Camera systems also have disadvantages. Unless expensive motor-driven systems are used, one can obtain only one photograph per animal visit. Some low-cost systems use a flashbar that may fail to flash, and the rate of failure seems to be related to weather conditions. Like other baited devices, camera detections are subject to biases associated with the attractiveness of bait and the inclination of the animal to approach a novel stimulus. In addition, these systems are subject to theft and must be camouflaged in areas frequented by people.

Design Considerations

The design of any survey or monitoring program depends on the program objectives. Once these objectives are clearly understood, an investigator must choose the detection device and sampling protocol. In addition, the investigator must make decisions about other design attributes of monitoring programs.

Spacing of Stations

The optimum distance between detection stations is a function of the study objective, the probability of detecting the target species, and economics or logistics. Stations should be close enough together to maximize the probability of detecting animals within a reasonably short time but far enough apart to assure that they are not recording the same animals (Roughton and Sweeny 1982). The longer the sampling period and the more stations located in an animal's range, the greater the chance that the animal will visit at least one station. Therefore, the spacing of stations is a trade-off between number of stations and number of days they are run.

Spacing should relate to the average size of the target species' home range. Assume that home range is circular with a radius W. Otis et al. (1978:77) recommended four stations per home range as a general guideline, so that the distance between stations (S) should be $S \leq W/2$ (Fig. 15.2). For an animal with a home range of 10 km^2, spacing would be about 0.9 km to assure four stations per home range (Fig. 15.2). Because home ranges of males are larger than those of females (Buskirk and McDonald 1989; Powell, this volume), optimal spacing for one sex may not be optimal for the other. Placing multiple stations within the home range of an animal increases the probability of

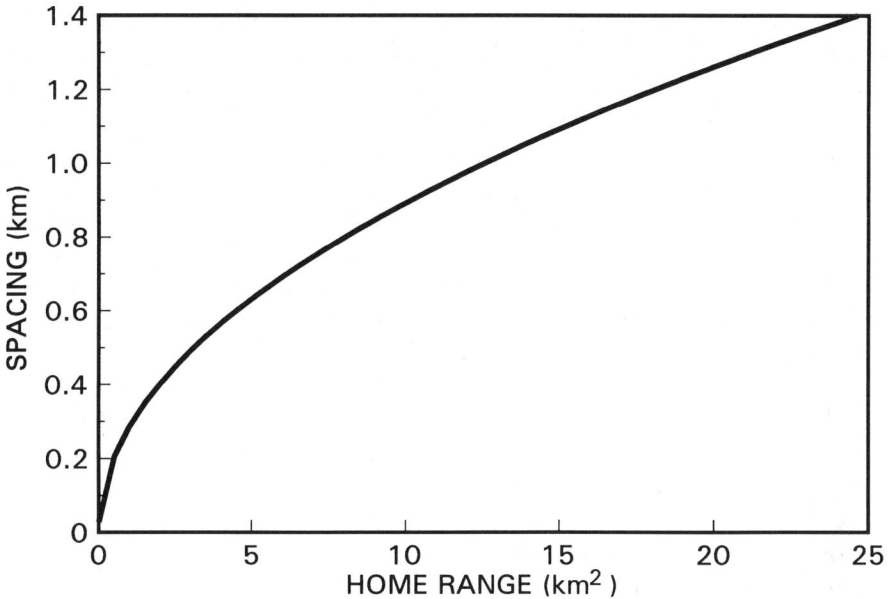

Figure 15.2. Spacing of sample stations needed to approximate an average of four stations per home range.

detection at the risk of decreased independence of observations. Where stations are deployed along transects, these spacing guidelines would yield an average of two stations per home range.

Sample Duration

   Length of a sampling session is important. A longer session leads to a higher probability of a station's being visited; however, the longer the session, the more likely the same animal will visit the same station more than once or visit adjacent stations (Roughton and Sweeny 1982). If the objective is to estimate relative abundance, multiple visits by the same animal will decrease both accuracy and precision. Interpretation of rates of detection as indices of abundance is difficult when one cannot clearly distinguish between detections of several individuals and repeated detections of the same individual. If, on the other hand, the objective is to record presence or absence, this should not be a problem. If the objective is to evaluate population trends, the variance of detection rate must be computed; variance is underestimated when the same animal is detected at multiple stations (Burnham et al. 1981, Roughton and Sweeny 1982).

The variable to be observed at each sampling unit is presence of the target species. The parameter likely to be estimated under most designs is the proportion (P) of occupied stations. Changes in presence can be tracked over time as an index of population trend. There is some probability of failing to detect an animal within the prescribed time interval or number of visits. The estimate ($\hat{P}$) will be biased in relation to the likelihood of missing animals. Azuma et al. (1990) developed a procedure for estimating this bias to adjust $\hat{P}$. Their method assumes a constant probability of detecting an animal at each visit and that each visit to a station is an independent and identical Bernoulli trial that can result in one of two outcomes: detected or undetected.

Timing of Sampling

Behavior of American martens and fishers undergoes seasonal changes associated with weather and reproductive status, and these changes can influence survey results. For example, I have found that American martens in my study site in Wyoming are two to three times more difficult to trap in summer than in other seasons. The number of trap-nights spent for each capture averaged 123 for three summers, 38 for two fall periods, and 55 for four winters. The presence of young-of-the-year in early fall is another important influence on detection rate. Males and females change their spacing behavior during mating and this too can affect detection rates.

Investigators should ascertain whether the proposed design adequately accounts for seasonal variation in behavior. Any study involving comparisons among years or among study areas should be standardized so that the same months are sampled in each case. Preliminary investigations should be undertaken to be sure that results are not confounded by unanticipated changes in behavior such as maternal care or the presence of foraging young.

Number of Stations

Reliable estimates of occurrence, relative abundance, or trends in population size over time require a sufficiently large sample. Sample size refers, in this context, to the number of sample units, which might be individual stations or closely spaced clusters of stations. Total detections will vary with detection probabilities, the spacing of stations, the duration of sampling bouts, and the number of stations per sample unit. The optimum number of stations depends on the average probability of detection for the target species and on the precision of the estimated detection rate.

In addition, if the objective is to estimate trends over successive years, the sample size must also be large enough to assure sufficient power—that is, the ability to detect a real trend (Toft and Shea 1983, Harris 1986, Gerrodette 1987, Link and Hatfield 1990). In general, a larger sample size will reduce the difference between the sample mean ($\hat{P}$) and the true (population) mean (P). Thus, a larger sample size increases the investigator's ability to detect a difference between populations or over time. Each additional sample, however, has its cost of supplies and time, and one must evaluate the benefits and costs of an additional sample.

A variety of procedures can be used to estimate appropriate sample sizes, given study objectives and known variabilities of mean detection rate. If estimates of variability are not available from previous studies, a pilot study should be conducted to estimate variance. Pilot studies have the added advantage of providing an opportunity to refine techniques.

Standardization

Any monitoring program to compare estimates of abundance over geographic areas or among years must be standardized. An investigator should be able to demonstrate that differences in detection rate reflect differences in population size, not differences in methodology. Care must therefore be taken to use a standardized bait; to run stations for the same length of time and during the same months each year; to visit each station at regular, consistent intervals; and to lay out stations with consistent spacing.

Large-scale surveys over broad geographic areas will require careful coordination to achieve reliable results. For example, there is a current need to determine the status and trend of fisher populations in the western United States. To accomplish such a survey, a large number of agencies and administrative jurisdictions must be willing to develop and use a standardized sampling protocol; and in this case, standardization is especially critical.

Conclusions

None of the techniques I have described will fully meet all monitoring objectives for both American martens and fishers (Table 15.4). Each has its advantages and disadvantages, and each needs further study to understand its utility and limitations. Research is needed to evaluate the efficiency of each method; to refine design of detection devices; to select the best baits or lures; and to evaluate effects of station placement, numbers of stations, sample duration, and timing of sampling.

For small-scale studies, live-trapping, snow transects (if weather conditions are appropriate), sooted plates, and cameras all appear to be suitable.

Table 15.4.   Application of various techniques to meet different monitoring objectives

Technique	Local presence	Geogr. dist.	Relative abund.[a]	Trend[b]	Density	Demographics[c]
Habitat survey		1	1	1		
Harvest records	2	1	1	1	1	
Live-trapping	2	2	2	2	2	2
Snow transects	2	2	2	1		
Sooted plates	2	2	2	1		
Hair snare	2	2	1	1		
Cameras	2	2	2	1	1	

Note: Numbers indicate relative efficacy of each technique: blank = not useful, 1 = moderately useful, 2 = very useful (if used in well-designed survey).

[a] Usually, an index of relative abundance for comparison of different study areas at the same time.

[b] Trend in an index of relative abundance over time.

[c] Estimates of birth and death rates, immigration, and emigration.

For larger-scale studies that require comparable results among sites, sooted plates and cameras offer the greatest promise. Both can be deployed in a design that can be carefully standardized. Although either type of device can be affected by moisture, both can be protected with inexpensive enclosures or protective covers. Both can yield recognizable images under most circumstances. I encourage further study of these devices to increase their reliability and maximize their ability to detect animals. I also encourage research to better understand appropriate study designs and data analyses to develop reliable monitoring systems. Martens and fishers are sensitive to loss of habitat (Buskirk and Powell, this volume), and resource management agencies need to implement sound monitoring strategies to know whether their land management practices are affecting habitat and populations in accordance with managers' objectives.

Acknowledgments

I thank the Rocky Mountain Forest and Range Experiment Station, the Pacific Northwest Research Station, and the Pacific Northwest Region of the USDA Forest Service for supporting portions of this work. In addition, I thank L. L. C. Jones and S. E. Henry for their many contributions, R. T. Golightly and R. A. Powell for comments on the manuscript, and J. L. Jones for administrative support.

PART IV

Status and Conservation of

Holarctic *Martes*

Introduction

Populations of most *Martes* species have undergone long-term declines in conjunction with human settlement, trapping, and habitat loss. This section shows that improved hunting regulation and habitat protection have led to recovery of some populations, but that in addition to other previously stable populations, some are now subjected to new threats from human activities.

The stone marten is unique among *Martes* in its use of warm, open, rocky sites and in its commensal relationship with humans. Bakeyev, describing the stone marten's status in the Commonwealth of Independent States, reports that populations are stable and high, and suggests that higher harvests could be sustained. In contrast, Bakeyev and Sinitsyn describe a less positive history of conservation of the sable, a species adapted to cold temperatures in boreal and taiga Asia. Records dating to the first century B.C. document the high value of sable pelts and heavy hunting pressure. By the late 1600s, the species was nearly extirpated. Populations remained low until hunting was temporarily banned in the mid-1930s and the price of pelts dropped. Massive reintroductions (>19,000 animals released) led to repopulation of much of the species' former range, but sables have apparently declined again in areas of high human density. Currently, improved hunting regulations and designation of protected refuges are intended to maintain healthy populations.

In China, the distribution and population size of sables have declined since the early 1900s owing to habitat loss and overtrapping. The sable is now designated as a Category I threatened and endangered species, the highest level of protection granted. Ma and Xu document the variable habitat associations of sables across China, ranging from larch-dominated coniferous forest in the west to mixed coniferous–broad-leaved forest in the east. As in the Commonwealth of Independent States, sables have undergone severe exploitation dating to ancient times. Unlike those in the CIS, however, sables in China have not recovered. An aggressive captive breeding program, establishment of eight large nature reserves, and protection from hunting may contribute to their eventual recovery.

Berg and Kuehn paint a brighter picture of the history of American martens and fishers in Minnesota. Following European settlement, both species declined severely as a result of logging, forest clearing and burning, and over-trapping through the 1920s. Now, however, forests have been reestablished, fur trapping is carefully controlled, and populations have largely recovered and occupy much of their former range. Because fishers use a diversity of forest types, Berg and Kuehn consider it unlikely that logging as currently practiced will have adverse effects on population-levels. Martens are more habitat-specialized, but public concern for maintenance of mature forest systems suggests that neither fishers nor American martens will again experience the population declines they did a century ago.

Finally, Tatara describes a population of the Tsushima marten, a subspecies of the Japanese marten that occurs only on the Tsushima Islands, off the northwest coast of Japan. The ecology of this subspecies is poorly known, but Tatara finds that conversion of native broad-leaved forest to conifer plantation, road kills, predation by feral dogs, and fragmentation of remaining habitat have all contributed to the subspecies' being considered one of Japan's vulnerable mammals.

MARTIN G. RAPHAEL

16 Stone Martens in the Commonwealth of Independent States

Yuri N. Bakeyev

Of the four *Martes* species in the Commonwealth of Independent States, the stone marten (*M. foina*) is the best adapted to warm climates and lacks morphological adaptations to survive severe winters with deep snow. Its preferred summer temperature is close to 18°C (Ponomarev 1944), although it inhabits some areas where summer temperatures average 10° higher. In the European republics it is never found north of 59° latitude. The bottoms of its feet are hairless with horny outgrowths, and the weight load per unit of area of foot is twice that of the Eurasian pine marten (*M. martes*), indicating a southern forest origin for *M. foina*.

The stone marten inhabits low- to medium-elevation forest-steppe and steppe zones of the Commonwealth of Independent States, but it penetrates into semidesert regions on the northwestern Caspian Sea lowland. Although it uses roadside forest shelter belts as corridors during dispersal, it prefers gully forests with rocky sites lacking forest canopy for permanent habitat. It readily uses small caves in taluses, red fox (*Vulpes vulpes*) and European badger (*Meles meles*) holes, and bird nests for resting sites. Because of its long association with humans in southern Europe, it has become commensal and often lives in attics of houses and in other buildings, even in towns and cities. Maternal nests have been found in farm buildings, heat lines, and irrigation pipes. In cities, stone martens are valued for killing rats and pigeons, and they are an aesthetic resource in parklands.

The stone marten population in urban areas of the European republics is large and stable. Populations are also found in the forests of the Carpathians, the Caucasus, the Crimea, and central Asia. In the Pamir Mountains of central Asia, stone martens are found along tourist routes at elevations up to 4900 m, sometimes occupying storage tents in tourist camps. Vegetation at these elevations is sparse and the only available plant food is the fruit of the Semenov honeysuckle (*Zonicera semenovii*). Prey are also scarce and the martens presumably survive by scavenging. During the 1980s, stone martens colonized new range as far east as the Volga River and are now found from Samara to Astrakhan. Populations are growing in areas where hunting is prohibited—for example, in the Ukraine near the site of the Chernobyl nuclear catastrophe. As populations have increased, stone martens have occupied mountain forests almost to the subalpine zone.

The ranges of stone martens and Eurasian pine martens overlap extensively. Population sizes of the two species on a site are inversely related, with stone martens more common in the Crimea, where there are no Eurasian pine martens, and in the forest-steppe zone. In the woodlands and in the Carpathians, the Eurasian pine marten is more common (Table 16.1). In the Caucasus, stone martens are more common in the xeric east, where there are few forests, and Eurasian pine martens are more common in the extensive forests to the west, where snow is deep and winters are long (Table 16.2). The ranges of stone martens and sables (*M. zibellina*) overlap only in the southwestern Altai, mostly in Mongolia and outside the Commonwealth of Independent States. The differences in preferred climate for these two species severely limits their area of overlap.

Although stone martens and Eurasian pine martens have similar diets, stones martens are more omnivorous. They eat mammals, birds, fruits, and insects, and their diets vary tremendously with the availability of food. In some areas, stomachs of harvested stone martens contain predominantly small mammals, especially murids, but in other areas they are predominantly full of fruits or birds.

Table 16.1.   Marten productivity in Ukraine forests, according to state pelt purchases in the 1960s

| | Pelts purchased | | | |
| | Stone martens (*Martes foina*) | | Eurasian pine martens (*Martes martes*) | |
Landscape/region	Total	Per 100 ha	Total	Per 100 ha
Woodlands	1174	0.38	2063	0.53
Forest-steppe	3544	2.26	766	0.48
Steppe	1555	1.79	129	0.14
Crimea	780	2.77	—	—
Carpathians	590	0.32	805	0.44
TOTAL	7643		3763	

Note: — = no data.

Table 16.2.   Number of *Martes* in the Caucasus during the period
1984–1987, according to interviews, and percentage of *Martes* pelt
purchases that were *M. foina* for 1968 and 1987

Region	No. of *Martes*, 1984–1987	% *M. foina* 1968	% *M. foina* 1987
Northern Caucasus	12,930	41	74
Krasnodar Krai	6,000	22	58
Stavropol Krai	2,000	33	74
Kabardino-Balkar Rep.	560	35	79
Severo-Osetin Rep.	800	50	71
Checheno-Ingush Rep.	1,400	57	85
Dagestan Rep.	2,600	95	95
Rostov Region	550	—	90
Transcaucasia	20,050	60	86
Georgia	13,000	44	73
Azerbaidzhan	5,600	89	94
Armenia	1,450	100	100

Stone martens have high reproductive rates for the genus *Martes*. Sexual maturity is reached at 14–15 months, and litter size averages between four and five (litter sizes from one to eight have been recorded). Male testes begin to enlarge and contain spermatozoa in May. Females come into estrus between mid-June and mid-August (Riabov 1982) and parturition occurs from late March to mid-May. Total gestation length varies from 234 to 279 days owing to variable delayed implantation.

Lobachev (1973) estimated that diameters of home ranges of live-trapped stone martens ranged from 2.3 to 4.5 km. Home ranges averaged about 9 km^2 in area and overlapped among individuals. Few individuals were resident and recaptured over a long period. Aggression between martens was not observed, and population density reached four per square kilometer. The age distribution of this lightly harvested population was 20% under 1 year of age, 26% between 1 and 2 years, and 54% 2 years or older.

The stone marten population in the Commonwealth of Independent States exceeds 100,000 and is not harvested to the extent possible. Trappers could take 30,000–35,000 animals a year, but the fur industry is not prepared to use large numbers of stone marten pelts. By better understanding stone marten biology, it should be possible to extend this species' range in the Commonwealth of Independent States.

17 Status and Conservation of Sables in the Commonwealth of Independent States

Nikolai N. Bakeyev and Andrei A. Sinitsyn

The sable (*Martes zibellina*), an occupant of the taiga, has a geographic range in the Commonwealth of Independent States covering about 7,000,000 km². As of the early 1990s, the species' range extends from the Severnaya Dvina and Mezen rivers eastward to the Pacific Coast and includes some of the Kurile Islands (Iturup, Kunashir), Sakhalin Island, and the Kamchatka Peninsula. Outside the Commonwealth of Independent States, the sable occurs on Hokkaido Island, Japan, and has a limited distribution in China, North Korea, and Mongolia. The range of sables is limited to the north by the boundaries of continuous forest, although in some areas it extends into the forest-tundra zone. To the south, the range reaches 55–60°N latitude in the taiga plain of west Siberia and 42° in the mountains of east Siberia (Fig. 17.1). The distribution of sables reflects their ability to withstand cold temperatures; they occur in Yakutia, which has the coldest winters in the Northern Hemisphere. Several adaptations, such as furred feet and use of protected microhabitats, enable sables to survive in the cold.

The distribution of sables also reflects their inability to occupy regions with high summer temperatures, even when the habitat appears otherwise suitable. The northern Maly Khingan of the Far East has apparently suitable habitat dominated by oak (*Quercus* spp.) and poplar (*Populus* spp.) but is devoid of sables. This area has abundant plant and animal foods, high habitat diversity, and large amounts of coarse woody debris that provide access to prey in winter. Yet sables only occasionally pass through this area during winter migrations. By contrast, in coniferous forest dominated by spruce (*Picea* spp.) and fir (*Abies* spp.) at higher elevations on the same slopes, sable densities are high (up to 1.5/km²). Although these forest zones are not

246

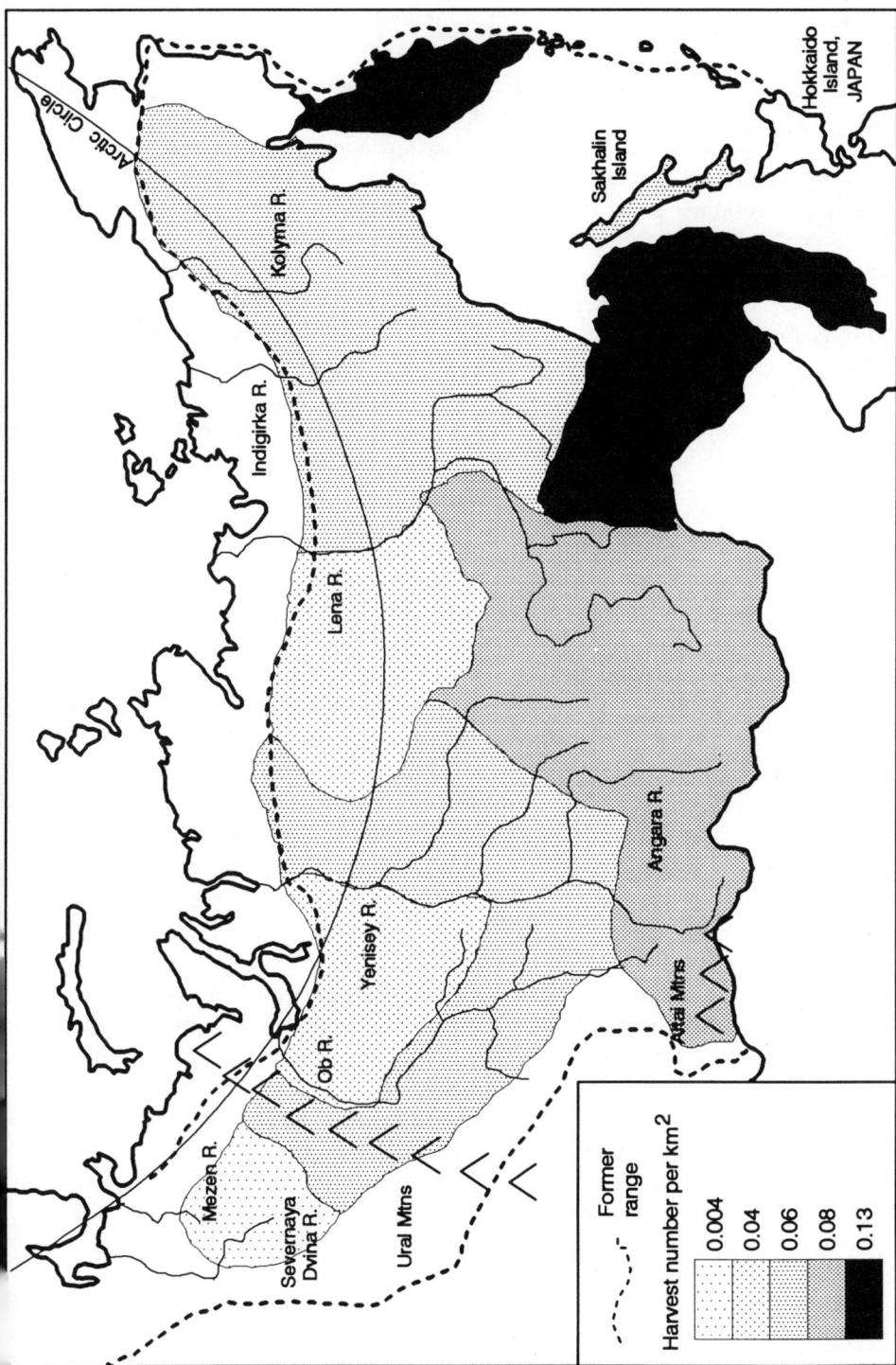

Figure 17.1. Distribution of sables (*Martes zibellina*) and pelt productivity (pelts/km²/year) in the Commonwealth of Independent States.

Hokkaido Island, JAPAN

Sakhalin Island

Arctic Circle

Kolyma R.

Indigirka R.

Lena R.

Angara R.

Yenisey R.

Altai Mtns

Ob R.

Ural Mtns

Mezen R.

Severnaya Dvina R.

Former range

Harvest number per km²

0.004

0.04

0.06

0.08

0.13

separated by any physical barriers, temperatures are colder in spruce-fir than in oak-poplar forests during all seasons.

Experimental releases have shown that lowland broad-leaved forest habitats are unsuitable for sables. Between 1951 and 1962, 806 sables captured from the Verkhnebureya region of Khabarovsk Territory were released in southern Sikhote Alin in an attempt to join a southern montane "island" with the main northern range. All 20 releases failed to establish sable populations (Abramov 1972). In the Suputinsk refuge, where 247 sables were released, all dispersed within several weeks, and none were reported harvested. Releases in extensive forests that included some spruce and fir in the Salair Mountains also failed. No experimental attempts to extend the range of sables southward succeeded. By contrast, all releases of sables into vacant range in the northeast, including Yakutia, the Magadan area, and mainland Kamchatka, were highly successful.

During the Pleistocene epoch, a time of major distributional changes for many vertebrate taxa, sables apparently persisted within parts of their current range. Sable bones of Pleistocene age have been found in the Far East, the Angara River Valley, middle Yenisei, Kuznetsk Alatau, the Altai Mountains, and the middle and north Ural Mountains (Kuzmina 1971, 1975; Yermolova 1978; Volodin et al. 1980). In these widely separated areas, several geographic races of sables developed, with gradations in size and color from west to east (Table 17.1). In the Ural Mountains a very light colored, large-bodied race occurs, whereas in the Far East the darkest and smallest races are found. Exceptions to this trend are the large sables that evolved independently in Kamchatka and on the Kuriles and Sakhalin Island.

Fur color, an important determinant of pelt value, differs greatly among geographic races, from an index of 1.7 (light) in the Ural Mountains to a mean of 3.6 (dark) in Yakutsk and Barguzin. Pelt color is also highly variable within regions and even within litters. One litter we examined had both very light and very dark kits. The genetic component of pelt color is clear from captive breeding studies (Table 17.2). When dark parents were crossed, offspring averaged 56% dark, 43% medium, and 1% light. Offspring of light parents averaged 31% dark, 46% medium, and 23% light.

The geographic range of sables has changed substantively because of natural and anthropogenic factors. To the west, the sable is generally limited by the Ural Mountains but occasionally penetrates far into the European republics. Here, the Eurasian pine marten (*M. martes*) predominates, presumably because it is better adapted to the warmer climate and associated habitats.

From the fifteenth through the seventeenth centuries, sables were common in the Severnaya Dvina River valley, in what is now northeastern Russia, and

Table 17.1.   Variability in the average weight of sables (*Martes zibellina*) from different geographic populations

Geographic population	Weight (g)	
	Males	Females
Southwest Altai	1427	1059
Tavda Trans-Urals	1330	910
Kondo-Sosva	1290	864
Turukhansk (Bachta River Valley)	1283	915
Chulym	1150	826
West Sayan	1126	825
Northeast Altai	1114	803
Angara region	1108	772
East Sayan	1100	787
South Trans-Baikal	1038	713
Northwest Baikal region	1087	782
North Baikal	1075	810
Barguzin	987	720
Bureya	847	722
Tyrma (Bureya River Valley)	835	594
Sikhote Alin	774	557
Sakhalin	1193	916
Kamchatka	1439	1119

reached west to northern Finland and Sweden at 50°N latitude (Geptner et al. 1967). Sables also extended southwestward into Byelorussia, Lithuania, and eastern Poland (Kirikov 1958, 1960). This extension of range coincided with a marked fall in temperature in 1550–1850, known as the "Little Ice Age." During this time northern spruce forests extended their range as far as 600 km southward, replacing broad-leaved forest. Ordynsk (Matt) annals of 1501 describe the levying of taxes in sable pelts in Vilno, Vitebsk, Polotsk, and Mogilyev. Pine martens were also widely distributed in this area. Later, as

Table 17.2.   Color of offspring from light, medium, and dark sables (*Martes zibellina*) crossbred on sable farms

Parents		Offspring color (%)			Color
Males	Females	Dark	Medium	Light	Index
Dark	Dark	55.7	42.9	1.4	3.53
Dark	Light	50.8	43.7	5.5	3.40
Medium	Medium	51.0	44.9	4.1	3.43
Light	Medium	45.7	42.1	12.2	3.21
Light	Light	31.0	46.4	22.6	2.86

the climate warmed and vegetation changed, the distributional limit of sables retreated eastward to the Urals, whereas pine martens continued to occupy the area west of the Urals. It is unlikely that humans affected this distributional change by selectively hunting sables, because where pine martens and sables occur sympatrically it is impossible to selectively harvest one species over the other.

Where they are sympatric, sables and Eurasian pine martens occasionally hybridize. Grakov (1981), created hybrids in captivity and found that male kidus (F_1 sable × Eurasian pine marten) were sterile. Female kidus back-crossed with male Eurasian pine martens produced kits in only one case, and female kidus bred with male sables produced no offspring. Therefore, although F_1 hybrids occur, reproductive barriers between sables and Eurasian pine martens are nearly complete. The narrow zone of sympatry of sables and pine martens extends from the Severnaya Dvina River Valley up to the Taz, Vach, and Vasyugan rivers (tributaries of the Ob River).

For several thousand years, humans have affected sable distribution and numbers through hunting. In the first century B.C. at the peak of Roman cultural development, sable fur adorned the necks of Roman women. Since then much history and lore has accumulated regarding sable fur. In 1020 A.D. Olaf, King of Norway, sent ships to fur auctions in the Perm lands and the Severnaya Dvina. Sable fur was known to the English in the Middle Ages, and the second bishop of Lincoln presented Henry I (1100–1135) with a cloak lined with black sable fur.

Supplying Europe with sable furs from Siberia was logistically difficult. From the fifteenth through the seventeenth centuries, Russian explorers traveled by river and sea to the mouths of the Yenisei and Lena rivers and ascended those rivers to as far as Lake Baikal. By land, they got as far as the Indigirka and Kolyma river valleys, where they found unharvested sable populations. With the colonization of Siberia in the seventeenth century, intensive harvesting of sables began. Thousands of working-class people went to Siberia in the hopes of lucrative trapping. One sable pelt was worth a draft horse or a milk cow, and by harvesting 50 sables, a hunter could become wealthy. The resulting sable harvest was enormous: an estimated 440,000 pelts were taken from the area now within the borders of Russia. Sable pelts were an important trade commodity and were almost a standard of currency for some purposes.

Because hunting was conducted from early autumn to late spring, by the late 1600s sables were nearly exterminated. This depression of sable populations lasted nearly three centuries. High prices for sable pelts, up to 600 rubles each, aggravated the problem. Government attempts to halt sable hunting were ineffective because of poaching.

Active restoration of sable populations did not begin until 1935, when a five-year ban on the hunting and sale of sable pelts was established. During the Great Patriotic War (1941–1945), professional hunters served in the military and so the harvest of sables was reduced without restrictive regulations. After the war, many small Siberian settlements merged into large ones, and access to distant hunting grounds became difficult because of poor roads. Further, the value of sable pelts fell sharply and a system of licensing was established, with the result that many sable hunters changed profession. The overall effect was to reduce hunting pressure to a point where numbers and distribution of sables began to increase.

Mass reintroductions reestablished sable populations in many areas. From 1940 to 1965 more than 19,000 animals were translocated into areas with low densities. Of these, 54% went to the Far East and Yakutia, 25% to west Siberia and the Urals, 18% to east Siberia, and 3% to middle Siberia. More than 1,400,000 km², 27% of the original sable range, was repopulated.

In the western part of the range, from the Urals to the Trans-Baikal, sables had persisted fairly well during the long period of overexploitation. There the range was restored with almost no human intervention during a 25-year period that ended in the 1950s, and an explosion in the number sables occurred. In mountainous forests of the Altai, Sayan, and Baikal regions, population densities reached three to four per square kilometer. Sables were reportedly so abundant that men could walk along sable trails in snow without skis and not break through the surface. By the end of the 1960s, overpopulation of sables was followed by the appearance of pathogens and parasites. Epizootics of a dermatitis, which had first been noted in the eighteenth century, caused decreases in some sable populations in the 1970s.

The recovery of sables was markedly different in Yakutia and the Far East. Here, they were driven to extinction over large areas, and the ban on hunting in 1935 had little effect on population size. Sables were reintroduced into this area from 1951 to 1962, which aided recovery. In Russia, the restoration of sable range now is nearly complete, as shown by harvest rates and purchases of pelts. The mean number of pelt purchases was approximately 170,000 in 1959–1964, 185,000 in 1964–1969, 135,000 in 1969–1974, 150,000 in 1974–1979, 155,000 in 1979–1984, and 250,000 in 1984–1989.

Besides reintroduction, natural dispersal of sables led to relatively rapid expansion of sable range over large areas. Sables marked in the Trans-Baikal Region dispersed as far as 200 km in one direction and as far as 100 km and over three treeless mountain ranges in another. The sable population near the research station in Barguzin was almost completely restored in two to three years through natural dispersal (Chernikin 1980).

Sables will travel as far as 300 km from translocation release sites. Dispersal begins in autumn, with movements first of young animals and then of adults. Sables also exhibit seasonal vertical migrations in the mountains. As snow depth increases in winter, sables move down in elevation, and as spring approaches they move back up, especially pregnant females seeking natal dens. Even migratory sables have stable home ranges 5–10 km in radius in summer. During this season they have strong homing instincts and will return from 20–25 km away when moved.

Sables are active throughout the day. When hungry, they leave resting sites to forage and may do so several times a day. During warm winter weather ($-10°$ to $-5°$C), activity is low. At temperatures of $-20°$ to $-25°$C, activity increases noticeably, but below $-30°$C sables remain in resting sites for as long as five to seven days. Sables also do not leave resting sites during heavy snowfalls (Rayevski 1947). Sables range from 2–19 km per day, depending on food availability and hunger.

Sables are omnivores, and major foods are mice (Arvicolidae), pikas (*Ochotona* spp.), pine nuts (*Pinus siberica, P. pumila*), and various fruits. In the Far East, sables often eat spawning salmon (Salmonidae) and grouse (Tetraonidae) that shelter in snow holes. Sables are more omnivorous in summer than in winter. In summer they eat insects, mollusks, earthworms, and berries, but they also prey on pikas and young birds and mice.

Foraging in winter is facilitated by coarse woody debris, which supports the snow and creates subnivean air pockets and passages. When snow is very deep and soft it is difficult for sables to enter subnivean spaces. Under these conditions, they can suffer food deprivation and may switch prey. For example, in Khabarovsk Territory in 1972, following a four-day snowfall that left 1.4 m of accumulated snow, sables resorted to feeding on musk deer (*Moschus moschiferus*), which were abundant. Within about 10 km of our field camp we found 11 carcasses of musk deer that had died of neck injuries inflicted by sables. Even though musk deer weigh 10 or more times as much as sables, they were usually killed by a single male. Several sables sometimes fed concurrently on a deer carcass, but we observed no agonistic behavior during these feeding episodes. After the snow had settled for several days, musk deer tried to run from pursuing sables but could still be caught. A sable would attack by biting the hind legs and belly of a musk deer and then, when the deer weakened, would kill it by biting through the base of the skull near the atlas.

Forest cutting and fire have not yet greatly influenced sable populations. During the 1980s, regenerating forest following fire and clear-cutting represented 12–15% of the total forested area in most regions. By the year 2000 that figure will rise to 17%, while about 13% will be mature forest. Although some areas have been clear-cut, mountainous taiga has fared better because many areas have slopes too steep to timber. In such areas, narrow bands of forest (300–400 m wide) have been selectively cut in accessible low-elevation mountains. This practice has created a mosaic of habitats favorable for sables. These cutovers are rich in mice and pikas, and snow thaws earlier there in the spring. Berry-producing shrubs are common, and many small passerines use the areas for nesting. In Khabarovsk Territory we have found high densities of mice near ripe raspberries (*Rubus* sp.) together with high numbers of sables, apparently exploiting this local food abundance.

Among sables sexual maturity occurs at age 2–3 years, and litter sizes range from one to four but are most commonly three. The pregnancy rate among adult females is 70–90%. Sables are extremely long-lived (15–20 years). Because of these reproductive characteristics, fluctuations in juvenile cohorts have little effect on the reproductive portion of the population. This facilitates population recovery after short-term declines in young cohorts.

Sable populations exhibit modest short (3–4 year) and long (7–10 year) population fluctuations, with amplitude ratios between high and low years of 1.2–1.5. As of 1993, sable numbers in the Commonwealth of Independent States are high and are estimated to be 1,000,000 to 1,300,000. During a hunting season, 300,000 to 350,000 sables are harvested. Production of pelts varies by area from 0.035 to 0.13 per square kilometer per year, and averages about 0.063 per square kilometer per year (Fig. 17.1). When population densities are between 0.2 and 0.26 per square kilometer, mean harvests are generally 25–30%. Harvests of up to 35% of preseason populations are permitted, however. Of harvested animals, 40–50% are juveniles and some-what more are females than males.

Although sable populations can sustain harvest rates of 25–30%, sables have apparently declined in some regions, especially those inhabited by humans. In the major taiga regions of sable fur production, hunting areas are assigned, and within each area are refuges where harvest is prohibited. The refuges allow sable populations to restore themselves after each hunting season. In mountainous regions, hunting areas are at least 250 km², whereas in lowlands accessible by snow tractors, they are much larger, between 700 and 1000 km². In many areas, hunting now is limited to as little as two months per winter. Hunting is closed on 1 January, because early in each hunting season many juveniles and males are harvested, whereas in late winter more females and adults are harvested. Hence, trapping early in the season conserves the reproductively most important segment of the population.

Current management goals for sable conservation call for maintenance of healthy populations and restoration of those that have been reduced or lost. A system of hunting areas and accompanying refuges has been created in proportions that can sustain populations overall. Because our most reliable index of sable populations is harvest per unit area harvests should be intensively monitored on a hunting-area basis. When a decrease in productivity occurs for three consecutive years, we recommend prohibition of hunting in that area for several years until the population recovers.

18 Distribution and Conservation of Sables in China

Yiqing Ma and Li Xu

Sables (*Martes zibellina*) are closely associated with coniferous forests and are a representative forest mammal of northern Asia. Since the 1890s, the distribution and numbers of sables in China have declined, primarily as a result of overharvesting and habitat destruction. The Chinese government has given sables the highest level of protection, Category I, for threatened wildlife. This chapter describes the distributional status, natural history, and history of exploitation and conservation of sables in the People's Republic of China.

Distribution and Taxonomy

Sables are widely distributed across northern Asia and nearby islands, with China being the southernmost part of their distribution. Morphology is highly variable; 20 subspecies have been described (Ognev 1925), of which 4 occur in China. *M. z. averini* (Bashanov 1943) is found primarily in the Altai Mountain region and the northern Xinjiang Uygur Autonomous Region of northwestern China. *M. z. princeps* lives in the northwestern Daxinganling Mountains of Heilongjiang and Inner Mongolia (north of 51°N) (Fig. 18.1). *M. z. linkouensis* (Ma and Wu 1981) occurs only in China, living in the Zhangguangchailing Mountains in south-central Heilongjiang Province and the southern Xiaoxinganling Mountains of Heilongjiang Province. *M. z. hamgyenensis* (Kishida 1927) occupies the Changbaishan Mountain region along the Korean border and southward into North Korea (Fig. 18.1, Table 18.1).

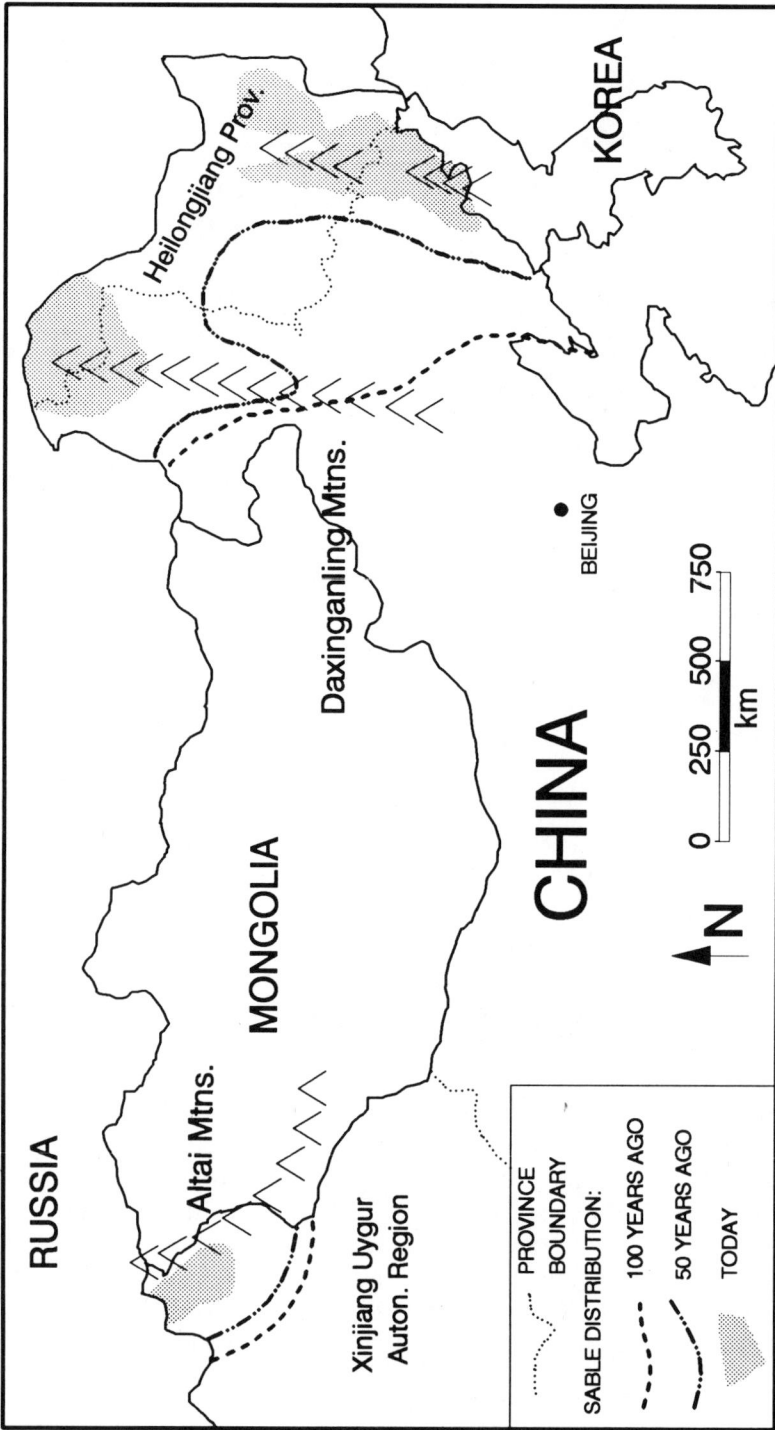

Figure 18.1. Historical and current distribution of sables (*Martes zibellina*) in China.

Table 18.1. Geographical variation of morphological characteristics of sables (*Martes zibellina*) in China

Subspecies	Color	Throat patch	Auditory ossicle length (mm)	Body length (mm)	Habitat
M. z. averini	Black-brown	Not visible, ashy brown	♂ 18.5	♂ 468 ♀ 409	Taiga
M. z. princeps	Black-brown	Not visible or reduced	♂ 18.3	♂ 412.5 ♀ 382	Taiga
M. z. linkouensis	Ashy brown	Not visible or reduced	♂ 17.3 ♀ 17.1	♂ 441.2 ♀ 388.6	Mixed forest
M. z. hamgyenesis	Fawn	Conspicuous salmon-yellow	♂ 18.3 ♀ 17.4	♂ 421.1 ♀ 378.3	Mixed forest

Habitat Ecology

Sables are most common in pristine montane forests far from human habitations. In China they occupy two major habitat types. Subarctic taiga forest is the primary habitat for the Altai and Daxinganling populations. In the northwestern Daxinganling Mountains, elevations range from 1000 to 1400 m and the topography is gentle, with wide valley bottoms. The climate is cold and dry, with seven-month-long winters and a frost-free period of 83–112 days. Mean annual temperature is about −9°C, and mean annual precipitation is about 46 cm. The dominant forest trees are larch (*Larix gmelini*) and camphor (*Pinus sylvestris* var. *mongolica*). Brushy pine (*Pinus pumila*) assumes a tree form farther north, but in the southern part of its range it is a medium to tall shrub on north-facing slopes, with long branches arching to form dense thicketlike mats up to 1.5 m deep. Larch-dominated forest has an understory of lingonberry (*Vaccinium vitis-idaea*), alder (*Alnus manshurica*), and mountain ash (*Sorbus amurensis*). Canopy cover in larch-dominated stands is typically about 60%. In autumn, sables often inhabit riparian areas where berries and nuts are abundant.

The second major habitat type occupied by sables is mixed coniferous–broad-leaved forest, which occurs at elevations of 300–700 m from the Daxinganling Mountains eastward. These eastern mountains are higher than the Daxinganling Mountains, and the highest peak, Changbaisan Mountain, reaches 2744 m. Because of the maritime influence, the climate is warmer and wetter than in the interior. The frost-free period is about 150 days, and mean annual precipitation is 500–700 mm. The dominant conifer is Korean pine (*Pinus koraiensis*), but dragon spruce (*Picea koraiensis*) and fir (*Abies holphylla*) occur in some places. The dominant broad-leaved species are maple (*Acer* spp.), linden (*Tilea* spp.), elm (*Ulmus* spp.), catalpa (*Juglans mandshurica*), northeast China ash (*Fraxinus mandshurica*), and Mongolian oak (*Quercus mongolica*). In both the taiga and mixed conifer–broad-leaved forests, food and cover conditions for sables appear to be best where the understory is dense.

Behavior and Diet

Sables are nimble and have keen hearing and vision. At the approach of danger, they climb trees and hide among the branches or seek refuge in underground dens. Sables generally are solitary, except during the mating season. Home ranges are thought to be 0.5–1 km in diameter, the largest more than 10 km in diameter during winter.

Although we have initiated studies of sables' diet, they are not complete. It appears that about 80% of the diet is mostly small birds and mammals. The remaining 20% is plant matter, including pine nuts, oak acorns, and berries. In the Daxinganling Mountains, sables commonly prey on Eurasian red squirrels (*Sciurus vulgaris*), voles (*Microtus* spp.), and red-backed voles (*Clethrionomys rutilus*). During the winter they often hunt for voles beneath the snow. In the eastern mountains of northeast China, they also prey on Siberian chipmunks (*Eutamias sibericus*), flying squirrels (*Pteromys volans*), and hares (*Lepus mandschuricus*). In spring and summer, sables eat bird eggs and nestlings, and occasionally capture adults of species as large as hazel grouse (*Tetrastes bonasia*) and ring-necked pheasants (*Phasianus colchicus*). Pine nuts are highly preferred, and in autumn sables eat many types of berries.

History of Hunting and Utilization

The fur of sables has been used in China for thousands of years. In ancient times, it was a highly prized commodity that ethnic minorities of northeastern China were required to produce. The Qing Dynasty stipulated that every official, soldier, or hunter who had attained 1.5 m in height must pay a tax of one sable pelt per year. These furs were used to make coats and other garments that could be worn only by high-ranking government officials, members of the imperial family, and aristocrats. According to historical records, in 1018 A.D., 65,000 sable furs originating from the lower Heilongjiang and Wushulijiang rivers, which form the northern and eastern boundaries of Heilongjiang Province, were submitted to the Liao Dynasty government. By about 1810, this annual harvest had fallen to 5400 furs from the northern Daxinganling Mountains. It continued to decline until only 4020 furs were submitted from all of northeastern China in 1882, and 3155 in 1889. Sables are now rare in China and are a minor component of the fur harvest.

Several methods were used to hunt sables. During winter, hunters looked for tracks in the snow, and if the tracks were fresh, followed them to the sable's den. The den was excavated, and when the sable ran, it was caught in a net erected around the den. Alternatively, leghold traps were set beneath the snow. Owenki hunters used dogs to drive sables up trees and then killed them with guns. This method was used infrequently, however, because it damaged the pelt.

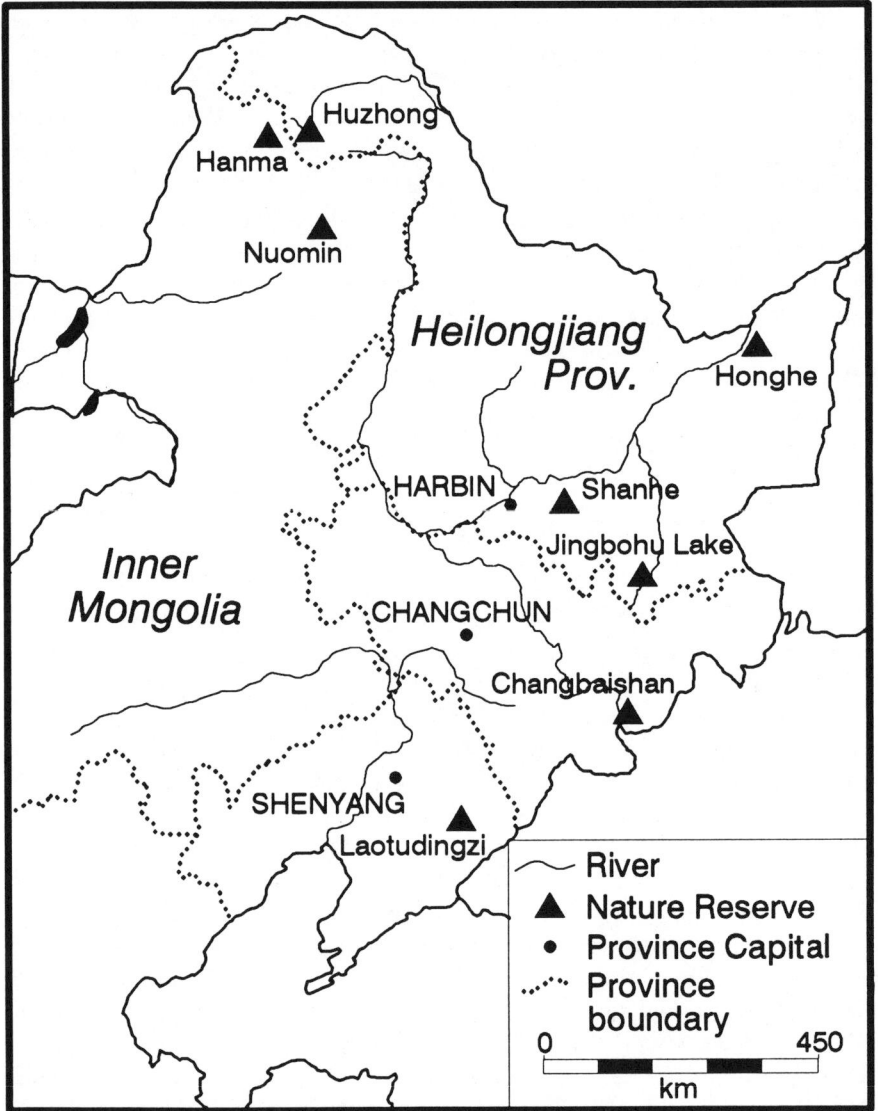

Figure 18.2. Nature reserves providing protection for sables (*Martes zibellina*) in north-eastern China.

Captive Breeding

Beginning in 1957, about 250 sables were captured from the Changbaishan and Daxinganling populations. They formed the core of a program for captive propagation that was aimed at restoring wild populations. The first reproduction from captive sables occurred in 1965, and as of 1993 the captive breeding population exceeded 500. Sexual maturity in captive sables usually occurs at 27–39 months but can be as early as 15 months and as late as 75. The breeding season extends from June to August, and total gestation, including embryonic diapause, lasts a mean of 257 days. Females have at most one litter per year, born from late March through May. Litters are usually two to three kits; the largest born in captivity is five. Neonate sables (*M. z. hamgyenensis*) weigh about 20 g, have incisors by 20–25 days, open their eyes by 30–35 days, and can be separated from their mothers by day 50, when weights reach about 500 g. Adult size is attained by five months of age, when males weigh about 1000 g and females about 800 g. Captive sables are active mostly between 6:00 and 10:00 A.M. and between 3:00 and 8:00 P.M. daily.

Conservation of Wild Populations

Sables have been hunted illegally for several decades; in the winter and spring of 1985–1986 more than 300 sables were hunted in and near the Huzhong Nature Reserve in the Daxinganling Mountains (Fig. 18.2). In an effort to reduce the illegal harvest and prevent extirpation, The Chinese government has taken measures to conserve wild sables and protect their habitat. The Wildlife Conservation Law of the People's Republic of China, which took effect in 1989, stipulates that all uses of sables, including research, are under government supervision. Penalties are provided for those who hunt sables inside or outside nature reserves or who trade sable pelts in violation of the law. By late 1990, one national and seven provincial nature reserves totalling 812,161 ha had been established for the protection of sables and their habitat (Fig. 18.2). In these areas, it is unlawful to hunt or disturb sables, or to harm their habitat.

19 Demography and Range of Fishers and American Martens in a Changing Minnesota Landscape

William E. Berg and David W. Kuehn

Fishers (*Martes pennanti*) and American martens (*M. americana*) are indigenous to Minnesota (Hagmeier 1956*a*), and at the time of settlement by Europeans their distributions were restricted to the forested zones, as mapped by F. J. Marschner (Heinselman 1974) (Fig. 19.1a). Fishers were distributed throughout the coniferous and deciduous forest zones (Fig. 19.1b), and martens occurred in the coniferous forest zone encompassing northeastern Minnesota (Fig. 19.1c). Various reports associating fishers with diverse forest habitats (Hagmeier 1956*a*, Powell 1982, Strickland et al. 1982*b*, Douglas and Strickland 1987) and martens with mature conifer forests (de Vos 1951*b*, Hagmeier 1956*a*, Strickland et al. 1982*a*, Strickland and Douglas 1987) corroborate Hagmeier's (1956*a*) historical distributions with presettlement forest zones. Hagmeier's (1956*a*) western and southern range extremes for fishers and martens (Fig. 19.1b,c) may have been due to these species' use of wooded portions of prairie, which may have been enhanced by climate conditions that were some 2°C cooler and 10% wetter in the mid-1800s than they are today (Bryson and Hare 1974).

Following the organization of the territory of Minnesota in 1849, indiscriminate human settlement, logging of pine forests (*Pinus* spp.) (Dana et al. 1960), and the fires of the early 1900s (Pyne 1982) drastically altered the forests and fauna. After logging peaked around 1900 (Dana et al. 1960), northern Minnesota's economy collapsed and many workers were forced into subsistence farming, hunting, and trapping. During this era, fishers and martens declined across their Great Lakes states range (Swanson et al. 1945, Jackson 1961, Powell 1982). As a result, fisher and marten trapping seasons were closed in Minnesota after 1929. During the next 50 years, aspen (*Populus* spp.) and balsam fir (*Abies balsamea*) reforested much of northern Min-

NORTHERN CONIFEROUS FOREST
EASTERN DECIDUOUS FOREST
TALLGRASS PRAIRIE

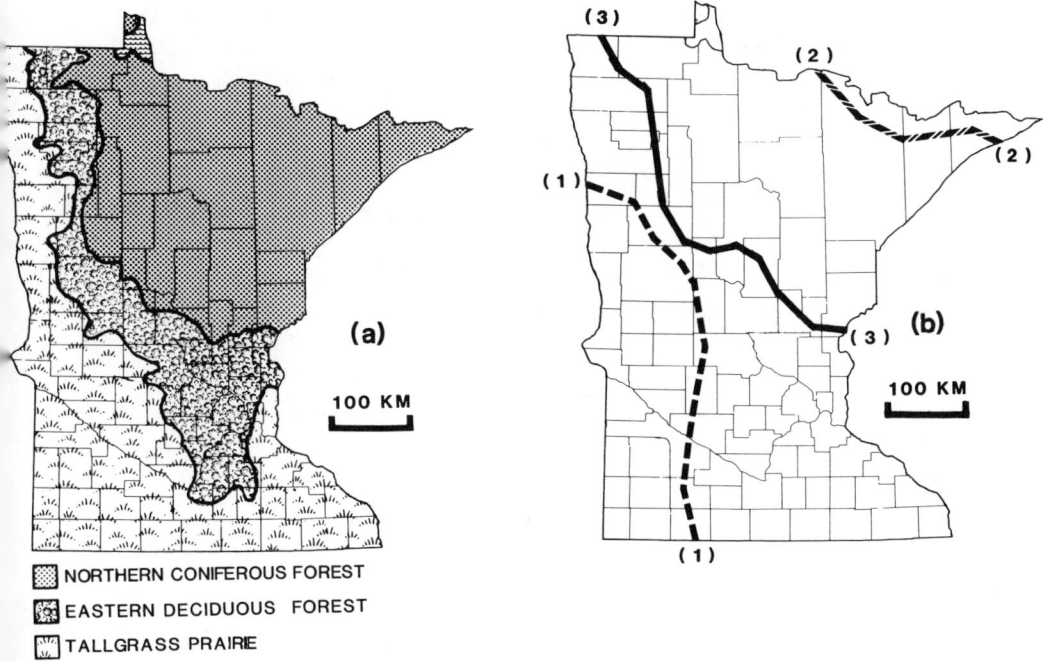

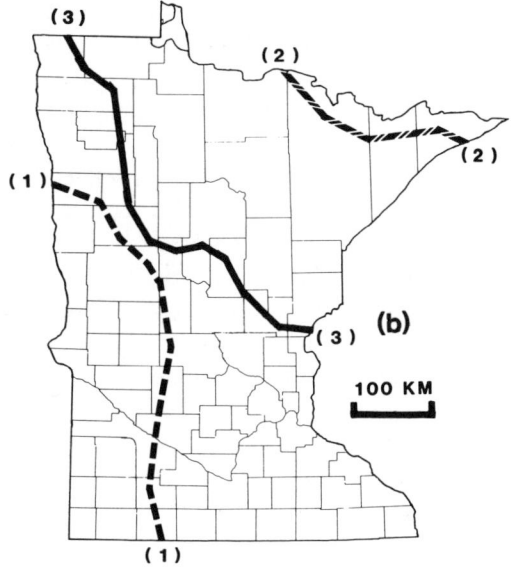

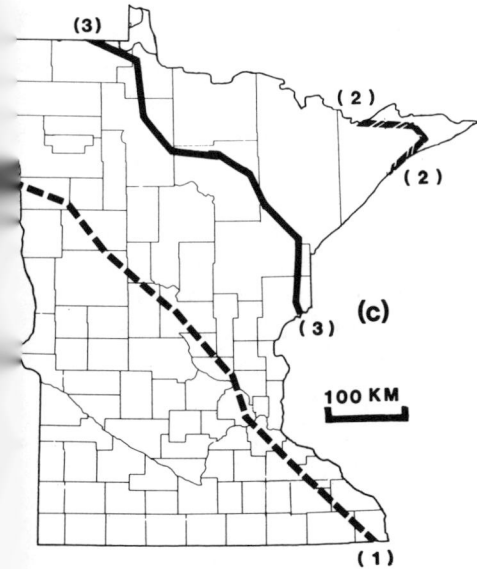

Figure 19.1. Distribution of vegetation, fishers (*Martes pennanti*), and American martens (*M. americana*) in Minnesota. (a) Natural vegetation at the time of the Public Land Survey, 1847–1907 (adapted from J. F. Marschner's 1930 map, as reported in Heinselman 1974). Southwestern limits of fisher (b) and American marten (c) distribution (1) before settlement, ca. pre-1800 (adapted from Hagmeier 1956a); (2) after settlement, ca. 1920; and (3) in 1990 (adapted from trapping registration data).

nesota, and as these forests matured, the protected fisher and marten populations began slowly to recover. Other species that inhabited Minnesota before logging, such as wolverine (*Gulo gulo*) and woodland caribou (*Rangifer tarandus*), were extirpated, however (Swanson et al. 1945).

Trapping seasons for fishers and martens opened again in 1977 and 1985, respectively, providing an opportunity to record data on distribution, harvest density, age, sex, and reproduction. Increased demand for forest products, which began in the 1980s, raises concern that even highly regulated logging may lead to excessive timber harvest and stand alteration (Zumeta 1990), which in turn may adversely affect some forest wildlife species. This chapter summarizes the range expansion and population growth of fishers and American martens in Minnesota through their post-1930s recovery, quantifies the demographic effects of exploitation, and offers a perspective on the impact of proposed timber harvests a century after fishers and martens were nearly extirpated.

Methods

Former distributions of fishers and American martens were determined from historical reports (Swanson et al. 1945, Hagmeier 1956a, Powell 1982). Current distributions were derived from confiscation records, registration data, and occasional observations through 1990 (Minnesota Department of Natural Resources [MDNR], unpubl. data).

Fisher and marten carcasses were obtained from fur harvesters and conservation officers. Juveniles (<1 year) were identified by open root canals in radiographs of canine teeth (Kuehn and Berg 1981, Dix and Strickland 1986b). Yearlings (1–2 years) and adults (>2 years) were identified by cementum annuli (Strickland et al. 1982a,b). In utero reproductive data were obtained from corpora lutea (Wright 1963). Population simulation models for both species using data on age, sex, in utero reproduction, harvest, and confiscations were developed from a modification of ONEPOP (Gross et al. 1973). The approximate rate of annual population growth from near extirpation (1920s) to abundance (1970s) was simulated using modeled finite rates of increase (Dixon 1981).

Results

Range and Distribution

Fishers. Fishers were common in Minnesota forests, even in the southeast, throughout the mid-1800s. Fur-trading posts in Minnesota collected 200 fishers in 1857 (Schorger 1942). Accounts suggest that by the late 1800s fishers were becoming rare (Swanson et al. 1945); by 1880 they were reported from only extreme northern Minnesota (Gunderson and Beer 1953). With fisher pelts bringing up to US$300 in the early 1900s (Balser 1960, Powell 1982), settlers trapping fishers stayed on the animal's trail until they caught it. The increasing rarity of fishers caused by overtrapping and habitat loss necessitated complete protection in Minnesota after 1929.

Remnant fisher populations were few and scattered, occurring mainly in extreme northeastern Minnesota (Fig. 19.1b). Gunderson and Beer (1953) hypothesized that the fisher was not extirpated because it was more difficult to trap than the marten. Between 1932 and 1942, 15 fishers were known to have been illegally or accidentally taken (Swanson et al. 1945). During the late 1940s, increasing numbers of accidentally trapped fishers were confiscated by conservation officers (Balser 1960). In the early 1950s, 36 to 41 fishers were confiscated annually; confiscations increased to approximately 100 annually by 1960. By the early 1960s, the main range was estimated (perhaps overestimated) to be 30,000 km^2 of northeastern Minnesota (Balser and Longley 1966). Because of this increase, fishers from Minnesota were used for reintroductions in Wisconsin and Michigan in 1958–1967, in Oregon in 1981 (Berg 1982), and in Montana in 1988–1990 (MDNR, unpubl. data).

Harvest distribution records from Minnesota's annual fisher trapping seasons since 1977 (except in 1980 when the season was closed) have documented the gradual expansion of range to the west and south. In 1990 fisher trapping was open in nearly all of the 87,000-km^2 fisher range (Fig. 19.1b); approximately 80% of the harvest occurs in the 52,000-km^2 "primary" range of northeast Minnesota. Harvests within this area generally take more than one fisher per 50 km^2, signs of fishers are common, and fishers are known to breed and raise young. In the late 1980s, fishers were observed and accidentally trapped on the edge of their range in northwestern and west-central Minnesota, suggesting continued range expansion (MDNR, unpubl. data).

American martens. Like fishers, American martens were depleted by the late 1800s. Martens ranked second only to beaver (*Castor canadensis*) in economic importance during the early fur-trade era (Yeager 1950). For example, 1600 martens were taken at five fur posts in northern Minnesota in 1857 (Schorger 1942), and in 1871, 96 were taken from Crow Wing County in north-central Minnesota (Swanson et al. 1945). The last known marten trapped in Minnesota for at least three decades was in the Northwest Angle (the northernmost point in Minnesota) in 1920 (Stenlund 1955). Swanson et al. (1945:63) stated that martens "may still occur rarely" along the Canadian border. Martens were listed as "probably now extinct" in Minnesota by

Hagmeier (1956a:156). In 1953 a marten was collected in northeastern Minnesota (Stenlund 1955); three were trapped accidentally in the early 1960s (Gunderson 1965), and one was live-trapped in 1969 (Maxam 1970). During 1972–1976, 50 sightings or captures were documented (Mech and Rogers 1977), and from 1977 to 1984, 1400 accidental marten captures were reported to conservation officers (MDNR, unpubl. data). Originating from remnant pockets of range in northeastern Minnesota in the 1960s, the marten range encompassed much of northeastern Minnesota by the late 1970s. In 1987–1990, 139 martens from northeastern Minnesota were translocated to northern Wisconsin (B. Kohn, Wisc. Dep. Nat. Resour., unpubl. data).

Annual American marten trapping seasons resumed in Minnesota in 1985. Registration data permitted accurate plotting of marten distribution in the 31,000-km² area open to trapping. The marten range continues to expand, and in 1990 encompassed 34,000 km² of northeastern Minnesota (Fig. 19.1c). Martens have been documented in Lake of the Woods County and Pine County, 51 km west and 80 km south respectively of the area open to trapping in 1990.

Harvests and Populations

Fishers. The fisher population was severely affected by the liberal trapping seasons of 1977–1979. The registered fisher harvest increased from 2150 in 1977 to 3032 in 1979 (Fig. 19.2), when the Minnesota fisher harvest was second only to Ontario's (Novak et al. 1987). The ratios of juveniles to adult females decreased from 8.4:1 in 1977 to 5.6:1 in 1979 (Fig. 19.3). The proportions of the available autumn population trapped (as estimated from the model) increased from 26% in 1977 to 42% in 1979. Illegal fisher trapping was rampant, and a large poaching ring was apprehended in 1983 (MDNR, unpubl. data).

Population modeling after 1979 indicated that fishers had declined 33% from 1977 to 1979 (Fig. 19.2); consequently, trapping was closed in 1980 (500 fishers were taken on Indian reservations in 1980). Since 1981, conservative trapping regulations (one per trapper per 10- to 16-day season) have permitted harvests ranging from 631 to 1642, which have taken an estimated 10–21% of the autumn population. Juvenile-to-adult-female ratios have varied from 10.5:1 (in 1981, one year after closure) to 4.7:1 in 1987 (Fig. 19.3). This restricted harvest has permitted fisher numbers to return to within 80% of 1977 levels (Fig. 19.2).

Despite dramatic changes in the juvenile to adult female ratios, the proportions of juveniles, yearlings, and adults in the harvests changed little through the population decline and recovery ($\bar{x} = 66\% \pm 3\%$ [SE] juveniles, 18% ±

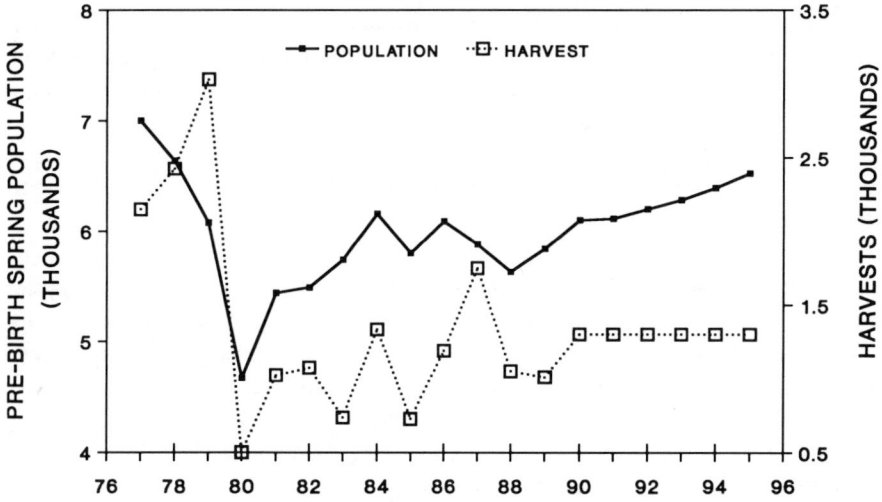

Figure 19.2. Fisher (*Martes pennanti*) population trends, estimated from modeling, 1977–1995, compared with registered harvests in 1977–1979 and 1981–1989 and projected harvests (at 1300 annually) in 1990–1995.

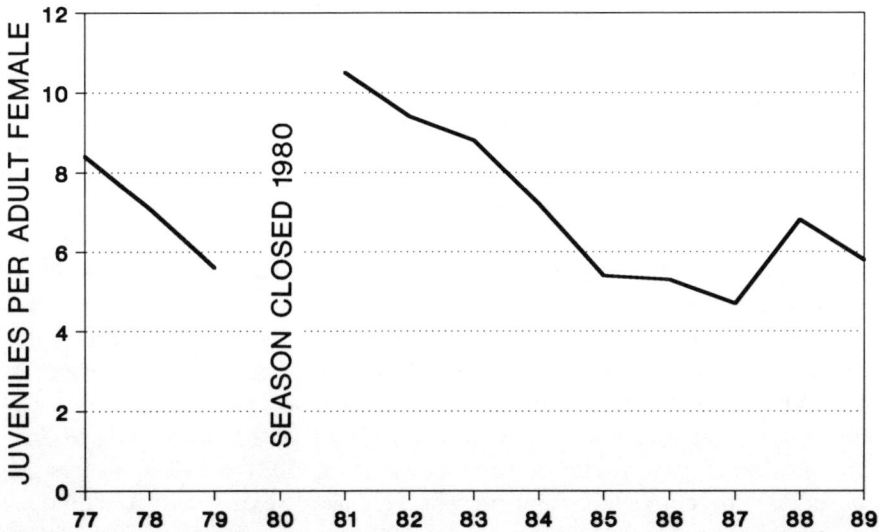

Figure 19.3. Ratios of fisher (*Martes pennanti*) juveniles (<1 year) to adult females (>2 years), 1977–1979 and 1981–1989.

3% yearlings, and 16% ± 3% adults). Males composed 48% ± 3% (range 44–54%), 41% ± 6% (28–50%), and 39% ± 6% (28–52%), respectively, of the juvenile, yearling, and adult cohorts. In utero reproductive data indicated a mean 96% ± 1% pregnancy rate ($n = 480$) and 2.9 ± 0.1 corpora lutea ($n = 459$) per pregnant yearling female, and 96% ± 1% pregnancy ($n = 308$) and 3.2 ± 0.2 corpora lutea ($n = 295$) per pregnant adult female (Kuehn 1989).

American martens. The lessons learned from the liberal fisher trapping seasons in 1977–1979 prompted a conservative approach to harvesting American martens. In 1985–1987, 10-day seasons allowed one marten per person to be trapped over a total of 17,800 km²; in 1988–1989, 16-day seasons allowed two martens per person over a total of 31,000 km². Harvests ranging from 430 in 1985 to 2072 in 1988 have taken 6–20% of the estimated autumn population. These conservative harvests have permitted the marten population to increase to an estimated 9100 in 1990 (Fig. 19.4) and their range to continue expanding.

Juvenile to adult female ratios initially showed the effects of the harvest; they declined from 17.2:1 in 1985 to 12.3:1 in 1986, 11.3:1 in 1987, and 8.6:1 in 1988. Overall harvest age structures have remained stable, with juveniles composing from 61 to 72% of the male harvest ($\bar{x} = 66\% \pm 4\%$ [SE]) and 67 to 76% ($\bar{x} = 71\% \pm 3\%$) of the female harvest. Overall, males have comprised 63% ± 5% of the juveniles harvested (range 57–69%), 64% ± 7% of the yearlings (50–71%), and 69% ± 7% of the adults (65–82%). A mean 83% of the females over 1 year ($n = 183$) were pregnant, with a mean 3.1 ± 0.8 corpora lutea per pregnant female.

Discussion

The decline of fisher and American marten populations during the settlement era and the gradual recovery during the mid-1900s in the western Great Lakes states are well documented (Jackson 1961, Strickland and Douglas 1981, Obbard et al. 1987). The rates of population recovery, however, are not. Whereas in Wisconsin and Michigan population recovery for both species has been augmented by reintroduction (Berg 1982), recovery in Minnesota has been from nontranslocated native stock. By the 1920s, fishers occurred only in remote areas and martens were nearly extirpated. Annual rates of increase can be simulated over 50–60 years to achieve the population levels in the 1970s and 1980s. If both fishers and martens numbered approximately 200 in 1920 in Minnesota, a mean annual rate of increase of 1.07 in

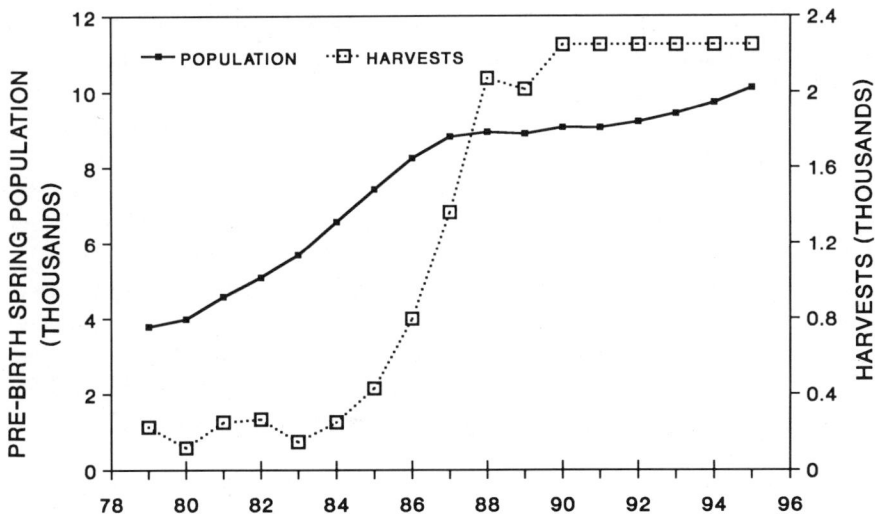

Figure 19.4. American marten (*Martes americana*) population trends, estimated from modeling, 1977–1995, compared with confiscations in 1979–1984, registered harvests 1985–1989, and projected harvests (at 2250 annually) in 1990–1995.

the model generates the approximately 7000 fishers estimated in 1977, and the 7400 martens estimated in 1985 (Figs. 19.2, 19.4).

Age, sex, and in utero reproductive data for fishers and American martens trapped in Minnesota resemble those found elsewhere in North America (Strickland et al. 1982*a*,*b*). Minnesota fisher harvests averaging 1 per 23 km² (but as high as 1 per 13 km²) during the harvests of 1977–1979 exceeded allowable recommended harvest densities of 1 per 26 km² in Ontario (Douglas and Strickland 1987). Using population models, Powell (1979*c*) calculated that fisher harvests exceeding 1 per 25 km² might cause the population to decline. Fisher harvest densities since 1981 of 1 per 31 km² to 1 per 83 km² in Minnesota fall well within the Ontario guidelines. Marten harvest densities approximating 1 per 16 km² in Minnesota are well below the 1 per 6 km² reported in Maine (Novak et al. 1987).

Considerable published and unpublished data (E. Orff, N.H. Fish and Game Dep., unpubl. data; Strickland and Douglas 1981; Strickland et al. 1982*a*,*b*; Douglas and Strickland 1987; Strickland and Douglas 1987) describe juvenile-to-adult-female ratios and their implications for fisher and marten harvest management. Although in Ontario juvenile-to-adult-female ratios of 5:1 to 7:1 for fishers were acceptable, similar ratios (Fig. 19.3) in Minnesota were concurrent with a population decline, perhaps because of the excessive harvest of a previously unexploited population, coupled with the somewhat lower fecundity of Minnesota fishers (Douglas and Strickland 1987). Strickland and Douglas (1987) considered a juvenile-to-adult-female harvest ratio of 6:1 acceptable for martens, less than the minimum reported in Minnesota ($\bar{x}$ = 8.6:1), where modeling indicates the marten population is increasing (Fig. 19.4).

Sex ratios are a less reliable (but still useful) means of assessing harvest intensity because both fisher males (Kelly 1977) and marten males (Yeager 1950) are more vulnerable to trapping. A too-high proportion of females in the harvest indicates overexploitation. Douglas and Strickland (1987) suggested that on the average, Ontario fishers declined when harvests contained less than 45% males and less than 5.7 juveniles per adult female. When fishers declined in Minnesota following the 1977–1979 seasons (Fig. 19.2), the overall proportion of males ($\bar{x}$ = 46%, range 40–50%) was similar to that in 1981–1989, when the population gradually recovered ($\bar{x}$ = 45%, range 43–49%). The juvenile-to-adult-female ratio was below Ontario's recommended 5.7:1 during 4 of the 12 fisher seasons. In Minnesota, males consistently dominated the marten harvest in all age classes.

Most sources attribute the decline of fishers and American martens in the early 1900s to extensive logging of mature and climax forests, unregulated trapping, and fire, and they attribute the recovery of these species to protection and gradual forest maturation (summarized for fishers by Powell [1982] and for martens by Nordquist and Birney [1988]). From the mid-1940s to the early 1970s, pulpwood logging in northern Minnesota took approximately 0.8–1.0 million cords annually (Dana et al. 1960) for use by five wood-products mills. This rate generally underharvested pulpwood timber, and many aspen forests gradually matured to conifer forests dominated by balsam fir. One indication of the fisher and marten responses to the forest succession is the decrease in fisher harvests in Cook County (the northeasternmost county and among the oldest, successionally) from 6% of the total state harvest in 1977 to 1–2% of the total state harvest throughout the 1980s. Conversely, marten harvests in Cook County increased from 9–10% of the total harvest in 1985–1986 to 15% in 1988. Although we lack data on trapping pressure, the concurrent fisher decline and marten increase in Cook County may have been tied to forest succession, as demonstrated by the presence of old (>60 years) aspen. In 1977, 24% of the aspen in northeastern Minnesota, the main range of American martens, exceeded 60 years in age, compared with 9% in north-central Minnesota (the main range of fishers) (Jakes 1980). In general, fishers are now common in most of forested northern Minnesota, and martens, while mainly associated with mature forests, have expanded their range into younger forests as well.

In 1990, 13 large wood-products mills existed in northern Minnesota. The consumption of pulpwood by these mills was 3.2 million cords in 1989 and is expected to approximate 4.9 million cords in 1995 and 7.0 million cords annually into the twenty-first century (Zumeta 1990). The total area affected will approximate 100,000 ha annually. Despite the increased logging, a repeat of the events following the extensive unregulated harvest of virgin forests is unlikely. Several controls are in place that regulate timber harvests. For example, the MDNR Division of Forestry attempts to restrict clear-cut area size to less than 8 ha in the northern forest, while maintaining generally irregular clear-cut boundaries. The Superior National Forest in northeastern Minnesota maintains a road density of less than 1 km of road per square kilometer of forest in order to minimize interactions between humans and wolves (*Canis lupus*). In an effort to maintain mature stands, 10% of old aspen will not be cut (MDNR, unpubl. data). Diverse forest ownership in northern Minnesota (21% state, 20% federal, 20% county and municipal, 39% private or other) poses a problem of timber harvest coordination that is still largely unsolved.

Because fishers prefer a diversity of forest types with a high degree of interspersion (Arthur et al. 1989*b*), it is unlikely that increased logging under present guidelines will adversely affect their numbers. Increased logging, however, has the potential to depress American marten populations by depleting mature forest stands. Because clear-cutting may reduce an area's carrying capacity for American martens, optimum marten habitat should include (1) 25% of the forest in mature age classes, (2) at least 50% conifers, and (3) uprooted blow-downs for subnivean rest sites (Soutiere 1979, Steventon and Major 1982, Buskirk et al. 1989). Present-day concerns for mature forest ecosystems, in concert with vastly improved logging practices and effective trapping regulations, suggest that neither American martens nor fishers will suffer the fate they did a century ago.

20 Ecology and Conservation Status of the Tsushima Marten

Masaya Tatara

The Japanese marten (*Martes melampus*) is distributed in the Japanese Islands and the southern Korean Peninsula (Anderson 1970; Fig. 20.1). Because it is primarily nocturnal and found only in forested areas, its natural history and ecology are poorly known. In Japan, the Japanese marten is trapped for its fur every hunting season (from 1 December to 31 January) except on Hokkaido Island, where it is sympatric with the fully protected sable (*M. zibellina*), and on the Tsushima Islands.

The Tsushima marten (*M. melampus tsuensis*), found only on the Tsushima Islands, has been protected from trapping since its designation as a vulnerable Natural Monument Species in 1971 by the Japanese Agency of Cultural Affairs. Conservation plans have not yet been developed, however, because little is known about the biology of this species. The study I report here was designed to collect basic information on the ecology of the Tsushima marten for use in conservation and management. Data were collected on distribution, habitat use, spacing patterns, and diets.

Study Areas and Methods

The Tsushima Islands (34° 23' N, 129° 20' E) are situated in the Korean Strait and comprise two main islands, North and South Tsushima, plus several smaller islands (Fig. 20.1). The total land area is approximately 710 km², of which 88% is forested. The climate is moderately warm, with temperatures ranging from a mean daily low of 4°C in January to a mean daily high of 26°C in August. Snowfall is uncommon and light. The human population in the islands was approximately 48,500 in 1990.

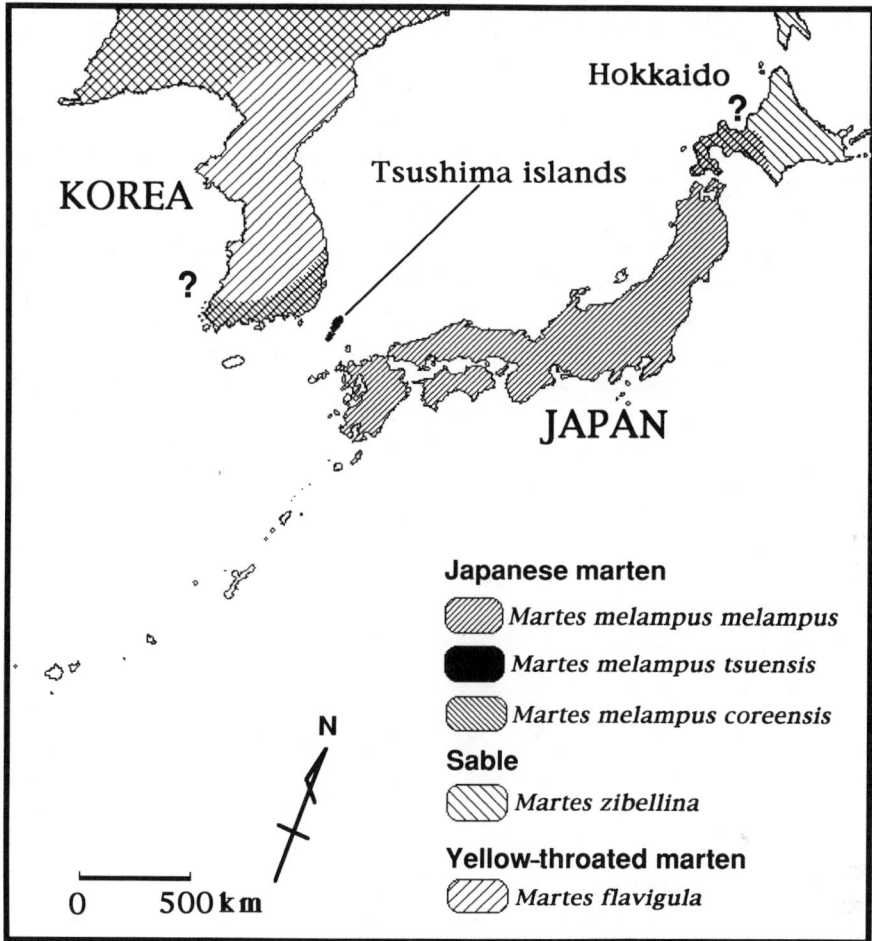

Figure 20.1. Geographic distribution of the Japanese marten (*Martes melampus*) and other *Martes* spp. in the Far East (modified from Anderson 1970.)

To determine the distribution of the Tsushima marten, I conducted scat surveys in September 1987, July 1988, and November 1988 in a total of 59 regions throughout most of the islands. Routes ranging in length from 1.1 to 7.0 km were established in each region, at between 0 and 558 m elevation. Trails and roads were searched and marten scats were sought along the census routes in transects approximately 4 m wide. Although the Tsushima Islands are inhabited by the Siberian weasel (*Mustela sibirica coreana*) and the leopard cat (*Felis bengalensis euptilura*), marten scats could be discriminated from those of other carnivores by size, shape, color, and smell.

Two areas were established for telemetry and diet studies. The Tanohama study area (TH) was located on the west coast of North Tsushima Island, and the Uchiyama study area (UY) was established in the center of South Tsushima Island. Both study areas were approximately 25 km^2 in area and were predominantly forested, but they also contained active and abandoned rice fields and residential areas. Natural forests were dominated by deciduous broad-leaved trees (e.g., *Quercus serrata*) in TH, and by evergreen broad-leaved trees (e.g., *Castanopsis cuspidata*) in UY.

Martens were captured in wooden box traps baited with chicken meat or live chicks from November 1987 to August 1990. Captured martens were immobilized with intramuscular injections of 50% ketamine hydrochloride (0.4–0.5 mg/kg body weight). They were then sexed, weighed, measured, and outfitted with 50-MHz transmitting radio collars weighing approximately 35 g for males and 25 g for females. Radio-collared martens were tracked for a few days after release to detect any detrimental effects of capture. For about two months following capture, locations of collared martens were estimated by triangulation from two or more points with two-element hand-held Yagi-type antennae at distances of 25–190 m. I assessed the accuracy of telemetry by locating transmitters placed by another observer.

Home range sizes were estimated with the convex polygon method. Martens were radio-tracked at about one- to five-hour intervals until convex polygon areas were saturated and at one- to 26-hour intervals following saturation. Areas of each habitat type were calculated from 1:5000 aerial photographs and were compared with percentages of telemetry locations fixed in each habitat type by chi-square goodness-of-fit tests and Z-statistic to determine habitat preferences. Furthermore, the scat census transects over-lapped home ranges, which enabled me to determine the distribution pattern of scats in the home range. Several home range areas were divided into 100-m × 100-m grids and the number of relatively new scats were counted along the lines.

In TH and adjacent areas, approximately 14.2 km of farm roads and pathways were established as scat sampling routes to determine diets, and scats were collected monthly from May 1986 through August 1987. The scats were soaked in water to soften them and washed through a 0.5-mm saran net. Remains were preserved in 80% alcohol for further identification. Animal remains were sorted to class level and identified to species when possible. Plant materials were sorted to species or genus. The percentage frequency of occurrence of each food type was used to show monthly variations of marten food habits.

Results

Marten scats were found along 57 of 59 routes. Density of scats varied among the routes (range = 0.3–8.2 scats/km), but martens appear sparsely distributed throughout the two main islands of Tsushima. The two routes that yielded no scats were on small islets (<0.7 km²) isolated from the main islands. Weight and total length of the nine male martens captured averaged (± SD) 1563 ± 148 g and 668 ± 21 mm. Four females averaged 1011 ± 125 g and 607 ± 13 mm. All martens captured were considered adults from tooth eruption and tooth wear. The sexes differed significantly in body measurements ($t = 7.03$, 24 df, $P < 0.001$).

The telemetry accuracy tests indicated that locations determined by triangulation were usually 5 m or less from the true locations. The greatest errors occurred when triangulation was attempted from 1 km or more; therefore, I usually tracked martens from 100 m or less. Numbers of telemetry locations varied from 33 to 427 per marten, and all independent locations were used to estimate the home range sizes. Home range size averaged 0.70 ± 0.26 km² for eight males and 0.63 ± 0.13 km² for three females. Between-sex differences in home range size were not significant ($P > 0.05$). In UY one male was captured twice in five months in the same area. In TH another male was captured twice in seven months in the same trap site. At least these two males appear to have been resident.

Habitat and Resting Site Use. Use of habitats differed significantly from what would be expected on the basis of spatial availability ($X^2 = 17.1$, $P < 0.05$). Broad-leaved forest composed 69% of the study area, but 88% of marten locations by telemetry were in this habitat. Martens used broad-leaved forest more than expected ($Z = 1.78$, $P < 0.05$), and conifer plantations significantly less than expected (5% of telemetry locations, 18% of study area). Open fields were also used significantly less than expected (6% of telemetry locations, 11% of study area).

Locations determined by triangulation on apparently resting martens were in limited areas within the home ranges. These areas, which I believe were resting sites, numbered from 4 to 10 per marten and were found in broad-leaved forest within each home range. Although I could not examine these resting sites at close range because of the possibility of disturbing the martens, I believe on the basis of radio signals that they were in tree dens or in ground burrows.

Spacing Patterns. Home ranges overlapped little: the 16 home ranges shared on average only 10.2% ± 9.4% of their area with other home ranges.

Tsushima martens appear to be territorial, judging from the exclusive spacing patterns. The line transects overlapped several marten home ranges, and on average 76% of the scats were found in the peripheral 50% of home ranges; the distribution of scats in each home range showed a concentration near the home range border. Therefore, scat concentrations can be used to determine not only areas occupied by martens but also maintenance, through scent marking, of their social system.

Food Habits. A total of 969 scats were collected in TH and neighboring areas for diet analyses (Fig. 20.2). Mammal remains in the scats comprised a relatively stable proportion (range 8.2–16.8%) of the diet throughout the year, whereas bird remains showed a peak from January (11.3%) to March (14.6%). Amphibians, mostly small adult frogs (*Rana tsushimensis*), were most common in the diet in February (9.7%). Insects were the most common animal class, but their proportion in the diet varied greatly from 10.5% in May to 27.8% in August; insects were consistently least common throughout winter. Centipedes, most of them large adult *Scolopendra subspinipes*, were frequently eaten during May and June (16.9–17.4%). Plant materials were the most common of all foods throughout the year (28.8–53.9%, Fig. 20.2). Berries and seeds, the most common plant remains, occurred at the highest frequency in April (41.0%, especially *Rubus hirsutus* and *Elaeagnus pungens*) and in September (47.3%, especially *Vitis ficifolia* and *Ficus electa*), following fruiting phenology. Relative occurrence of plant matter other than berries and seeds reached a maximum of 17.5% in December and January, when pollen and stamen parts of camellia (*Camellia japonica*) were eaten.

Discussion

Tsushima martens are found mostly in broad-leaved forests along valleys. Their important foods (small mammals, insects, and fruits) may be less abundant in other habitats. Although about 88% of the area of the Tsushima Islands is forested, the forests are harvested. As of 1993, 34% of the forest was in conifer plantation, and the amount of land planted with conifers will likely increase in the future. The important foods of Tsushima martens may not be common in such plantations. Martens appear sparsely distributed on North and South Tsushima Islands, and the population is expected to decline as its preferred habitat is destroyed.

Home range size was similar between sexes, about 0.5–1.0 km^2. Home ranges of American martens (*M. americana*) determined by the same methods are larger (3.5–9.6 km^2: Raine 1982, Taylor and Abrey 1982), as are

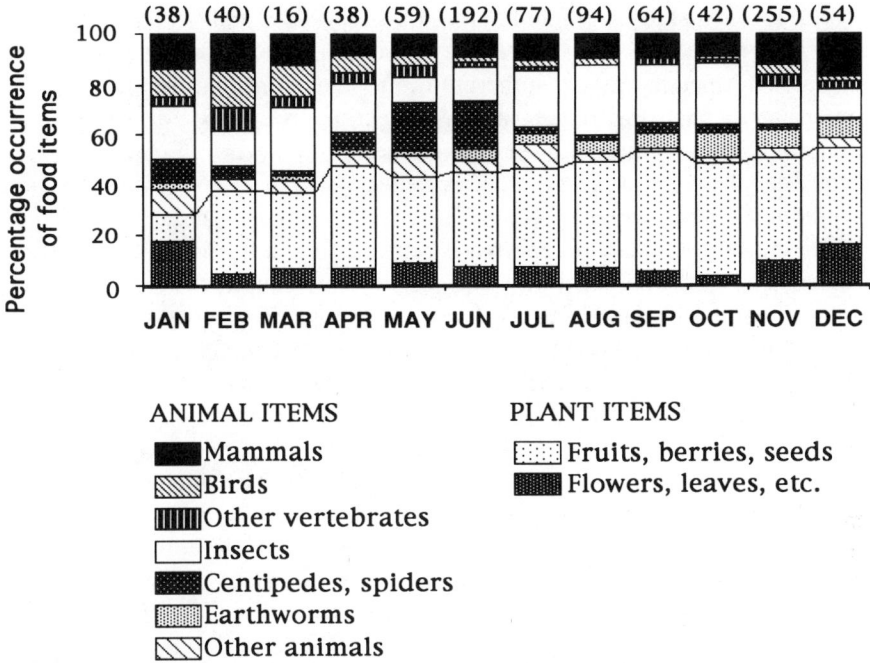

Figure 20.2. Seasonal changes in diet of the Tsushima marten, determined from 969 marten scats collected monthly in the TH area from May 1986 through August 1987. Numbers in parentheses are sample sizes for each month.

those of fishers (*M. pennanti*; 15.0–35.0 km^2: Powell 1982, Raine 1982). The small home ranges found in this study may be due to small sample sizes and short monitoring periods. Tsushima martens were followed for 15–37 days, whereas animals in other studies were followed for 31–198 days. Home range sizes for Tsushima martens, however, may also depend on the greater availability and distribution of foods such as fruits and berries within their preferred habitats.

Tsushima martens appear to be territorial, as do fishers (Powell 1982). Many solitary carnivores scent-mark with scats (Sandell 1989*a*), and Tsushima martens may scent-mark their territories in specific locations to facilitate intraspecific communication. The "doughnut" distribution pattern of scats in home ranges suggests that scats are used as scent marks to maintain home ranges with little overlap from conspecifics.

Each marten studied chose at least four resting sites in broad-leaved forests. The resting sites were in tree dens and ground dens. Buskirk and Powell (this volume) consider resting sites to be important for American martens, both to lower resting thermoregulatory costs in winter and to provide protection from predators. The climate in Tsushima is much milder than in the habitats of American martens; therefore, thermoregulation is probably not as important in choosing resting sites. Protection from predators, especially feral dogs, may be more important than thermal protection. Feral dogs were distributed throughout most of the forested areas on the two main islands, and five adult martens killed by feral dogs have been found since 1986.

The diet of the Tsushima marten is rather different from that of American martens and stone martens (*M. foina*). Small mammals are usually the primary foods for martens in North America and Europe. These mammals include voles in the United States (Murie 1961, Weckwerth and Hawley 1962), mice and shrews in Spain (Delibes 1978), and lagomorphs in Czechoslovakia (Holisova and Obrtel 1982). Small mammals were not the dominant foods for Tsushima martens in this study, but small rodents were eaten fairly constantly throughout the year and likely were important foods. The small rodents eaten by Tsushima martens could not be identified to species, but two species of wood mice, *Apodemus speciosus* and *A. argenteus*, were both found in the study area and presumably were eaten. The mammal fauna in Tsushima is unusual in lacking squirrels and lagomorphs, both of which are common in other areas of Japan. The diets of *Martes* spp. appear to reflect the variation and abundance of local food types. Insects and fruits were the most important foods for Tsushima martens. Goszczynski (1976) and Arthur et al. (1989*b*) reported that fruits were very common in the diets of stone martens and fishers but were seasonal foods. In Tsushima, fruits were found in marten scats throughout the year. At some times of the year, fruits appear to be the only foods. Food sources other than small mammals, insects or fruits may be secondary and complementary.

Natural habitats on the Tsushima Islands have been altered to conifer plantations that have few fruiting trees or grasses and probably have poor food resources for martens. Because the remaining preferred forests are patchily distributed, it will be difficult to preserve the subspecies. The Tsushima marten was listed as one of the vulnerable mammals in the 1991 Red Data Book of Japan.

Fifty-three dead martens were found on the islands since 1986, 38 (72%) of which were killed by vehicles and 5 (9%) by feral dogs. Paved roads not only isolate forested habitats, but they also present a significant mortality factor for the current marten population. Although no information is available on the number of feral dogs in Tsushima, the feral dog population should be controlled for the survival of the Tsushima marten. A conservation plan for the Tsushima marten should consider the impacts of continuing habitat loss, habitat segregation due to the road system, road kills, and mortality caused by feral dogs.

Acknowledgments

Special thanks go to Y. Ono and T. Doi, Department of Biology, Kyushu University, for critical comments on my study. I also received valuable comments from two anonymous reviewers. The Japan Wildlife Research Center assisted with the survey on marten distribution. The study was partly supported by the Studies for Natural Monuments in Tsushima through the Ministry of Education, Science, and Culture, Japan.

Introduction

Habitat ecology of martens and fishers is interlinked with other aspects of their biology. A species' body size and shape, for example, impose constraints on its ecological needs and capabilities. Each population manifests the outcome of the interplay of these constraints with adaptation and opportunism. Buskirk and Powell begin this section with a synthesis and contrast of the habitat ecology of American martens and fishers. They point out the assumptions that are central to most habitat selection studies and reveal the challenge faced by all—that is, the importance of fitness to defining habitat quality, and the use of behavior and population attributes as surrogates for fitness. American martens tend to associate with mature coniferous forests, whereas fishers are sustained by a broader range of broad-leaved and mixed coniferous forests. Both species seldom venture into large openings, although the reason that the larger-bodied fisher shows a stronger avoidance for such areas is unclear.

Buskirk and Powell contend that habitat structure is important to American martens and fishers for different reasons. Standing and downed trees are beneficial to American martens because they provide access to subnivean areas where the animals forage and seek thermal cover during winter. Fishers, on the other hand, hunt their prey on top of the snow. Habitat structure for them is favorable when it reduces sinking into the snowpack, ensuring that less energy is expended for locomotion. Buskirk and Powell emphasize that "a mouse in the mouth" is not synonymous with prey density. A predator's success at capturing prey depends not only on prey density but on factors that affect prey availability. Buskirk and Powell propose that prey availability, not prey density, is likely the foremost factor in the selection of foraging habitat by American martens and fishers. Prey availability depends strongly on habitat structure.

Differences in habitat use by American martens and fishers are, in part, linked to their feeding ecology. Martin searches for patterns in diet studies of American martens and fishers across a wide geographic range. American martens feed on smaller prey and have less diverse diets than do fishers. These diet differences may constrain martens' choices of suitable habitat more than fishers'. Martin goes further and hypothesizes that martens are more sensitive to habitat alterations than are fishers.

Clevenger extends this theme of interdependency between feeding ecology and habitat use to Eurasian pine martens and stone martens. While both species are similar in size and shape, they exhibit striking dietary and habitat differences. Both Martin and Clevenger examine variation in *Martes* species' diets and the relationship of this variation to geography and habitat. Dietary variation within and across sites shows the adaptability of each species and helps to identify the range of tolerable ecological conditions.

Although complete knowledge of habitat ecology is not yet available, we clearly have both a foundation and a need for formal frameworks for habitat ecology. Thomasma et al. create such a framework for fishers by incorporating habitat attributes, food, and other factors into a habitat suitability model. After testing this model, they conclude that our notions of fisher habitat ecology appear to be generally correct, but that some aspects need to be reexamined and refined.

We need not only additional information about habitat ecology of martens and fishers, but, more important, we need different approaches. Buskirk and Powell assert that feeding ecology is vital to understanding habitat ecology, but both Martin and Clevenger note that diet studies seldom report abundances and availabilities of prey; we often know little about prey selection. Habitat use, while easy to measure and describe, appears to have reached its limits of insight. Habitat selectivity studies provide a better perspective of habitat ecology, but the relationships between habitat use and fitness remain elusive. The thrust of habitat research needs to be toward understanding the factors affecting choices made by *Martes* species within specific ecological contexts.

ALTON S. HARESTAD

21 Habitat Ecology of Fishers and American Martens

Steven W. Buskirk and Roger A. Powell

Several lines of evidence suggest that the genus *Martes* evolved in forested habitats (Anderson, this volume). These include the habitat associations of most extant species (Brainerd, this volume; Tatara, this volume; Bakeyev and Sinitsyn, this volume) and the association of fossil and subfossil remains with those of forest mammals (Voorhies 1990) and plants (Anderson 1984, Grayson 1984). Extant close relatives of *Martes* (Holmes and Powell, this volume) are good tree climbers. Additionally, the paucity of fossils of early *Martes* argues for a long association with forest habitats (Anderson 1970), since forest environments are not conducive to fossil formation and preservation.

Fishers (*M. pennanti*) and American martens (*M. americana*), like most of their Old World congeners, have been recognized as forest associates since the time of early naturalists (Seton 1929). Some recent researchers have considered them specialists for certain forest types (Bissonette et al. 1989), whereas others have deemed them more adaptable to different habitats (Strickland et al. 1982*a*). Harris et al. (1982) characterized fishers and American martens as the most obligate for late-successional habitats of 68 species of forest mammals in the Pacific Northwest. These contradictory viewpoints reflect inconsistent results of field studies, interpretational and semantic differences among biologists, and a basic lack of understanding of the selective factors affecting habitat choices by these two species.

Habitat quality is defined in terms of the fitness a habitat confers on its occupants (Fretwell 1972). In the case of fishers and American martens, measurement of total fitness, and even of some fitness correlates such as survival and reproduction, is difficult, even by standards for mammals. As a result, attempts to compare quality of habitats for these two species have bypassed the need to measure fitness relative to habitat type by assuming that habitat preferences, as indicated by time- or distance-based proportional use, maximize fitness of the animals that exhibit those preferences. In this review we try to show how this assumption can be misconstrued or tacitly modified to suggest that all animals show consistent preferential use of high-quality habitat throughout their lives. We also try to use the most appropriate scales for understanding choices made by individual animals in local areas, for geographic comparisons within species, and for comparisons between species. Because there is no evidence that kin- or group-selective forces act strongly on *Martes*, our basis for all comparisons are the choices or preferences of individuals.

The issue of scale enters into our review in several ways and may have influenced our conclusions. No universally appropriate scale has been determined for habitat analyses because the scale must match the goals of the analyst; choosing the incorrect scale can lead to reverse conclusions (Rahel 1990). First, habitat variables used in comparisons must be appropriate to the geographic scale of the comparisons. Information on cover types used by animals in different areas may be misleading or not at all useful when comparisons are made only between cover types identified by their component species. We gain little by knowing that cover types, identified by tree dominants, used by American martens in California are different from those in Maine; the forests share no tree species and martens therefore cannot possibly use the same cover types in both places. Cover types can be useful, however, when considered in nontaxonomic comparisons involving successional stages, structure, and similar factors.

Second, analyses of habitat preferences by animals within a population may yield different results when done on different scales. Kelly (1977) reported opposite results for choice of cover type by fishers, using analyses of cover types within home ranges and analyses of time spent in forests of different cover types. Knowing the composition of the forest used by the fisher population studied by Kelly in the White Mountains of New Hampshire is different from knowing the preferences of the individual fishers for particular cover types within the forest.

Finally, an individual animal's choice of habitat will vary with its activities. Foraging behavior may depend on the scale of physical structure of logs, branches, and rocks on the forest floor. Wilbert (1992) showed that choices of resting places were made on the scale of the individual piece of coarse woody debris and were relatively nonselective at larger scales within the forest. Choice of where to establish a home range may be made on the landscape scale.

In this chapter we review the commonly measured habitat factors that affect where American martens and fishers spend their lives and that we assume determine individual fitness. We compare preferences for cover types and for specific habitat features by the North American *Martes*, and we try to identify the biological principles that might explain both habitat selection and variation in habitat selection by these species. We conclude with some generalizations that can be used as starting hypotheses for future research. We acknowledge that, in our attempt to generalize, we may have excluded important exceptions and that our generalizations will not apply to fishers or American martens in some places and under some circumstances.

Our discussion pertains to established habitat types and to successional sequences of stand types and habitat characters. We mention only incidentally the effects of major disturbances. Fire and logging are dominant causes of habitat change in landscape mosaics throughout the North American range of fishers and American martens, and many studies have included recent burns and clear-cuts among the habitat types investigated. Thompson and Harestad (this volume) consider the effects of major habitat disturbances on fishers and American martens.

Cover Types

In general, the habitats used by fishers and American martens are forest or woodland landscape mosaics that include conifer-dominated stands. These landscape mosaics have variable amounts of broad-leaved (hardwood) forests, ranging from none to as high as 80% composition for fishers (Powell 1993) and 100% for American martens (Allen 1984). In the parts of their geographic ranges with a mixture of mesic and xeric forests, both species prefer mesic forests over xeric ones. For example, American martens in the Rocky Mountains prefer high-elevation stands dominated by spruce (*Picea* spp.) and fir (*Abies* spp.) over stands dominated by dry-site species such as ponderosa pine (*Pinus ponderosa*) and lodgepole pine (*P. contorta*) (Campbell 1979, Buskirk et al. 1989). In some forests, for example the taiga, this preference is not seen because xeric stands are absent.

At temperate latitudes, mesic forests are commonly riparian. American martens in the Sierra Nevada of California selected riparian forests for foraging (Spencer et al. 1983, Martin 1987), and martens in the Rocky Mountains of Wyoming selected riparian forests for resting during winter (Buskirk et al. 1989). Riparian forests were considered by Buck et al. (this volume) to be very important for fishers in the coastal mountains of northern California, and in Idaho, Jones (1991) found fishers close to open water, primarily streams, more often than expected.

For both species, physical structure of the stand seems to be more important than species composition. Both prefer forests tending toward low and closed (although not uniform) canopies. American martens in Yosemite National Park in California and fishers in the White Mountains of New Hampshire preferred forests with canopies that were low in height (Kelly 1977, Hargis and McCullough 1984). American martens in Idaho preferred forests with more than 30% canopy closure, and those in the Sierra Nevada of California preferred forests with 40–60% canopy closure and avoided those with less than 30% canopy closure (Koehler and Hornocker 1977, Spencer et al. 1983). Canopy closure is a major component of the Habitat Suitability Index model developed for fishers by Allen (1983), and fishers in the Upper Peninsula of Michigan preferentially used forests with high index values (Thomasma et al., this volume). Studies have generally shown that fishers and American martens avoid open areas where no overstory or shrub cover exists. Some animals studied never used open or nonforested areas (Hawley and Newby 1957a; Kelly 1977; Spencer et al. 1983; Arthur et al. 1989b; Jones, this volume), whereas others would occasionally travel along edges of open areas or cross narrow open areas (Robinson 1953, Koehler and Hornocker 1977, Soutiere 1979, Simon 1980, Buskirk 1983, Raine 1983, Hargis and McCullough 1984). Even these animals still showed a marked, general avoidance of open areas. Some studies (e.g., Streeter and Braun 1968) have reported that American martens occasionally traveled as far as 3 km from forested areas in summer, provided that a substitute for vegetative cover (e.g., talus fields) was available.

Snow cover is also associated with broad-scale and seasonal habitat selection, but in different ways for the two species. Where deep snow accumulates, American martens prefer cover types that prevent snow from packing hard (e.g., from wind) and that have structures near the ground that provide access to subnivean spaces (Hargis and McCullough 1984, Corn and Raphael 1992). In contrast, fishers may pay higher energetic costs to travel in soft, deep snow, which may explain why they concentrate winter foraging in cover types where snow is shallow or becomes packed (Raine 1983).

Throughout the distributions of these two species, the predominant late-successional forest types are dominated by conifers. Where these late-successional forests are characterized by spruce and fir, American martens are more closely associated with old forests than are fishers, which also use forests with a significant broad-leaved component. Both fishers and American martens disappeared from forests in the northeastern and midwestern United States early in this century (Strickland et al. 1982a). When protected from trapping, fishers recolonized midsuccessional second-growth forested areas in New York, Maine, and Minnesota that had continuous canopy of mixed broad-leaved and conifer species (Powell 1982). Fishers also were successfully reintroduced to forests of these types in Michigan, Wisconsin, and New England (Powell 1982). Martens did not recolonize these forest types and reintroductions did not establish viable populations (Churchill et al. 1981; Berg and Kuehn, this volume; Slough, this volume). Martens did recolonize forests with a significant late-successional component, however, and reintroductions to such forests were successful.

In coniferous forests of the Pacific Northwest, American martens are associated with high-elevation spruce-fir forests, whereas fishers are associated with lower elevation forests dominated by late-successional Douglas-fir (*Pseudotsuga menziesii*) associations (Buck et al. 1983; Raphael, this volume). It is well known that structural components and prey populations in late-successional spruce associations are important to American martens. Thus martens meet their needs within a single category of cover types. There is no such consistent association for fishers; fishers are able to meet their needs in several very different cover types within each region. We hypothesize that, in fulfilling the life needs of fishers, structural components and prey associations in midsuccessional mixed broad-leaved and coniferous forests of the midwestern and northeastern United States are equivalent to those in late-successional Douglas-fir forests in the Pacific Northwest.

Habitat preferences of martens and fishers parallel those of their preferred prey (Buskirk and MacDonald 1984, Koehler et al. 1990). American martens prefer *Microtus* spp. and *Synaptomys* spp. (mesic-site species) over *Peromyscus* spp. and *Eutamias* spp. (xeric-site species) (Koehler and Hornocker 1977). The bases of these preferences are not well understood, but we hypothesize that habitat preferences of American martens are generated by relative difficulty of catching prey. Still, it is not clear whether American martens prefer cover types that are occupied by prey whose behaviors make them easy to catch, or cover types with physical structure that renders prey more vulnerable. The linkage between habitat preferences of fishers and their prey may be clearer. Dense lowland forests in northeastern North America are preferred by snowshoe hares (*Lepus americanus*), a major prey species for fishers (Powell 1982, Arthur et al. 1989*b*). In the Pacific Northwest, the range of the snowshoe hare coincided with areas that were originally in late-successional Douglas-fir forests (Bittner and Rongstad 1982).

In Eurasia (Stroganov 1969, Degn and Jensen 1977, Pulliainen 1981, Syrjanov 1989), sables (*Martes zibellina*) and to a lesser extent Eurasian pine martens (*M. martes*) and Japanese martens (*M. melampus*) associate with late-successional coniferous forests, often of spruce, fir, or larch (*Larix* sp.). These forests are similar in structure to those occupied by American martens. There is no living Old World equivalent to the fisher. Eurasia has no *Martes* species that is adapted to midsuccessional second-growth forests. North America in turn has no *Martes* species that is adapted to agricultural areas and low-density urban areas, as is the stone marten (*M. foina*). This latter niche is, however, well filled by raccoons (*Procyon lotor*) and striped skunks (*Mephitis mephitis*).

Effects of Edge, Area, and Insularity

Evidence for preference or avoidance of edges is scarce, likely because most of the habitat preference studies for American martens and fishers have used radiotelemetry. By definition, edges are small in at least one dimension and often they are small in total area. The error associated with radiotelemetry estimates for locations of American martens and fishers usually is greater than the areas of edge habitats. This has caused some researchers to include within edges much habitat that would better be included in adjacent cover types (Johnson 1984), or vice versa. Nonetheless, some data on edge use are available. Studies in the northern Sierra Nevada of California reported strong preference by American martens for forest-meadow edges (Simon 1980, Spencer et al. 1983, Martin 1987). Hawley (1955) reported heavy use of the interfaces between mature and regenerating coniferous stands by American martens. Kelly (1977) found that fishers preferred edges between forested and nonforested areas of the White Mountains of New Hampshire during winter, but he found no such preference in summer. On the other hand, Rosenburg and Raphael (1986), working only during summer in northwestern California, found that fishers' use of stands was negatively correlated with the amount of edge. This relationship, however, may have been confounded by patch size and stand fragmentation, both of which are generally related to amount of edge.

Although these findings may fail to support a unifying principle of edge use, we hypothesize that edge preference or avoidance is a primary function of the individual values of the habitats composing the edge. An interface between a graminoid meadow and a mesic late-successional coniferous forest is likely to be more used by American martens than are stand interiors, whereas the edge between a recent clear-cut and a xeric early-successional forest type will likely be avoided. Fishers avoid open areas such as open bogs, meadows, clear-cuts with no overhead cover, and rights-of-way for roads, pipelines, and electric lines (Kelly 1977; Powell 1977; Arthur et al. 1989*b*; Jones and Garton, this volume) but have been tracked along the edges of such areas (Powell 1977, 1982). Fishers are more likely to approach an interface between an open area and a dense lowland conifer stand with a large snowshoe hare population and to avoid an interface between a recent clear-cut and a second-growth hardwood forest. The near-universal avoidance of open areas by fishers and American martens (Kelly 1977; Koehler and Hornocker 1977; Powell 1977; Simon 1979; Arthur et al. 1989*b*; Thomasma et al. 1991; Jones and Garton, this volume) suggests that the areas of patches of good habitat and the distribution of open areas with respect to these patches may be critically important to the distribution and abundance of these species. Both species will travel through forested areas that are not of preferred compositions. Therefore, patches of preferred habitat that are interconnected by other forest types are used by American martens and fishers. But patches of preferred habitat that are separated by open areas of sufficient size are not likely to be used at all.

On a large scale, this behavior explains the distribution of American martens in the Rocky Mountains. Many small mountain ranges with preferred habitat for American martens have lost their populations in postglacial time (Patterson 1984) and cannot be colonized because of large open areas surrounding the mountains (Gibilisco, this volume). On a smaller scale, highly fragmented forests may contain preferred habitats for fishers and American martens that are so separated by open areas that these species cannot make use of the habitat that is available. The amount of insularity that can be tolerated by these species, especially American martens, is not well understood.

Seasonality of Habitat Use

Both North American *Martes* species show seasonal variation in their use of habitats. Generally, fishers (Kelly 1977; Arthur et al. 1989*b*; Jones and Garton, this volume) and American martens (Campbell, 1979, Soutiere 1979, Steventon and Major 1982, Wilbert 1992) use a wider range of cover types in summer than in winter. During winter both tend to prefer conifer-dominated over hardwood-dominated forests and old conifer-dominated stands over young ones. Thus, one must be cautious in generalizing that any particular habitat is best for either species; what is best in one season may not be in another.

Fishers in the White Mountains of New Hampshire preferred lowland and conifer forests and avoided hardwood forests and clear-cuts in winter; they were less selective during summer (Kelly 1977). Fishers in Maine show similar preferences, with conifer-dominated forests the most important winter type (Arthur et al. 1989*b*). At least some documented seasonal change in habitat use by fishers is caused by a change in avoidance of open habitats.

Some open habitats are densely covered with deciduous young trees or shrubs. These habitats are truly open during winter and provide no overhead cover and little subnivean structure. During summer, however, the same habitats may have a low but dense canopy and therefore be used to some degree by fishers (Kelly 1977).

American martens show seasonal differences in habitat use that are at least as pronounced as those of fishers (Koehler and Hornocker 1977, Campbell 1979, Soutiere 1979, Steventon and Major 1982). In areas where martens limit their use of nonforested habitats in summer, they generally avoid those types completely in winter (Koehler and Hornocker 1977, Simon 1979), although some anecdotes (Buskirk 1983) place martens more than 300 m from trees even in winter. Martens may occupy large recent burns in Alaska in winter, but these areas have complex structures formed by fallen boles beneath the snow. Studies investigating preference among martens for conifer-dominated stands consistently show increased preference for conifers and increased avoidance of open areas from summer to winter (Kelly 1977, Koehler and Hornocker 1977, Soutiere 1979, Steventon and Major 1982). Seasonal differences in habitat selection are important because they may account for reported differences between or within geographic areas. In other words, seasonal variation easily may be mistaken for habitat flexibility or lack of selectivity.

Activity

Habitat differences of fishers and American martens associated with activity and inactivity correspond roughly to differences attributed to resting and foraging behaviors. Fishers (Powell 1977) and American martens have two principal nonreproductive activities: sleeping and traveling. They travel to maintain territories, to forage, and to find resting sites. Fishers seldom climb trees except to enter resting sites (Jones 1991). They are solitary except for a brief mating period and have a small number of daily active periods separated by rest periods (Powell 1979b). With the exception of more frequent tree climbing, the same is largely true of American martens.

Data on differences in habitat use as a function of activity level are not consistent for American martens. Major (1979) found that martens in Maine did not rest preferentially in any habitat type. But Steventon and Major (1982) reported that resting martens and active martens showed similar preference hierarchies for habitat types and favored conifer-dominated stands. Buskirk (1983) found that in winter, active martens preferred stands dominated by black spruce (*Picea mariana*), whereas they rested almost exclusively in stands dominated by white spruce (*P. glauca*). These differences were associated with differences in the availability of red squirrel (*Tamiasciurus hudsonicus*) middens for resting, which were concentrated in stands of white spruce (Buskirk 1984).

Kelly (1977) reported that fishers in the White Mountains of New Hampshire used most habitats with no significant preference but that they rested more than expected in softwood-dominated and riparian habitats and were active more than expected in recent clear-cuts. Arthur et al. (1989*b*) in Maine and Jones and Garton (this volume) in Idaho reported greater habitat selectivity by fishers for resting (conifers preferred) than for foraging. By comparing snow-tracking data (active fishers) to radiotelemetry data (both active and inactive), Buck et al. (1983) found that fishers were more likely to be active in open areas than to rest there and that they are more likely to rest in habitats with dense canopy near the ground. Buck et al. (1983) suspected, however, that these differences might be caused by sampling biases. Although these findings are not entirely consistent, they suggest that fishers are more selective of habitats used for resting than for foraging. American martens are consistent in their preference for late-successional coniferous forests, but they vary in unpredictable ways in their relative preference of certain subtypes for foraging as opposed to resting.

Demography and Habitat Selection

When researchers have been able to compare habitat preference of adult and juvenile fishers and American martens, juveniles have been less selective (Weckwerth and Hawley 1962; Burnett 1981; Buskirk et al. 1989; Buck et al., this volume). This may be because juveniles have not yet learned the best habitats to choose, because adults are socially dominant and exclude juveniles from good habitats, or both. Despite spending more of their time in what are considered suboptimal habitats, juveniles may nonetheless be maximizing their fitness by choosing the best of a set of suboptimal choices. As in the case of seasonal shifts in habitat choices, demographic variation in habitat choices can mislead researchers trying to infer the best-quality habitats. The assumption that animals will choose the best habitats available to them does not mean that they all have equal access to high-quality habitats from which to choose.

Several authors have also found that habitat choices may differ by sex. The adult male fishers studied by Buck et al. in the coastal mountains of northern California strongly selected habitats with no broad-leaved component, whereas juveniles of both sexes exhibited no significant selection for any habitats. Buck et al. infer, but present no direct evidence, that adult males somehow prevented females and juvenile males from using preferred habitats. Kelly (1977) found statistically significant selection for mixed conifer-hardwood and alder lowland habitats and against open areas by females during winter and by males over the whole year. These differences may have been an artifact of sample sizes. Kelly also found that male and female fishers partitioned their home ranges by elevation and hypothesized that this reduced intersexual competition for food, especially during winter. He did not propose a mechanism for this elevational partitioning. Baker (1992) reported stronger selection for habitat types by female American martens than by males. These authors have not posed a common mechanism underlying such sexual differences.

Life Needs and Habitat Use

Several behavioral traits or life needs of fishers and American martens have been invoked to explain the associations between these species and their habitats. American martens are presumed to associate with structurally complex forests to avoid becoming another predator's prey. Accounts of predation on martens are few (Grinnell et al. 1937:200, Nelson 1973:219), yet biologists often have inferred that avoidance of areas without overhead cover evolved in response to fitness costs of exposure to predation in these areas—the "psychological need" for overhead cover described by Hawley and Newby (1957a). Support for the predator-avoidance hypothesis is provided by seasonal differences in the distances that American martens will venture into nonforested areas (Koehler and Hornocker 1977, Soutiere 1979). On the snow surface during winter, American martens are more visible to predators than when the ground is snow-free, and martens show a coincidental increased avoidance of treeless areas in winter. No data are available, however, that allow differentiation between choices made to avoid predators or to avoid places with few available prey.

The evidence for predator avoidance as a factor important to fishers is even weaker, yet fishers appear even more likely than martens to avoid open areas. Tracking studies have shown that fishers will go out of their way to cross open areas at the narrowest sites possible (Powell 1977), yet only Roy (1991), Buck et al. (this volume), and Krohn et al. (this volume) have documented predation on fishers. *Martes* species do not use the "mouse pounce" hunting behavior that is so successful for canids for catching mice in herbaceous habitats. It may be that the capture behaviors evolved by martens in forests are not effective in open-country foraging, but this still does not explain fishers' extreme avoidance of open areas. It is interesting to note that stone martens, morphologically similar to but behaviorally different from the boreal forest martens, are often seen in open areas (Jensen and Jensen 1970; Degn and Jensen 1977; Herrmann, this volume).

For both American martens and fishers, openings differ in important ways. Clear-cuts, tornado blow-downs, or burned areas with large amounts of coarse woody debris may have no overstory or shrub cover, yet still provide good cover. These nonforested areas differ critically from open grasslands, alpine tundra, or other areas of short herbaceous vegetation with regard to the structural features that are so important to *Martes*.

A second need invoked to explain the association between these species, particularly American martens, and structurally complex forests is access to subnivean spaces. By midwinter in most areas, American martens prey mostly on small mammals that live in subnivean spaces formed by vegetation and coarse woody debris near the snow-ground interface (Francis and Stephenson 1972; Martin, this volume). Snow settles around horizontal and diagonal structures, leaving air pockets that connect microtine tunnels with the snow surface (Pruitt 1957). Martens use these spaces mostly to capture prey (Simon 1980, Hargis and McCullough 1984) but also to find resting sites (Buskirk et al. 1989). Although not specifically investigated, it has generally been assumed that digging through deep snow in the absence of these passageways, especially for foraging, would be prohibitively inefficient. Presumably, the high energetic costs of tunneling through deep snow is not repaid by small packets of food (small mammals) captured with low success rates when there are no passageways to subnivean pockets. American martens generally hunt in deep snow by investigating subnivean access points (Bateman 1986, Corn and Raphael 1992) and perhaps picking up the scent of prey at the upper openings of these passageways. This should increase the efficiency of exploratory trips that are made into some passageways. Foraging "blind" in the absence of olfactory cues provided by the passageways would be approximately random with respect to prey positions and likely would be extremely inefficient. Support for this hypothesis also comes from seasonal differences in habitat selection for foraging. American martens forage most closely to structurally complex stands when snow depths are greatest (Koehler and Hornocker 1977). Where prey that live above the snow (e.g., snowshoe hares) are highly available, however, subnivean structure may be less important for foraging, but still we see strong habitat associations with late-successional forest (Thompson and Colgan 1990).

Access to subnivean spaces seems unimportant in explaining habitat associations of fishers. Fishers eat mostly foods that occur above the snow in winter (Powell 1982) and, being three to five times the size of American martens, are able to use far fewer passageways through the snow. Few resting sites of fishers are subnivean, and subnivean foraging has seldom been reported for fishers.

American martens also appear to choose habitats with structure that helps them reduce energetic costs during rest. Much evidence suggests that martens are energetically constrained during winter (Buskirk et al. 1988; Harlow, this volume), and Buskirk et al. (1989) and Taylor (1993) showed that martens rested preferentially in the warmest microenvironments in winter. During cold weather they rested in subnivean sites associated with coarse woody debris, but switched to above-snow sites when temperatures there were higher than those below the snow surface. Wilbert (1992) reported the same position shift for resting over a longer term, from winter to summer. The resting sites used during the coldest weather were in forests containing the highest densities of coarse woody debris and having the most complex overall structure. It has not been shown that martens actually lower energy expenditure by choosing these rest sites, but R. W. Threader (Ontario Hydro, Toronto, pers. commun.) has found that at ambient temperatures below thermoneutrality, American martens preferred artificial resting sites that were thermally efficient and that could be warmed to thermoneutrality by the loss of body heat. Metabolic costs during resting, thermal efficiency of resting sites, and forest structure appear functionally linked to behavioral choices by American martens. It is possible, however, that other factors, such as proximity to foraging areas, may be important in choosing resting sites as well.

Because fishers are so large, thermal losses while resting are likely not as important as they are for American martens. But snow conditions affect the energy costs of locomotion for fishers more than for American martens. Raine (1983) presented data suggesting that fishers in Manitoba avoided deep, soft snow when foraging, whereas sympatric American martens did not. Although morphologically similar, fishers sink much deeper into soft snow and drag their bellies in very deep snow, whereas martens rarely do (Raine 1983). When snow in the woods is deep, fishers travel on frozen streams and lakes, where snow is shallow and where ice provides a hard surface; Raine inferred that this reduced the cost of locomotion. Snow depth and hardness often varies with moisture content, forest structure, canopy closure, and cover type (Campbell 1979). Powell (unpubl. data.) never noted a belly drag in following more than 150 km of fisher tracks in the Upper Peninsula of Michigan and northern Minnesota, where snow depth exceeded 1 m during each winter of research, but snow penetrability was low relative to that in Raine's 1983 study area. These fishers sometimes ran along their own tracks, tracks of snowshoe hares, and researchers' snowshoe tracks when snow was deep. Physical characteristics of snow cover explain patterns of habitat use by fishers more than such patterns for American martens, the more so in areas with deep, soft snow than in areas with shallow or hard snow.

Physical complexity near the forest floor may affect fisher habitat choices indirectly. Snowshoe hares are most common in habitats with dense physical structure near the ground (Litvaitis et al. 1985). This structure may come from low conifer branches or from coarse woody debris, rocks, and small trees. Because snowshoe hares are important prey for fishers wherever the two species are sympatric (Powell 1982), and because fishers hunt for hares in the dense habitats where they are abundant, complex physical structure in forests affects fisher habitat choices through the habitat choices of snowshoe hares.

Most researchers assume that prey abundance explains habitat use for foraging to some extent, but evidence is inconclusive. In some studies, fishers and American martens selected habitats that held the highest numerical or biomass densities of potential prey (More 1978), but in most studies they did not (Bateman 1986; Martin, this volume). The best examples of American martens selecting habitat patches with high prey numbers were in herbaceous or low shrub meadows during summer or fall (Spencer et al. 1983, Buskirk and MacDonald 1984). The best examples for fishers were in dense conifer habitats with high hare densities (Powell 1981*b*, Arthur et al. 1989*b*). Habitats with high prey numbers may not have high numbers of available prey, however, and prey size, palatability, and behavior all influence what fishers and American martens can catch and eat. Potential prey that are seldom or never eaten, but that occur within a predator's home range, may or may not be considered available. Prey choice should change as populations of preferred prey change (Charnov 1976) and thus abundant and available prey may be avoided if highly preferred prey are abundant. In addition, the conventional method of ranking prey by energy content or rate of energy acquisition (MacArthur and Pianka 1966, Charnov 1976, Pyke et al. 1977, Powell 1981*b*) does not consider the risks involved in capturing prey. New foraging models that include risks of injury have not been applied to *Martes* species. Although prey density does not appear to be the foremost factor explaining habitat use in most areas, Harlow (this volume) predicts that American martens can only survive in home ranges where food can be captured on a predictable basis all winter. How do we reconcile these two ideas? We believe that American martens and fishers optimize their rates of prey capture by sensing patch structure, physical structure near the ground, prey abundance, and probably prey behaviors. They must choose habitat for foraging on the basis of prey "catchability," which involves more than prey abundance. Relating prey catchability to forest structure, prey abundance, and prey behavior has yet to be done.

Conclusions

American martens and fishers appear to be among the most habitat-specialized mammals in North America. We believe that changes in habitat availability, more than any other factor, will affect the geographic distributions of these two species over the next several decades. Trapping is currently well enough regulated to prevent broad-scale depletion in virtually all North American jurisdictions. Therefore, we believe that, although trapping may affect density and demography in many areas (Powell, this volume), the effects will be local and temporary. Do American martens and fishers require particular forest types—for example, old-growth conifers—for survival? We think they do. Ecological dependency has been defined in terms of viability of populations (Ruggiero et al. 1988), and distributional losses of marten and fisher populations in response to habitat change provide evidence that populations require the habitats that individuals, especially reproductive adults, behaviorally prefer. These losses include natural extinctions that have occurred in insular habitats in postglacial time (Patterson 1984) and anthropogenic ones now in progress (Buskirk 1992; Gibilisco, this volume). Past extinctions and future extinctions that cannot be prevented must be studied to understand which habitats are important and why, and how much is required to assure population persistence.

Although there is logic in our use of specific life needs to explain habitat selection behaviors of fishers and American martens, there is also speculation and teleology involved. We and others believe that habitat selection patterns are probably most easily explained in terms of fitness or its currencies. We therefore propose explanations, infer the currency, and leave the "explanations" largely untested. Our guesses may not be correct. We offer them as hypotheses to be tested and likely rejected in the process of learning more about these species.

Acknowledgments

G. M. Koehler provided helpful comments on an outline of this chapter. We appreciate constructive reviews of drafts by J. A. Bissonette and an anonymous reviewer.

22 Feeding Ecology of American Martens and Fishers

Sandra K. Martin

Over 35 published studies describe the feeding ecology of American martens (*Martes americana*) and fishers (*M. pennanti*). All report on diet, and some contain additional information on abundance of prey or on behavior of the predators. This literature gives insight on the geographic variation in diet and on the influence of relative abundance of potential prey on food selection by American martens and fishers. A few authors have explored seasonal patterns and potential sex- and age-related differences in the diet. The relationship of sexual dimorphism exhibited by American martens and fishers to differences in males' and females' feeding ecology has also been examined. Investigating diet increases our understanding of the food resources available to American martens and fishers, the predatory capabilities of these mustelids, and the limitations their environment places on their acquisition of food. Responses of mustelid populations to prey fluctuations is difficult to research, and to date the published data are few. This aspect of feeding ecology is critical for understanding relevant population and community ecology and for translating scientific understanding into effective population and habitat management.

The literature on the feeding ecology of American martens and fishers poses as many questions as it answers. The mechanisms of coexistence for these closely related mustelids, sympatric throughout much of their ranges, remain largely unknown. How food resources affect American marten and fisher populations is also largely unknown. Knowledge of the effects of habitat alteration on feeding ecology is critical for understanding and predicting the effects of forest management on these species.

Research on feeding ecology ranges from prosaic scat and stomach analyses to intricate theoretical models of the interrelationships of diet and sexual selection, predator-prey interactions, and population parameters. Further research needs to focus on the causes of the geographical variation found in the diets of American martens and fishers, to explore relationships between feeding ecology and population ecology of these species, and to relate these findings to the ecology of the communities these mustelids inhabit.

American Martens

In studies of American marten diets conducted across their range, voles (*Microtus* spp. and *Clethrionomys* spp.) recurred as important food items and were dominant in half of the diets (Table 22.1). Only 2 of 22 studies reviewed (Table 22.1) did not find voles to be at least 10% of the identifiable diet sample. Simon (1980), working in California, found voles (*Microtus* spp.) to be the second most commonly consumed mammalian prey, after Douglas squirrels (*Tamiasciurus douglasii*), but did not conclude that voles were important prey. Nagorsen et al. (1989) reported a diverse diet for American martens studied on Vancouver Island, British Columbia, which included vegetation, birds, fish, ungulate carrion, and deer mice (*Peromyscus maniculatus*), but they found very little consumption of voles.

Avian, invertebrate, and vegetative items were grouped by most authors, and are totaled by category in Table 22.1. Total number of identifiable food items ranged from 6 to 27 in the 22 diet studies reviewed. The number of food items representing at least 10% (frequency of occurrence) of the identifiable sample ranged from 2 to 5. In 14 studies, the dominant food item represented at least 30% of the identifiable diet.

Vegetation (fruit, berries, nuts, fungi, lichens, grass, conifer needles, leaves, twigs, and bark; Table 22.1) was an important diet item in over half of the studies examined. Vegetation dominated one-fourth of the studies (6 of 22) by percent occurrence. Frequency in percent of occurrence is the most commonly calculated measure of food item importance and allows comparisons to be made between data sets. However, because a relatively small amount of material occurring in a scat or gastrointestinal tract has the same importance in the measure as a much larger amount of another food item, this measure may be misleading (Day 1966). Additionally, vegetation may be consumed secondarily by carnivores when they ingest the stomachs of herbivorous prey.

In four of six studies, vegetation occurred in a high percentage of scats or intestinal tracts, but the volume of vegetative material was low. These studies included three from California. Simon (1980) reported a 44% occurrence and 17% volume in the identifiable diet sample; Hargis and McCullough (1984)

Table 22.1. Diversity of American marten (*Martes americana*) diet

Study (location)	Material analyzed	No. items	No. categories Total	No. categories ≥10%[a]	H'[b]	Diet categories ≥ 10% in sample (% occurrence)
Bateman 1986 (Newfoundland)	56 scats	63	6	4	0.61	Snowshoe hares (46%), meadow voles (27%), masked shrews (10%), birds (10%)
Soutiere 1979 (Maine)	412 scats	562	18	3	0.89	Short-tailed meadow voles (28%), red-backed voles (27%), birds (13%)
Clem 1977b (Ontario)	183 GIs	72	13	4	0.99	Birds (22%), red-backed voles (19%), snowshoe hares (11%), meadow voles (11%)
Thompson & Colgan 1987a (Ontario)	918 scats	849	9	4	0.85	Red-backed voles (30%), deer mice (18%), vegetation (13%), shrews[c] (10%)
Raine 1987 (Manitoba)	107 scats	120	7	4	0.65	Snowshoe hares (50%), birds (15%), red squirrels (13%), vegetation (10%)
Douglas et al. 1983 (Northwest Territories)	172 scats	381	10	2	0.59	Vegetation (65%), insects (13%)
Lensink et al. 1955 (Alaska)	402 scats, 64 GIs	576	7	3	0.36	Voles[c] (73%), vegetation (17%), birds (10%)
Buskirk & MacDonald 1984 (Alaska)	467 scats	622	9	2	0.50	Voles[c] (66%), vegetation (15%)
Cowan & MacKay 1950 (Alberta, British Columbia)	197 scats, 3 stomachs	254	18	3	1.04	Red-backed voles (30%), short-tailed meadow voles (12%), heather voles (10%)
Murie 1961 (Wyoming)	528 scats	675	18	2	0.66	Voles[c] (43%), vegetation (37%)
Marshall 1946 (Montana)	64 scats	66	8	3	0.75	Red squirrels (38%), flying squirrels (21%), red-backed voles (14%)

(*continued*)

Table 22.1. (Continued)

Study (location)	Material analyzed	No. items	No. categories Total	No. categories ≥10%[a]	H'[b]	Diet categories ≥ 10% in sample (% occurrence)
Weckwerth & Hawley 1962 (Montana)	1758 scats	3103	27	4	1.06	Vegetation (22%), short-tailed meadow voles (15%), insects (12%), red-backed voles (12%)
Koehler & Hornocker 1977 (Idaho)	129 scats	151	15	3	0.89	Red-backed voles (30%), *Microtus* spp. (26%), vegetation (10%)
Campbell 1979 (Colorado)	145 scats	188	12	3	0.83	Red-backed voles (28%), northern pocket gophers (25%), vegetation (18%)
Gordon 1986 (Colorado)	32 GIs	95	11	2	0.82	*Microtus* spp. (40%), shrews[c] (21%)
Quick 1955 (British Columbia)	250 GIs	235	9	2	0.68	Red-backed voles (42%), birds (29%)
Nagorsen et al. 1989 (British Columbia)	701 GIs	1025	15	5	0.84	Vegetation (29%), birds (20%), fish (15%), ungulates (14%), deer mice (11%)
Newby 1951 (Washington)	95 scats, 17 stomachs	164	19	2	0.93	Insects (45%), pine squirrels (10%)
Simon 1980 (California)	99 scats	409	16	2	0.75	Vegetation (44%), insects (27%)
Zielinski et al. 1983 (California)	300 scats	265	20	2	1.14	*Microtus* spp. (18%), birds (10%)
Hargis & McCullough 1984 (California)	91 scats	217	17	4	1.02	Vegetation (24%), *Microtus* spp. (14%), red squirrels (14%), birds (11%)
Martin 1987 (California)	100 scats	91	16	3	0.92	Vegetation (34%), insects (18%), *Microtus* spp. (12%)

Note: Data from 22 studies are presented by geographic area, from east to west. Diversity indices were calculated; data labeled "unidentified species" were not included in the calculations. All occurrences of birds, insects, and vegetation were totaled into three groups, respectively. GIs = gastrointestinal tracts.

[a]Percent of occurrence in the sample used to calculated the diversity indices.

[b]Shannon diversity index.

[c]Data were not differentiated to species.

noted a 24% occurrence and 6% volume; and Martin (1987) reported a 34% occurrence and 18% volume. One study from British Columbia (Nagorsen et al. 1989) found 29% occurrence and 9% volume. Douglas et al. (1983) also found high occurrence of vegetation in the Northwest Territories but did not provide other data for interpretation. Weckwerth and Hawley (1962) examined a large sample of scats ($n = 1758$) from Montana and recorded a high percent occurrence of vegetation, primarily seeds from fruits. They believed that these were important seasonal foods for American martens, eaten in proportion to their availability.

Vegetation provides an important food resource for American martens but likely is secondary to mammalian prey in dietary importance. In some cases, vegetation may be used as a substitute when preferred prey cannot be obtained. Use of local high-density patches of vegetative foods, such as some berries, may provide lower energetic cost/benefit ratios than would hunting mammalian prey. The relative availability of vegetative and animal foods and the resulting energy budgets for American martens have not yet been explored.

Birds were significant dietary items in nearly half of the studies examined (Table 22.1). The majority were passerines, but some grouse (*Bonasa umbellus*, *Canachites canadensis*) were also found (Marshall 1946, Lensink et al. 1955, Murie 1961, Bateman 1986, Raine 1987, Thompson and Colgan 1987*a*). Birds, however, may be overrepresented in scat contents. Zielinski (1986) found birds were highly overestimated in comparison with mammals, as measured by percent occurrence in the diet of a captive European ferret (*Mustela putorius furo*). Feathers are more easily observed in scats than are mammalian hairs, and birds also have higher percentages of undigestible materials per unit weight than do similarly sized mammals, generating more scats per unit weight of food for avian prey than for mammalian prey.

Insects made considerable contributions to American marten diets in five studies reviewed (Table 22.1), and were the dominant item in the identifiable sample in one of these (Newby 1951). As with vegetation, percent occurrence may be a misleading indicator of true importance in American marten diets for this food category, inasmuch as the proportion of insects by volume of the sample was usually low. Also, as with vegetation, martens may have consumed insects in bird crops. Newby (1951) found ants, hornets, and yellow jackets to dominate a high proportion of scats collected on his study site in Washington, however, and concluded that they were a staple in the summer diet. Weckwerth and Hawley (1962) also considered insects important in the American marten diets they investigated in Montana. Hornets, yellow jackets, and ants were the insect species found most often in scats they collected, and they concluded that insects provided a seasonally available diet item for American martens in their study. Simon (1980) found relatively high percentage of occurrence for insects but low relative volume, suggesting that their importance in the diet may be overestimated. Martin's 1987 data were similar. Douglas et al. (1983) did not provide any interpretation for the high occurrence of insects they found.

Snowshoe hares (*Lepus americanus*) contributed importantly to American marten diets in three studies, and dominated the diet in two of these (Table 22.1). All these studies were conducted in eastern and midwestern North America, specifically Newfoundland (Bateman 1986), Ontario (Clem 1977*b*), and Manitoba (Raine 1987). This geographic pattern may have some meaning, but without data on the availability of hares versus other prey, we cannot know why American martens eat this relatively large prey more in the east than elsewhere.

Most studies note the catholic nature of the American marten's diet. It often includes shrews (*Sorex* spp.), deer mice, red squirrels (*Tamiasciurus hudsonicus*), mountain phenacomys (*Phenacomys intermedius*), flying squirrels (*Glaucomys sabrinus*), northern pocket gophers (*Thomomys talpoides*), fish, ungulates (carrion), and Douglas squirrels (Table 22.1). Harlow (this volume) suggests that one factor important to the energetic balance of American martens is their ability to adjust predatory patterns and prey type, exhibited by the opportunistic diets observed across their range. The prevalence of voles as a major diet item throughout the range of American martens, however, is notable.

Geographic Variation in Dietary Diversity

Shannon diversity indices (Shannon and Weaver 1949) were calculated for 22 American marten studies (Table 22.1). For these data, H' increases as the number of diet categories increases, and as the distribution of proportion of the total diet among the categories becomes more even. When the studies are arranged in descending order of H', several patterns emerge (Table 22.2). Investigations conducted in the subarctic had the lowest diet diversity (H' < 0.60; Lensink et al. 1955, Douglas et al. 1983, Buskirk and MacDonald 1984). High latitude ecosystems are less complex in terms of community structure than temperate zone ecosystems, and less dietary diversity for American martens in these environments may be expected. Also, diets dominated by large prey such as snowshoe hares (Bateman 1986, Raine 1987) and red squirrels (Marshall 1946) had relatively low diversity (H' < 0.77). Larger prey would provide more meals per carcass, necessitating fewer kills per unit time and ultimately equating with a less diverse diet. Most studies from

Table 22.2. Studies of American marten (*Martes americana*) diet, arranged in descending order of H', the Shannon diversity index

Location	Reference	H'
California	Zielinski et al. 1983	1.14
Montana	Weckwerth & Hawley 1962	1.06
Alberta, British Columbia	Cowan & MacKay 1950	1.04
California	Hargis & McCullough 1984	1.02
Ontario	Clem 1977*b*	0.99
Washington	Newby 1951	0.93
California	Martin 1987	0.92
Idaho	Koehler & Hornocker 1977	0.89
Maine	Soutiere 1979	0.89
Ontario	Thompson & Colgan 1987*a*	0.85
British Columbia	Nagorsen et al. 1989	0.84
Colorado	Campbell 1979	0.83
Colorado	Gordon 1986	0.82
California	Simon 1980	0.75
Montana	Marshall 1946	0.75
British Columbia	Quick 1955	0.70
Wyoming	Murie 1961	0.66
Manitoba	Raine 1987	0.65
Manitoba	Bateman 1986	0.61
Northwest Territories	Douglas et al. 1983	0.59
Alaska	Buskirk & McDonald 1984	0.50
Alaska	Lensink et al. 1955	0.40

the Pacific states had relatively high dietary diversity (Newby 1951, Zielinski et al. 1983, Hargis and McCullough 1984, Martin 1987).

To test the effect of changes in importance of vegetation in the data sets for both American martens and fishers, I halved the number of occurrences of vegetation in each data set where vegetation represented at least 10% of the identifiable diet, and recalculated H' for each. The values of H' changed, but the geographic and food item patterns did not. Reducing the relative importance of vegetation in those data sets already exhibiting high diet diversity increased H'—and decreased its value in those data sets already exhibiting low diet diversity.

Seasonal Variation in Diet

Where seasonal diets have been examined, patterns have been similar in different areas of the continent. Seasonally abundant foods, such as insects and fruit, are eaten when available (Koehler and Hornocker 1977, Simon

1980). Buskirk and MacDonald (1984) reported that voles and fruit were important autumn foods for martens in Alaska, but that they gradually declined in importance over the winter. Zielinski et al. (1983) suggested that deepening winter snows made voles (*Microtus* spp.) less available to American martens on their study site in California, and that American martens then switched to Douglas squirrels and snowshoe hares. Voles appeared with decreasing frequency in late winter and spring diets, and were replaced in midsummer by ground squirrels (*Spermophilus lateralis*) and chipmunks (*Eutamias* spp.). The authors suggested that the appearance of juveniles of these species resulted in a more attractive prey source for American martens. The factors involved in prey switching, including seasonal variation in relative prey abundance, have not been investigated in much depth across the range of American martens.

Relative Prey Abundance

Examination of relative prey abundance in relation to the diet of American martens can illuminate an important aspect of feeding ecology: do American martens respond opportunistically to densities of prey, or do they take prey selectively from the environment, thereby suggesting a more finely tuned, and possibly more vulnerable, relationship with the community they inhabit? Three studies examined prey abundance as well as diet and found no evidence for prey selection (Campbell 1979, Soutiere 1979, Gordon 1986). Seven other investigations did find evidence for prey selection (Weckwerth and Hawley 1962, Francis and Stephenson 1972, Douglas et al. 1983, Buskirk and MacDonald 1984, Bateman 1986, Martin 1987, Thompson and Colgan 1990). All of the latter studies, except Bateman's, found that American martens selected for voles (*Microtus* spp.). Bateman suggested that American martens selected for snowshoe hares at her study site in Newfoundland, but she could not support this conclusion with field data.

The American marten inhabits the forest interior, as do red-backed voles (Maser et al. 1981), so American martens' predation of this species is easily understood. *Microtus* spp. can occur in forest openings but are most abundant in meadows, primarily those associated with riparian systems within a forest landscape. Their prevalence in the diet of American martens, even unsupported by accompanying prey availability data, alerts us to the possible occurrence of prey selection. This is further supported by five of the field studies cited above. The localized nature of *Microtus* spp. abundance may partially explain American marten preference (Buskirk and MacDonald 1984). Predator use of prey patches can increase overall predation success, and American martens could quickly identify and locate rich habitat patches, such as riparian meadows. American marten dependence on such localized prey might increase the vulnerability of individuals to local vole population decline or even extinction. This vulnerability would be minimized by an ability to switch prey when required.

The ultimate measure of success for a species is the reproductive output of groups of individuals and long-term population viability, both the result of the sum of the group members' levels of fitness. Two studies found that American marten populations declined when faced with local decreases in mammalian prey (Weckwerth and Hawley 1962, Thompson and Colgan 1987a). Although American martens can shift prey preference seasonally, the effects on long-term American marten population viability have not been investigated where this phenomenon has been documented.

Predatory Behavior

American martens have been observed to avoid open areas, and this behavior has been hypothesized as a predator avoidance mechanism (Hargis and McCullough 1984). Koehler and Hornocker (1977) observed American martens passing through, but not hunting in, openings less than 100 m wide in Idaho. They assumed that American martens were using open meadows in summer, however, because of the occurrence of fruit, insects, and ground squirrels in the diet at that time of year. Raine (1987) noted the circuitous hunting path American martens used in his Manitoba study area, with frequent investigations of logs, roots, and stumps. Spencer and Zielinski (1983) also recorded zigzag search paths and frequent investigation of coarse woody debris and other potential prey locations in their winter tracking of American martens in California.

Sexual Dimorphism

The sexual dimorphism of American martens has been hypothesized as an evolutionary mechanism for sexual separation along the food axis of the niche, reducing intersexual competition (Brown and Lasiewski 1972). In contrast, Moors (1980) advanced the hypothesis that sexual dimorphism in mustelids is a result of different selective pressures on males and females, although secondary benefits may accrue from partitioning food resources between sexes.

The relationship of sexual dimorphism in American martens and hypothesized intersexual differences in diet was investigated by Nagorsen et al. (1989) on Vancouver Island, British Columbia. They concluded that females consumed more small mammal prey, and a greater proportion of small birds

($<$10 g) than did males, but they generally observed extensive dietary over-lap between sexes. Nagorsen (this volume) presents data that provide general support for a hypothesis of greater sexual dimorphism on islands, although differences in sexual dimorphism indices between insular and adjacent main-land populations were not significant. Nagorsen suggests that competition for food may be an important factor for larger body size of island martens. Specifically, body size may be in part determined by the presence of larger competitors, few of which were sympatric with Nagorsen's island popula-tions.

Thompson and Colgan (1990), working in Ontario, also investigated hypo-thetical sexual diet differences, but found none. This important hypothesis warrants further collection of field data, especially inasmuch as the diet observed in the Vancouver Island study was unusual (voles did not signifi-cantly contribute to the diet), and the conclusions drawn are suspect in their general applicability to American martens.

Holmes and Powell (this volume) compared sexual dimorphism indices for American martens and fishers, using cranial and dental measures with trophic significance. They found no evidence for resource partitioning as selective pressure for sexual dimorphism for these mustelids. Rather they suggest that *Martes* populations may respond to increased competition for food resources by shifting their behavior, habitat use patterns, diet, or even demography. They conclude that resource partitioning between sexes of mustelids is a result of sexual dimorphism, and not a driving factor resulting in sexual dimorphism. This hypothesis deserves testing with the long-term diet studies they suggest are needed.

Fishers

Most investigations of the diet and feeding ecology of fishers have been conducted in the eastern United States; only a few are from the midwestern United States or from Canada, and one study was published from California (Table 22.3). This paucity of information for the West represents a serious deficiency in our knowledge, especially because fisher distribution is more limited and densities appear to be much lower in these regions than in eastern and midwestern zones (Gibilisco, this volume).

The occurrence of a food item in one scat, or even one gastrointestinal tract or stomach (for larger prey) does not necessarily equate to one prey animal killed and eaten (Floyd et al. 1978, Korschgen 1980, Zielinski 1986). Occur-rence data can be used as an index of actual diet, however. Until captive feeding trials with fishers are conducted, no credible correction factors exist

Table 22.3. Diversity of fisher (*Martes pennanti*) diet

Study (location)	Material analyzed	No. items	No. categories Total	No. categories ≥10%[a]	H'[b]	Diet categories ≥ 10% in sample (% occurrence)
Coulter 1966 (Maine)	242 GIs, 127 scats	563	20	4	1.09	Deer (22%), snowshoe hares (14%), porcupines (13%), passerine birds (11%)
Arthur et al. 1989b (Maine)	69 scats	92	11	5	0.92	Fruit (27%), porcupines (15%), snowshoe hares (13%), gray squirrels (12%), birds (10%)
Stevens 1968 (New Hampshire)	153 stomachs, 337 scats	775	18	3	1.07	Short-tailed shrews (23%), vegetation (17%), white-footed mice (10%)
Kelly 1977 (New Hampshire)	40 GIs	54	15	2	1.06	Red-backed voles (22%), white-footed mice (13%)
Giulano et al. 1989 (New Hampshire)	158 GIs[c] 173 GIs[d]	220[c] 173[d]	11[c] 11[d]	3[c] 4[d]	0.94[c] 0.93[d]	Vegetation (23%), birds (22%),[c] vegetation (18%), deer (17%), voles (10%)[e]
Hamilton & Cook 1955 (New York)	60 GIs	86	13	4	1.00	Vegetation (22%), deer (13%), red-backed voles (12%), red squirrels (12%)
Brown & Will 1979 (New York)	332 GIs	531	19	3	0.95	Vegetation (38%), red squirrels (12%), deer (10%)
de Vos 1952 (Ontario)	57 GIs	84	13	4	0.87	Snowshoe hares (30%), porcupines (24%), insects (14%), deer (10%)

(*continued*)

Table 22.3. (*Continued*)

Study (location)	Material analyzed	No. items	No. categories Total	No. categories ≥10%[a]	H'[b]	Diet categories ≥ 10% in sample (% occurrence)
Clem 1977*b* (Ontario)	270 GIs	307	14	5	1.11	Porcupines (22%), snowshoe hares (15%), muskrats (14%), passerine birds (10%), ruffed grouse (10%)
Raine 1987 (Manitoba)	159 scats	191	10	2	0.50	Snowshoe hares (70%), vegetation (11%)
Powell 1977 (Michigan)	35 scats	42	7	5	0.80	Snowshoe hares (26%), porcupines (17%), deer (17%), mice (17%),[e] squirrels (1%)[e]
Grenfell and Fasenfest 1979 (California)	8 stomachs	18	7	4	0.71	Vegetation (44%), insects (17%), deer (11%), white-footed mice (11%)

Note: Data from 12 studies are presented by geographic area, from east to west. Diversity indices were calculated; data labeled "unidentified species" and species used as bait were not included in the calculations. All insect and vegetation occurrences were totaled into two groups, respectively. Ruffed grouse occurrences were counted separately, but all other occurrences of birds (primarily passerines) were grouped together. GIs = gastrointestinal tracts.

[a] Percent of occurrence in the sample used to calculate the diversity.

[b] Shannon diversity index.

[c] Data from sample collected 1965–1967.

[d] Data from sample collected 1987.

[e] Data were not differentiated to species.

with which to treat these data. The general patterns in the data discussed below do not appear to be tremendously influenced by fine gradations in numbers of occurrences.

Fishers eat a variety of foods throughout their range, including mammalian and avian prey, ungulate carrion, vegetation, and insects. Reported food items include snowshoe hares (de Vos 1952, Powell 1977, Raine 1987, Kuehn 1989), red-backed voles (Kelly 1977), red squirrels (Brown and Will 1979), porcupines (*Erethizon dorsatum*: Clem 1977*b*), short-tailed shrews (*Blarina brevicauda*: Stevens 1968), passerine birds (Giulano et al. 1989), and a variety of vegetation (apples: Arthur et al. 1989*b*; false truffle: Grenfell and Fasenfest 1979; unidentified plant matter: Hamilton and Cook 1955).

The total number of identifiable food items ranged from 5 to 18 in 13 studies that examined diet (avian, invertebrate, and vegetative items are grouped; Table 22.3). The number of food items that represented at least 10% (frequency of occurrence) of the identifiable diet (most samples include a percentage of material that could not be identified to species or even genus), ranged from 2 to 5. In 4 studies, the dominant food item represented at least 30% of the identifiable diet.

There is a great deal of commonality in the diets reported in all 13 studies. Five food items were repeatedly reported as important parts of fisher diets: snowshoe hares, porcupine, deer, passerine birds, and vegetation. All studies found at least two of these items to contribute 5% or more to fisher diet. All studies except one (Grenfell and Fasenfest 1979) reported snowshoe hares, porcupines, or both as important food items. The relative contribution of these prey to fisher diets appears to vary geographically.

Geographic Variation in Dietary Diversity

Shannon diversity indices (Shannon and Weaver 1949) were calculated for 12 of the 13 studies reviewed (Table 22.3). When the studies are examined in descending order of their diversity indices (H'; Table 22.4), two patterns emerge: geographic location and dominant food items. Fisher diets in the eastern United States had the highest diversity (H' > 0.90, Table 22.4; Hamilton and Cook 1955, Coulter 1966, Stevens 1968, Kelly 1977, Brown and Will 1979, Arthur et al. 1989*b*, Giulano et al. 1989), and those studies conducted in the Midwest and West had the lowest diet diversity (H' < 0.80, Table 22.4; de Vos 1952, Grenfell and Fasenfest 1975, Powell 1977, Raine 1987). An exception to this geographic pattern, Clem's 1977*b* study from Ontario, had a diversity index of 1.1.

The second pattern involves major food types. Fisher diets documented in eastern studies, which exhibited the highest diet diversity indices, were com-

Table 22.4.   Studies of fisher (*Martes pennanti*) diet, arranged in descending order of H′, the Shannon diversity index

Location	Reference	H′
Ontario	Clem 1977*b*	1.11
Maine	Coulter 1966	1.09
New Hampshire	Stevens 1968	1.07
New Hampshire	Kelly 1977	1.06
New York	Hamilton & Cook 1955	1.00
New York	Brown & Will 1979	0.95
New Hampshire	Giulano et al. 1989	0.94[a]
		0.93[b]
Maine	Arthur et al. 1989*b*	0.92
Ontario	de Vos 1952	0.87
Michigan	Powell 1977	0.80
California	Grenfell & Fasenfest 1979	0.71
Manitoba	Raine 1987	0.50

[a]Data collected in 1987.
[b]Data collected in 1965–1967.

posed of comparatively large proportions (33–77% of the identifiable sample) of small food items or deer (*Odocoileus virginianus*) carrion. Hamilton and Cook (1955) reported that vegetation, red-backed voles, red squirrels, and deer formed 59% of fisher diets in New York. Brown and Will (1979), also in New York, found that vegetation, red squirrels, and deer made up 62% of fisher diets. Stevens (1968) reported that short-tailed shrews, white-footed mice (*Peromyscus leucopus*), and vegetation composed 50% of fisher diets in New Hampshire. Kelly (1977), also in New Hampshire, reported that red-backed voles and white-footed mice constituted 35% of fisher diets. Authors of a third New Hampshire study reported that birds, fruit, and mice made up 70% of fisher diets in 1965–1967, and birds, fruit, voles, and deer made up 66% of the diets in 1987 (Guilano et al. 1989). Arthur et al. (1989*b*), studying fisher diets in Maine, found fruit, gray squirrels (*Sciurus carolinensis*), and birds to form 49% of the diets. Coulter (1966), also in Maine, found that deer and birds comprised 33% of fisher diets.

Relatively undiverse fisher diets in midwestern and western studies had comparatively high proportions (37–70% of the identifiable sample) of large prey. Snowshoe hares were the major prey item in three of the four studies: in Manitoba, snowshoe hares, 70% (Raine 1987); in Ontario, snowshoe hares, 30%, and porcupines, 24% (de Vos 1952); in Michigan, snowshoe hares, 26%, and porcupines, 17% (Powell 1977). Kuehn's (1989) data are not directly comparable, but he also found snowshoe hares were an important part of the fisher diet in Minnesota when hare populations were high. As snowshoe hares declined over his eight-year study, consumption of small mammals, such as deer mice, voles, lemmings (*Synaptomys* spp.), moles (*Condylura cristata*), and shrews, increased. The author suggested these species may be important alternative prey. Consumption of deer, other mammals, birds, reptiles, and fruit were not correlated with consumption of hares. Body fat stores in fishers were correlated with snowshoe hare availability, but no reproductive parameters showed such a relationship. Fishers are one of the few regular predators of porcupines (Powell 1982). Biologists in Michigan determined that fishers caused a major reduction in porcupine populations in an area where fishers had been reintroduced 40–60 years after they had been extirpated (Earle and Kramm 1982).

The only study of fisher diet reported from western North America exhibited a low diversity index (H′ = 0.71: Grenfell and Fasenfest 1975), but the food items and their proportions in the diet would place this sample in the same category as the eastern studies reported above. The very small sample size that Grenfell and Fasenfest examined (eight fisher stomachs), however, may warrant exclusion of this study from discussions of the general patterns of geography and food types. The study is unique in its findings of false truffles (hypogeous fungi) in the stomachs of fishers. The authors suspected, but could not prove, that fishers were directly feeding on the fungi. The importance of fungi as a food for many forest vertebrates has only been widely noted in the past 15 years (Maser et al. 1978, 1986), and earlier research may have overlooked the possible importance of this vegetation to fisher diets.

Clem's results from Ontario (1977*b*) also do not fit the patterns described above. His fisher diet sample exhibited a diversity index of 1.1, yet the primary prey were relatively large: porcupines, muskrats (*Ondatra zibethica*), and snowshoe hares, together representing 50%. The diet he observed was more evenly distributed among the food categories than others from the midwest. These differences might be more interpretable if they could be coupled with information on relative prey abundance. Local variations in abundance of snowshoe hare and alternative prey no doubt affect fisher diets.

Voles, shrews, mice, squirrels, and passerine birds yield fewer calories per carcass than do the larger hare and porcupine. Vegetation similarly provides a small part of the fisher diet, although domestic fruits (apples) can be an exception. Ungulate carrion provides a rich food source, but its availability is unpredictable. Fishers must require more kills per unit time to maintain themselves on small prey, resulting in a higher variety of prey taken in their overall diet.

Body size and predatory abilities of fishers are presumably relatively constant over their range, but habitat type, variety, and interspersion vary strongly across the continent (Buskirk and Powell, this volume). In the eastern part of the species' range, a variety of forest types are available, often with a high degree of interspersion. This variety is reflected in the diet of fishers. In the West, fishers feed heavily on snowshoe hares in coniferous forest habitat, and on porcupines in hardwood-dominated habitat. Although little data on relative prey abundance are presented in the studies I reviewed, if fishers do prefer larger prey such as snowshoe hares and porcupines, with higher caloric value per kill, their greater use of a variety of small prey in the East may reflect an inability to acquire adequate amounts of larger prey.

Arthur et al. (1989*b*) suspected a preference for hares in their study in Maine, based on snow tracking. They assessed relative prey abundance on snow transects, and found all prey species present in all habitat types. But they found changes in fisher movement patterns when dense patches of hare tracks were encountered. The authors suspected that fishers specifically searched for hares, even though they found no difference between encounters of prey tracks along fisher trails and along randomly placed transects. Although no data are available, fishers in the western part of their range may also rely heavily on snowshoe hares, a common inhabitant of the western coniferous forests.

Seasonal Variation in Diet

We know little about seasonal changes in fisher diets. Coulter (1966) concluded that there were seasonal differences in his diet data, but this has been questioned (Powell 1977). Stevens (1968) found that in summer, fishers made high use of fruits and nuts, which were plentiful in his study area. Clem (1977*b*) also looked for, but did not detect, seasonal differences. His data, however, were collected only between November and February, arguably too short a period to test for seasonal changes in diet. It seems likely that the diet of this opportunistic species would change as food availability fluctuates seasonally.

Predatory Behavior

Powell (1982) reported that fisher foraging behavior and patterns were correlated with prey density and habitat. He found that fishers traveled along straight lines for considerable distances when they hunted in open, upland hardwood forests in Michigan, which were good winter porcupine habitats. In areas where prey were found in dense local patches, foraging patterns were characterized by frequent changes in direction.

Sexual Dimorphism

Like American martens, fishers exhibit sexual dimorphism. Several authors have investigated intersexual differences in diets of fishers, but most data sets do not exhibit them (Coulter 1966, Stevens 1968, Clem 1977*b*, Kelly 1977, Giulano et al. 1989). Kuehn (1989) found that deer carrion was an important winter food for fishers in Minnesota and that males consumed more than females. Pittaway (1983) observed a male fisher defend a deer carcass from a red fox (*Vulpes fulva*) in Ontario, and suggested that the smaller female fishers would have more difficulty fighting off competitors.

Giulano et al. (1989) examined fisher diets in New Hampshire for sex- and age-related differences. They found no sexual differences in diets, although females were in poor condition more often than males. Juvenile fishers consumed fruit more often than did older fishers, and large males consumed birds less often than did smaller fishers of both sexes.

Comparison of American Marten and Fisher Diets

Over the ranges of the two species, fishers have a more diverse diet than do American martens. The average Shannon diversity index for fisher diets was 0.92 (SD = 0.17), and for American martens was 0.81 (SD = 0.20). A Mann-Whitney test indicated that the difference between these averages was marginally significant ($U = 197$, $P = 0.06$).

In the Midwest, where studies indicate fishers eat proportionally more large prey (snowshoe hares, porcupines), their diets were not as diverse as those exhibited by American martens ($H' = 0.72$, $n = 3$). Similarly, American martens had relatively low diet diversity where their diets were dominated by large prey. There are few data for fisher diets from western North America, and diet diversity may be as low as, or lower than, that recorded in the Midwest. The average fisher diet may therefore be much less diverse over the range of the species than available data indicate.

Raine (1987) examined sympatric American marten and fisher diets in Manitoba, and found that fishers took proportionally more snowshoe hares than did American martens, although hares represented a greater proportion of the diet for both species than reported elsewhere. Raine's data showed a lower diet diversity for fishers ($H' = 0.50$), than for American martens ($H' = 0.65$). The opposite relationship was found with Clem's 1977*b* data from Ontario, (American martens: $H' = 0.99$, fishers: $H' = 1.11$). Until more research is conducted, these relationships must remain tentative.

Another difference between the two North American *Martes* species may be their demographic response to prey fluctuations. The one study that investigated potential numerical responses in fisher populations to prey population fluctuations found no adverse effects on reproductive output as snowshoe hare populations declined (Kuehn 1989). Two teams have researched American marten population response to fluctuating prey levels (Weckwerth and Hawley 1962, Thompson and Colgan 1987a), and both concluded that prey declines affected the American martens. Weckwerth and Hawley (1962) determined that a decrease in small mammal density resulted in a loss of female martens in the resident population, increased dispersal rate of juveniles away from the study area, and reduced reproductive success of resident American martens. Thompson and Colgan (1987a) found reduced ovulation rates in young female American martens, reduced production of young, and dispersal of formerly resident adults from their study area in response to a decline of most prey species over a five-year period.

It would be enlightening to examine fisher population response to prey population declines in an area where fishers do not depend on large prey. The geographic variation observed in fisher diets suggests they can use a range of prey sizes from shrews to porcupines. Measuring the effect of this variability on population attributes and long-term viability will allow us to gauge the relative quality of the habitats providing these diets for fishers.

Diversity in diet will be influenced by what foods are available to an animal—an immediate reflection of diversity of the environment—and by the food-procuring abilities of the animal. Conversely, habitat preferences may be generated by the availability of food, which is a function of both prey density and of "catchability" (Buskirk and Powell, this volume). Comparisons of American marten and fisher diets may illuminate differences in the types of communities these species inhabit and their relative predatory abilities. Such studies would help predict the effects of habitat change.

I hypothesize that American martens, with smaller home ranges and less diverse diets, are more negatively affected than sympatric fishers by habitat alterations that reduce prey populations. I suspect that the American marten's propensity for voles may result in a vulnerability to habitat change that the fisher does not have. Where fishers rely more heavily on larger prey (snowshoe hares and porcupines), they exhibit less diet diversity than American martens. In some areas, however, their diet is more diverse than that of American martens.

The larger body size of fishers may make available a greater range of potential prey than is the case for American martens. Also, there is some evidence that fishers acquire seasonal body fat reserves, have much greater thermal tolerance, and may have greater short-term ability to withstand food shortages than American martens (Harlow, this volume). The potential for greater plasticity in the diet, and the larger home ranges of fishers, could dampen the effect of prey population declines caused by habitat alterations. The limited data available also suggest that fisher populations may be more resilient than American martens to declines in the availability of important prey.

Conversion of late to early seral stages induced by intensive timber harvesting causes increases in some small mammal populations, but these species are not the dominant prey of the American marten over most of its range. The local reduction of mature forest habitat by timber harvest probably causes declines in species, such as the red-backed vole, which depend on coarse woody debris and dense forest canopy. The open, xeric conditions that occur after timber harvest could similarly reduce the optimal habitat for *Microtus* spp. in upland sites. While American martens, like many faunal species, appear to prefer habitat edges, the spatial distribution of forest/opening interfaces and mature forest stands is probably a critical component in maintaining habitat quality (Buskirk and Powell, this volume). Effects of forest harvest on adjacent riparian meadow systems are probably more complex and currently are more obscure. At the very least, the changes in local hydrologic balances in the forest-meadow ecosystem resulting from substantial logging of forest stands must jeopardize the maintenance of a meadow in its preharvest state. The immediate results of timber harvest, therefore, are likely to be reductions in vole populations within the forest and also in the high-density vole populations in adjacent riparian meadows.

Weckwerth and Hawley (1962) and Thompson and Colgan (1987a) report lowered fecundity and general population decline for American martens when vole populations were reduced. These data support the concern that timber harvest negatively affects American martens in the harvest area through reduction of a critical component of the prey base. A more thorough understanding of the relationships between prey fluctuations and changes in mustelid populations will improve our ability to predict the effects of habitat change on these predators and deserves priority in future research.

23 Modeling Habitat Selection by Fishers

Linda E. Thomasma, Thomas D. Drummer,
and Rolf O. Peterson

The long-term survival of free-ranging animal populations requires that a certain amount of optimum habitat be available. Recommendations regarding the type and amount of habitat are based in part on the premise that an animal will "prefer" habitats that fulfill its life requisites. In other words, animals should prefer habitats that maximize their own fitness (Buskirk and Powell, this volume).

The U.S. Fish and Wildlife Service developed Habitat Suitability Index (HSI) models to quantify habitat quality for selected fish and wildlife species. Thomasma et al. (1991) tested the HSI model for fishers (*Martes pennanti*) (Allen 1983) in the Ottawa National Forest, Michigan, against the hypothesis that fishers used habitat in proportion to its availability. We rejected that hypothesis because fishers used the habitats with higher HSI value with greater frequency than expected, based on the habitats' availability, and this preference was found to increase with increasing HSI value (Thomasma et al. 1991).

To determine habitat preference of a species, researchers have used a variety of procedures to compare the relative amount of a resource or habitat used by a species to the relative availability of that resource or habitat. Neu et al. (1974) compared the proportions of different habitats available to moose (*Alces alces*) to the proportion of moose sighted in the various habitats. Johnson (1980) proposed a method based on use and availability data for individuals.

These analyses determine animals' preference for or avoidance of categories of habitat, such as cover type. The HSI and three of its components, however, are continuous rather than categorical values. Rather than assign arbitrary cut-points to categorize these variables, we examined habitat selec-

316

tion as a smooth function, called a selection function, of the value of the variable. Our previous study (Thomasma et al. 1991) used logistic regression functions to estimate these selection functions. Logistic regression, however, imposes shape restrictions on the functions (Seber 1984:308). Our non-parametric estimates of the selection functions impose no shape restrictions and therefore provide additional insight into habitat selection patterns.

In this chapter we amplify our earlier analysis to further examine fisher habitat selection and the utility of habitat models, calculating nonparametric estimates of HSI and habitat selection functions. A secondary objective is to evaluate fisher preference for and avoidance of different cover types.

Study Area

The 200-km² study area was in the Kenton and Ontonagon Ranger districts of the Ottawa National Forest in the Upper Peninsula of Michigan. Lake Superior gives rise to moderate temperatures (average annual temperature is 6°C) and increased cloudiness and snowfall (236–381 cm/year). Landforms and soils within the study area vary as a result of the last glaciation. Vegetation is dominated by maple and oak (*Acer saccharum*, *A. rubrum*, and *Quercus rubra*). Minor components include aspen (*Populus grandidentata* and *P. tremuloides*), birch (*Betula alleghaniensis* and *B. papyrifera*), eastern hemlock (*Tsuga canadensis*), white pine (*Pinus strobus*), spruce (*Picea mariana* and *P. glauca*), fir (*Abies balsamea*), larch (*Larix laricina*), and plantations of jack and red pine (*Pinus banksiana* and *P. resinosa*).

Methods

The Fisher HSI Model

HSI models are used in planning and environmental impact assessment to predict the effects of land-use actions on species and their habitats. The HSI is a numerical index ranging from 0 (unsuitable habitat) to a maximum value of 1 (optimal habitat). The index is assumed to have a positive linear relationship with the potential carrying capacity of the habitat (U.S. Fish and Wildlife Service 1981).

Based upon a review of the literature, Allen (1983) constructed the HSI model for fishers, assuming that prey availability and foraging strategies determine habitat use. He also assumed that winter and early spring habitat is the most restrictive component of the annual habitat requirements of fishers and that dense, old-growth forest stands provide suitable winter cover for fishers and their prey (Allen 1983). The life requisite of the fisher HSI model is winter cover.

The four habitat variables in the fisher HSI model are the following: V_1 = percent tree canopy closure; V_2 = mean overstory tree diameter at 1.4 m height (dbh); V_3 = tree canopy diversity; and V_4 = percent of overstory canopy composed of deciduous species (Allen 1983). The relationships of these variables to the suitability indices (SI) are shown in Figure 23.1. The HSI is determined by combining the SI scores with the following equation:

$$(SIV_1 \cdot SIV_2 \cdot SIV_3)^{1/3} \, SIV_4 = HSI$$

Measurements

During the winters of 1985–1986 and 1986–1987 HSI data were collected on 132 plots where fisher tracks crossed line transects. The transects were also sampled at 180-m intervals ($n = 497$) to determine available habitat. On each 263-m² plot we measured the four habitat variables and computed the HSI value for the plot. The cover type for each plot was determined based on the predominant overstory vegetation. Thomasma et al. 1991 provided complete methodology.

Statistical Analyses

Model Performance. In the previous analysis (Thomasma et al. 1991) we tested the HSI model in part by performing discriminant analyses using logistic regression functions (Seber 1984). The HSI model was not invalidated because the HSI variable discriminated between used and available plots ($P = 0.0002$). In the multivariate analysis to determine the contributions of individual habitat variables, both the mean dbh of overstory trees and the percent of overstory canopy composed of deciduous species contributed to the discriminant function ($P = 0.019$, $P = 0.033$, respectively). Percent tree canopy closure did not contribute to the model in the presence of the other habitat variables ($P = 0.101$). The diversity variable was unable to differentiate between used and available plots at any level of analysis and is not addressed further in our work.

Habitat Selection Functions. A selection function is a mathematical function of a habitat attribute. It is proportional to the probability that a species uses habitat with that attribute. Let X denote a habitat attribute measured on each sampling unit (e.g., dbh of overstory trees) and let $s(X)$ denote the selection function for X. Let $f_u(X)$ and $f_a(X)$ represent the density functions,

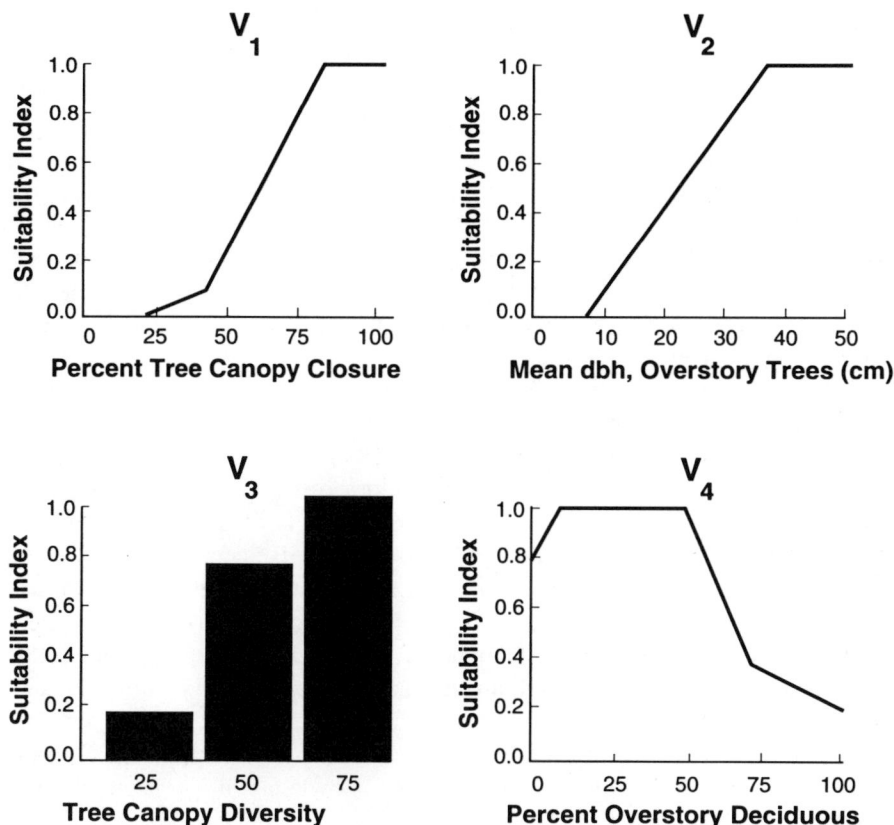

Figure 23.1. The four habitat variables and their relationships with the suitability indices (SI) for the fisher HSI model. From Allen 1983.

or probability distributions, of X for the used and available units, respectively. For example, $f_u(X)$ and $f_a(X)$ could represent normal distributions, perhaps with different parameters. The species "selects" from the available distribution, $f_a(X)$, and "produces" the used distribution, $f_u(X)$. Using the theory of weighted distributions (Mahfoud and Patil 1982) it can be shown that:

$$f_u(X) = s(X)f_a(X)/\theta$$

where θ is a constant equal to the expected value of the selection function $s(X)$ over the density function $f_a(X)$. Thus, if the animal selects habitats randomly with respect to X (i.e., the animal neither prefers nor avoids habitats based on X), then $f_u(X) = f_a(X)$, and the ratio of the two functions is constant with respect to X. If the animal selects habitats as a function of X, then the two distributions are different. If we knew the functions $f_u(X)$ and $f_a(X)$, we could plot the ratio $f_u(X)/f_a(X)$ versus X to examine the selection function. Since the density functions for the habitat variables are unknown, we estimated them and computed the ratio of the estimates at each value X to estimate the various selection functions.

The numerical value of $f_u(X)/f_a(X) = s(X)/\theta$ can be interpreted. The constant θ, which is unknown, represents the average value of the selection function over all values of X. If $s(X)/\theta > 1.0$ at a value X, this implies above-average selection for that value of X, whereas a value of $s(X)/\theta < 1.0$ implies below-average selection.

We used kernel estimators (Silverman 1986), a nonparametric procedure, to estimate used and available density functions for the HSI variable and other continuous habitat variables in the HSI model. Kernel estimates of density functions can be envisioned as smoothed histograms. A kernel is a function that controls exactly how the smoothing is done. The bandwidth determines the range over which the smoothing will occur. A small bandwidth results in an "irregular" estimate of $f(X)$ that is sensitive to sampling error because it contains excessive detail. It is desirable to use a large bandwidth, yielding a "smoother" estimate of $f(X)$, but too large a bandwidth may result in excessive loss of detail. Both the kernel and bandwidth are user-selected, although it is possible to determine optimal kernels and bandwidths if enough is known about the underlying density function. Optimality criteria were not used here because we had no prior knowledge of any properties of any of the density functions we estimated.

Because of the possible effects of subjective kernel and bandwidth selection, we tried several kernels and bandwidths for each analysis. The general shape (increasing or decreasing) of the selection functions and the general location of turning points were robust to choice of kernel and bandwidth. The bandwidths used to produce Figures 23.2 and 23.3 were: for HSI, 0.1; for mean dbh of overstory trees, 5 cm; for percent tree canopy closure, 10; and for percent overstory deciduous, 7.5. All estimates presented here were constructed with the kernel $K(X) = \exp(-X^2/2)$, similar to a standard normal density function. Values near X received the most weight in estimating $f(X)$.

Cover Type Analysis. The test of the HSI model and estimation of habitat selection functions do not yield a qualitative analysis of fisher habitat selection. To obtain such an analysis, we categorized the fisher-used plots and available plots into 1 of 11 cover types (Table 23.1). The distributions of used and available cover types were compared with a chi-square test of homogeneity (Conover 1971:149). We then compared the used and available propor-

Table 23.1. The cover type composition of the used and available plots and Bonferroni pairwise comparisons, Ottawa National Forest, Michigan

Type	% Use (n)	% Available (n)	Z	P
Hardwoods	37.1 (49)	42.4 (205)	−1.08	0.28
Mixed hardwoods				
$\leq$ 25% conifer closure	9.9 (13)	13.6 (66)	−1.15	0.25
25% < conifer closure $\leq$ 50%	9.9 (13)	4.9 (24)	2.09	0.04
> 50% conifer closure	15.9 (21)	3.5 (17)	5.25	<0.0001[a]
Mixed birch	0.8 (1)	1.2 (6)	−0.46	0.65
Mixed aspen	8.3 (11)	7.2 (35)	0.43	0.67
Swamp conifer	6.1 (8)	4.6 (22)	0.72	0.47
Open/brush	3.8 (5)	8.9 (43)	−1.94	0.05
Pine plantations	8.3 (11)	13.6 (66)	−1.63	0.10
Mixed pine[b]	—	—	—	—
Fir/spruce[b]	—	—	—	—
TOTAL	100 (132)	100 (484)		

[a]Indicates an overall 0.05 level of significance.
[b]Cover types eliminated from analyses because of small sample size.

tions for each of the 11 cover types with two-sample Z-tests for proportions (Conover 1971:141). We used the Bonferonni approach (Miller 1981:15–16) to maintain an overall level of significance of $P = 0.05$. Two proportions must be different at the $0.05/11 = 0.00454$ level of significance to be declared statistically significantly different.

Results

Estimates of Density Functions

The univariate density functions were estimated on available plots and plots used by fishers for the HSI variable and its three components: mean dbh of overstory trees, percent tree canopy closure, and percent of overstory that is deciduous (Fig. 23.2). Both the used and available HSI distributions were primarily composed of low values (Fig. 23.2A). The distributions were most different at high HSI, where $f_u(\text{HSI}) > f_a(\text{HSI})$, implying fisher preference for large HSI values. The distributions for mean dbh of overstory trees both appeared to be Gaussian, with the used distribution shifted slightly towards higher dbh values than the available distribution (Fig. 23.2B). The tree canopy closure distributions were similar, except in the 0–40% range, where $f_a(\text{percent closure}) > f_u(\text{percent closure})$ (Fig. 23.2C). The overstory de-

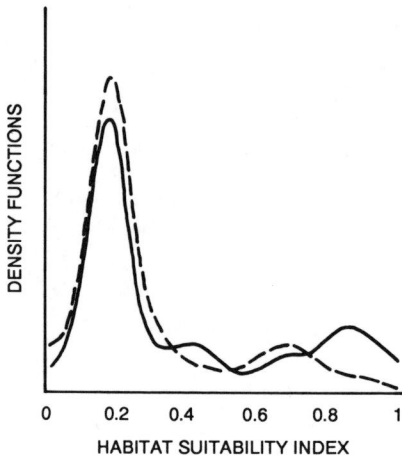

HABITAT SUITABILITY INDEX

A

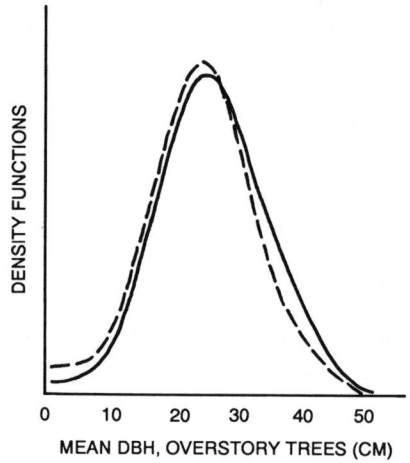

MEAN DBH, OVERSTORY TREES (CM)

B

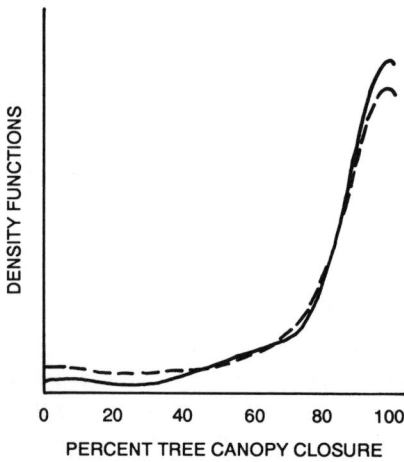

PERCENT TREE CANOPY CLOSURE

C

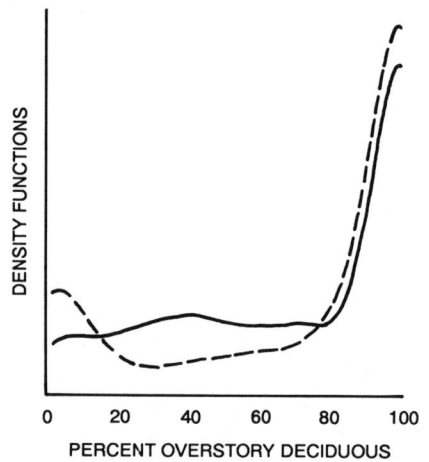

PERCENT OVERSTORY DECIDUOUS

D

Figure 23.2. Estimated available (*dashed line*) and used (*solid line*) distributions for the HSI and three habitat variables for fishers (*Martes pennanti*) in Michigan.

ciduous distributions were both left-skewed, with f_u(percent deciduous) $>$ f_a(percent deciduous) in the 20–80% range (Fig. 23.2D).

Habitat Selection Functions

The selection function (the ratio of the estimated used and available densi-ty functions) of each variable was plotted on the variable value (Fig. 23.3), and these plots were used to determine fisher habitat selection patterns and to assess the nonlinear transformations used to convert raw habitat data into suitability indices (Fig. 23.1). The HSI selection function increased from 0 to 0.1, declined slightly from 0.1 to 0.3, and increased over the HSI range 0.3 to 0.45, exceeding 1.0 at HSI = 0.36 (Fig. 23.3A). The selection function then decreased over the range 0.45–0.6, dropping below 1.0 at HSI = 0.53. The selection function increases for HSI $>$ 0.6, exceeding 1.0 at HSI = 0.75.

The selection function for mean dbh of overstory trees increased over the range 0–40 cm and exceeded 1 at 27 cm (Fig. 23.3B). The function de-creased when mean dbh of overstory trees exceeded 40 cm, although data were sparse in this region.

The percent tree canopy closure selection function increased in the 0–9% range, then decreased in the 9–23% range (Fig. 23.3C). The selection func-tion increased in the 23–54% range, exceeding 1.0 at 47%. The function then decreased in the 54–73% range, dropping below 1.0 at 62%, and increased thereafter, exceeding 1.0 at 85%. For percent overstory deciduous, the selec-tion function increased from 0–32% overstory deciduous, exceeding 1.0 at 14% (Fig. 23.3D). The function declined from 33–86% overstory deciduous, dropping below 1.0 at 76% and increasing slightly in the 86–100% range.

Cover Type Analysis

From the chi-square analysis, fishers did not use cover types in proportion to their availability (X^2 = 38.5 [8], $P <$ 0.0001). The mixed hardwoods with conifer closure greater than 50% and the open/brush cover type contributed the most to the chi-square statistic.

Mixed hardwoods with conifer closure greater than 50% was the only cover type for which use significantly exceeded availability ($P <$ 0.0001) (Table 23.1). Other cover types indicated possible preference or avoidance, but the differences in the used and available proportions were not statistically significant. Mixed hardwoods with conifer closure greater than 25% but no more than 50% indicated possible preference (P = 0.037) and open/brush areas indicated possible avoidance (P = 0.052).

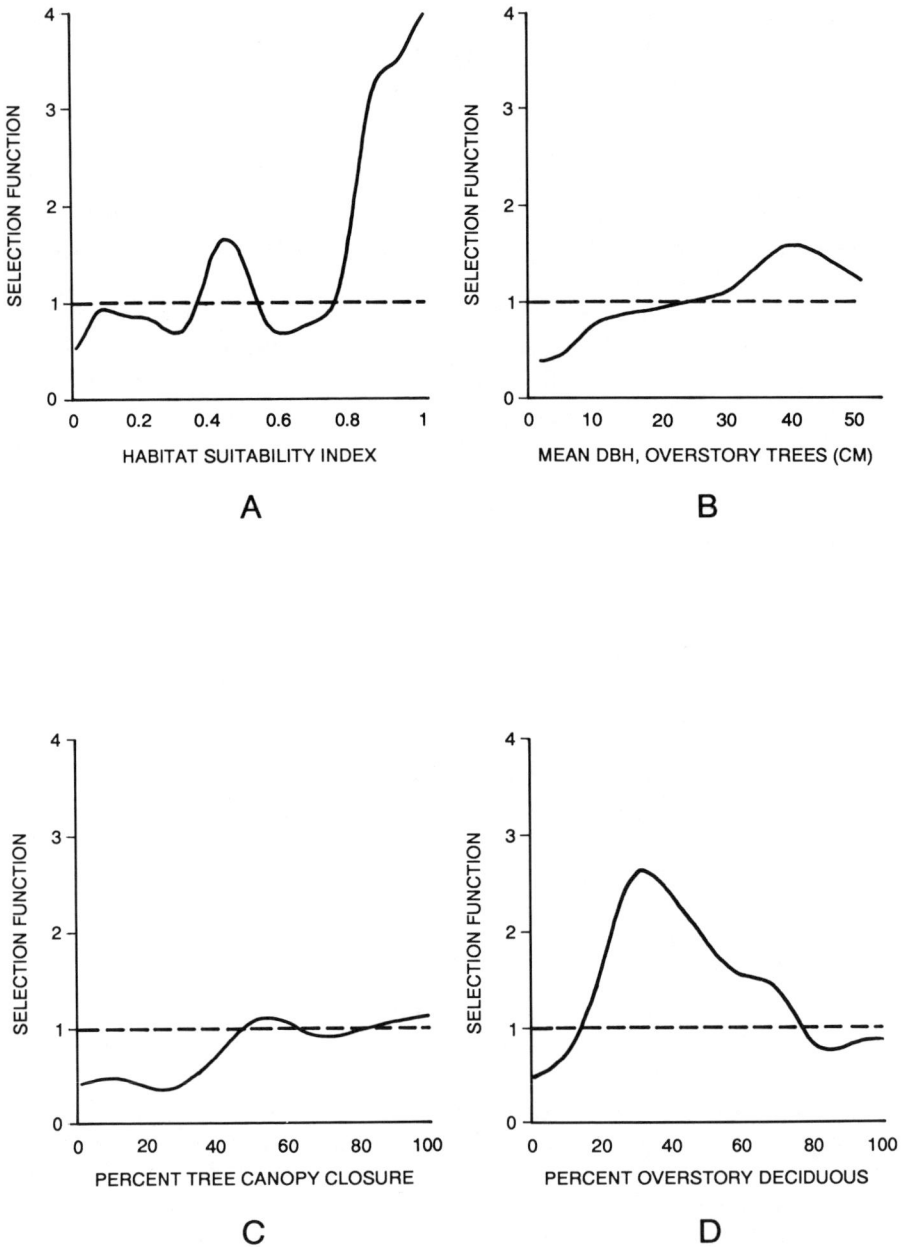

Figure 23.3. Estimated selection functions for the HSI and three habitat variables for fishers (*Martes pennanti*) in Michigan. A value greater than 1.0 indicates above-average selection.

Discussion

Habitat Selection Functions

The estimated selection function for HSI did not increase monotonically. Most significantly, a reversal occurred in the 0.4–0.6 range, with the selection function dropping below 1.0 at HSI = 0.53, and remaining below 1.0 up to HSI = 0.75. Of the available plots, 73 had HSI in the range 0.45–0.75, and 48 (65.8%) of these were pine plantations. Seventeen fisher-used plots were in this HSI range but only 6 (35.3%) were pine plantations. These proportions differed significantly (Z test, $P = 0.021$) and we concluded that the HSI model placed excessively high values on pine plantations.

For other habitat variables, the estimated selection functions were in rough agreement with the original transformations proposed by Allen (1983). The selection function for dbh increased with dbh and exceeded 1.0 at 27 cm. This implies that fishers prefer habitats with trees greater than 27 cm dbh. The percent tree canopy closure selection function increased with increasing canopy closure, exceeding 1.0 at 47%. Fishers were also found to prefer an overstory tree canopy with deciduous closure between 14 and 76%. The maximum selection function value for percent overstory deciduous occurred at 33%.

It should be noted that this univariate analysis ignores the relationships among habitat variables. A similar multivariate analysis is therefore preferable but more difficult to interpret.

Cover Type Analysis

The conifer component of the preferred cover types consisted primarily of eastern hemlock. The mean dbh of hemlock trees within 21 fisher-used plots with conifer closure greater than 50% was 39.1 cm. Hemlock growing in groups or stands produces a dense canopy, providing winter shelter for a variety of wildlife. Some of the larger hemlock stands within the study area were known to be used by wintering white-tailed deer (*Odocoileus virginianus*) for yarding. Of the 21 fisher-used plots, 32.9% had signs of white-tailed deer, 23.8% had small mammals, 19.0% had squirrels (*Tamiasciurus hudsonicus*), 14.3% had snowshoe hares (*Lepus americanus*), and 4.8% had ruffed grouse (*Bonasa umbellus*). These species are known fisher prey (de Vos 1951*b*; Clem 1977*b*; Raine 1987; Arthur et al. 1989*b*; Giuliano et al. 1989; Martin, this volume). Porcupines (*Erethizon dorsatum*) are also associated with this habitat type (Brander 1973), but were not seen in our study plots. Raine (1983) hypothesized that prey availability and snow conditions are possible factors that affect winter habitat use by fishers. Lacking comparable prey data on other cover types and on any available plots, however, we cannot make valid statistical inferences concerning prey availability and use.

24 Feeding Ecology of Eurasian Pine Martens and Stone Martens in Europe

Anthony P. Clevenger

Eurasian pine martens (*Martes martes*) and stone martens (*M. foina*) occupy biotopes ranging from Mediterranean to Fennoscandian taiga. Their distributions overlap extensively in mainland Europe, extending from Poland to the northern Iberian Peninsula (Fig. 24.1). Similar in size, both have slender, elongated bodies and are agile and well adapted for arboreal pursuit. Unlike sables (*M. zibellina*) and Eurasian pine martens, these two species are distinct species and do not hybridize where sympatric (Anderson, this volume).

Sympatry among similar mustelids has previously been explained in terms of body size differences that reduce competition for food (Rosenzweig 1966, 1968; McNab 1971; King and Moors 1979; Erlinge 1986). In general, similarly sized Carnivora often differ in terms of diet and foraging strategies (Erlinge 1972, Jenkins and Harper 1980). Overlapping diets have been reported for pine martens and stone martens in several studies (Baud 1981, Braña and del Campo 1982, Debieve et al. 1987, Jedrzejewski et al. 1989, Marchesi et al. 1989), but mechanisms for the two species' coexistence have only been proposed, not well explored. Powell and Zielinski (1983) noted the large diet overlap of these two species, and they suggested habitat differences as a structuring mechanism.

More than 25 published studies have described the diets of Eurasian pine martens and stone martens, most of which present quantitative data. But less than half provide year-round data and even fewer were multiyear studies. Sex- and age-related differences in marten diets have scarcely been investigated in Europe. Important aspects of foraging ecology, such as relative prey abundance, population responses to prey fluctuations, and predatory behavior of martens, are largely unstudied as well. These topics should be addressed in future studies so that data eventually become available to conduct critical tests of the two species' use of food resources and foraging strategies.

326

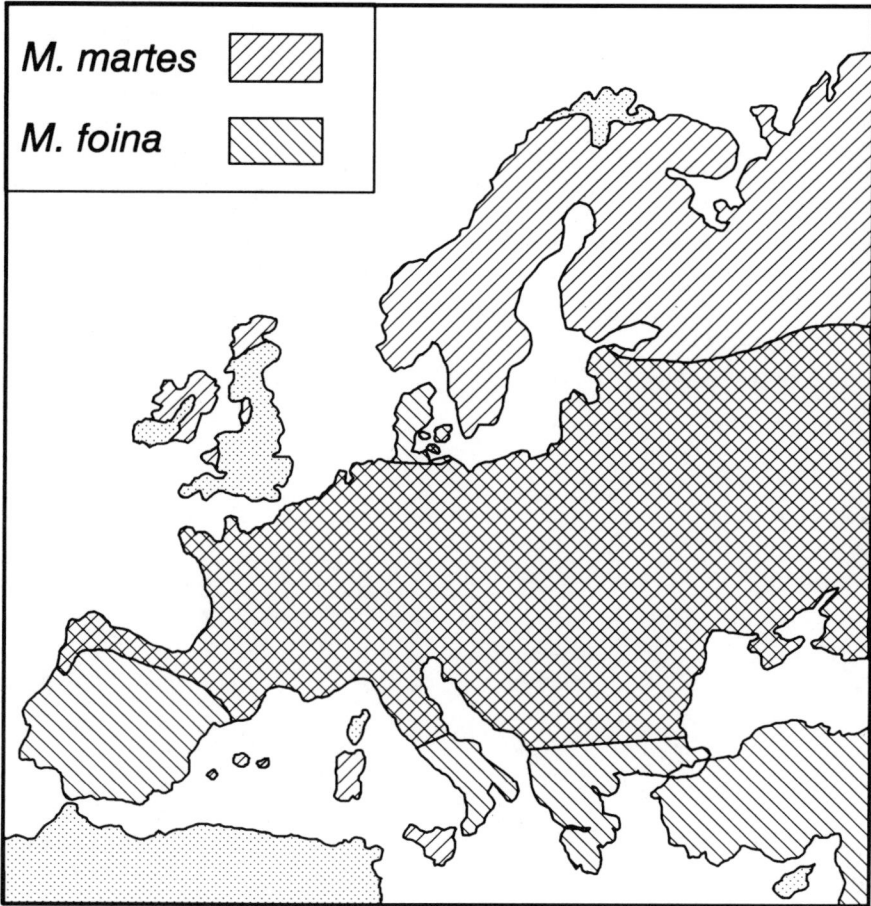

Figure 24.1. Present distribution of Eurasian pine martens (*Martes martes*) and stone martens (*M. foina*) in Eurasia. Data from Anderson 1970 and Corbet 1978.

In this chapter I summarize and compare the most important quantitative studies of diets of Eurasian pine martens and stone martens. I review the diets of both species by principal food categories and use feeding niche breadths and prey size indices to contrast the species' diets and possible foraging behaviors. Finally, I discuss food resource use and the possible importance of interspecific competition among sympatric *Martes*.

Eurasian Pine Martens

I used the results from 15 studies (Table 24.1). Of these, only 7 reported year-round data, 6 described winter diets, and 2 were fall-winter studies. For

Table 24.1. Diets of Eurasian pine martens (*Martes martes*) from 15 studies

Study (location)	Latitude	Material analyzed	Seasons (duration)	B_s[a]	PSI[b]	Diet categories $\geq$ 10% in diet (% occurrence)[c]
Lockie 1961 (Ireland)	54°N	337 scats	S, Su, F, W (2 yr)	0.50	—	M (36%), I (24%), V (19%)
Warner & O'Sullivan 1982 (Ireland)	53°N	609 scats	S, Su, F, W (4 yr)	0.61	35.5	Wood mice (13%), hares (12%); M (31%), B (46%), I (52%), V (46%)
Marchesi & Mermod 1989 (Switzerland)	47°N	823 scats, 31 GIs	S, Su, F, W (4 yr)	0.53	24.7	Bank voles (15%), wood mice (13%), *Microtus*/*Pitymys* (10%); M (38%), I (14%), V (30%)
Labrid 1987 (France)	46°N	171 scats	S, Su, F, W (1 yr)	0.51	—	Wood mice (48%), bank voles (39%); M (59%), B (21%), I (36%), V (57%)
Ruiz-Olmo & Lopez-Martin, in press (Pyrenees)	43°N	445 scats	S, Su, F, W (1 yr)	0.03	12.0	Rowenberry (70%); M (39%), I (33%), V (93%)
Clevenger 1993*b* (Cantabrian Mts.)	43°N	193 scats	S, Su, F (1 yr)	0.38	17.6	Wood mice (34%), field vole (18%); M (75%), I (32%), V (35%)
Clevenger 1993*b* (Minorca)	40°N	1180 scats	S, Su, F, W (1 yr)	0.64	53.5	Rats (12%), wood mice (10%); M (39%), B (31%), I (44%)

				[a]	[b]	[c]
Pulliainen 1981 (Finland)	68°N	2700 scats	W (4 yr)	—	—	M (77%), B (14%), V (21%)
Nyholm 1970 (Finland)	67°N	134 GI	W (?)	—	—	M (37%), B (26%), V (14%)
Storch et al. 1990 (Sweden)	60°N	94 scats	W (3 yr)	—	—	Shrews (35%), red squirrels (31%); M (81%), B (24%)
de Jounge 1981 (Sweden)	60°N	51 scats, 32 PR	W (2 yr)	—	—	Red squirrels (56%)
Reig & Jedrzejewski 1988 (Poland)	52°N	62 scats	W, Sp (1 yr)	—	—	Bank voles (28%), wood mice (25%), field voles (15%); M (>90%), A (11%), I (11%)
Ansorge 1989b (SE Germany)	49°N	66 GI	W (4 yr)	—	—	Bank voles (17%), field voles (15%); M (47%), B (17%), I (19%)
Garzon et al. 1980 (Pyrenees and Cantabrian Mts.)	43°N	53 GI	F, W (10 yr)	—	—	Wood mice (33%), field voles (30%); M (79%), B (30%), I (34%), V (35%)
Braña & del Campo 1982 (Cantabrian Mts.)	43°N	55 scats, 20 GI	F, W (3 yr)	—	—	Wood mice (16%), field voles (11%); M (62%), B (22%), I (32%), V (77%)

Note: GI = gastrointestinal tracts, PR = prey remains, M = mammals, B = birds, R = reptiles, A = amphibians, I = insects, V = vegetation. — = could not be calculated from data reported.

[a] Food niche breadth (Hespenheide 1975).

[b] Prey size index (Erlinge 1987).

[c] Percent occurrence of six food categories and mammal prey species ≥ 10%.

both Eurasian pine marten and stone marten diets, data were divided into six food categories: mammal, bird, reptile, amphibian, insect, and vegetation. Studies based on sample sizes of less than 50 were not considered.

Diet measures from the studies were expressed variously. All studies reported frequency data (*n*) for each food item; some, however, did not calculate frequency of occurrence (*n* food item/*n* scats). To standardize the results for the analyses in this chapter, I calculated frequency of occurrence for those studies that did not include it. Generally, frequency of occurrence overestimates the importance of undigestible foods (birds, insects, vegetable material) and small food items (Zielinski 1986, Reynolds and Aebischer 1991).

Mammals were the dominant food in 11 of the 15 studies. Wood mice (*Apodemus* spp.), field voles (*Microtus* and *Pitymys* spp.), and bank voles (*Clethrionomys* spp.) were the principal food items, constituting more than 10% of the diet in 8, 7, and 4 studies, respectively. The highest proportion of voles in the diet was reported from Finland (>60% occurrence) (Pulliainen 1981) and from Sweden (de Jounge 1981). In these studies, voles, Norwegian lemmings (*Lemmus lemmus*), and red squirrels (*Sciurus vulgaris*) were the most important foods. Only in the Spanish Pyrenees (Ruiz-Olmo and Lopez-Martín, in press) did neither voles or wood mice appear in more than 10% of the diet.

Dominant prey species other than voles included common moles (*Talpa europaea*) in Switzerland (Marchesi and Mermod 1989) and rats (*Rattus* spp.) on the Balearic island of Minorca (Clevenger 1993*a,b*). Uncommon mammal prey included hedgehogs (*Erinaceus europaeus* and *Atelerix algirus*) (Braña and del Campo 1982; Moreno et al. 1988; Reig and Jedrzejewski 1988; Clevenger 1993*a*) and bats (Clevenger 1993*a*).

In 2 studies, vegetation (fruits, berries, lichens, grass, nuts, mushrooms, etc.) was the most important food, appearing in more than 90% of the samples from the Pyrenees (Ruiz-Olmo and Lopez-Martín, in press) and 75% of those from the Cantabrian Mountains (Braña and del Campo 1982) (Table 24.1). Rowanberry (*Sorbus aucuparia*) was the leading fruit eaten. Ruiz-Olmo and Lopez-Martín (in press) found rowanberry in 70% of the scats collected over the year. Seasonal sample sizes were not included with their data, however, and this figure may represent sampling biases rather than a real pattern. Rowanberry also was found in more than 10% of the diets reported in Germany (Ansorge 1989*b*) and Switzerland (Marchesi and Mermod 1989). Other fruits represented in more than 10% of samples were blackberries (*Rubus* spp.) and carob fruit (*Ceratonia siliqua*) in Minorca (Clevenger 1993*b*), and rose-hips (*Rosa* spp.) in Switzerland (Marchesi and Mermod 1989).

All studies reported that the contribution of insects to the diet of Eurasian pine martens was high over the year. But in only 2 of the 15 studies were they the predominant food (Table 24.1). Insects occurred in more than 10% of the diet in 10 studies, but only in 4 did they form 33% or more of the diet, which suggests that their importance may be overestimated. Warner and O'Sullivan (1982) found that beetles (Coleoptera) were an important staple year-round, in addition to the bees and wasps (Hymenoptera) eaten during summer. In Minorca, I found the same annual trend (Clevenger 1993*b*) except that in summer, martens preyed heavily on grasshoppers (Orthoptera). Both studies indicated that high insect consumption corresponded with periods of abundance.

Birds composed more than 10% (frequency of occurrence) of the diet in 8 studies and 33% or more in only 1 report. In Scandinavia, tetraonids are common prey (Nyholm 1970, Pulliainen 1981), whereas forest passerines are the main avian prey elsewhere in Europe (Lockie 1961, Garzon et al. 1980, Warner and O'Sullivan 1982, Ansorge 1989*b*, Marchesi and Mermod 1989). Recent results from Minorca differ markedly, however; Eurasian pine martens preyed nearly exclusively on shrubland bird species (15 of 16 species identified in 46 scats with avian prey) during the passerine breeding season (Clevenger 1993*a*). Predation on cavity-nesting birds is also common and widespread, although of lesser frequency (Baudvin et al. 1985; Sonerud 1985*a,b*).

Reptiles and amphibians were scarcely present in the 15 studies. Reptiles were more frequently consumed than amphibians; the former were more than 5% of the diet in 4 studies, whereas amphibians represented more than 5% of the diet in only one study. Reig and Jedrzejewski (1988), working in Bialowieza National Park, Poland, found toads in 11% (frequency of occurrence) of the pine martens' diet in winter and early spring.

Diets varied seasonally in several studies. Seasonally abundant foods such as fruits and insects were important during summer and fall. Marchesi and Mermod (1989) documented significant seasonal diet differences in Switzerland. In Minorca, grasshoppers were in more than half of the summer scats, whereas during the fall, carob fruit predominated (Clevenger 1993*b*). Variations in diet from year to year were also reported. Pulliainen (1981) found that Eurasian pine martens in Finland switched to alternative winter foods such as berries and mushrooms when microtine populations were low.

To assess geographic patterns in pine marten diets in Europe, I plotted arcsine transformed percent frequency of occurrence of food categories against latitude for studies with data from three seasons or more (Lockie 1961; Warner and O'Sullivan 1982; Labrid 1987; Marchesi and Mermod 1989; Clevenger 1993*b*; Ruiz-Olmo and Lopez-Martín, in press). Results showed that there was no significant trend in pine marten food selection with latitude ($P > 0.10$).

Food Niche Breadth

Standardized feeding niche breadth (Hespenheide 1975) was calculated for 7 Eurasian pine marten studies (Table 24.1), using the six food categories and the niche breadth formula of Levins (1968), where Pi is the proportion of food item i in the total diet. Breadth values increase as the diet becomes more generalized, and reaches 1.0 when all foods are exploited equally. Only studies with year-round data were used in the analysis, since niche breadths from seasonal diets are not equivalent (Marchesi and Mermod 1989). An exception was made for the three-season Cantabrian Mountain data (Clevenger, 1993b). When compared with year-round data from Minorca, the difference in sampling intensity and duration had no effect on niche breadth value.

Sample sizes for the 7 studies ranged from 171 to 1180 ($\bar{x} = 541$, SE = 139); however, there was no significant relation between food niche breadth and sample size ($r = 0.41$, $F = 1.0$, $P = 0.36$). The mean niche breadth for the 7 studies was 0.45 (SE = 0.07) and ranged from 0.027 (Ruiz-Olmo and Lopez-Martín, in press) in the Spanish Pyrenees to a high of 0.64 in Minorca (Clevenger 1993b). Latitude also had no significant effect on food niche breadth ($r = 0.30$, $F = 0.5$, $P = 0.52$).

Excluding the Minorcan study, the 4 from northern Europe (Lockie 1961, Warner and O'Sullivan 1982, Labrid 1987, Marchesi and Mermod 1989) had wider mean niche breadth values ($\bar{x} = 0.534$, SD = 0.05) than the 2 from the south (Clevenger 1993b; Ruiz-Olmo and Lopez-Martín, in press) ($\bar{x} = 0.205$, SD = 0.25). A Mann-Whitney test, however, indicated that the difference between these means was not significant ($U = 8$, $P = 0.11$). If latitude is used as a general measure of dietary diversity for Carnivora (Gompper and Gittleman 1991), one would expect the opposite, since the diversity of animal communities increases with decreasing latitudes (Pianka 1966). The wide food niche breadth in the Minorcan study may be explained by the insular environment (only *Mustela nivalis* is a potential competitor), and the insular population's large body size (Alcover et al. 1986).

Prey Size Selection

An index of mean available prey size was calculated for 5 of the 15 studies (Lockie [1961] and Labrid's [1987] data did not permit inclusion) according to the formula and prey weights of Erlinge (1987) (Table 24.1). This index was chosen over other methods because avian prey are included, as opposed to strictly mammalian prey (Jaksic and Braker 1983, Iriarte et al. 1990). All mammalian prey of Eurasian pine martens were classified according to their average weights, and all avian prey in the diet were assigned a standard weight of 40 g.

Sample size had no effect on prey size index ($r = 0.83$, $F = 6.4$, $P = 0.08$). The mean prey size index for the 5 studies was 21 (SE = 2.7) and ranged from 12 from the Spanish Pyrenees (Ruiz-Olmo and Lopez-Martín, in press) to 54 in Minorca (Clevenger 1993*b*). The relationship between latitude and prey size index was not significant ($r = 0.07$, $F = 0.01$, $P = 0.91$).

Stone Martens

A total of 14 studies were used in the review of stone marten diets (Table 24.2). Nine of the 14 studies reviewed year-round data, 1 reported three-season information, and 1 provided summary data from an annual diet. Three single-season diet studies (summer, fall, and winter) were also reviewed. Stone marten diets consist of a broad spectrum of foods, including mammals, birds, vegetation, invertebrates, and reptiles. Vegetation (primarily fruits and berries) was the leading food category in 7 of 14 studies, mammals predominated in 4, and birds and insects in 1.

Wild and cultivated fruits were the most common vegetation in the diets reviewed. In southern France, Cheylan and Bayle (1988) found that more than 80% (frequency of occurrence) of the stone marten diet consisted of two varieties of fruits, mostly juniper berries (*Juniperus* sp.). Rowanberries were also important seasonal foods in diets in Switzerland and central Italy (Marchesi et al. 1989, Serafini et al. 1992). The lowest proportions of vegetation (<2% occurrence) were reported from Romania (Romanowski and Lesinski 1991) and southwestern Spain (Amores 1980).

Mammals occurred in more than 10% of samples in 12 studies. Field voles were the main prey, constituting more than 10% of the diet in 6 studies, followed by wood mice and shrews (*Crocidura russula*) in 1 study each. Marchesi et al. (1989) reported that stone martens in Switzerland preyed mostly on field voles and water voles. Wood mice and shrews were also identified as important foods in a deciduous woodland study area in northern Spain (Delibes 1978) and in northeastern France (Waechter 1975).

Several stone marten diets revealed a high proportion of birds. In Denmark, nearly half of the birds consumed by stone martens were passerines (Rasmussen and Madsen 1985), whereas Columbiformes predominated in an urban Swiss stone marten diet (Tester 1986). Birds comprising 12 species appeared in 23% of the stone marten diets in northern Spain (Delibes 1978).

Invertebrates, primarily insects, were important foods in all studies in all seasons. In southwestern Spain, insects were the principal food; for example,

Table 24.2. Diets of stone martens (*Martes foina*) from 14 studies

Study (location)	Latitude	Material analyzed	Seasons (duration)	$B_s{}^a$	PSI[b]	Diet categories ≥ 10% in diet (% occurrence)[c]
Rasmussen & Madsen 1985 (Denmark)	57°N	178 scats, 44 GI	S, Su, F, W (1 yr)	0.52	—	Field voles (32%); M (46%), B (58%), I (36%), V (15%)
Skirnisson 1986 (northern Germany)	54°N	246 scats	S, Su, F, W (1 yr)	0.52	49.9	*Microtus* (24%), rabbits (13%), wood mice (11%), earthworms (24%); M (51%), B (36%), I (18%), V (26%)
Skirnisson 1986 (northern Germany)	54°N	348 scats	S, Su, F, W (1 yr)	0.57	21.9	*Microtus* (32%), wood mice (12%), earthworms (29%); M (58%), B (28%), I (10%), V (65%)
Marchesi et al. 1989 (Switzerland)	47°N	882 scats	S, Su, F, W (2 yr)	0.38	19.5	Field voles (14%), water voles (12%); M (34%), I (30%), V (61%)
Tester 1986 (rural Switzerland)	47°N	440 scats	S, Su, F, W (1 yr)	0.29	—	M (23%), B (18%), I (13%), V (62%)
Tester 1986 (urban Switzerland)	47°N	407 scats	Su, F, W (9 mo)	0.23	—	B (54%), V (61%)
Cheylan & Bayle 1988 (southern France)	43°N	393 scats, 15 GI	S, Su, F, W (1 yr)	0.20	12.9	M (16%), I (10%), V (68%)

334

Serafini et al. 1992 (central Italy)	43°N	378 scats	S, Su, F, W (1 yr)	0.43	—	M (22%), B (14%), I (43%), V (77%)
Delibes 1978 (northern Spain)	43°N	148 scats, 9 GI	S, Su, F, W (1 yr)	0.59	26.7	Wood mice (15%), shrews (13%); M (53%), B (23%), I (42%), V (50%)
Amores 1980 (southwest Spain)	38°N	539 scats	S, Su, F, W (1 yr)	0.47	17.5	M (35%), B (18%), R (24%), I (78%)
Ansorge 1989a (southeast Germany)	49°N	1159 GI	W (2 yr)	—	—	Field voles (23%); M (12%), B (16%), V (11%), farm animals and refuse (>50%)
Waechter 1975 (northeast France)	48°N	431 scats	Su (2 yr)	—	—	Field voles (20%), shrews (11%); M (45%), I (13%), V (61%)
Leger 1979, cited in Libois 1991 (France)	48°N	146 scats	S, Su, F, W (1 yr)	—	—	M (>25%), B (>40%), V (>50%)
Romanowski & Lesinski 1991 (Romania)	45°N	103 scats	F (1 yr)	—	—	Microtus (30%), Spermophilus (15%); M (57%), B (49%), R (11%), I (16%)

Note: GI = gastrointestinal tracts, M = mammals, B = birds, R = reptiles, A = amphibians, I = insects, V = vegetation. — = could not be calculated from data reported.

[a]Food niche breadth (Hespenheide 1975).

[b]Prey size index (Erlinge 1987).

[c]Percent occurrence of six food categories and mammal prey species ≥ 10%.

335

beetles appeared in 30% of the diets (Amores 1980). Serafini et al. (1992) indicated that beetles and grasshoppers are consistently important stone marten foods in central Italy. Besides insects, 8 studies noted earthworms (Lumbricidae) in the diet of stone martens. Trace amounts (<5% occurrence) of these annelids appeared in 5 diets, but in 3 they constituted more than 25% of the sample. Although stone martens are capable diggers, earthworms may have been indirectly consumed when martens preyed on earthworm predators (birds, shrews, and wood mice).

Reptiles and amphibians were rarely eaten. They appeared in 5 and 4 of the 14 diets I reviewed, respectively. In southwestern Spain (Amores 1980), 12 or more species of reptiles, ranging in size from 10 g (*Psammodromus algirus*) to 200 g (*Malpolon monspesulanum*) made up 24% of the diet. Reptiles, primarily grass snakes (*Natrix* sp.), were common foods eaten by stone martens in Romania during the fall (Romanowski and Lesinski 1991).

Stone martens living in cities, villages and their vicinities often consume human refuse and other foods outside their normal diets. Nine of the 14 stone marten diets reviewed included human refuse. Tester (1986) reported that in Basel, Switzerland, 18% of the diet of urban stone martens was made up of domestic refuse, as opposed to 3% in nearby rural villages. Skirnisson (1986) and Ansorge (1989a) also found that human refuse commonly occurred in German stone marten diets.

I used 10 studies reporting annual diets of stone martens to examine dietary variations in relation to latitude. Except for reptiles, none of the food categories showed a significant trend with latitude ($r = 0.682$, $F = 7.0$, $P = 0.03$).

Food Niche Breadth. Feeding niche breadths were calculated for 10 studies of stone marten diets consisting of year-round data (Table 24.2). Sample sizes ranged from 157 to 580 ($\bar{x} = 372$, SE = 42), but there was no significant relation between food niche breadth and sample size ($r = 0.543$, $F = 3.3$, $P = 0.10$). The mean niche breadth for the studies was 0.42 (SE = 0.04) and ranged from 0.20 from southern France (Cheylan and Bayle 1988) to 0.59 in northern Spain (Delibes 1978). Latitude also had no significant effect on food niche breadth ($r = 0.170$, $F = 0.1$, $P = 0.36$).

The narrow niche breadth of stone martens in southern France can be explained by the high proportion of wild fruits in the diet. The diet from northern Spain had a more balanced composition of food categories, which Delibes (1978) attributed to the large area and long period over which his sample was collected.

Prey Size Selection. Prey size indices were calculated for 6 of the studies (Table 24.2). The data from Serafini et al. (1992), Rasmussen and Madsen

(1985), and Tester's (1986) could not be included in these calculations. Again, sample size had no effect on prey size index ($r = 0.58$, $F = 2.0$, $P = 0.22$). The mean prey size index for the 6 studies was 25 (SE = 5.4) and ranged from 13 from southern France (Cheylan and Bayle 1988) to 50 in northern Germany (Skirnisson 1986). No significant relationship was observed between latitude and prey size index ($r = 0.618$, $F = 2.4$, $P = 0.18$).

Diets of European Martens

Eurasian pine martens and stone martens are opportunistic predators with generalized diets. Although none of the studies analyzed food selection in terms of availability, diets of both species appear to reflect local food availabilities. The diets of the two martens overlapped considerably, but mammals were the most important prey in 11 of the 15 Eurasian pine marten studies and in only 4 stone marten studies. On the other hand, vegetation was the most important food category in 7 out of 14 studies of stone marten diets but in only 2 reports on Eurasian pine marten diet.

To analyze differences in diet composition, I compared the mean proportion (percent occurrence) of each of the six food categories in the 7 Eurasian pine marten and 10 stone marten diets. Diets for the two species were significantly different (X^2 contingency table test, $X^2 = 10.5$, $P = 0.03$). Eurasian pine marten diets appeared to have higher proportions of mammals, whereas vegetation and insects occurred in higher proportions for stone martens. But when pair-wise comparisons were made among the six food types, Mann-Whitney tests showed no significant species differences ($P > 0.10$ for all tests). Sample sizes were small, however, and the results are not conclusive.

Among-species differences were also found in a Swiss study (Marchesi et al. 1989) in which sympatric Eurasian pine marten and stone marten diets were studied concurrently. Diets of the two species differed from each other both seasonally, except in spring, and over the entire year. Pine martens preyed more than stone martens on mammals and birds, whereas both species preyed equally on insects.

Diet Diversity and Prey Selection

Seven Eurasian pine marten and 10 stone marten studies showed no significant species difference between mean food niche breadths (Mann-Whitney test, $U = 44$, $P = 0.40$). Likewise, Marchesi et al. (1989) documented high niche overlap (0.94) between the two martens in Switzerland. But they found that although prey in the two diets were nearly identical, how prey were exploited differed markedly. Pine martens preyed primarily on rodents from forest habitats (*Clethrionomys glareolus* and *Talpa europaea*), whereas stone martens selected those from open habitats (*Microtus* spp., *Pitymys* spp., and *Arvicola terrestris*).

A species' food niche breadth is simply the range of food resources it uses (Hespenheide 1975), but my calculation of it refers to the contribution of food categories rather than of individual foods to the diet. My niche breadths therefore may not reflect the total number of plant and animal species exploited by European martens nor the complexity of the community they live in. Islands are species-depauperate compared with mainland areas (Lack 1942, Sondaar 1977, Reed 1982), but predator food niche breadths can be wider on islands as a result of reduced competition and expanded resource bases (Lawlor 1982). Of the 29 studies I reviewed, the widest food niche breadth was from the Minorcan pine marten diet, which had foods nearly equally distributed among categories (Clevenger 1993*b*). The range of food resources used may, therefore, be influenced more by environmental variability and the distribution and abundance of nonprey populations than by simple prey species diversity or abundance.

The mean prey size indices for 5 studies of Eurasian pine martens and 6 of stone martens did not differ significantly (Mann-Whitney test, $U = 17$, $P = 0.78$). But some prey size differences between diets of the two species merit attention. In 3 Eurasian pine marten studies (Lockie 1961; Warner and O'Sullivan 1982; Clevenger 1993*b*) large prey items such as hares, rabbits (*Oryctolagus cuniculus*), and rats were more than 10% of the diet. Of the same three prey species, only rabbits formed more than 10% of the stone marten diets in Germany (Skirnisson 1986). In 1 study of stone martens in Switzerland (Marchesi et al. 1989) a medium-sized prey (*Arvicola terrestris*) was equally important. The same study reported that prey size indices for pine martens (24.7) were also slightly larger than those for stone martens (19.5).

Martin (this volume) shows that for American martens (*M. americana*) diets low in diversity were dominated by large prey. My results differed. Eurasian pine marten diets with wide food niche breadths ($B_s > 0.60$; Warner and O'Sullivan 1982; Clevenger, 1993b) also had high proportions of large prey and correspondingly high prey size indices (PSI > 35, Table 24.1). The results were not as clear for stone martens. Studies characterized by large prey items (Skirnisson 1986, Marchesi et al. 1989) had disparate food niche breadths (range $= 0.37–0.57$), and the indices for prey size were also varied (range $= 19–50$, Table 24.2). Nevertheless, my data sets are smaller than Martin's and caution should be taken in interpreting my results.

Several authors (Rosenzweig 1966, Schoener 1967, Ashmole 1968,

Erlinge 1969) have proposed that predator size within a hunting set correlates to prey size, which assumes that larger predators have a wider spectrum of prey sizes (and food types) available to them than smaller predators. Martin (this volume) documents that diet diversity indices for the larger-bodied fisher (*M. pennanti*) were significantly different, although marginally so, than for American martens. I found interspecific differences in neither prey size indices nor food niche breadths, as would be expected for similar-sized predators. The intraspecific analysis, however, revealed the opposite: there was no significant relationship between food niche breadths and prey size indices for pine martens ($r = 0.811$, $F = 5.8$, $P = 0.09$) and for stone martens ($r = 0.514$, $F = 1.4$, $P = 0.30$). I tend to believe that the intraspecific results were more subject to error inasmuch as the regression was performed with few data ($n = 5$ and 6, respectively), and more data will be needed to adequately test this interesting relationship.

Discussion

This analysis shows that Eurasian pine marten and stone marten diets are similar in composition and breadth. The diets of both species are generalized and include foods that are seasonally most abundant and accessible. But more important, their diets appear to be defined by each species' foraging habitats.

Although diets overlapped greatly, they show ecological separation. Pine martens are associated with forest habitats and prey principally on forest-dwelling rodents (*Clethrionomys glareolus* and *Apodemus* spp.), whereas stone martens are more catholic, occupying open fields, shrublands, villages, and woodlands (Marchesi et al. 1989; Kruger 1990; Herrmann, this volume).

Dietary niche breadths of Eurasian pine martens and stone martens did not differ, but intraspecific values varied greatly. The small sample size used in this study most likely contributed to the nonsignificant results. Also, diet breadth may be affected by competing sympatric predators (Van Valen 1965, Roughgarden 1972), which were not considered in my analysis. To obtain more accurate information on food niche breadth, one should quantify diets over several years to examine year-to-year variations in prey and other environmental factors. Marchesi et al. (1989) documented significant differences in diets of both marten species, but because field vole populations were high in their study, they predicted greater food sharing between martens in most years.

Competition may not have influenced the evolution of these two species on mainland Europe. For competition to occur between species with similar ecological requirements, resources must be limiting (Cody and Diamond

1975). Food does not seem to be limiting for either species; it seems unlikely that dietary partitioning has been selected for. Stone martens probably evolved in the Middle East and later moved north into Europe (Anderson, this volume). Their infiltration northward through the range of Eurasian pine martens was possible only because stone martens were adaptable to the environmental variability they encountered. The highly heterogenous environment of Europe probably provided alternative habitats for stone martens when coexistence with Eurasian pine martens in woodlands was difficult.

Pine martens and stone martens are highly adaptable, opportunistic feeders. Living in and around villages, stone martens encounter a more stable and diverse food supply than that in the nearby woodland habitats occupied by pine martens (Herrmann, this volume). This plasticity in habitat use likely makes stone martens less vulnerable to loss of habitat than Eurasian pine martens.

Compared with those for North American *Martes*, data for European *Martes* are scarce. Much basic research is still needed before databases will permit rigorous testing of hypotheses. Studies are needed throughout the range of the two species that collect basic diet data and investigate numerical responses of marten populations. Of particular importance is the quantification of available prey to determine martens' food preferences, as well as the identification of prey to species level. Both will provide much-needed and valuable information on marten foraging ecology.

Introduction

Over much of the distribution of martens and fishers, logging has been implicated as a factor in the reduction of these species' abundance and range. Does the association of the boreal forest martens with older forests mean that martens and fishers do not use younger forests and cannot tolerate habitat changes resulting from logging? In the following four chapters, the authors address this broad issue and provide direction for development of habitat management prescriptions.

Brainerd et al. examine use of seral stages by Eurasian pine martens in an intensively managed forest. They caution that their definition of "old forest" includes forests more than 70 years old that have been subjected to silvicultural practices. Contradicting the belief that Eurasian pine martens require old-growth forest, they found these animals using forest stands of differing ages. Brainerd et al. conclude that habitat use depends more on size and placement of clear-cuts than on the age of forest stands, although they recognize that within forest stands, structural attributes are important to martens.

These patterns of habitat use in managed landscapes are also evident for American martens. Thompson and Harestad note that three main hypotheses explain the American martens' preference for old forests: predator avoidance, special habitat features (e.g., coarse woody debris and large-diameter trees), and prey abundance. Our review of habitat use in relation to disturbance reveals that American martens select older stands over early seres, but that younger stands receive considerable use. We propose a model for relating carrying capacity for the American marten to the extent of logging. Our conclusions support those of Brainerd et al.: small dispersed cuts within a forested matrix should have less impact on marten populations than large

341

contiguous clear-cuts. We suggest that up to 25% forest removal may even increase population density of American martens.

Jones and Garton report the results of their study of habitat ecology of fishers in Idaho, one of the few such efforts for the western states or provinces. They found that fishers used all but the nonforested seral stage and this use was primarily in young, mature, and old-growth forests. They emphasize that management of fisher habitat must be considered at a landscape rather than a stand level. Availabile and connected patches of mature and old-growth stands is critical to maintaining fisher populations in managed forests. Jones and Garton note that fishers evolved in forested systems subjected to fire and other natural disturbances and suggest that these disturbances and human-caused ones, if of small scale, likely benefit fisher populations.

Buck et al. also compare fisher habitat use in lightly and heavily harvested western forests. They found that habitat use differed between treatments, but more important they conclude that lightly harvested forests more closely approach suitable fisher habitat than do heavily harvested forests. Buck et al. hypothesize that selective cutting, which maintains continuity of forest cover and structure, is likely to ensure retention of habitat features important to fishers.

For both martens and fishers, forest fragmentation and loss of connectivity are important problems for which management solutions are not now available. But the lack of solutions to these problems reaffirms the centrality of understanding habitat ecology to successful habitat management. The spectre facing habitat managers is that options and opportunities for maintenance of marten and fisher populations are lost as forest harvesting proceeds and remaining large intact forest stands are modified. Managers rely largely on the assumption that habitats preferred by martens and fishers are the best habitats and that preservation of these best habitats will secure the future of marten and fisher populations. Perhaps it would be more prudent to vary approaches to habitat management by jurisdiction and to monitor the effects of these practices on marten and fisher populations. As we learn more, successful practices can then be extended to other areas. The challenge is to ensure that large-scale habitat management practices are successful enough to compensate for local failures.

ALTON S. HARESTAD

25 Eurasian Pine Martens and Old Industrial Forest in Southern Boreal Scandinavia

Scott M. Brainerd, J.-O. Helldin,
Erik Lindström, and Jørund Rolstad

The rapid conversion of Fennoscandian boreal forests from relatively natural forests to a mosaic of uniform monocultures of coniferous trees has awakened concern over the fate of a variety of forest-adapted species (Punkari 1984, Helle 1985, Gamlin 1988, Baskin 1990). In boreal Eurasia, pine martens (*Martes martes*) are often associated with old stands of coniferous forest (Aspisov 1959, Grakov 1972, Bjärvall et al. 1977, Wabakken 1985, Storch et al. 1990). In Sweden and Russia, snow-tracking censuses indicated that densities of marten tracks were higher in tracts of undisturbed, old virgin forests than in surrounding areas fragmented by large-scale clear-cutting (Aspisov 1959, Grakov 1972, Bjärvall et al. 1977, Jonsson 1992). These studies support the notion that old coniferous forests are important habitat for pine martens. In Scandinavia, popular accounts (e.g., Selås 1990*a*) imply that this species is an old-forest specialist, on the basis of these studies and research on the closely related American marten (*M. americana*: see reviews by Clark et al. 1987; Strickland and Douglas 1987; Bissonette et al. 1989; Thompson 1991; Buskirk and Powell, this volume; Thompson and Harested, this volume) and sable (*M. zibellina*: review in Baskin 1990). Since the 1960s, however, Eurasian pine marten numbers have increased markedly in Russia (Grakov 1972, Baskin 1990) and Fennoscandia (Krott and Lampio 1983, Storch et al. 1990, Helldin and Lindström 1991) despite continued clear-cutting and intensified forest management. Indeed, increased prey biomass in clear-cut areas (Hansson 1978, Sonerud 1986, Henttonen 1989) may have benefited martens and other medium-sized predators, thus offsetting the negative effects of forest removal to some extent (Romanov 1956, Krasovsky 1970, Henttonen 1989, Baskin 1990, Brainerd 1990).

The structural characteristics associated with old virgin forests (snags, woody debris, large trees with cavities, abundant shrub layer, diverse vertical structure) may be less abundant in managed forests. In southern boreal Scandinavia, however, forestry practices often create an abundance of dead and rotting stems, branches, and stumps in younger forests (Hansson 1978; Majewski and Rolstad, in preparation). Such structural components may be analogs of the physical structure of virgin forest.

As part of a broader ecological study on Eurasian pine martens in the industrial forests of southern boreal Scandinavia, we investigated the use of different age classes of coniferous forests during 1987–1991. In this chapter we address the widely held notion that the Eurasian pine marten is an old-forest specialist by testing for its preference for old seminatural stands of industrial coniferous forest.

Study Areas

Our research was conducted in two forested areas near the southern limit of the boreal zone (Ahti et al. 1968) of Sweden and Norway. Grimsö Wildlife Research Station (59°40′N, 15°25′E) is situated in south-central Sweden, and our efforts were restricted to the southern portion (50 km²) of the study area. Varaldskogen Wildlife Research Area (60°10′N, 12°30′E) is located on the Norwegian-Swedish border 175 km west of Grimsö and covers 100 km². Grimsö is relatively flat (75–125 m elevation), whereas the topography at Varaldskogen is more hilly, varying between 200 and 400 m elevation. Both areas are dominated by commercial stands of Scots pine (*Pinus sylvestris*) and Norway spruce (*Picea abies*). Stands dominated by deciduous trees are rare, but birch (*Betula pubescens*), alder (*Alnus incana*), and aspen (*Populus tremula*) are sometimes present in coniferous stands. Bogs and agricultural fields are rare at Varaldskogen but compose 21% of the Grimsö study area. Lakes and rivers compose between 5 and 15% of both study areas. The substrate in both areas is glacial till, dominated in many places by fields of large boulders.

Modern forestry practices have created a fine-grained mosaic (Pielou 1974, Rolstad 1991) of clear-cuts, plantations, and older-forest stands for martens in both areas. The majority (>80%) of clear-cuts are 10 ha or less, and rarely exceed 50 ha. At Grimsö these cuts are generally interspersed in a mosaic of older forest; in Varaldskogen, clear-cuts and plantations often adjoin, creating contiguous cutover areas of several hundred hectares. Two-thirds of the Grimsö study area is in spruce-dominated (>50%) stands, whereas about half of the Varaldskogen forests are dominated by spruce.

Average January temperatures are −4.4°C for Grimsö and −7.3°C at

Varaldskogen, with average snowfalls about 40–50 cm. The winters of 1988–1989, 1989–1990, and 1990–1991, however, were unusually mild and virtually snow-free. Average July temperatures are 16.2°C in both study areas. Detailed descriptions of Grimsö and Varaldskogen are given by Ceder-lund (1981) and Rolstad et al. (1988), respectively.

Materials and Methods

For this analysis, we used data collected on Eurasian pine martens cap-tured during the winters of 1986–1987, 1988–1989, 1989–1990 and 1990–1991 at Grimsö ($n = 12$: 7 males and 5 females) and the winters of 1989–1990 and 1990–1991 at Varaldskogen ($n = 6$: 3 males and 3 females). Two females at Grimsö and two males at Varaldskogen were monitored for more than a year. Data on three of the martens included in the Grimsö data were also used in Storch et al. 1990. The basic methods apply to both study areas.

Wooden box traps ($40 \times 40 \times 60$ cm) were baited with either honey or the viscera of cervids from the area. Martens were drugged with a combination of ketamine hydrochloride (Ketalar, 10 mg/kg body weight) and xylazine (Rompun, 2 mg/kg body weight). Each animal was marked with a small plastic rototag and a metal earclip, and with a radio collar (Televilt AB, Ramsberg, Sweden) equipped with either a metal loop or whip antenna. Radio collars weighed 40 g for males and 25 g for females, about 2.5% of body weight. Martens were released at their trap sites after recovery and monitored from forest roads in each area. They were also visually located with hand-held receiving equipment, primarily at resting sites or more rarely while they were moving.

A subsample of marten radiolocations at Grimsö ($n = 887$) and Var-aldskogen ($n = 394$) were selected for this analysis. Active locations com-posed 52% and 22% of the Grimsö and Varaldskogen samples, respectively. Radiolocations included triangulations, biangulations, close-tracking loca-tions (100 m or less from the animal), and visual observations. In this analysis, we included only triangulations with the longest side 250 m or less and 1000 m or less from the furthest tracking station and cross bearings with the longest bearing 200 m or less from the marten. Each triangulation was also evaluated in terms of the angle between respective bearings, which was greater than 45°. The angle between cross bearings was 45–135°. Since martens often were continuously radio-tracked over 3- to 12-hour periods, only independent locations (Swihart and Slade 1985) were used in this analy-sis to avoid autocorrelation and to standardize data across individuals and study areas. This method was applied to locations less than 12 hours apart. For some martens the independence criterion could not be satisfied. In such instances, locations more than 12 hours apart were considered independent. Home range centers for independence analyses were computed on the RANGES IV program (Kenward 1987). Martens were not sampled evenly, and possible influences associated with varying sample sizes were compensated for by weighting expected values for each individual with its number of locations.

Habitat use was determined by plotting locations on detailed forest habitat maps provided by the Swedish and Norwegian Forest Services. Habitat availability was measured within the home range (95% minimum area convex polygon, Mohr 1947) for each individual. This analysis was split into two seasons: winter (16 Nov–15 Apr), and spring-summer (16 Apr–15 Sep) for all years. Most transmitters failed before autumn, and this season was excluded from the analysis because of low sample sizes.

Four forest age classes were considered: 0–8 years, 9–30 years, 31–70 years, and more than 70 years. These categories roughly corresponded to clear-cuts, young plantations, middle-aged plantations or seminatural forests, and old seminatural forests, respectively. Only locations in forest habitats (including clear-cuts), which compose more than 95% of the Grimsö data and 100% of those at Varaldskogen, were used in this analysis; nonforest habitat types such as bogs and agricultural fields were not included.

We tested for preference for old forest in three ways. First, we compared the proportion of each marten's locations in old forest against the proportion of this forest class within the marten's home range. Within each season, only martens with 20 or more locations were included in this analysis. Active selection, indicated by plots above the 45° median line, was tested with a 1-tailed, 1-group Wilcoxon's signed-rank test within each season and study area. (Points exactly on the median line were excluded.) Possible differences among study areas and seasons were tested with a 1-way, 2-group Wilcoxon signed-rank test. Next, we compared the total number of marten locations with their weighted expected values between old and younger forest age classes (1-tailed chi-square goodness-of-fit test). Active, inactive, and total locations were analyzed within seasons and study areas, and all martens were included. Finally, chi-square tests were used to further compare use and availability among the four forest age classes for this sample. If use varied from that expected, the Bonferonni-Z simultaneous confidence interval test (Neu et al. 1974) identified forest classes which were preferred or avoided within each season and study area. Statistical significance was inferred if $P < 0.05$.

Results

Use of Old Forest by Individual Martens

Habitat use nearly differed between seasons ($Z = -1.92$, $P = 0.054$) but not between study areas ($Z = -0.80$, $P = 0.43$ (Fig. 25.1). During winter, martens selected old forest at Varaldskogen ($T^+ = 0$, $n = 4$, $P < 0.05$) but not at Grimsö ($T^+ = 11$, $n = 9$, $P = 0.10$). When the winter samples from both areas were combined, martens preferred old forest over habitat available within their home ranges ($T^+ = 15$, $P < 0.02$, $n = 13$). During spring and summer, martens did not use old forest more than expected at Grimsö ($T^+ = 21$, $n = 9$, $P > 0.40$) or Varaldskogen ($T^+ = 2$, $n = 4$, $P > 0.15$), or when these area samples were combined ($T^+ = 31$, $n = 13$, $P > 0.10$).

Use of Old Forest by the Sample Population

Eurasian pine martens did not use old forest more than forests 70 years old or younger at Grimsö during the winter or spring-summer seasons, or when these seasons were combined (Table 25.1). At Varaldskogen, martens preferred resting sites in old forest during winter, but active locations during both seasons and resting sites during the spring-summer season were distributed randomly between old and younger forests. When study areas were combined, martens used old forests more than expected in general and for resting. During spring and summer, martens tended to select old forests while moving but not while resting. When seasons and study areas were combined, active locations occurred in old forest more than expected: the same was true when active and inactive locations were pooled.

Use of Forest Age Classes by the Sample Population

During winter, martens preferred old forest for resting at Varaldskogen, but not at Grimsö (Fig. 25.2). In both study areas, martens avoided forests 0–8 years old while resting, and when active and inactive locations were pooled within study areas ($P < 0.001$). Otherwise, their use of forest age classes did not differ from that expected in either study area.

During the spring and summer months, martens did not show a preference for old forests (Fig. 25.3). In both study areas, martens avoided forests 0–8 years old for resting and moving, a trend that continued when active and inactive locations were combined ($P < 0.001$). At Grimsö, martens strongly avoided forests 9–30 years old for resting but not for movement. This avoidance remained when active and inactive locations were pooled ($P < 0.01$).

Grimsö

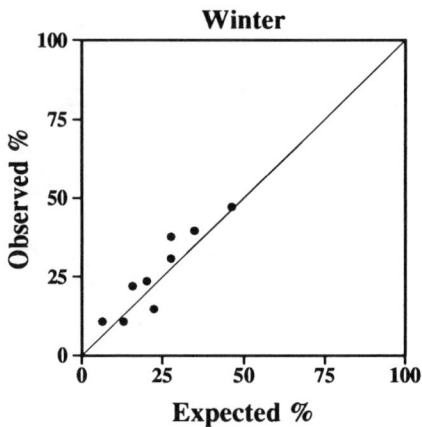

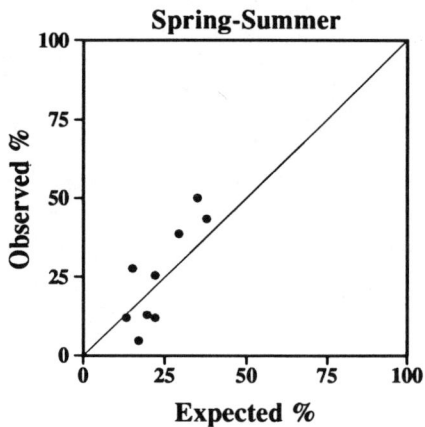

Varaldskogen

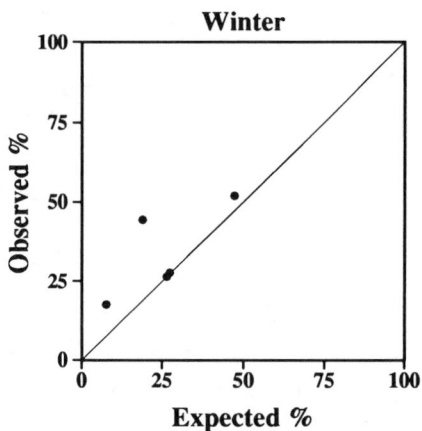

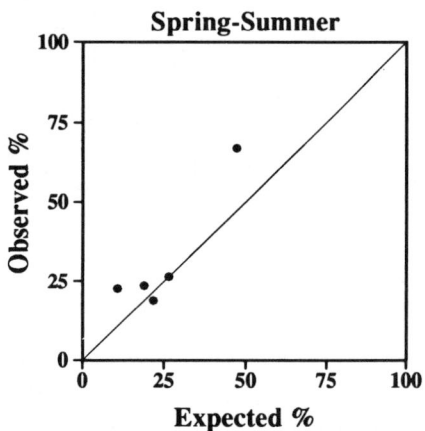

Figure 25.1. Observed/expected ratios of use of forests more than 70 years old by Eurasian pine martens (*Martes martes*) at Grimsö Wildlife Research Station, Sweden, and Varaldskogen Wildlife Research Area, Norway, during winter and spring-summer seasons, 1986–1991. Each point represents proportionate use of this forest class by each individual within its home range.

Table 25.1.   Comparison of observed and expected Eurasian pine marten (*Martes martes*) radiolocations in forests ≤70 years old and >70 years old at Grimsö Wildlife Research Station, Sweden, and Varaldskogen Wildlife Research Area, Norway by season, 1986–1991

Study area and season[a]	Activity	≤70 yr		>70 yr		*n*	*P*
		Obs.	Exp.	Obs.	Exp.		
Grimsö							
winter (11)	Active	148	152	53	49	201	0.26
	Inactive	167	170	59	56	226	0.32
	All	315	322	112	105	427	0.22
spr-sum (8)	Active	175	185	72	62	247	0.07
	Inactive	166	159	47	54	213	0.14
	All	341	344	119	116	460	0.37
TOTAL (12)	Active	323	337	125	111	448	0.06
	Inactive	333	329	106	110	439	0.33
	All	656	666	231	221	887	0.22
Varaldskogen							
winter (6)	Active	36	39	15	12	51	0.16
	Inactive	98	114	50	34	148	0.001
	All	134	153	65	56	199	0.03
spr-sum (6)	Active	26	28	11	9	37	0.22
	Inactive	112	118	46	40	158	0.14
	All	138	146	57	49	195	0.09
TOTAL (6)	Active	62	67	26	21	88	0.11
	Inactive	210	232	96	74	306	0.002
	All	272	299	122	95	394	<0.001
Combined							
winter (17)	Active	184	191	68	61	252	0.15
	Inactive	265	284	109	90	374	0.01
	All	449	475	177	151	626	0.008
spr-sum (14)	Active	201	213	83	71	284	0.05
	Inactive	278	277	93	94	371	0.45
	All	479	490	176	165	655	0.16
TOTAL (18)	Active	385	404	151	132	536	0.03
	Inactive	543	561	202	184	745	0.06
	All	928	965	353	316	1281	0.008

Note: *P* values are for a 1-tailed chi-square goodness of fit test with 1 df.
[a]Numbers in parentheses are the number of martens studied for that season.

Martens at Grimsö preferred forests 31–70 years old for resting and movement, and this preference increased when these locations were pooled ($P <$ 0.001). At Varaldskogen, forests 9–70 years old were used as expected, even when seasons were combined for active, inactive, and pooled locations in this study area.

Grimsö

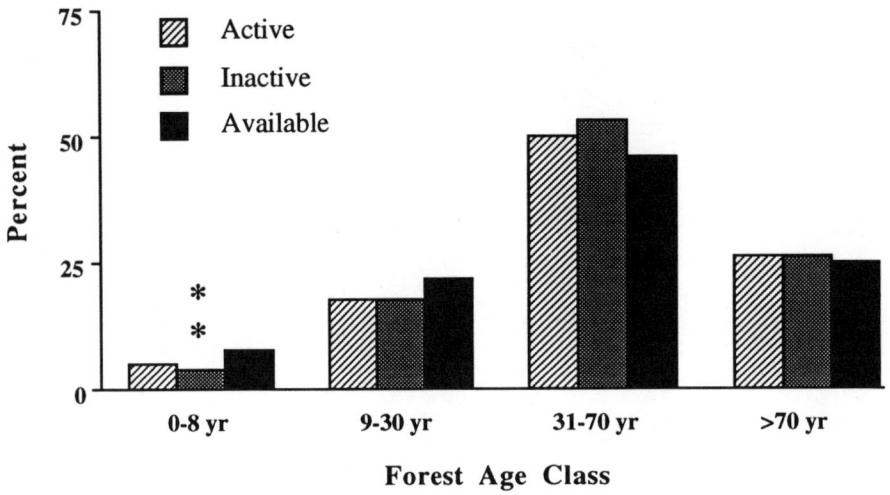

Varaldskogen

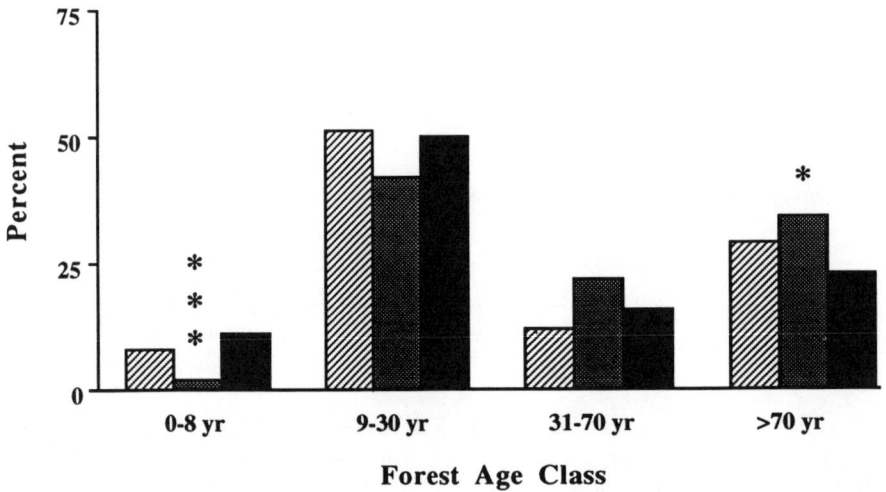

Figure 25.2. Eurasian pine marten (*Martes martes*) use of forest age classes during winter at Grimsö Wildlife Research Station, Sweden, and Varaldskogen Wildlife Research Area, Norway, 1986–1991. Sample sizes are given in Table 25.1. *P*-values for Bonferonni-Z tests: *$P < 0.05$; **$P < 0.01$; ***$P < 0.001$.

Grimsö

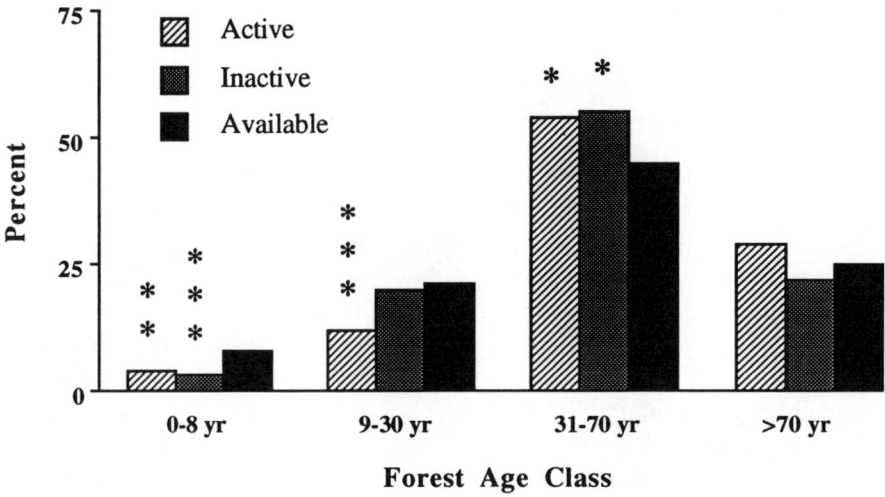

Varaldskogen

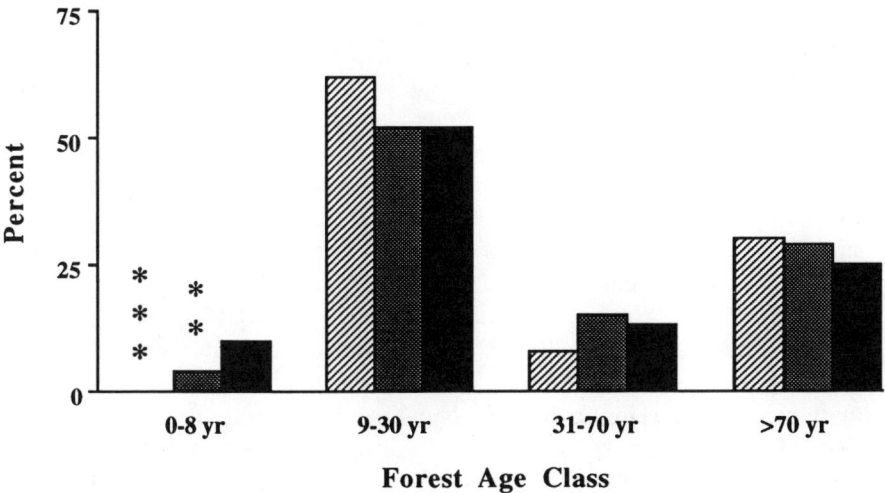

Figure 25.3. Eurasian pine marten (*Martes martes*) use of forest age classes during spring-summer at Grimsö Wildlife Research Station, Sweden, and Varaldskogen Wildlife Research Area, Norway, 1986–1991. Sample sizes are given in Table 25.1. *P*-values for Bonferonni-Z tests: *P* < 0.05; **P* < 0.01; ***P* < 0.001.

Discussion

Snow-tracking studies in Eurasia (Aspisov 1959, Grakov 1972, Wabakken 1985), including Grimsö (Storch et al. 1990) have indicated that martens prefer old-forest stands in winter. We found this preference held for that season but could not confirm it for spring-summer (Storch et al. 1990). The rather broad use of forest age classes in our study suggests that Eurasian pine martens were able to meet their life requirements in forests altered by modern forestry practices. What can explain this broad use of younger forests?

Eurasian pine martens probably select forests where foraging efficiency is maximized (Thompson 1986; Thompson and Harested, this volume) and predation threats are reduced (Brainerd 1990, Storch et al. 1990). They are diet generalists (Höglund 1960, Lockie 1961, Nyholm 1970, Morozov 1976, Pulliainen 1981*b*, Wabakken 1985, Storch et al. 1990) and thus can find food in a wide variety of habitats. Foraging may be concentrated where martens can obtain the most prey (Thompson 1986), which may be in early seres (e.g., voles: Hansson 1978, Sonerud 1986, Henttonen 1989). Martens have many enemies (Nyholm 1970, Pulliainen 1981*a*, Jonsson 1983, Korpimäki and Norrdahl 1989, Storch et al. 1990) including the fox (*Vulpes vulpes*), and overhead escape cover probably influences habitat selection (Pulliainen 1981*a*, Wabakken 1985, Storch et al. 1990). Many young forest stands should provide adequate escape cover for martens.

Structure near the ground may facilitate foraging, resting, and escaping predators. In our study areas, the abundance of dead stems from thinning and cutting practices in younger forests may provide cover for a variety of prey species, including microtines (Hansson 1978). The broken, rocky substrate randomly distributed in both areas may also facilitate access to prey, as well as providing cover for resting, thermoregulation, and escaping enemies.

In our study areas, many younger forests were spruce-dominated. Spruce forests, with their dense crowns, higher site productivity, lighter and softer snow cover, and diversity of prey species (Wabakken 1985) were favored by Eurasian pine martens in other studies (Grakov 1972, Pulliainen 1981*a*, Wabakken 1985). Squirrels (*Sciurus vulgaris*) prefer old spruce-dominated forests (Andrén and Lemnell 1992), and martens often rest in squirrel nests in spruce trees (Pulliainen 1981*a*, Storch 1988). Squirrels are important to martens as food (Höglund 1960; Nyholm 1970; Pulliainen 1981*a*, 1984; Wabakken 1985; Storch et al. 1990), and martens hunt them heavily when they are abundant and other prey species are relatively scarce (Nyholm 1970), preferring old spruce forests during these periods.

Large trees and snags with cavities excavated by black woodpeckers (*Dryocopus martius*), important to martens as natal dens (Selås 1990*b*), are found in a variety of forest types and ages in commercial forests (Majewski and Rolstad, in preparation). In our study, female martens used such structures in all forest classes more than nine years old (Brainerd et al., in preparation). Snow cover, which restricts access to prey and impedes martens' movements (Grakov 1972, Wabakken 1985), was generally lacking during most of our study and may have allowed martens to use a wider variety of forest types than in winters of normal snowfall (Storch et al. 1990).

Eurasian pine martens were consistently located in old-forest stands within their home ranges during winter, but this did not indicate a strong preference for such stands (Fig. 25.1). If martens depend on old forests, they should exhibit a strong preference for such forests when they are scarce within home ranges and prefer them less as their availability increases. During spring-summer, the effect was very nearly the opposite of what we might expect for an old-forest specialist. The literature implies, however, that old forests are important to pine martens in Eurasia. How should we interpret our results in the light of such studies?

Clear-cutting practices may adversely affect Eurasian pine marten populations (Grakov 1972, Brainerd 1990). Martens generally avoided clear-cuts and young plantations 0–8 years old in our study, as in other studies (Grakov 1972, Pulliainen 1981a, Wabakken 1985, Storch et al. 1990). At Grimsö and Varaldskogen, cutting mosaics were fine-grained (from the perspective of pine martens), unlike the coarse-grained, contiguous clear-cutting (>1000 ha) of northern Sweden and Russia. The lower marten densities observed in the fragmented forests of these regions (Aspisov 1959, Bjärvall et al. 1977, Jonsson 1992) may be related to the intensity and scale of cutting. Large-scale clear-cutting may adversely affect marten populations (Grakov 1972; Thompson and Colgan 1987b; Thompson 1991; Thompson and Harested, this volume), whereas small (<10 ha) cuts may benefit marten populations through increasing their food supplies (Brainerd 1990).

Martens in cutover areas may experience higher predation by foxes, which prefer clear-cuts (Christensen 1985) for hunting field voles (*Microtus agrestis*), a favored prey (Lindström 1982). Increased fox densities in cutover areas may reduce marten densities through direct predation, particularly on juveniles, although published data supporting this hypothesis are lacking (see Jonsson 1992). In regenerating forests lacking snags and cavity trees, natal dens may be located on the ground (D. H. Bakka, pers. commun.), perhaps exposing newborn martens to higher predation risks.

Old virgin forests may have higher structural diversity than commercial forests of similar age. Complex physical structure may increase martens' foraging success (Thompson 1986; Thompson and Harested, this volume), with woody debris providing access to small mammals, overwintering insects, and other food in winters of high snowfall (Grakov 1972). Such debris might also provide shelter for resting and denning (Buskirk et al. 1988, 1989; Buskirk and Powell, this volume) or for escaping from enemies.

Old stands of industrial coniferous forest had some importance to Eurasian pine martens, particularly during winter months. Such stands varied in structure from monocultures devoid of understory and vertical structure to semi-natural stands with diverse structure. Younger forests, including many monocultures of spruce and pine, were generally used in proportion to their availability, and forests 31–70 years old were even favored. This implies that Eurasian pine martens are capable of using a wide variety of forest types in commercial forests and are therefore not old-forest specialists. Structural elements such as site productivity, tree species composition, canopy coverage, vertical structure, and rockiness and the presence of dead wood and tree cavities probably influence marten habitat selection, and we plan to examine their importance (Brainerd et al., in preparation). As Buskirk and Powell emphasized in this volume, habitat selection patterns may be best explained in terms of fitness (Fretwell 1972). At the individual level, martens can exploit a wide variety of forest ages and types. At the population level, however, densities should be highest where reproduction and survival are maximized. The ideal habitat for martens is probably in landscapes dominated by expanses of mature, natural forests, where foraging opportunity, cover, and choice of denning sites may be increased and predation risks reduced.

Acknowledgments

We thank the Swedish Sportsmen's Association, the Swedish Agency for Environmental Protection, the Norwegian Scientific Research Council, and the Norwegian Agricultural Research Council for funding. D. H. Bakka, E. Rolstad, and K. Sköld contributed immeasurably to the success of this study by assisting in the capture and monitoring of martens. We also thank A. Harestad, J. Swenson, P. Wegge, and two anonymous referees for comments on earlier versions of the chapter. P. Majewski kindly interpreted Russian-language articles.

26   Effects of Logging on American Martens, and Models for Habitat Management

Ian D. Thompson and Alton S. Harestad

The fate of American martens (*Martes americana*) is of particular concern in several areas of North America because of loss of preferred habitat through logging of old forests. Habitat destruction was suggested as the primary cause of population decline in areas where American martens have become extinct or are currently threatened (Yeager 1950, Bergerud 1969, Dodds and Martell 1971*a*, Davis 1983, Thompson 1991). In some jurisdictions—for example, New Brunswick (New Brunswick Dep. Nat. Resour. 1990), and U.S. National Forests in Oregon and Washington (M. Raphael, USDA For. Serv. Pac. Northwest Res. Stn., pers. commun.)—habitat criteria for American martens are used to establish area objectives for mature and old forest age classes. Loss of habitat and decline of American marten populations are indicative of broader problems in forest management: an inability to regenerate forest ecosystems with all their complex community processes, and a lack of long-term large-scale planning.

In this chapter we review information on the effects of habitat change through commercial logging on American martens, propose generic models of selection of forest age class, and suggest management protocols to maintain American marten populations in areas where commercial logging is conducted. We propose our models as hypotheses with the suggestion that they be tested and improved through research and management-level experiments.

Effects of Timber Harvesting on American Martens

Why American Martens Choose Old Forest

Buskirk and Powell (this volume) discuss habitat selection by American martens; we provide only a summary here as a basis for discussion of the effects of logging. There are three main hypotheses for why American martens choose old forests:

Predator avoidance. American martens may prefer to live under a mature tree canopy that provides protection from avian predators (Herman and Fuller 1974, Pulliainen 1981*a*, Hargis and McCullough 1984). Further, mature forests are used infrequently by possible mammalian predators such as red fox (*Vulpes vulpes*) and lynx (*Felis lynx*).

Special habitat features. Coarse woody debris and large-diameter trees found in old forests, and not in younger ones, are needed for winter resting, maternal denning and to provide access beneath the snow surface for American martens to hunt small mammals in winter (Steventon and Major 1982; Wynne and Sherburne 1983; Thompson 1986; Buskirk et al. 1988, 1989). Winter dens are usually associated with coarse woody debris. Natal dens have been found in large-diameter partially hollow trees (Wynne and Sherburne 1983), particularly white cedar (*Thuja occidentalis*) in the East. Bergerud (1969) suggested that the extensive logging of white pine (*Pinus glauca*) could have been partly responsible for the decline of American martens in Newfoundland. Pine is the only tree species large enough to accommodate maternal dens of martens on that island.

Prey abundance. Thompson and Colgan (1987*a*) suggested that American marten populations are proximally food-regulated; therefore they should forage where they can obtain the most prey, not necessarily where most prey occur. Foraging success by American martens was found to be 21–119% greater in old-forest stands compared with successional forest, depending on prey density (Thompson 1986). American martens are highly catholic in diet, and although preferences have been found occasionally (Martin, this volume), we doubt that the occurrence of individual prey species influences overall habitat selection.

Effects of Habitat Change

American martens probably select habitat for all three reasons—to avoid predators, to locate needed shelter and to achieve maximum foraging success. Disturbances such as fire, insect infestation, blow-down, and logging alter forest structure and habitat features on a broad scale, with consequent effects on prey abundance and the presence of competitors and predators. The extent of habitat change can be put in context only at a landscape level, which is the appropriate scale for managing the species. The effects of logging on habitat depend on how much suitable habitat remains and the amount that will redevelop quickly enough on a large enough scale for martens to persist, given the particular forest type.

The immediate impact of timber harvesting is a loss of habitat, coupled with changes in ecosystem processes and structures to which American martens are not well adapted. The effects of timber harvesting on American marten habitat can be grouped into two categories: short-term, small-scale effects, acting on local populations (e.g., the effects of a five-year logging operation), and long-term effects at the population and landscape level.

Short-Term Effects. Clear-cut logging results in successional forest types for 70–200 years, depending on the forest type, original stand composition, type of logging system, and silvicultural actions. Where the forest ultimately regenerates to provide suitable habitat for American martens, the local effects can be short-term. Studies have shown that American martens generally avoid recent clear-cuts, and some have suggested that martens will either not cross large areas with little canopy cover, or will use direct-line travel between uncut edges (Buskirk and Powell, this volume). In Maine, Soutiere (1979) and Steventon and Major (1982) found that American marten habitat could be maintained by selective logging. In that study, use of minimum diameter restrictions for balsam fir (*Abies balsamea*) (15 cm) and for spruce (*Picea* spp.) and hardwoods (40 cm) produced a 16–43% reduction in stem basal area. American martens apparently prefer a canopy cover of mature trees greater than 30% (Koehler and Hornocker 1977, Spencer et al. 1983), although some authors have suggested that 50% canopy or greater is required (Hargis and McCullough 1984, Snyder 1984). Thompson (1994) found a winter canopy of 50% in his most heavily used uncut stands, whereas stands regenerating after logging and dominated by deciduous trees and shrubs had a canopy of about 20% and were little used by American martens.

Regenerating (<45 years after clear-cut) successional forest supported 0–33% of the population levels compared with nearby uncut forest, depending on type of regeneration and amount of original forest removed (Soutiere 1979, Snyder and Bissonette 1987, Thompson et al. 1989). In the only published study to report consistent use of young forest, Thompson (1994) found that some American martens occupied home ranges within cutovers (10–40 years old, with <10% residual old forest) but that these animals were subjected to high predation rates and were particularly susceptible to commercial trapping. Furthermore, animals in cutovers were young (<3 years) and were likely forced to disperse from better uncut habitats nearby. There was no reproduction by martens in second-growth forest during the five years of the study because individuals did not live long enough to reproduce. Home range sizes in successional forest were significantly larger for both sexes compared with those in old forest. Male ranges were 3.4 ± 1.4 km² in uncut forest and 5.0 ± 1.0 km² in cutover areas, whereas females occupied 1.0 ±

0.4 km² and 3.1 ± 1.1 km², respectively). The distance between core areas in home ranges was substantially greater in uncut areas (260 m) than in cutover areas (780 m) (Thompson and Colgan 1987a,b).

Commercial timber harvesting often leaves residual stands of timber not large enough to be commercially valuable or on sites where conditions preclude logging. Depending on these stands' size and isolation, American martens may use them for foraging (Soutiere 1979, Snyder and Bissonette 1987). Such stands were core areas for American martens living in cutovers in Ontario (Thompson 1994). Snyder and Bissonette (1987) showed that residual stands smaller than 16 ha were used less than expected. American martens used shrub-stage stands in Maine to feed on berries in summer and fall (Soutiere 1979, Steventon and Major 1982), and were reported to forage along edges of meadows for *Microtus* spp. in California (Simon 1980, Spencer et al. 1983). All these findings have implications for management programs that attempt to maintain a local population of American martens during timber extraction.

Long-Term Effects. Under two conditions, timber harvesting has long-term (more than one rotation) effects on American martens: first, when the second-growth forest type is not favorable to American martens even in the mature stage; and second, when logging proceeds at an unsustainable rate so that insufficient habitat is available over a long enough time to support American martens, resulting in their extirpation from a local fauna. An example of the first condition is the conversion of mature and old forest to short-rotation plantation or precommercially thinned forest. In boreal systems, thinned and planted stands are often harvested before they can develop attributes that are important to American martens.

Large areas of boreal and central Ontario, Quebec, and Maine provide examples of conversion forest, where logged-over mixed-wood stands have regenerated to primarily deciduous forests. These stands will not support American martens owing to low prey populations and unfavorable structural conditions (Buskirk and Powell, this volume). Winter canopy in deciduous forest is thin and likely provides less protection from avian predators than do conifer-dominated stands (Thompson 1994). Furthermore, deciduous stands lack the structural diversity of older mixed forest. Lack of structure is partly responsible for lower small mammal densities in successional and hardwood boreal forests, compared with those in uncut old forest (Martell 1983).

In Newfoundland and Nova Scotia, overlogging combined with insect damage has resulted in fragmented American marten habitat and an unbalanced age structure of second-growth forest. Insufficient old forest will be available to maintain viable American marten populations in those provinces during the next 50–100 years (Thompson 1991). New Brunswick may soon face a situation similar to that in other Atlantic provinces, as timber managers move to a short-rotation forest and rapidly liquidate the remaining old-growth stands. Forest management must be planned at the landscape level (several thousand square kilometers) to ensure an adequate supply of all age classes of all ecosystem and community types. Planning should include maintaining sufficient old-growth forest to enable the survival of species, including American martens, that are associated with that stage of forest in each ecosystem.

General Management Models

For studies in which American martens were found to use several age classes of forest, we calculated use-availability ratios for time spent (or proportion of tracks recorded) in each age class (Fig. 26.1, Table 26.1). The index values were then used to suggest the form for our habitat model. Because the various studies (Table 26.1) used different methods for defining habitat use, assigning habitat types to successional categories had to involve some subjective judgments. Therefore, we simply defined use/availability indices as the proportion of use in a habitat type divided by the proportion of that habitat type available.

The only density figures for American martens in logged areas come from Soutiere (1979), who showed a reduction in population of about 67% with removal of approximately 60% of the timber; and from Thompson (1994), who found a 90% reduction in numbers of American martens over several areas where more than 90% of the forest was cut and the remaining 10% occurred in small isolated patches.

Our generalized models for the management of American marten habitat are based on the following assumptions:

1. American martens will use forest with a canopy of at least 30%—and prefer a canopy of 50–70%—in which more than half is provided by mature or old softwood trees.
2. American martens prefer large-diameter downed wood for winter dens and as habitat for their major prey species. This feature is characteristic of old conifer-dominated forest.
3. American martens require large-diameter standing old or dead trees for natal denning and these trees are most abundant in old forest.
4. American martens prefer forest with a complex understory or forest with fine-grained patches (gaps).
5. In the short term, large cutovers (contiguous cuts over a 40-year period of 2 km²/year) eliminate American marten habitat. The more highly fragmented

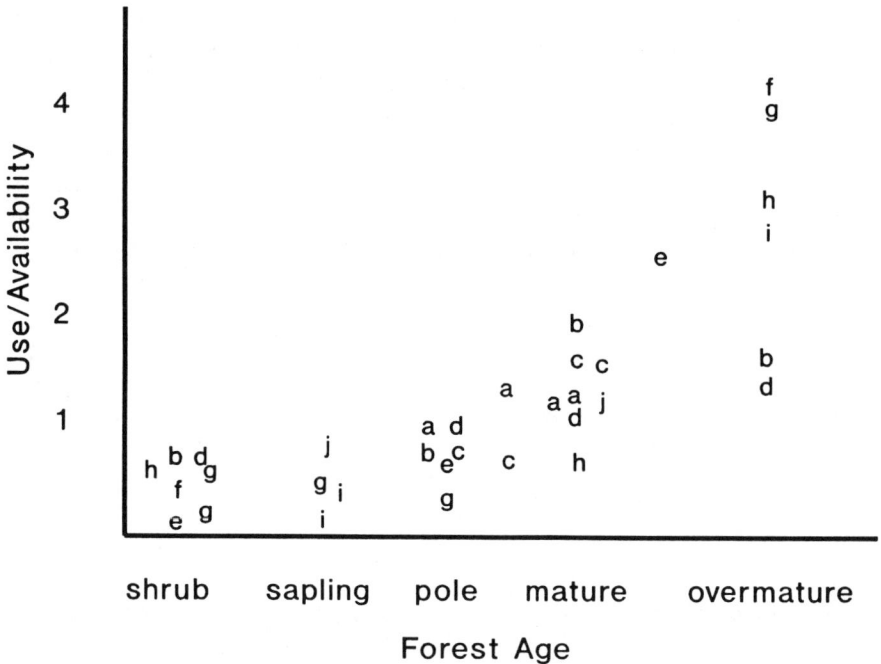

Figure 26.1. Use/availability data derived from published studies of American martens (*Martes americana*). a, Francis and Stephenson 1972: b, Soutiere 1978; c, Taylor and Abrey 1982; d, Kelly 1982; e, Spencer et al. 1983; f, Bateman 1986; g, Thompson et al. 1989; h, Raine 1983; i, Snyder 1984; and j, Major 1979. See text and Table 26.1 for further explanation.

the mature forest becomes, the lower will be its carrying capacity for American martens.

6. Carrying capacity of an even-aged mature conifer forest could be improved in the short term by evenly dispersing a harvest of 20–25% of the stem basal area in 0.5 to 3.0-ha patches. We fully recognize that an even-aged management strategy and patch cutting are mutually exclusive.

7. Selective logging, including using a shelterwood system, will not reduce a habitat's carrying capacity for American martens if removals are kept below 30% of the stem basal area every 50 years in boreal and montane forests, or every 100 years in temperate rain forests.

8. Second-growth conifer-dominated forest that has attained maturity or old-growth condition can be equivalent to natural forest of the same age in ability to support American martens. (Actually, we do not believe this to be the case under current forest management regimes and suggest that a substantial amount of research is needed on this question.)

Table 26.1. Use/availability indices calculated from published studies for which it was possible to discern proportion of time spent, or proportion of tracks of American martens (*Martes americana*) occurring, in forest stands of different ages

Reference	Forest age and type	Use/availability index
Francis & Stephenson 1972	Pole	0.83
	Pole-mature	1.22
	Mixed mature	1.15
	Conifer mature	1.07
Soutiere 1978	Shrub	0.64
	Pole	0.67
	Mature	1.86
	Overmature	1.39
Taylor & Abrey 1982	Pole	0.76
	Pole-mature	0.60
	Mixed mature	1.38
	Conifer mature	1.30
Kelly 1982	Shrub	0.65
	Pole	0.84
	Mature	0.97
	Overmature	1.23
Spencer et al. 1983	Shrub	0.00
	Pole	0.68
	Mature/overmature	2.31
Bateman 1986	Shrub	0.23
	Overmature	3.88
Thompson et al. 1989	Shrub (<5 yr)	0.02
	Shrub (10 yr)	0.55
	Sapling	0.40
	Pole	0.20
	Overmature	3.75
Raine 1983	Shrub	0.58
	Mature	0.51
	Overmature	2.84
Snyder 1984	Shrub	0.84
	Sapling (15 yr)	0.28
	Sapling (23 yr)	0.00
	Overmature	2.84

Model of Carrying Capacity and Stand Development

The forest depicted in the model (Fig. 26.2) is conifer-dominated northern boreal and is predominantly even-aged (i.e., ±20 years) until maturity. In the older stages, it becomes uneven in age for a time that depends on the forest type, as trees die and the stand slowly converts to a young stand. We suggest

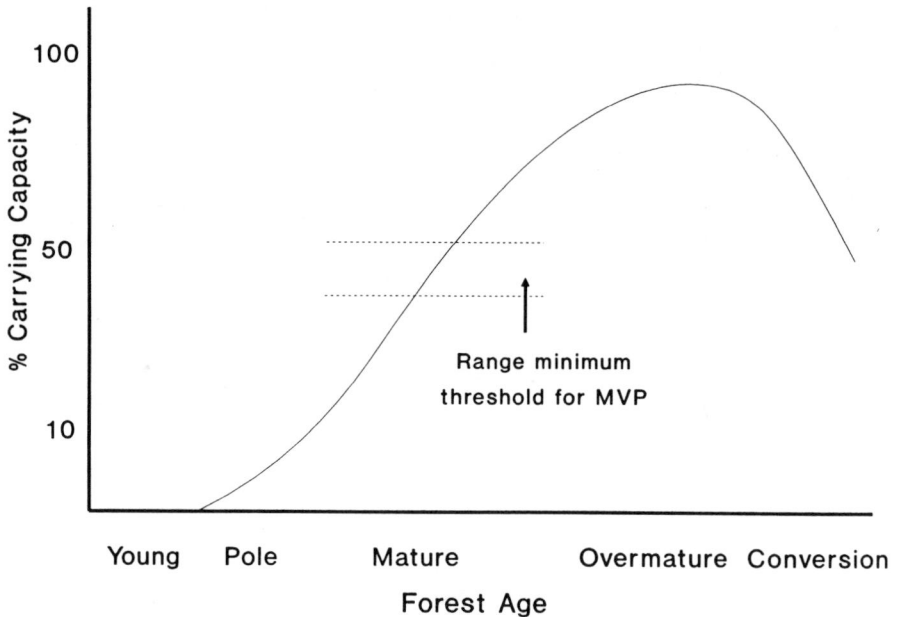

Figure 26.2. Proposed model for the relationship between carrying capacity for American martens (*Martes americana*) and forest stand age in boreal conifer-dominated forest. Management for minimum viable population (MVP) should be at a stand age for which carrying capacity is at least at or near 50% of maximum.

that the model also represents temperate rain forest, except that forest type maintains a shifting steady-state mosaic of old growth for a protracted period (several hundred years).

The model predicts a low carrying capacity for American martens in boreal forest for the first 70–80 years. Carrying capacity increases dramatically in late maturity and finally declines as the stand converts to a younger forest through windthrow, insect infestation, and fire. Actual time periods obviously would vary with forest type. We predict that carrying capacity for American martens is highest when the forest is overmature. Specific characteristics that have developed by late maturity and that enhance habitat quality for American martens include the following: a diverse understory in openings created where large white spruce, black spruce (*Picea mariana*), or white pine have fallen; numerous downed logs; trees suitable for denning; and diverse ground covers with deep moss carpets.

We suggest a threshold of 40–60% of maximum carrying capacity, below which a minimum viable population (MVP) could not be maintained. We predict that the threshold in boreal conifer and mixed woods occurs at about

80 years, in lodgepole pine (*Pinus contorta*) forest at approximately 100 years, and for temperate rain forest at about 60 years. Below these thresholds, martens would inhabit suboptimal habitat and therefore spend excessive energy on hunting and have less time available for social interaction and reproduction. Furthermore, we suggest that the population would be subjected to a high rate of predation that would continually alter the social structure, further reducing production. In practical terms, the threshold indicates that managers cannot rely on young-forest stands to maintain American marten populations.

Model of Carrying Capacity and Forest Removal

Our proposed model for the relationship between forest harvesting and American marten carrying capacity (Fig. 26.3) expresses a dependence on the scale and dispersion of cutting in a tract of forest during a period of 20–30 years. The model represents management of a mature conifer-dominated forest. We believe that the relationship between American marten density and logging over the short term (i.e., one rotation and with a low probability of stochastic population events) is linearly declining if clear-cuts are not dispersed and if 75–90% of the timber is removed in each stand.

On the other hand, if timber removal is dispersed over a large area in approximately 1.0- to 3.0-ha cuts, with less than 25% of total stems removed, then the carrying capacity of an area is predicted to increase. Even with a system of small cuts, as the number of cuts increases during several passes, the total undisturbed forest remaining obviously declines. Once logging reaches a particular threshold, which we predict at about 20–30% removal, the forest increasingly takes on the coarse-grained pattern of islands of trees in a predominantly logged area, and carrying capacity for American martens declines precipitously. The rapid decline would be the result of increased predation on American martens in open areas, high edge-to-forest ratio, and a reduction in the size of forest interior. The model has clear implications for objectives with respect to maintaining American martens in a local area.

In boreal forests under normal operating conditions, a series of extremely small cuts is neither economical nor desirable at the landscape level given that large areas of habitat are needed for American martens. At the landscape level, we suggest that the relationship between American marten carrying capacity and forest area is probably sine-wave form, with the amplitude dependent on habitat quality and the wavelength dependent on forest type, site conditions, and silvicultural practices. We suggest that landscape-scale management for American martens is possible, if sufficient attention is paid

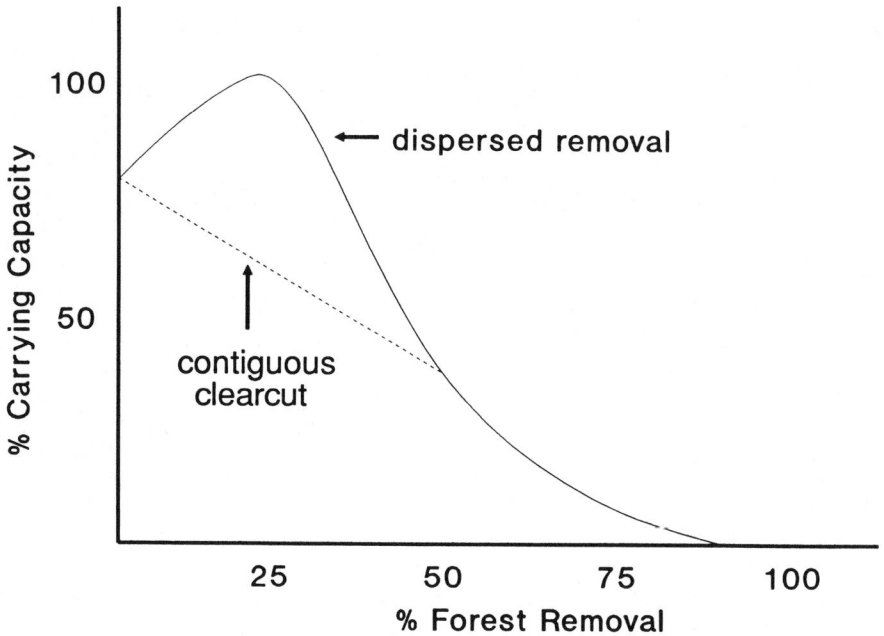

Figure 26.3. Proposed model for the effect of forest removal over a 30-year period on American marten (*Martes americana*) carrying capacity. Dispersed removal refers to selective cutting or patch-cutting blocks of less than 3 ha in boreal conifer-dominated forest.

to developing large blocks of old-growth forest with attributes required by American martens.

How Much Old Forest?

We calculated MVP for American martens using Lehmkuhl's (1984) model and following the method of Clark (1989):

1. We assumed that a minimum effective population size (N_e) of 50 is needed to counteract the effects of genetic drift and possible inbreeding.
2. The N required to produce an N_e of 50 is derived by adjusting various factors that affect the population. N_e is adjusted upward by 40% to account for variation in kits per litter in cases where data variance in litter size is unknown. $N = 70$.
3. The number is adjusted for an adult sex ratio of 1.2 males per 1 female (Thompson and Colgan 1987a); therefore, $N = 71$.

4. Overlapping generations reduce the effective population size. Therefore, this value is further adjusted ($\times 2$) for overlapping generations. $N = 142$.
5. The value is also corrected for population fluctuation based on Thompson and Colgan 1987a: $142 \cdot 1.67 = 237$.

Using home range and density data from boreal mixed woods in Ontario, one can calculate that approximately 600 km² of old-forest habitat is needed to maintain an MVP of 237 American martens. If we assume that population density in early mature forest is about half that in old-growth forest (Fig. 26.2) and is about 75% of maximum density in mature forest, then management for a contiguous area of 1000 km² with 400 km² in each of the 100–140-year and 140–180-year classes and 200 km² in the 180+-year class would be a suitable objective for an area of about 5000 km² of boreal coniferous and mixed-wood forest. That is, at the landscape management level, 20% of the forest should be mature and old at any time. Some or all of the mature forest could occur within parks or other protected areas. However, even protected forests age and die. Commercially managed areas surrounding parks must be planned in context with the protected lands to ensure that a supply of suitable mature and overmature forest habitat is always available.

Management agencies should decide on a protocol for the number of MVPs or total American marten population, particularly in small areas where concern for American martens may drive the planning system in some years. For example, small jurisdictions may wish to consider managing their forests in separate areas to maintain at least three viable American marten populations. That strategy would protect against stochastic population events in any one population during periods when numbers might be low. Such a strategy is currently being examined for New Brunswick (New Brunswick Dep. Nat. Resour. 1990) and Newfoundland, although in the latter case finding sufficient habitat for the three populations is problematic (Thompson 1991).

Management Regimes

We suggest two management regimes to protect or maintain American marten habitat. Our proposed management plans depend on the available forest and its age structure within a particular jurisdiction. The first regime is for areas in which a limited amount of suitable forest remains; the second is for areas in which available habitat is not yet a critical problem in the long-term survival of American martens.

Where American martens occupy limited areas of old forest (e.g., Newfoundland, New Brunswick, the Gaspé Peninsula in Quebec, and some U.S.

National Forests), we suggest that the objective for short-term management (e.g., 30–50 years) should be to maintain as much old forest as possible for as long as possible. Therefore, even-aged management of current habitat is not an option. Long-term and landscape objectives should include the development of sufficient future potential habitat in surrounding areas to perpetuate American martens. The program should have the following components: an improved database that refines information for individual stands, including age information more specific than pooling 20- to 40-year age groups; some measure of decadence, site type, and predicted regeneration trajectory; a carefully designed integrated pest management program to protect against outbreak of insect pests; a harvesting strategy to remove only the most decadent stands in small patches and in a manner that maintains connectivity among uncut areas, and possibly to improve the food base by management for habitats favorable to hares and grouse; retention of snags and coarse woody debris during logging; GIS mapping of surrounding landscape to project future stand development, enabling the planning for American marten habitat through models and habitat supply analysis; use of silviculture to speed stand development (thinning, fertilization, herbicides); and monitoring of American marten populations and their habitat.

In areas where habitat supply is not immediately critical, short-term planning to maintain remaining habitat would not be the overriding immediate objective. Long-term planning still is a necessity, and even-aged management is a viable option if appropriate to the forest type. The key to management planning of the forest is ecosystem supply analysis. These analyses will permit long-term assessment of the availability of forest types and age-class development at temporal (more than one rotation) and spatial (several thousand square kilometers) scales appropriate to management priorities, which should include the preservation of biotic diversity. Such a program might be based on species such as the American marten, which requires expanses of forest in particular age classes and with certain habitat attributes. Components of ecosystem supply analysis include mapping forest classes that are predictive of wildlife species associations; creating and implementing timber production models for planned harvesting operations, which will result in sufficient ecosystem availability to meet biodiversity objectives; and a monitoring system to ensure objectives are being met.

Research Needs

Individual hypotheses into the question of why American martens require old forest should be examined experimentally, with a view to using the knowledge to improve management of areas for this species. Management-level experimentation is required to verify our proposed models. In particular, our second model should be tested to determine whether forest management can be carried out to enhance American marten populations in areas of mature forest.

There also is a need to research the development of old-growth forest in commercially managed areas. In particular, second-growth forests should be examined for their habitat capability for American martens. Furthermore, we need intensive efforts to redevelop poor and understocked second-growth forests.

Ecologically based forest classification systems are now available for many jurisdictions. Such systems use soil nutrient content and moisture content as prime variables in the classification of forest plant communities. Research should test these classification systems as predictors of American marten habitat and carrying capacity.

27 Habitat Use by Fishers in Adjoining Heavily and Lightly Harvested Forest

Slader G. Buck, Curt Mullis, Archie S. Mossman, Ivan Show, and Craig Coolahan

Habitat use by fishers (*Martes pennanti*) has been studied in eastern and central North America (de Vos 1952; Coulter 1966; Clem 1977*a*; Kelly 1977; Powell 1977; Raine 1983; Arthur et al. 1989*b*; Thomasma et al., this volume) and in the northern Rockies (Hash and Hornocker 1978; Jones and Garton, this volume). Other work has focused on the Pacific states (Schempf and White 1977, Raphael 1984, Rosenberg and Raphael 1986, Raphael 1988, Criss and Kerns 1990), but even with these studies, habitat ecology has received little attention. Pacific fishers (*M. p. pacifica*: Hall 1981) are widely regarded as old-growth forest associates and have received increased attention from conservation groups. In 1990 an environmental coalition petitioned the U.S. Fish and Wildlife Service to list Pacific fishers as endangered, citing the lack of recent sightings and the absence of large uncut areas within suspected fisher range. The petition was denied on the grounds that insufficient information existed to warrant listing (*Federal Register* 56[8]:1159–1161). Interest in Pacific fishers and in wildlife of old-growth forests remains high, however. Additional efforts to secure federal protection for Pacific fishers are likely. We especially need information on how current timber harvest practices affect fishers—information critical to the conservation of fisher populations. Our goal was to gain some of this information by comparing habitat use by fishers in an area of substantial timber harvest with that in an area of lighter harvest.

Study Area

The 150-km² study area was located in the coastal mountains of Shasta-Trinity National Forest in northwestern California, just south of Big Bar.

Within this area, fishers were studied concurrently at two sites: Corral Bottom, where timber was heavily harvested (HH), and Hayfork Bally, which was lightly harvested (LH) (Table 27.1). The main differences between the two sites were total length of logging roads and amount of forest harvested by presalvage logging.

Summers are mild and almost all precipitation falls during the cool winters. Winter snow is infrequent and is often followed by rain and rising temperatures, which melt the snow rapidly. Vegetation in the study area is mostly coniferous forest of almost equal proportions of Douglas-fir (*Pseudotsuga menziesii*) and white fir (*Abies concolor*), with lesser amounts of ponderosa pine (*Pinus ponderosa*), sugar pine (*P. lambertiana*), and incense cedar (*Calocedrus decurrens*).

Douglas-fir dominates at lower elevations on north-facing slopes and co-dominates with white fir along lower drainages and at most elevations above 1372 m. White fir dominates at the highest elevations on sites with moist north-facing slopes. Mixtures of Douglas-fir, ponderosa pine, and madrone (*Arbutus menziesii*) occur on lower southern exposures. Canyon live oak (*Quercus chrysolepis*) and black oak (*Q. kellogii*) are found on xeric sites up to 1524 m. Canyon live oak and giant chinquapin (*Castanopsis chrysophylla*) also occur frequently in the understory. Montane chaparral occurs on southern exposures at the highest elevations.

Presalvage logging in the study area was conducted by selective removal, usually of the largest trees in a stand, in anticipation of future mortality from any cause (B. Wulf, USDA Forest Service pers. commun.). Over much of the HH site, many of the largest commercial conifers had been cut by pre-

Table 27.1. Characteristics of heavily harvested and lightly harvested study sites in Shasta-Trinity National Forest, California, 1977–1980

	Heavily harvested	Lightly harvested
Size	72.5 km²	77.7 km²
Elevation range	853–1280 m	838–1916 m
USFS land	80%	97%
Private land	20%	3%
Forested	87%	93%
Clear-cut	7%	5%
Presalvage logging[a]	25%	12%
Seasonally maintained roads	115 km	72 km
Four-wheel drive roads	224 km	8 km

[a]Presalvage logging is a form of selective cutting, defined as estimated harvest of timber anticipating future mortality from any cause. Quantitative data were unavailable from USDA Forest Service (USFS) silvicultural records.

salvage logging. Presalvage logging decreased the vertical diversity, decreased canopy closure, and made the stands more xeric (Lee 1978). More important, it retained most of the preharvest midstory hardwood component, changing the stand from conifer-dominated to hardwood-dominated. In the HH site, presalvage logging resulted in larger amounts of hardwoods and mixed conifers and hardwoods than in the LH site. Although trapping was restricted to the original study area, any radio-collared fisher that traveled beyond this area was monitored and the study area was expanded to include the animal's home range.

Methods

Between November 1977 and March 1979, fishers were captured using single-door wire cage traps. They were immobilized with a mixture of ketamine hydrochloride (11–17.6 mg/kg body weight) and acepromazine (1.1 mg/kg body weight) administered intramuscularly. The animals were weighed, measured, and fitted with a 110-g or 60-g radio collar (Telonics Inc., Mesa, Ariz.). All fishers were ear-tagged with numbered plastic rototags (Nasco West, Inc., Modesto, Calif.). The lips of large fishers were tattooed using a domestic rabbit tattoo kit (Valentine Equipment Co., Bridgeville, Ill.). Age was estimated by palpation of the sagittal crest and inspection of teeth. Fishers with little or no sagittal crest development, minimal tooth wear, and no tooth discoloration were classified as juveniles (≤ 12 months). Fishers with obvious sagittal crests and worn teeth, with or without discoloration, were classified as adults (>12 months).

Fishers were located using telemetry from vehicles, from fixed-wing aircraft, and on foot. All ground locations obtained with telemetry were determined using the null-peak method (Beaty 1978). All locations, including trap sites, were plotted on USFS 1:1320 blueline maps. Home ranges were plotted by the minimum convex polygon method (Mohr 1947) for all fishers that were located at least four times. Trap locations were included in a fisher's home range if the location occurred on the edge of a range. We excluded trap locations that were obviously the result of breeding-season wandering by males.

USDA Forest Service timber-type maps (1977 edition) were used to determine habitat use within the study area. To determine habitat availability, we used a Calcomp 2300 digitizer to measure the areas of timber-type polygons within each fisher's home range. Within fisher home ranges, 109 timber types occurred; 38 occurred only in the HH site, 39 occurred only in the LH site, and 32 were found in both sites. Timber types were grouped into four timber type groups according to similarities in physiognomy, crown size, crown cover, and USFS timber-type modifiers (Table 27.2). Chi-square goodness-of-fit tests were used to compare fisher habitat use with habitat availability and to test for fisher habitat selectivity among timber-type groups. The threshold for statistical significance was $P < 0.05$. Because traps were placed where we thought we would capture fishers, timber-type data from trap sites was excluded from this analysis. And because we investigated habitat use in an area where many individual timber-type polygons often occurred within small areas, we included only telemetry locations from nonmoving fishers whose locations were determined over a short period.

For the statistical analysis, adult males, adult females, and juveniles were considered separately. Each fisher was treated independently. Percent of telemetry locations in each timber-type group was our measure of habitat use. Percent of each timber-type group in a fisher's home range was used as a measure of habitat availability. All percentages were based only on the area

Table 27.2. Timber-type groups in heavily harvested and lightly harvested sites of the Shasta-Trinity National Forest, California, 1977–1980

Timber-type group	No. individual timber types	Description
Open	14	Areas with no standing trees; clear-cuts, grass, barren areas. Conifer stands with crown size 4–12 m, crown cover <40%, with open areas among conifers.
Brush-pole	36	Montane shrub, chaparral, plantations. No large trees. Conifer stands with crown size 2–12 m, crown cover <40%, with shrubs among conifers. Small conifer stands with crown size <7.3 m, crown cover >20%.
Hardwoods	26	Stands of live oak, Oregon white oak, Pacific madrone, mixed commercial hardwoods. Conifer stands with crown size 2–12 m, crown cover <40%, with hardwoods.
Mature closed conifer forest	26	Conifer stands with crown size 4–12 m, crown cover >40%. Includes conifer stands with crown cover <40% but associated with riparian areas, *Salix* sp., *Alnus* sp., *Corylus* sp.
Unlabeled	1	Unlabeled polygons on USFS timber-type maps.

encompassed within a fisher's home range. Using home range alone is conservative because it reduces the number of assumptions involving the availability of areas in which individual fishers were not located. It also reduces the assumptions involving sampling effectiveness.

Our null hypothesis was that fishers showed no difference between habitat availability and habitat use. Our logic was that the null hypothesis should be tested first for all observed animals. If no significant differences were found, no further statistical analysis was possible and we could conclude that there was no statistical difference between habitat availability and use. Where significant differences were found, we partitioned the fishers into meaningful groups, such as LH and HH, and repeated the analyses on each group.

Results

We captured five male and three female fishers in LH and nine male and four female fishers in HH. All, except one adult male in HH (no. 22), were radio-collared. A total of 169 telemetry locations were obtained in LH and 260 locations in HH. Overall, fishers showed a highly significant difference between use and availability of habitats ($X^2 = 82.9$, 48 df, $P < 0.001$). Therefore, the data were disaggregated by age and the chi-square analysis repeated. Juveniles showed no significant difference ($X^2 = 17.9$, 15 df, $P = 0.27$), indicating that they used habitat in proportion to its availability. Adult fishers, however, showed a highly significant difference ($X^2 = 65.0$, 33 df, $P < 0.001$), so we further partitioned adults into adult males and adult females. No significant habitat selection was found for adult males ($X^2 = 81.1$, 21 df, $P = 0.07$), but highly significant selection was found for adult females ($X^2 = 33.9$, 12 df, $P < 0.001$). So, differences between habitat availability and use were primarily the result of selection by adult females.

When the data were partitioned into HH and LH groups, no significant selection was found for the LH site ($X^2 = 35.2$, 30 df, $P = 0.24$). Fishers in the HH site, however, showed highly significant habitat selection ($X^2 = 78.0$, 36 df, $P < 0.0001$). When the data were disaggregated by age, adults showed highly significant selection ($X^2 = 57.3$, 21 df, $P < 0.0001$), but juveniles showed no difference between habitat availability and use ($X^2 = 20.8$, 15 df, $P = 0.15$). Further partitioning of adults revealed that adult females in the HH site showed highly significant differences between habitat availability and use ($X^2 = 31.9$, 6 df, $P < 0.0001$), and adult males showed significant differences ($X^2 = 25.4$, 15 df, $P = 0.045$).

Within the LH area, fishers appeared to use timber-type groups in proportion to their availability. Within the HH area, however, adult fishers, particularly adult females, showed some form of selection or avoidance of timber-type groups.

Discussion

Our analyses indicate that fishers used habitats differently in the LH and HH sites. Within the LH site, fishers used TTGs in proportion to their availability and we conclude that, during our study, timber harvest had not detectably affected their habitat use. Within the HH site, fisher use of timber-type groups differed as a result of timber harvest practices. We hypothesize that management practices that reduce mature conifer habitat and increase the proportion of hardwoods created suboptimal fisher habitat.

Almost all investigators have documented fisher selection for closed-forest types (Buskirk and Powell, this volume; Jones and Garton, this volume). Our results indicate that within the LH site, fishers select habitats with overhead canopy. Although a mature hardwood component within coniferous forest may be valuable by increasing prey diversity, fishers may avoid areas with an overabundance of hardwoods (Kelly 1977, Hash and Hornocker 1978). Our results suggest that the mixture of timber types in the LH site more closely met the habitat needs of all fisher sex and age classes and that, within the HH area, the level of harvesting influenced habitat use.

Differences in habitat use between the two study sites may also be explained, in part, by the social behavior of fishers. As mature forest is removed, we would expect territorial interactions among fishers to increase. We would also expect the number of fishers to decrease. If timber management practices create timber types that are suboptimal, then survival and reproduction of fishers should decrease within these timber types.

Some evidence supports this hypothesis: 7 radio-collared fishers died during our study—2 adult males, 1 adult female, and 4 juveniles. All were recovered in habitats considered sub-optimal by our analysis: clear-cuts, areas without overhead canopy, and hardwood-dominated stands. Four of the 12 fishers collared within the HH site died there. Three of the 8 fishers collared in the LH site died, 2 within the HH site. Four mortalities were caused by predation by other carnivores. One juvenile from the HH site was almost certainly killed by another fisher.

As closed-forest types are removed, fishers found in the remaining mature forest would likely be resident animals with well-established home ranges or very strong individuals from disturbed areas that succeeded in displacing previous residents. Adult male fishers would have the best chance of occupying these habitats because they are the largest and strongest and possibly the most aggressive (Powell 1982). Male fishers have much larger home ranges than females (Powell, this volume). Therefore, they may be less susceptible to proportional changes in the habitats within their range; that is, a 10% reduction of mature conifer habitat may alter habitat use by an adult female more than that of an adult male.

Management Implications

Harris (1984) and Jones and Garton (this volume) suggest that habitat management for fishers be carried out on a landscape scale. This would allow for a mixture of young, mid-, and late-successional stages that would provide both prey and resting sites. The optimal proportions of these stages are unknown and most likely vary geographically. A critical unknown is the amount and type of timber removal that may be beneficial to fishers—and at what level removal and the resulting fragmentation of habitat becomes detrimental to fishers. Our data suggest that within our area, the level may differ between sexes and perhaps between adults and juveniles. Additional research on fisher habitat use in heavily managed forests is needed to clarify these relationships. To gain this information, one could radio-collar fishers in an area where no harvest has occurred, then monitor them as different harvest strategies are implemented.

Managing for an optimum level of hardwoods may be more difficult than managing for conifer successional stages. Thomasma et al. (this volume) found that in Michigan the maximum habitat selection function for percent overstory of hardwoods was 32%. Applying data from one area directly to another based on only cover types can be misleading (Buskirk and Powell, this volume). Buskirk and Powell (this volume) suggested, however, that structure can be useful when considering broad cover types. An example of this is riparian areas. Regardless of the geographical location or the specific timber management practice, the landscape approach is enhanced by incorporating riparian areas as connecting corridors between areas of mature closed conifer forest. Fishers in our area, particularly within the HH site, frequently used riparian "stringers" as travel corridors. Jones and Garton (this volume) reported similar results.

Timber harvest practices influence the distribution and possibly the abundance of fishers. Both clear-cutting and presalvage thinning modify fisher habitats. Clear-cuts are poor habitat and a large new clear-cut has little value to fishers (Powell, this volume; Jones and Garton, this volume). Clear-cutting alters site conditions by removing the insulating canopy and exposing the site to the drying effects of sun and wind. Additionally, the composition of tree species in northwestern California is often changed by replanting with Ponderosa pine rather than firs. Ponderosa pine grows faster and is better suited to the new site conditions.

Once a dense cover of shrubby vegetation is established, fishers may use clear-cuts for hunting, as suggested by Jones and Garton (this volume). Snow tracks indicated that fishers within our study area used clear-cuts with dense brush, such as whitethorn (*Ceanothus cordulatus*). But these early-successional stages are temporary and their value may be reduced by herbicides or brush removal to enhance growth of conifers. Leaving brushy areas may be beneficial to fishers.

The effects of presalvage or selective logging are less apparent than those of clear-cutting, but these practices may have a greater impact on fishers over time, especially if the harvest involves large areas. Selective logging results in many of the same site effects as clear-cutting. Selectively cut sites and clear-cuts often are replanted with Ponderosa pine. In addition to the removal of large commercial conifers and the increase in relative abundance of hardwoods, snags are often removed, thereby reducing the number of potential den and rest sites for fishers and their prey.

The regional effect of clear-cutting and of selective cutting followed by replanting with pines is to change vegetation composition from a mesic closed forest of predominantly Douglas-fir and white fir to a drier open forest dominated by Ponderosa pine. The long-term effects of this change are unknown, but they may be detrimental to fishers. Schempf and White (1977) found that only 8% of 206 fisher sightings occurred in forests composed of greater than 80% *Pinus* spp. This difference likely is underestimated because of visibility biases associated with habitat. Pine forests generally are more open than fir forests; therefore, fishers should be more easily seen in pine forests.

If our hypotheses prove correct, timber management practices that result in open stands, an abundance of hardwoods, and xeric conditions over large areas create conditions unsuitable for the maintenance of fisher populations. For fisher populations to be maintained, extensive clear-cutting of mature closed conifer forest should be minimized and selective cutting conducted so that adequate habitat is provided for all fisher sex and age classes.

Acknowledgments

The original idea and early planning for this study came from R. L. Garrett. We thank R. A. Powell for insight into capture and handling techniques and for advice throughout the study. K. A. Nelson assisted in collec-

376 *Slader G. Buck et al.*

tion of field data. Many other people helped in many ways. We particularly thank T. S. Burton, J. Gordon, J. Kahl, R. Escano, D. Wright, B. Wulf, C. Powell, R. M. Ensminger, R. G. Botzler, D. E. Craggie, J. M. Allen, P. Collins and J. Dixon. This research was funded by the USDA Forest Service. The California Department of Fish and Game provided traps, immobilization equipment, and postmortem carcass examination. The U.S. Fish and Wildlife Services Division of Animal Damage Control provided additional traps. We thank Humboldt State University for logistical support.

28 Selection of Successional Stages by Fishers in North-Central Idaho

Jeffrey L. Jones and Edward O. Garton

Predicting the effects of forest management on fisher (*Martes pennanti*) populations requires an understanding of their habitat relationships, as well as such characteristics as movements, size of home ranges, and food habits. Allen's 1983 habitat suitability index model was designed to aid managers in evaluating the effects of habitat alteration on fishers, but the model was based on data from eastern North America and is probably not appropriate for western habitats. Little is known about the ecology and behavior of fishers in western North America. Only two studies have described habitat relationships of fishers in California (Buck 1982, Mullis 1985), and none have been conducted in the northern Rocky Mountains. Our study investigated and attempted to explain habitat-use patterns of fishers in north-central Idaho.

Study Area

We conducted our study in the Nez Perce National Forest, Idaho County, Idaho. Boundaries of the area were roughly the South Fork Clearwater River to the south and west, Meadow Creek to the east, and the Selway River to the north. The specific study area of about 1010 km² was defined by the home ranges of 13 radio-collared fishers. Elevations within this study area range from 1006 m to 2165 m.

Most forests within the area are in the grand fir (*Abies grandis*) and subalpine fir (*A. lasiocarpa*) vegetation zones (Cooper et al. 1987). Grand fir habitat types (Cooper et al. 1987) dominate the area (75.9%), whereas subalpine fir, Douglas-fir (*Pseudotsuga menziesii*), ponderosa pine (*Pinus ponderosa*), and lodgepole pine (*P. contorta*) habitat types occur on approx-

377

imately 16.7%, 5.0%, 1.8%, and 0.6% of the area, respectively. Other tree species present in the area include western larch (*Larix occidentalis*), Engelmann spruce (*Picea engelmannii*), and a few western red cedar (*Thuja plicata*). Pacific yew (*Taxus brevifolia*) was often a major component of the grand fir–ginger (*Asarum caudatum*) and grand fir–queencup beadlily (*Clintonia uniflora*) habitat types and commonly reached heights of 10 m.

Annual precipitation and snowfall at nearby Elk City, Idaho (1230 m), average 85 cm and 353 cm, respectively; annual maximum and minimum temperatures average 13°C and −3°C, respectively (Pierce 1983). Winter snowpacks during our study ranged from about 0.5 m at lower elevations to more than 2.0 m at higher elevations. The study area is generally covered by snow from early November through mid-April.

Methods

Capturing Fishers. Fifty cage-type livetraps baited with meat scraps and scented with a commercial attractant were placed along a trapline at locations expected to have a high trapping success. Trap intervals varied from about 3 km to 12 km. We trapped from 1 September through 15 April, although some trapping occurred during summer. We checked traps daily except when snow conditions delayed inspection. We sometimes set traps in areas where noncollared fishers had been observed.

Captured fishers were immobilized with ketamine hydrochloride (Hash and Hornocker 1980). Anesthetized animals were weighed, measured, sexed, tattooed, aged according to sagittal crest development (Wright and Coulter 1967), examined for external parasites and physical abnormalities, and fitted with 78-g radio collars.

Locating Fishers. We located radio-marked animals from fixed-wing aircraft and from the ground, but to ensure accuracy we used only ground locations for habitat analyses. Owing to difficulties in locating animals, a precise sampling design was not used. We tried to locate animals twice each week and to obtain 30 or more observations per animal per season. To increase the independence of observations, we did not record locations if the animal was relocated within eight hours of its previous observation. We approached to within 10 m of resting animals before observations were recorded. Consequently, error polygons (Mech 1983) for resting fishers were generally less than 0.05 ha. In summer, active animals were generally approached to within 80 m, resulting in an error polygon of less than 1 ha. Winter locations were also determined by back-tracking fisher tracks in snow. When back-tracking, use sites were recorded at 500-m intervals, and were assumed generally to be hunting observations.

Habitat Availability. The study area was defined by pooling all observations of the fishers and circumscribing them within a minimum convex polygon. An adequate map depicting successional stages was not available for the study area. Consequently, habitat availability on a broad scale was estimated by randomly distributing points throughout the study area as described by Marcum and Loftsgaarden (1980). Habitat availability for individual animals was estimated by using those random plots falling within an individual's home range determined by the minimum convex polygon technique (Hayne 1949). Random points were distributed within individual fishers' home ranges to ensure that each animal had at least as many random plots as plots at used sites.

Field Methods. Each fisher location and random point was classified into one of six successional stages as described by Thomas et al. (1979)—grass-forb, shrub-seedling, pole-sapling, young forest, mature forest, and old-growth forest—on the basis of dominant and codominant tree heights, distribution of tree size classes, stand decadence, and presence of snags and logs.

We distinguished two seasonal periods, based on whether snow covered more than or less than 50% of the study area at about 1230 m. We refer to these periods as winter and summer.

Statistical Analyses. We did not use trap sites for habitat analyses because of the potential bias attributable to baiting, unless a fisher had been previously observed at a location before traps were set.

To minimize the potential Type II error rate when testing for fisher selection of successional stages, we reduced the number of habitats by combining the grass-forb and shrub-seedling types into a nonforest category (Alldredge and Ratti 1986). Chi-square tests were used to determine whether habitat use differed between sexes, seasons (summer and winter), and activities (resting and hunting). Macrohabitat selection was determined by comparing habitat use with habitat availability following the procedure of Marcum and Loftsgaarden (1980); tests were conducted with $\alpha = 0.10$ owing to the conservative nature of the Bonferroni Z statistic (Alldredge and Ratti 1986). We frequently did not have adequate sample sizes for individual animals to approximate a chi-square distribution (Roscoe and Byars 1971). Consequently, chi-square heterogeneity tests (Zar 1984) were conducted to ensure that radio-collared animals could be pooled, which effectively increased sample sizes and allowed adequate approximations of the chi-square distribution.

Results

Nine male and seven female fishers were captured and radio-collared. Of these, five males and four females had sufficient locations for macrohabitat analyses. We obtained 153 summer and 93 winter locations from these animals.

Successional Stages: Vegetation Structure

We determined habitat characteristics of successional stages by measuring overstory canopy cover, tree and snag density by size class, log volumes by size class, and understory cover of trees, shrubs, and herbs (Jones 1991). Old-growth stands were characterized by dense canopies; high densities of large-diameter trees, snags, and logs; high coniferous understory cover; and moderate deciduous understory cover. Mature stands had the highest densities of moderately large trees (34.3–47.0 cm diameter at breast height [dbh]), snags 24.1–34.3 cm dbh, and logs 14.0–34.3 cm in diameter. Ground cover of logs was also highest in mature stands. The highest densities of trees 11.4–34.3 cm dbh and of snags 14.0–24.1 cm dbh were found in young forest stands. Relatively high volumes of 14.0- to 21.6-cm diameter logs also were found in young forest stands. Young forests had the highest understory cover of deciduous shrubs. Large-diameter trees and snags were rare in pole-sapling, shrub-seedling, and grass-forb stands. The pole-sapling stands had the greatest availability of trees 1.3–11.4 cm dbh and the lowest canopy densities of forested sites. Canopy cover in shrub-seedling and grass-forb stands never exceeded 15%.

Seasonal Use and Selection of Successional Stages

Five animals were observed nine or more times in each season (summer and winter) and were evaluated for seasonal differences in cover type use. Use of successional stages shifted significantly between summer and winter ($X^2 = 29.8$, df $= 3$, $P \leq 0.0001$; Table 28.1). Fishers used mature forests more in summer ($X^2 = 4.8$, df $= 1$, $P = 0.028$), whereas young forests were used more in winter ($X^2 = 20.7$, df $= 1$, $P \leq 0.0001$). Use of the other types did not differ between seasons ($P > 0.10$).

Sexual differences in use of successional stages during summer were analyzed for five male and four female fishers and during winter for three males and two females. Use of successional stage did not vary significantly by gender during summer ($X^2 = 3.80$, df $= 3$, $P = 0.28$) or winter ($X^2 = 2.2$, df $= 2$, $P = 0.34$). Therefore, sexes were pooled for all further analyses of successional-stage selection.

Table 28.1. Selection of successional stages by fishers (*Martes pennanti*; n = 9) near Elk City, Idaho

Successional stage	Summer observations			Winter observations		
	Use	Random	CI[a]	Use	Random	CI
Nonforest	0	35	0.04−	0	20	0.04−
Pole-sapling	3	35	0.05−	0	24	0.05−
Young forest	12	88	0.08−	39	61	0.14+
Mature forest	114	135	0.10+	39	107	0.15
Old-growth	24	23	0.08+	6	11	0.07
TOTAL	153	316	—	84	223	—

[a]90% CI = (% of random locations − % fisher use locations) ± indicated value. + = preference, − = avoidance at $P < 0.10$ (Z-test).

Nine fishers had 10 or more summer-use observations. Of these summer locations, 90% occurred in either mature or old-growth forest (Table 28.1). No observations of fishers occurred in the nonforest habitat type. Bonferroni confidence intervals showed significant selection or avoidance in each of the five successional stages during summer (Table 28.1). Fishers preferred the old-growth and mature forest types, and avoided the nonforest, pole-sapling, and young forest successional stages.

We observed no winter use in either the nonforest or pole-sapling successional stages (Table 28.1). In winter, fishers used young and mature forest cover types at the same intensity (46%). Use of old-growth forests dropped to less than half that of summer use. Bonferroni confidence intervals indicated that in winter, fishers preferred young forests and avoided nonforest and pole-sapling areas. We detected no selection (preference or avoidance) for mature or old-growth forests. The observed seasonal shift in use of successional stages was readily apparent when the habitat selection patterns were compared between seasons; the most preferred successional stage in winter (young forests) was avoided in summer.

Use of Successional Stages for Resting and Hunting

During summer, only six fishers were located during both resting and hunting bouts. No fishers were located in nonforested sites while resting or hunting. Summer use of successional stages differed significantly ($X^2 = 13.5$, df = 3, $P = 0.004$; Table 28.2) between resting and hunting sites for all six animals. Use of pole-sapling forests for hunting was significantly greater than for resting ($X^2 = 11.5$, df = 1, P 0.001). Significant differences between resting and hunting use of young ($X^2 = 0.5$, df = 1, $P = 0.47$),

Table 28.2. Selection of successional stages by fishers (*Martes pennanti*; $n = 6$) at resting and hunting sites during summer near Elk City, Idaho

Successional stage	Resting observations			Hunting observations		
	Use	Random	CI[a]	Use	Random	CI
Nonforest	0	31	0.05−	0	31	0.05−
Pole-sapling	0	19	0.04−	3	19	0.17
Young forest	7	64	0.10−	3	64	0.18
Mature forest	69	99	0.13+	16	99	0.24+
Old-growth	12	15	0.09	1	15	0.11
TOTAL	88	228	—	23	228	—

[a]90% CI = (% of random locations − % fisher use locations) ± indicated value. + = preference, − = avoidance at $P < 0.10$ (Z-test).

mature ($X^2 = 0.2$, df $= 1$, $P = 0.67$), and old-growth forests ($X^2 = 1.3$, df $= 1$, $P = 0.25$) were not detected.

About 92% of summer resting-site observations occurred in mature or old-growth forest, whereas no such observations occurred in the nonforest or pole-sapling types (Table 28.2). Bonferroni confidence intervals indicated that fishers chose mature forests for resting, avoiding nonforest, pole-sapling, and young forest types (Table 28.2). A significant difference between availability and resting use of old-growth forests was not detected.

Of the summer hunting observations of six fishers, about 74% occurred in mature or old-growth forests, whereas none occurred in the nonforest type (Table 28.2). Fishers used a broader range of successional stages for hunting than for resting, even though we collected fewer hunting observations. Specifically, hunting observations included the pole-sapling cover type, whereas resting observations did not. Bonferroni confidence intervals for summer hunting-site observations (Table 28.2) showed that mature forests were preferred and nonforests were avoided. Use did not differ from availability for the other successional stages.

Comparing selection patterns for resting versus hunting suggested that mature and old-growth forests were used more for resting, whereas pole-sapling and young forests were used more for hunting. Fishers avoided pole-sapling and young forests for resting sites, whereas differences in use and availability were not detected for these types for fishers while hunting. Although old-growth stands were used less for hunting than they were for resting, they were used in proportion to their availability for both activities. Mature stands were preferred and nonforest types were avoided for both activities.

During winter, only four fishers were observed in both resting sites ($n =$

52) and hunting sites ($n = 19$). These four animals used only three successional stages (young, mature, and old-growth forests) during winter for both
hunting and resting activities. We could not detect a difference in use of
successional stages by activity type (hunting versus resting, $X^2 = 0.5$, df $=$
2, $P = 0.80$).

Discussion

Fishers in north-central Idaho did not use habitats in proportion to their
spatial availability. Our findings regarding habitat use concur with those of
other studies (Quick 1953; Coulter 1966; Kelly 1977; Powell 1977, 1978;
Buck 1982; Mullis 1985; Arthur et al. 1989*b*) in that fishers did not use
nonforested habitats. Evidence of microtines, yellow-bellied marmots (*Marmota flaviventris*), and ground squirrels (*Spermophilus* spp.) in the diet of
fishers in our study area, suggested, however, that fishers may have made
forays into nonforested or sparsely forested habitats for hunting (Jones 1991).
Mature to old-growth coniferous forests have commonly been considered
optimal or preferred fisher habitat (de Vos 1951*b*, Coulter 1966, Ingram
1973, Kelly 1977, Schempf and White 1977, Buck 1982, Allen 1983,
Raphael 1984, Mullis 1985, Rosenberg and Raphael 1986) especially in
areas with deep snow (Arthur et al. 1989*b*). Our results suggest, however,
that although fishers preferred mature and old-growth forests during summer,
young forest was the most preferred successional stage in winter. Even
though we did not detect significant selection of mature or old-growth forest
in winter, these types were represented by 53% of the winter-use locations
and should still be deemed important.

The observed seasonal shift in use of successional stages is further supported by analyses in which the microhabitat structure and vegetative composition also differed between summer and winter habitat (Jones 1991). Although the physical characteristics of snow cover may result in seasonal
variations in habitat-use patterns (Buskirk and Powell, this volume), we
believe the most plausible explanation for the seasonal shift in habitat use by
fishers is a concurrent shift in prey use. Jones (1991) reported that snowshoe
hares (*Lepus americanus*), voles (*Microtus* spp. and *Clethrionomys gapperi*),
and red squirrels (*Tamiasciurus hudsonicus*) were the primary prey for fishers
in north-central Idaho. The importance of voles in the diet may decrease over
the winter with a concomitant increase in consumption of red squirrels and
possibly snowshoe hares. A similar shift in prey use has been reported for
American martens (*Martes americana*) (Zielinski et al. 1983). Additional

research on the habitat relationships of important prey of the fisher is needed to fully understand seasonal variation in habitat use by fishers. Until the completion of additional studies, the observed seasonal variation should not be mistaken for habitat flexibility (Buskirk and Powell, this volume).

In general, sites used in winter differed less from random sites than did sites used in summer. Compared with summer, for which we found significant selection or avoidance of all five successional stages, use in winter differed from availability for only three of five stages. Furthermore, one less successional stage (young forests) was avoided in winter. This suggests that fishers use a more diverse array of habitats and are less selective of habitats in winter than in summer. In contrast, Buskirk and Powell (this volume) suggested that fishers use a wider range of cover types in summer than in winter. These apparent contradictions in habitat-use observations may be due to differences in thermoregulatory costs, prey availability, and the effects of snow cover on habitat use among study areas from widely separated geographic areas (i.e., the northeastern United States and the northern Rocky Mountains).

Similarly, fishers appeared to use a wider variety of habitats when hunting than when resting, at least in summer. The apparently random use of the pole-sapling, young, and old-growth forests for hunting may, however, have been due to inadequate sample sizes (Dixon and Massey 1969, Alldredge and Ratti 1986). Arthur et al. (1989*b*) also reported that active fishers probably used a wider variety of forest types than resting fishers and found little evidence to suggest that hunting fishers strongly selected for particular forest types. After a review of several fisher studies, Buskirk and Powell (this volume) similarly suggested that fishers were more selective of habitats used for resting than for foraging. We found that younger-aged forests appeared suitable for hunting but were rarely used for summer resting sites. More structurally complex forests seemed to have been preferred for both activities, but simpler stand structures were used for hunting (Jones 1991).

Although fishers preferred young forests in winter, they selected localities with higher availability of large-diameter trees ($\geq$47 cm dbh), snags (>52 cm dbh), and logs ($\geq$47 cm) relative to sites 50 m distant (Jones 1991). When using young forest stands, fishers often sought areas with at least one large tree, snag, or log that had survived the stand replacement fires from earlier in the century. Because large-diameter logs often were used as temporary dens in winter (Jones 1991), it is not surprising that fishers selected winter sites with many available logs. Thus, even though many sites used in winter were classified as young forests, they contained several characteristics commonly associated with older forests.

Management Implications

Landscape Management

Although fishers in north-central Idaho preferentially selected mature to old-growth forests, their population density and stability most likely respond to overall resource abundance (i.e., macrohabitat structure; Morris 1987, Adler 1988). Therefore, as Harris (1984) suggested, fisher habitat management must involve the management of a system of mature forests as opposed to the management of individual stands. Management at a landscape scale should incorporate a variety of young- to midsuccessional stages to promote a diversity of prey species, in conjunction with late-successional stages to provide key resting habitat. In a managed forest, the habitat factor we believe most likely to limit fisher populations is the availability and connectivity of mature and old-growth forests that provide optimal resting habitat.

Fishers in the northern Rocky Mountains have evolved under a fire regime that created numerous small openings within a matrix of mature-forested habitats. Mean fire-free intervals (mostly between surface fires) in north-central Idaho range from six years in ponderosa pine–Douglas-fir/bunchgrass areas to 40 years or more in subalpine-fir habitat areas (Arno and Petersen 1983). Consequently, timber harvest practices that mimic natural landscape patterns and processes may not be detrimental to fisher populations. In fact, conversion of some areas of older age classes to younger age classes may promote a diversity of prey species and thus have long-term benefits for fishers. On the other hand, Rosenberg and Raphael (1986) reported that fishers were very sensitive to forest fragmentation in northwestern California. Additional research on the relationships among forest fragmentation, timber management, and fishers in the northern Rockies is needed to develop a conservation strategy for this species.

In our study, fishers avoided openings and forested areas with 40% or less canopy cover (Jones 1991). Preferred resting habitat patches should therefore be linked by travel corridors of closed-canopy forest. High connectivity of preferred habitats would allow the landscape to support such wide-ranging species as the fisher (Harris 1984; Buskirk and Powell, this volume). Some evidence from our study area suggests that fishers preferred forested riparian areas for resting sites and used them extensively for traveling (Jones 1991). In addition, forested riparian sites likely provide optimal habitat for two preferred prey in our study area: snowshoe hares (Bookout 1965, Bittner and Rongstad 1982, Pietz and Tester 1983) and southern red-backed voles (Koehler et al. 1975, Koehler and Hornocker 1977, Campbell 1979). Thus, riparian forests would likely make excellent corridors to connect preferred habitats.

Stand Management

Fishers seemed to prefer large-diameter Engelmann spruce trees and hollow grand fir logs as resting sites in north-central Idaho (Jones 1991). These two species should therefore dominate stands to be managed for fisher habitat in this region. Stands containing, or adjacent to, riparian areas seem to be particularly important to fishers during all seasons (Jones 1991), and should be managed conservatively if maintaining fisher habitat is a goal.

Fishers' tolerance of habitat islands is not well understood (Buskirk and Powell, this volume). Large isolated stands probably have a lower probability of fisher presence than smaller, less insular stands. We recommend that mature to old-growth forest stands, to be considered effective fisher habitat, should be at least 51 ha and have 50% or more of their perimeter in contact with pole-sized or older forests. Stands with these attributes should have about a 70% probability of fisher occurrence (Rosenberg and Raphael 1986).

At the stand scale, fisher habitat capability would be degraded in the short term by clear-cut logging. Although we did not evaluate fisher habitat selection with respect to stand age, fishers likely would avoid clear-cut areas for at least 50 years (through the pole stage), use them occasionally for another 60–100 years, and likely not preferentially select them until the trees were 80–100 years old in the case of lodgepole pine (during winter) or 120–160 years old in the cases of mixed-conifer forests. Although we found that fishers prefer young forest in winter, it is important to note that these stands regenerated under natural circumstances, after fires. Consequently, they retained several structural characteristics—a few residual large-diameter live trees, snags, and logs—that would not be expected in most recently harvested stands.

The process of recovery of a clear-cut stand, from the standpoint of fisher habitat, could be accelerated by the following practices:

1. Retaining of an abundance ($\geq$12.3 trees/ha) of cull grand fir trees for future den logs. The objective would be to have trees at least 45.7 cm dbh that would begin to fall 80–100 years after logging.
2. Retaining at least 54 but no more than 109 metric tons/ha of large-diameter logs. An abundance of logs should aid the recovery of southern red-backed voles, providing prey that fishers may begin to use once the regenerated stand has reached the pole stage.
3. Retaining decks of cull logs and a few slash piles for potential fisher resting sites and for habitat for snowshoe hares.

Uneven-aged management would better maintain fisher habitat at the stand level. Harvesting individual trees or small ($\leq$5 ha) plots would likely not reduce fisher habitat capability, and could in fact increase within-stand diversity, which might improve prey diversity and abundance.

We currently lack the information needed to develop a conservation plan for fishers in the northern Rockies. Therefore, adequate management of fishers and their habitats may require adoption of a landscape-based approach. Two advantages of a broader strategy are that it has the ability to maintain the integrity of ecological systems and that it can operate with relatively little information (Hunter 1991). Applying such an approach would require land managers to adopt a long-term, large-scale plan (Thompson and Harestad, this volume), one that would mimic natural landscape patterns and processes. This in turn would involve management that would keep certain proportions of a forest in various successional stages, together with a specific frequency distribution of various patch sizes and linkages across the landscape. Such an approach would help insure the viability of fisher populations within a managed landscape.

Acknowledgments

This research was funded by the Idaho Department of Fish and Game through Federal Aid in Wildlife Restoration Project W-160-R, and the USDA Forest Service. Additional support was provided by the University of Idaho Forest, Wildlife and Range Experiment Station (Contribution no. 583) and the Idaho Trappers' Association. We thank D. D. Gale, A. Hubbs, and M. Wright for help with the fieldwork.

Introduction

This section explores physiological adaptation of *Martes* in its two most intensively studied forms: the timing of reproductive events through reflex ovulation and embryonic diapause, and the balancing of energy budgets in cold environments through the coordinated use of morphology, physiology, and behavior.

Harlow discusses the factors that affect the risk of starvation, a factor that probably contributes to mortality in winter. American marten, and probably other *Martes*, have extremely limited body fat stores and must balance their energy budgets over brief periods, perhaps only a few days. Harlow challenges the common assertion that mustelids pay high mass-specific costs for thermoregulation but recover those costs through increased foraging efficiency. A surprising number of mustelids, including members of three subfamilies, have been reported to have resting metabolic rates lower than predicted on the basis of body size. The increased foraging efficiency of long, thin-bodied mustelids is assumed to relate to capture of prey in spaces that are narrow and tortuous, but the costs may not be as high as once thought. For small mustelids, Harlow states, deposition of body fat represents a trade-off between the risk of poor tolerance of food scarcity and that of poor access to suitable sites for foraging and resting. Factors that confound these relationships include food-caching, which has been reported on rare occasions for North American *Martes*, and controlled hypothermia, both of which would tend to prolong fasting endurance. Finally, Harlow discusses the physiology of fasting, showing that coordinated catabolism of fat and muscle is a mechanism for which there is evidence and adaptive value in lean-bodied carnivores.

Females of all species of *Martes* generally breed first at 12–15 months of

389

age; the precise age varies with environmental conditions. Breeding during a brief estrus occurs in all but two species from late June to early August. Fishers exhibit postpartum estrus in March and April, and the poorly studied yellow-throated marten may breed in October. Ovulation induced by copulation has been documented in the sable but is thought likely for other species. Six of the seven *Martes* species are known to exhibit embryonic diapause of species-specific durations, during which blastocysts increase in size but not in cell number and the function of corpora lutea is reduced. Implantation in the endometrium is accompanied by enlargement and activation of the corpora lutea and is followed by an active gestation of 24–35 days.

Mead summarizes the postulated reasons for the evolution of delayed implantation in Carnivora and suggests plausible mechanisms accompanying the evolution of the delay. In addition, he elaborates on the possible linkage between seasonal ecological energetics and reproductive cycling. Because starvation causes endocrine-mediated anestrus, the evolution of summer estrus may be related to the physiological inability of some *Martes* to breed at more energetically demanding times of year—for example, while lactating or during winter.

Mead summarizes the methods by which reproductive condition and performance may be inferred from *Martes* and points out that the later the reproductive stage examined (corpus luteum, blastocyst, placental scar), the more reliable the index of natality.

STEVEN W. BUSKIRK

29 Trade-Offs Associated with the Size and Shape of American Martens

Henry J. Harlow

The size and shape of an animal's body most likely has evolved to optimize its success within its particular niche (Casey and Casey 1979). Size and shape may act in concert to influence energetic income from, and losses to, the environment. The activity of an animal may therefore be limited by both the "energetic ceiling" associated with body size (Drent and Daan 1980, Kirkwood 1983, King 1989*a*) and physical characteristics such as body shape. Some homeotherms have unusual shapes associated with specific habitats and behaviors. This chapter considers the energetic trade-offs and compensatory mechanisms that go with the long, lean bodies of mustelids, with specific reference to American martens (*Martes americana*), which occur only in areas with cold winters (Gibilisco, this volume).

Since the early 1970s, general references have been made to costs and benefits of a small, elongated, lean body in a cold environment. King (1989*a*) presented an excellent treatment of advantages and disadvantages of small size to weasels, dealing with the interplay of profit and loss. Mustelids' relatively short fur length and high surface-to-volume ratio lead to higher metabolic rates (Brown and Lasiewski 1972, Casey and Casey 1979, Chappel 1980, Sandell 1989*b*). Brown and Lasiewski (1972) interpreted this lack of morphological adjustment to cold in small mustelids as part of a trade-off. Because thermoregulation by many small mustelids requires more energy than in "normally" shaped mammals of the same mass, body size and shape must yield energetic benefits in other ways. A long, thin shape allows these animals to enter burrows and dens of small rodents, which compose most of their diet (Simms 1979; Erlinge 1980; Martin, this volume), as well as allowing them to enter their own resting sites (Buskirk et al. 1989). However, a long, thin shape influences more than surface-to-volume ratios; it also

391

severely limits body fat reserves and results in a suite of morphological, physiological and behavioral adjustments that reduce the risk of starvation in *Martes* and other mustelids.

Recently, Lima (1986) and Rogers (1987) proposed models of optimal body mass in birds. These models viewed winter body mass, especially body fat stores, as a trade-off between predation risk and the risk of starvation during periods of inclement weather. In the following discussion the basic concept of optimal body mass will be broadened to include not only the relation of mass to body fat, but also to shape. The risks faced by long, thin carnivores are not simply those of being preyed on but those of not having access to small prey and insulated resting sites. This discussion will show how a model of optimal body size and contour that maximizes overwinter survival represents the best trade-off between the need to obtain prey and rest sites and the need to minimize starvation risk.

On one side of the ledger in such a proposed model, adding body fat would decrease the availability of food and resting sites and increase exposure to predation. This assumes that (1) fattening increases body girth and limits access to small prey and resting sites, (2) fattening decreases agility and the ability to capture prey and escape predators, and (3) maintaining high body fat during winter requires high intake of prey and increased exposure to predation while foraging. On the other side of the ledger, fat storage and increased body mass would supply metabolic fuel during periods of forced inactivity, for example, in bad weather.

Many temperate vertebrates do not maintain their fat reserves at their physiological maximums (King and Farner 1966, Blem 1975), which implies that a significant cost is associated with maintaining large fat reserves (Stuebe and Ketterson 1982, Lima 1986). Fat stores would benefit animals with unreliable food resources but may be a liability for animals for which food availability is constant and predictable (Ekman and Hake 1990). When resource predictability is high, body fat should be low. For small mustelids, then, winter body fat manifests a trade-off between the risk of poor tolerance of food scarcity and that of poor access to suitable resting sites and subnivean prey.

This ledger-book model requires the identification of other physiological factors that may influence the energy budget and needed fat reserves. Low energy demands limit the need for stored fat and could result from several factors: low basal metabolic rate (BMR), low cost of locomotion, fur insulation, low cost of reproduction, high assimilation efficiency of food, or ability to enter torpor. Also, food caching, high fasting tolerance, and perhaps the ability to store alternate reserves such as labile protein could combine with low energetic demands to influence an animal's stored energy needs, regardless of food predictability.

Factors That Determine Body Fat Stores

Mrosovsky and associates hypothesized that many mammals maintain fat content at a level that varies over the year (Mrosovsky and Powley 1977, Mrosovsky and Sherry 1980). After an increase in summer and fall, a programmed decrease in body fat occurs during winter (Barnes and Mrosovsky 1974, Mortensen and Blix 1985). This depletion of stored fat ensures that animals are rarely below a specified level for body fat. As they use their fat reserves, their appetite drops, lowering the energetic costs of foraging and exposure to cold (Mortensen and Blix 1985).

But while some animals show increased body fat at the onset of winter (Harlow 1981*a,b*; Bartness and Wade 1984), others reduce their fat stores at the end of fall (Hoffman 1973, Petterborg 1978), and still others, including American martens, have low body fat content (less than 5%) in the fall with no recorded seasonal alteration (Buskirk and Harlow 1989). Pitts and Bullard (1968) found that fat mass of terrestrial eutherians is proportional to fat-free body mass raised to the 1.19 power. Martens are lean throughout the year by this standard. This finding corroborates those of Thompson and Colgan (1987*a*) for American martens, and Simms (1979) for other small mustelids. Larger mustelids such as the American badger (Harlow 1981*a,b*) are different; their body fat content in the fall is far greater (33%) than predicted and decreases markedly from autumn to March. Seasonal body fat stores may also be present in fishers (*Martes pennanti*: Powell 1979*b*).

The lack of a seasonal fat cycle in martens may be seen as the result of short-term balancing of energy budgets. Factors contributing to this budget include size and frequency of meals, amount and type of dietary fat, and activity and energy expenditure by animals. We know that animals may maintain neutral or positive energetic balances without altering food intake by decreasing activity levels. For example, badgers (Harlow 1981*a,b*) and some rodent species (Cornish and Mrosovsky 1965, Hirsh 1973, Schemmel 1976, Mrosovsky and Powley 1977) can regain a prefasting weight after a fast by decreasing activity levels and increasing the efficiency of food assimilation. Likewise, the size and frequency of meals can influence fat deposition. When gorged, many animals deposit more fat than when they eat the same amount over many frequent meals. If forced to consume few but large meals containing long-chain fatty acids, many animals transfer a large portion of the ingested energy to adipose tissue, where it is stored as triglycerides (Cohn et al. 1965, Leveille 1970, Schemmel 1976). Furthermore, different protein metabolism under different rates of ingestion influences the extent of fat deposition. A sudden large intake of protein can result in lower protein synthesis, increased fat deposition, and higher urinary nitrogen loss (Cohn et al. 1965). Therefore, an increase in the frequency of protein ingestion and lower loads of amino acids per unit time can enhance the efficiency of protein synthesis in mammals generally (Cohn 1963, Mochrie 1964). While few studies have dealt with small carnivores, empirical data strongly suggest that single large meals, high in either fat or protein, should result in greater fat deposition and lower protein synthesis in *Martes*.

American martens, like weasels (Gillingham 1984, King 1989*a*), have a limited ability to eat large meals. Unlike their metabolic rate, the gut mass of Carnivora varies linearly with body mass. Small mustelids, then, have high mass-specific metabolism, but they have a limited capacity to consume and assimilate food. They compensate by ingesting many small meals per day. Microtine rodents are the most common prey of American martens (Martin, this volume); each is a small high-protein meal. When American martens feed on larger prey such as squirrels and hares, the volume of the meal may be restricted by the gut capacity. As a result, fat deposition from eating a single large daily meal may be impossible. A high-protein diet consumed in small amounts over the day could therefore result in body protein anabolism or stasis and a lean body contour. R. Powell (North Carolina State Univ., pers. commun.), however, has noted that weasels (*Mustela* spp.) can gain weight with ad libitum feeding while in captivity, a finding consistent with observations of other animals (King 1974, Rogers 1987) for which fat mass in winter clearly is less than the physiological maximum. Thus, low seasonal fat levels in small mustelids such as American martens may result from limitations in rates of food intake and assimilation, balanced by high energetic costs for finding and capturing prey.

Food caching allows many animals to cope with unpredictable or variable food resources (Roberts 1979, Lea and Tarpy 1986) in a way that is functionally analogous to storage of body fat (McNamara et al. 1990). Variability in food supply induces hoarding in some species, similar to deposition of body reserves in others (Lima 1986, McNamara et al. 1990). Food caching, therefore, can either augment or act as an alternate to body fat reserves in an unpredictable environment. Hoarding is likely to occur when the costs of carrying reserves on the body are high (McNamara et al. 1990). This suggests that lean mustelids like American martens should depend on food caches in environments with unpredictable and variable food supplies. Documentation of food caching is rare for American martens, however. Of the many diet studies (Martin, this volume), only three have mentioned the possibility of caching food (Murie 1961, Zielinski et al. 1983, Henry et al. 1990). This contrasts with the well-documented caching behaviors of weasels (Powell 1975, Oksanen et al. 1985). The apparent absence of food caching by American martens, like the scarcity of body fat, supports the view that they must have predictable access to adequate food. The extent to which a stable food resource dictates the level of fat reserves would depend, however, on physiological and behavioral factors influencing energy use.

Basal Metabolic Rate

Variations in total energy expenditure appear to correlate with variations in basal metabolic rate (BMR) (McNab 1989). Body mass is generally acknowledged to be the most important factor determining the BMR in mammals (McNab 1980, 1986), but the correlation with mass accounts for only 67% of the variation in BMR. For example, mustelids weighing less than 1 kg generally have BMRs about 20% higher than other eutherians of the same size. This may be because BMR is functionally associated with factors other than mass, such as diet, activity level, and climate (McNab 1989). Predators consuming vertebrates have BMRs equal to or greater than those predicted by the Kleiber curve, whereas animals that depend on invertebrates, fruits, and vegetation have BMRs less than expected from the Kleiber relationship (McNab 1969, 1983, 1986). Martens prey on vertebrates and, like other mustelids, would be expected to have a BMR higher than predicted. Additionally, martens are highly active and live in areas with cold winters, both of which contribute to an elevated BMR in mammals (McNab 1986, 1989).

The presence of ecological mavericks within carnivore families, however, throws doubt on the concept that phylogeny has the predominant influence on the level of energy expenditure independent of body size and diet (McNab 1989). For example, many mustelids such as ferrets (*Mustela putorius*: Morrison and Ryser 1951), American badgers (Harlow 1981*a*), and spotted skunks (*Spilogale putorius*) have BMRs that are not elevated above the predicted values for nonmustelid eutherians. The generalization that mustelids have elevated BMRs that adapt them to cold climates (Scholander et al. 1950, Iverson 1972) may therefore have important exceptions.

The BMR of American martens (Buskirk et al. 1988) is below that predicted for a 1-kg mustelid. In fact, many mustelids have metabolic rates close to those of other mammals of similar size if care is taken to determine true resting metabolism (Farrell and Wood 1968). For example, in the study by Buskirk et al. (1988) the "respirometer" used for determining oxygen consumption was the American marten's usual resting box. To measure metabolic rate, the researchers gently closed and sealed the entrance door and monitored air ventilated through the respirometer for oxygen content. As a result, the study animals were not disturbed and often slept through measurement periods. The lack of high BMR observed in this study—a finding that may be more general among small mustelids—should reduce the need for fat reserves in martens. It must be remembered, however, that while the BMR is not as high as that predicted for a small mustelid, the weight-specific BMR of martens is still high. To suggest that a lean body form is simply the result of a low BMR by martens is not supportable.

Thermoregulation

Short fur and the high surface-to-volume ratio of small mustelids are associated with high thermal energetic costs resulting from a lower critical temperature (T_{lc}) that is higher than predicted for body size (Morrison 1960, King 1989a) and from increased thermal conductance (Herreid and Kessel 1967). The T_{lc} is the lowest ambient temperature at which passive mechanisms requiring little energy can maintain normal body-core temperature (T_b). Below the T_{lc}, active heat production (e.g., shivering) is necessary. Minimum thermal conductance is expressed by the absolute value of the slope of the regression line through metabolic rates below the T_{lc}. The high thermal conductance associated with short fur and a high surface-to-volume ratio results in a T_{lc} of small mustelids that is so high that ambient temperatures generally fall below it in winter (Casey and Casey 1979, Buskirk et al. 1988). And energetic costs to stay warm below the T_{lc} are high. Conductance and T_{lc} have been measured for American martens and model predictions are available for fishers (Powell 1979b). The conductance value calculated for martens (0.0527 ml O_2/g · hr · °C) by Buskirk et al. (1988) was greater than the 0.0334 ml O_2/g · hr · °C predicted for an animal of its body mass ($C = 1.05M^{-0.51}$, where M is in grams; Bartholomew 1977), but closer to the 0.046 ml O_2/g · hr · °C reported for American martens by Worthen and Kilgore (1981).

The T_{lc} of martens reported by Buskirk et al. (1988) was 16°C and approximates the predicted T_{lc} of 20°C ($T_{lc} = T_b - 4M^{0.23}$, where M is in grams; Bartholomew 1977) for similar-sized mammals. Although the T_{lc} of American martens in winter was lower than predicted by body size, Buskirk et al. (1988) showed that this species is not adapted to severe cold by having highly efficient fur, especially when compared with other cold-climate mammals. For example, the T_{lc} of winter-acclimated varying hares (*Lepus americanus*) was −5°C (Hart et al. 1965), that of the red fox (*Vulpes*) was −13°C (Irving et al. 1955), and that of the arctic fox (*Alopex lagopus*) was −40°C (Underwood 1971), all far below values predicted for body size. Powell's (1979b) model of daily energy expenditure estimated the T_{lc} for a male fisher by adjusting the protocol of Porter and Gates (1969) to account for other temperate mammals. Powell (1979*b*) estimated T_{lc} to be $-30°C$ for active fishers, a lower value than that predicted for a 4-kg mammal ($T_{lc} = T_b - 4M^{0.23}$; Bartholomew 1977). Powell's (1979*b*) estimate suggests that fishers rarely experience ambient temperatures below T_{lc}. In contrast, American martens are generally, and almost always while active, exposed to ambient temperatures during the winter that are below T_{lc} (Buskirk et al. 1988, 1989). Clearly, the morphological adaptations of American martens to cold are not as effective as those of some ecologically similar mammals. As a result, martens would be expected to pay high energetic costs to rest at ambient temperatures in winter. Body reserves would therefore be a benefit to accommodate resting periods, especially those prolonged by bad weather, in protected sites.

Torpor

Hypothermic torpor, another potential physiological response to balance energy budgets (Wang 1978), is uncommon among temperate mammals, especially mustelids, (King 1989*a*) which maintain T_b within a very narrow range. Perhaps one reason is that mammals using this strategy must accumulate large reserves of white adipose tissue (Lyman and Chatfield 1955, Mrosovsky 1966) and brown adipose tissue (BAT) (Heldmaier et al. 1981). The latter is needed for nonshivering thermogenesis during arousal from hypothermia (Smith and Hock 1963). The only mustelids believed capable of reducing T_b appreciably are the American badger (Harlow 1981*a*) and perhaps the European badger (*Meles meles*) (Paget 1980). Harlow (1981*a*) found a 9°C depression in T_b in the American badger; however, this species has body fat stores of 33% of total weight in winter and nonshivering thermogenic capability indicative of BAT reserves (Harlow and Miller 1985). No small mustelid has been shown to have BAT. In spite of this, free-ranging American martens monitored by Buskirk et al. (1988) depressed T_b by a mean of 2.9°C during resting episodes. Assuming a Q_{10} of 2.0, martens would achieve a 14.5% reduction in metabolic energy consumption while hypothermic. T_b depression of this magnitude is characteristic of deep sleep but not of more profound torpor (French 1988). Buskirk et al. (1988) estimated daily energy savings from T_b depression by assuming that martens rested for 12 hour daily, depressed T_b by a maximum of 2.9°C during each resting episode, and conformed to characteristic cooling and rewarming curves. Estimated daily savings of energy were 4%. Taken alone, this cannot be considered an important energetic strategy.

Locomotor Costs

Foraging by small mustelids requires high energy expenditures (King 1989a). Taylor et al. (1970) and Wunder (1975) calculated that metabolic rate increases linearly with running speed as follows:

$$\text{Running metabolic rate} = k_2 \, W^{0.75} + k_3 \, W^{0.6} \cdot S$$

where S is running speed, W is weight of the animal, and k_2 and k_3 are constants. Powell's (1979b) model for daily energy expenditures of fishers includes a prediction of the cost of locomotion during prey capture. It assumes that the majority of prey capture is associated with running and conforms with the formula above. Powell (1979b) found k_2 and k_3 for fishers running on a treadmill to be above values for other mammals. Therefore, the cost of locomotion by an elongated mammal with short legs is higher than predicted by body mass. Added to this is the energy required for bounding through deep snow. American martens have less difficulty moving through deep snow than fishers (Raine 1983), but Powell's (1979b) equations suggest that high demands of locomotion may place additional needs on martens for stored energy.

Reproductive Costs

Reproductive costs in the form of ovulation, gestation, and lactation have been linked to fat reserves in many mammals. Lactation is generally considered the most energetically demanding period, with costs two to four times those of gestation (Nelson and Evans 1961, Migula 1969, Myrcha et al. 1969). But reproductive costs for many mustelids are proportioned differently and may be even higher than those of other eutherians. Lilligraven (1975) pointed out the possible energetic advantages of prolonged gestation and short lactation over short gestation and prolonged lactation. If gestation is defined as the period of rapid development after implantation of the blastocyst, then delayed implanters, including many mustelids, have short gestation and long lactation. For example, American martens undergo a 7- to 8.5-month embryonic diapause, followed by a 27-day active gestation and birth in April or May (Mead, this volume). Few empirical data are available on the maternal costs of gestation and lactation for mustelid mothers. But for American badgers, Harlow et al. (1985) found that length of active gestation was shorter than predicted by body weight (Blueweiss et al. 1978). From data on fetal tissue growth, energy requirements and the length of gestation, Harlow et al. (1985) calculated that producing two American badger cubs from implantation to parturition would cost the mother only 2% in energy expenditure above normal requirements. By measuring caloric intake of hand-reared American badger pups and cost of milk production by the mother, and with estimates of assimilation efficiency and growth of neonates, these researchers estimated energetic cost to the mother of nursing two pups during a 40-day lactation period to be 16 times that of gestation. The greater cost of transferring maternal energy into pup growth compared with fetal growth is due to the additional energy losses associated with energy transfer into milk by the mother and then milk digestion, assimilation, and tissue growth by the pups (Harlow et al. 1985). As a result, the overall cost of reproduction from implantation to weaning would be expected to be greater for mammals that have a short gestation and lengthy lactation period. Lactation by Eurasian pine martens (*Martes martes*) and stone martens (*M. foina*) was reported by Schmidt (1943) to be about six weeks, twice that for mink (*Mustela vison*) and polecats. Allometric equations (Trillmich 1986) would therefore underestimate energetic costs for lactating martens and other mustelids. Indeed, Powell's model (Powell and Leonard 1983) for daily energy expenditure predicts the cost of lactation in fishers at 2500–2900 kJ/day, compared with 1140 kJ/day using allometric predictions for a 3-kg mammal (Trillmich 1986).

It is apparent that considerable food or stored energy is required to raise mustelids to weaning. Given that marten litter size is slightly larger than predicted for mammals ($Y = 2.45 \ M^{-0.136}$; Sacher and Staffeld 1974) and that lactation is prolonged, energetic costs for reproduction must be high. While mustelids such as American badgers and (perhaps) fishers (Powell 1979*b*) may have enough fat to pay these costs, American martens clearly do not. For this species, reproduction must be paid for with other body reserves or abundant and predictable foods.

Assimilation Efficiency

Carnivores living with food scarcity maximize their use of the available calories in several ways. Young (1944) and Whitney (1948) suggested that the stomachs of some carnivores can act as storage organs, allowing food to pass a little at a time into the intestine. This would provide the benefit of continuous nutrient uptake during periods of food deprivation, thereby reducing variation in blood glucose, but it requires a distensible stomach and pyloric sphincter control not yet described for any mustelid.

Another mechanism for maximizing use of calories involves variation in assimilation efficiency. American mink seasonally eat more food than required for maintenance, producing feces containing only partially digested food (Errington 1967). Likewise, Lampe (1976) found assimilation efficiency in American badgers to be negatively correlated with biomass consumed. Gorging by a carnivore, therefore, can result in a low assimilation efficiency and poor nutrient uptake.

As already discussed, the volume of food that can be consumed by small mustelids is severely limited (Gillingham 1984), so small meals must be followed by rapid gut passage to maximize the daily ingestion rate. But assimilating nutrients is positively related to gut passage time: the longer the exposure of food to absorptive surfaces of the gut, the more efficient the nutrient uptake. The gut passage time of martens (eight hours: More 1978) is about four times that for mink (Sibbald et al. 1962), but it is the same as that of American badgers, which are an order of magnitude heavier (Harlow 1981c). These data suggest that the slow rate of food passage in American martens should result in more efficient food assimilation. Additionally, Harlow (1981c) found that when badgers were fed after a fast, their food passage rate decreased, with a concomitant increase in assimilation efficiency. In contrast, when Gillingham (1984) put weasels on a fasting diet, they were unable to replace lost body mass, even when fed ad libitum for eight hours daily. For American martens, Harlow and Buskirk (1991) found that fasting for five days caused a mean loss of 24% of body mass, which was replaced by three weeks of ad libitum feeding. It is possible that martens have optimized the number of meals consumed per day in association with gut passage time, which could reduce the need for body fat reserves.

Foraging Efficiency

An animal should reduce its foraging time when nutritional benefits are less than costs. For American martens, this may occur when ambient temperature is low and wind speed is high (Pruitt and Lucier 1958, Pauls 1978). In fact, movements of weasels (Kraft 1975, King 1989a) and American martens (Buskirk et al. 1988) decrease when ambient conditions are cold and windy. Extended time in an insulated nest (MacLean et al. 1974, Erlinge 1980, King 1989a) should reduce thermoregulatory costs and perhaps the need for stored energy. With limited fat reserves, however, the need for food forces activity. Marten movement is greatest when food is least available (Thompson and Colgan 1987a, 1990). To increase predatory efficiency and limit the need for stored fats, carnivores can restrict their foraging to coincide with the most vulnerable (generally the most active) phase of prey activity rhythms, thereby increasing their capture efficiency, compared with hunting at random (Kaufman 1974, Curio 1976, Daan 1981). Martin (this volume) portrays American martens as food generalists that demonstrate just such an adjustment in response to prey type and availability.

Any prey eaten by a predator has its costs in terms of search energy and time, and handling energy and time, and its benefits in terms of its net food value (Krebs and Davies 1984). Martens may, therefore, optimize prey choice on the basis of net profitability, which may vary with snow depth and relative abundance of prey types, especially types that live above instead of below the snow. Data presented by Thompson and Colgan (1990) support the optimal foraging prediction that as prey abundance decreases, the diet niche widens. They concluded that American martens behave like dingoes (*Canis dingo*), in that their hunting strategy ensures that small prey are captured incidentally between less frequent kills of large prey (Corbett and Newsome 1987). An efficient foraging strategy could therefore be expressed in a reduced need for fat reserves.

Fasting Tolerance and Fat Reserves

The fasting survival time (t_s) for 1.1-kg mammal at 5°C is 6.5 days, as predicted by Lindstedt and Boyce (1985):

$$t_s = \frac{2948M^{1.19}}{317M^{0.75} + [(\dot{E}_t + |\dot{E}_t|)/2]}$$

where M is in kilograms and $\dot{E}_t$ is the rate of energy use (kJ/day)

$$\dot{E}_t = (16M^{0.5}) [38 - (22.5M^{0.25} - T_a)]$$

After five days of no food or water, however, American martens in one study were in excellent health and very aggressive (Harlow and Buskirk 1991). These animals appeared to be far from energetically in extremis. Martens, therefore, have a higher capacity for enduring fasting than predicted for a mammal of their size and certainly of their leanness. Harlow and Buskirk (1991) examined how this might be accomplished by investigating biochemical changes in American martens during three phases of fasting. Phase I for most mammals occurs during the first several hours or days of a fast and is characterized by the retention of normal blood glucose levels through the catabolism of glycogen and protein reserves. Phase II is identified by elevated blood ketone bodies and fatty acids from triglyceride breakdown and reduced protein catabolism, with a concomitant drop in blood glucose. This protein-conserving phase reduces skeletal protein degradation, which would lead to loss of strength and mobility. The depletion of fat reserves marks the onset of Phase III, as protein is again the major catabolic substrate and blood glucose levels rise (Le Maho et al. 1981).

Harlow and Buskirk (1991) asked whether American martens behave like lean laboratory rats, with little resistance to fasting and no Phase II fast (Li and Goldberg 1976, Goodman and Ruderman 1980), or whether they use fat and labile protein tissue to tolerate food and water deprivation. They compared the fasting response of lean martens with that of a similarly sized but obese mammal, black-tailed prairie dogs (*Cynomys lucovicianus*). Both species are adapted to a food and water fast of up to five days, despite great differences in fat stores. Unlike laboratory rats, both martens and prairie dogs entered a state of protein conservation. This conservation was lower in martens than in prairie dogs, as evidenced by higher daily loss of nitrogen in the urine and greater concentration of amino acids in the plasma. Harlow and Buskirk (1991) proposed that the differential inhibitory effect of ketone bodies on protein catabolism may cause this difference. Ketone bodies may have even less of an inhibitory role on protein catabolism in rats, which appear to use only one substrate type at a time while fasting.

Several potential advantages of combining fat and protein catabolism during a fast are apparent. First, Harlow and Buskirk (1991) calculated that by using such combinations, martens can maintain fat stores about three days longer than if they were the sole substrate, and muscle tone can be prolonged. Second, Bintz et al. (1979, 1980) hypothesized that catabolism of fat and protein in a ratio of 1.9:1 helps maintain water balance, owing to the high water content of protein tissue. In fact, fasting American martens used fat and protein in about this ratio. Additionally, Harlow and Buskirk (1991) found that American martens hydrolyzed urea by intestinal urealytic microbes. The potential advantage of such a mechanism is to reduce urinary urea and urinary water loss. It would also conserve nitrogen by recycling it into amino acids for protein synthesis (Nelson et al. 1975, Harlow 1987). Therefore, American martens may be adapted to fasting by combined use of energy substrates—a way not seen in other lean animals.

Labile Proteins as Alternatives to Fat Stores

Although American martens conserve protein during a fast, their relatively high metabolic rate places considerable demands on protein tissues, compared with animals that store fat (Harlow and Buskirk 1991). An adaptive advantage could be realized if martens had labile protein reserves. Many vertebrates, including birds (Jones and Ward 1976, Le Maho et al. 1981), fish (Kendall et al. 1973), and mammals (Millward and Waterlow 1978,

Yacoe 1983, Torbit et al. 1985) use protein as an energy source and store protein seasonally for this purpose. These protein reserves appear to be in specific compartments, including the skin and viscera (Millward and Waterlow 1978, Torbit et al. 1985) as well as blood albumins (Garcia-Rodrigues et al. 1987). Additionally, when skeletal muscle is catabolized for energy, effects on locomotor capacity may be minimized by using specific muscle groups (Le Maho et al. 1981) and labile proteins stored in the sarcoplasm between the myofibrils, without affecting the sarcotubular system (Kendall et al. 1973).

Conclusions

Carnivores can make substrates available for metabolic pathways either through food intake or from mobilized fat reserves. Starvation risk thus depends on both foraging success and the size of body reserves. Large body size due to fat reserves carries the risks of impaired access to food and to resting sites as well as loss of agility in catching prey. For American martens, these risks may outweigh those of starvation. Martens have a relatively high mass-specific BMR. Their thermal efficiency is relatively low, with a T_{lc} above most ambient winter temperatures. Unlike many small mammals facing winter cold, they do not undergo deep torpor. While movement over and under the snow may be less costly for martens than for fishers (Raine 1983), their locomotor efficiency is still low compared with that of other mammals. As for many mustelids, lactation is lengthy, which places an additional reproductive energy burden on females.

The costs associated with all these attributes are high; they seem to add more to the cost side than the benefit side of the energy ledger. My premise is that if BMR, locomotor, thermoregulatory, and reproductive costs are high and do not explain the lack of fat reserves, then American martens must possess additional adaptive mechanisms to maintain energy balance. The following factors may set martens apart as high maximizers of derived energy: (1) shallow daily torpor, especially in response to inclement weather; (2) behavioral thermoregulation; (3) potentially high nutrient assimilation as a result of slow gut passage time; (4) optimization of food intake by adjusting predatory patterns and prey type; (5) combining fat and protein as metabolic substrates during fasting; and (6) storage of protein to augment small fat reserves. These features may allow martens to cope with periods of cold-induced food scarcity without relying on fat reserves, while at the same time allowing a long, lean body shape to exploit an otherwise unavailable food resource.

30   Reproduction in *Martes*

Rodney A. Mead

The outline of the female reproductive cycle is now known for six of the seven species belonging to the genus *Martes*, but the reproductive physiology of male *Martes* is less well documented. Little is known about reproduction in Japanese martens (*M. melampus*) of either sex. This chapter summarizes the literature and points out where additional information on reproductive physiology is needed. A considerable amount of research on mustelids has been conducted in the Commonwealth of Independent States and China. Much of this work remains untranslated and thus unavailable to most western scientists.

Reproduction in Female *Martes*

The gross anatomy of the female reproductive tract of *Martes* is essentially the same as that of other mustelids. The ovaries are encircled by the convoluted oviducts and completely covered by the mesosalpinx. The ostium of the oviduct is surrounded by numerous fingerlike projections (fimbrae), which are covered with ciliated epithelium and function to remove the ovulated oocytes and enveloping cumulus masses from the surface of the ovary. The uterus is bipartite, with a short internal septum located within the body (corpus) near the junction of the horns (cornua). The cervix is located at the distal end of the corpus. The caudal end of the urethra unites with the ventral surface of the vagina and opens into the vaginal lumen through the urethral orifice. The common chamber posterior to this opening is called the urogenital sinus. The integumentary folds that surround the orifice of the urogenital sinus and contribute to the external genitalia are called the vulva.

404

The age at first mating has been reported for five species of *Martes*. Females of some populations of all species (Table 30.1) have been reported to breed when they are 12–15 months old. There are even reports of pregnancy in juvenile stone martens (*M. foina*), but such observations are rare (Stubbe 1968; S. Broekhuizen, Res. Inst. Nat. Manage., Arnhem, Netherlands pers. commun.). Schmidt (1943) reported that only 43% of captive martens bred at 1 year of age. Most female sables (*M. zibellina*), American martens (*M. americana*), stone martens, and Eurasian pine martens (*M. martes*) do not breed in captivity until they are 24–27 months of age and some reportedly do not breed until their third year. Although several female American martens were observed to mate when they were about 15 months old, the youngest female was 3 years old when she gave birth (Markley and Bassett 1942). Krott (1973) reported that both sexes of *M. martes* were mature at 14 months and matings were fruitful in the wild. Several possible explanations may account for these seemingly contradictory reports. Puberty may be delayed in both captive and wild populations as a result of nutritional deficiencies (Fitzgerald et al. 1982, Fitzgerald 1984, Dubey et al. 1986), although the

Table 30.1. Approximate age at first mating of males and females of five species of *Martes*

Species	Sex	Age at first mating	Source
M. americana (American marten)	Male	15–39 months	1–6
	Female	15–39 months	1–6
M. foina (stone marten)	Male	15–27 months	7–9
	Female	15–27 months	8,9
M. martes (Eurasian pine marten)	Male	14–27 months	9–11
	Female	14–39 months	9,11,12
M. pennanti (fisher)	Male	12 months	13,14
	Female	12 months	13–15
M. zibellina (sable)	Male	27 months	9,16
	Female	15–39 months	9,16,17

Sources:
1. Yerbury 1947
2. Orsborn 1953
3. Jonkel & Weckwerth 1963
4. Weckwerth 1957
5. Hawley 1955
6. Markley & Bassett 1942
7. Stubbe 1968
8. Madsen & Rasmussen 1985
9. Schmidt 1934, 1943
10. Danilov & Tumanov 1980a
11. Krott 1973
12. Danilov & Tumanov 1980b
13. Wright & Coulter 1967
14. Leonard 1986
15. Eadie & Hamilton 1958
16. Ponomarev 1938
17. Song et al. 1988.

effect of nutrition on reproductive performance has not been investigated in *Martes*. Reduced fecundity in wild martens, however, has been correlated with declines in major prey populations (Strickland 1981, Strickland and Douglas 1987). Altered behavior or stress in captive animals could be responsible for failures to mate. Errors in aging wild martens could also account for some of the conflicting reports. A report by Kartashov (1989), who studied changes in age-related fertility of wild sables for seven consecutive years, suggests, however, that the age at which sexual maturity is reached varies with environmental conditions. In some years more than 70% of the 15-month-old females appeared to breed, as denoted by the presence of corpora lutea in their ovaries. In other years less than 20% of the females in this age class reproduced. Moreover, the percentage of females breeding in any given year varied for all age classes, but reproduction in sables in the ≥1- and ≥2-year-old age classes was considerably more sensitive to environmental effects than that of older age classes. This difference may be a result of young animals being forced by older ones to reside in less favorable habitat. Similar fluctuations in the percentage of yearling American martens breeding have been reported (Strickland and Douglas 1987, Thompson and Colgan 1987*a*). The latter investigators reported lower ovulation rates and reduced production of young in 1- and 2-year-old martens in response to scarce prey. Failure of the younger age classes to reproduce should have a pronounced effect on recruitment because they compose a substantial proportion of the population. Nearly all female fishers (*M. pennanti*) are sexually mature at 12 months.

Most *Martes*, with two notable exceptions, commence mating in late June and the breeding season continues through July and early August (Fig. 30.1), with most matings occurring in July (Schmidt 1934, Markley and Bassett 1942). Slight variations in timing of the breeding season and other events in the reproductive cycle appear to occur in species that have wide geographic distributions. Yellow-throated martens (*M. flavigula*) have been bred in the Republican Zoo of Kaunas in Lithuania in October through early November (A. Andruiskevicius, Kaunas Zoo, Lithuania, pers. commun.). Fishers mate in late March through April (Lowe 1930, Hodgson 1937, Hall 1942, Wright and Coulter 1967). Although the duration of the mating season has not been well documented for wild *Martes*, it would appear that most matings occur within a 30- to 45-day period (Fig. 30.1). The time of year of sable matings is controlled by seasonal changes in the light/dark cycle (Song et al. 1988), as is the case for other mustelids (Bissonette 1932; Herbert 1969, 1989; Mead and Neirinckx 1990). Photoperiod is therefore assumed to be the primary environmental cue for timing of breeding in all *Martes*.

The first visible sign of estrus in *M. americana*, *M. martes*, *M. foina*, and *M. pennanti* is the enlargement of the vulva (Enders and Leekley 1941;

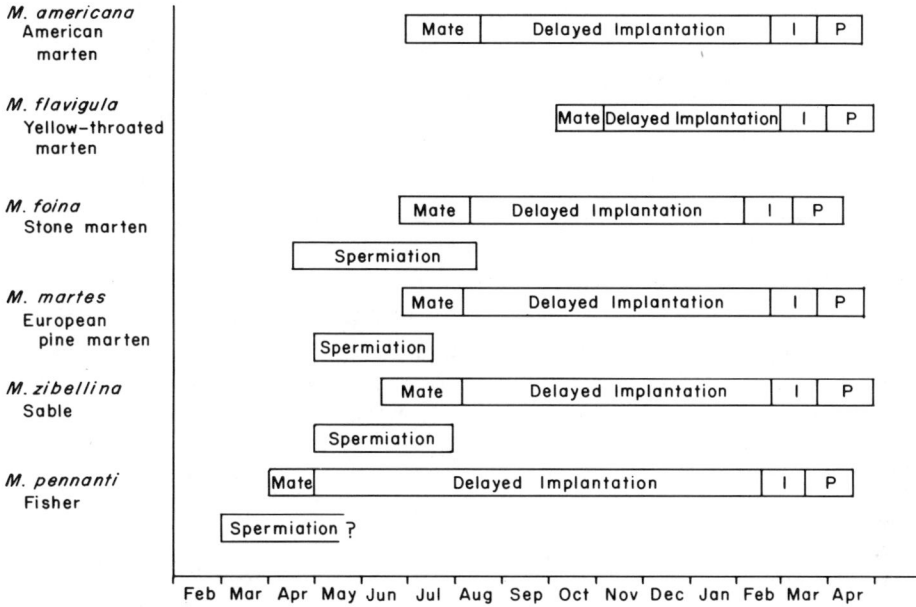

Figure 30.1. Reproductive cycles of six of the seven species of *Martes*. Most breed in July, and all exhibit delayed implantation and give birth in March and April. I = implantation, P = parturition.

Labree 1941; R. Cochrane, West Virginia Univ., pers. commun.). Early regression of vulval swelling does not occur after mating; the vulva of a mated American marten remains swollen for 6–39 days (Enders and Leekley 1941). Most matings occur when vulval swelling is maximal. Estrus and the breeding season have been successfully detected by monitoring vaginal cytology in several genera of mustelids (Hamilton and Gould 1940, Hansson 1947, Travis et al. 1978, Stenson 1988, Mead et al. 1990). Enders and Leekley (1941) reported taking vaginal smears from *M. americana* but did not describe their results. Elevated blood levels of estrogens prior to estrus are responsible for the thickening and cornification of the vaginal epithelium and vulval swelling in American mink (*Mustela vison*) (Hammond 1951, Enders and Enders 1963). Blood levels of estradiol in sables averaged 41.5 pg/ml immediately before mating (Polyntsev et al. 1985).

Estrus is reportedly brief, lasting 2–8 days and occurring 3–9 days after parturition in fishers (Hodgson 1937, Labree 1941, Hall 1942). Some fishers may exhibit a second estrus within 10 days of the first mating (Hodgson 1937). Captive *M. americana, M. foina, M. martes,* and *M. zibellina* exhibit one to four periods of sexual receptivity, which usually last 1–4 days and recur at 6- to 17-day intervals during the breeding season (Ashbrook and Hanson 1930; Schmidt 1934, 1943; Enders and Leekley 1941; Markley and Bassett 1942; R. Cochrane, West Virginia Univ., pers. commun.). It seems likely, however, that the duration of sexual receptivity in unmated martens and fishers would be longer than that reported by these authors. If ovulation was induced by coitus, one would predict that ovulation and the accompanying increase in progesterone secretion would have altered the sexual behavior of the females in these studies and would have led to a shortening of the estrous period. Whereas most *M. martes* and *M. foina* (>70%) exhibit multiple heat periods (Schmidt 1934, 1943), most *M. zibellina* (52%) exhibit a single estrous period (Bernatskii et al. 1976). Krott (1973) reported that estrus appeared to last about 10 days in wild *M. martes*, and females that were bred did not return to heat. Multiple copulations with one or more males occurred during each period of receptivity under captive conditions (Schmidt 1943). It is therefore assumed that wild females would also copulate with more than one male if the opportunity arose.

Female American mink permit several matings at irregular intervals and will ovulate again if bred within 7–28 days after the first mating (Hansson 1947). About 75–90% of such repeated matings result in all kits being sired by the male of the second mating (Hansson 1947, Venge 1956). Mink kits were sired by both males in about 15% of the females that were known to have ovulated during both matings (superfetation), but less then 10% of the kits were sired by the first male (Shackelford 1952). Thus the mating of mink during successive heats generally leads to the loss of most of the first set of embryos, although the exact mechanism responsible for this loss remains obscure (Adams 1981). If additional matings during subsequent estrous periods in *Martes* result in further ovulations, corpora lutea from both matings should be retained, as occurred when a second ovulation was induced with hormones in *M. martes* (Canivenc et al. 1969). One would therefore expect numbers of corpora lutea and corpora albicantia to substantially exceed those of embryos, but this does not seem to be the case in *Martes* (Douglas and Strickland 1987, Strickland and Douglas 1987). Bernatskii et al. (1976) found the incidence of ovulation in once-mated sables of known fertility to be quite low (13 of 47, or 28%) and the number of corpora lutea between once- and twice-mated females was not significantly different. They concluded that most females ovulated only after the second mating. The incidence of ovulation at each mating has not been investigated in other *Martes* and the occurrence of superfetation in this genus remains questionable. The relatively low and constant number of corpora lutea within all *Martes*, however, suggests that estrus may be recurring because the female failed to ovulate during previous matings. Alternatively, matings during successive estrous cycles do not lead to further ovulations.

Although ovulation is presumed to be induced by coitus, this has been documented only for sables (Bernatskii et al. 1976). Ovulation occurs 73–84 hours after coitus (Bernatskii et al. 1976). Peripheral blood levels of progesterone, ranging from 600 to 850 pg/ml at the time of mating, increase three- to fourfold 80–90 hours after copulation (Polyntsev et al. 1975, 1985). Copulation has been observed to occur on the ground (Markley and Bassett 1942, Schmidt 1943) and in trees (Henry and Raphael 1989) and is prolonged, averaging 30–50 minutes for sables, stone martens, and Eurasian pine martens (Schmidt 1934, 1943). Most matings of *M. americana* were more than one hour long (Markley and Bassett 1942) and about four hours long for captive fishers (Labree 1941, Hall 1942). Such lengthy copulations are not required to induce ovulation or ejaculation in mink or ferrets (*Mustela putorius*), but fertilization usually fails to occur if copulation is interrupted 5–12 minutes after intromission (Venge 1956, Carroll et al. 1985, Miller and Anderson 1989).

The time required for embryos to traverse the oviducts in *Martes* is unknown. Tubal transport of embryos is completed by the sixth day after coitus in ferrets (Robinson 1918) and by the eighth day in mink (Hansson 1947), but more than 11 days are required in long-tailed weasels (*Mustela frenata*) (Wright 1948). Upon reaching the uterus, blastocysts of all *Martes* remain encased by an acellular membrane, the zona pellucida, and undergo a prolonged period of developmental arrest known as embryonic diapause. At least six of the seven species of *Martes* exhibit an obligate delay of implantation, and embryonic diapause probably also occurs in Japanese martens. M. Tatara (Kyushu Univ., pers. commun.) reported that *M. melampus* breeds from mid-July through early August and gives birth from late April to mid-May. Resumption of embryonic development and implantation are delayed for periods ranging from 172 days in *M. flavigula* to 358 days in *M. pennanti* (Fig. 30.1; A. Andruiskevicius, Kaunas Zoo, Lithuania, pers. commun.; Hall 1942; Douglas 1943). Delayed implanting blastocysts have been described for most species of *Martes* (Marshall and Enders 1942, Enders and Pearson 1943*a*, Baevsky 1963, Aubert and Canivenc 1986). The diameter of the blastocysts gradually increases throughout the prolonged preimplantation period, but there is little or no accompanying increase in cell numbers in sables (Baevsky 1963, 1971). Cytological changes in the uterus and changes in uterine protein synthesis and secretion during delayed implantation have been reported for several species of mustelids (Enders and Enders 1963, Mead et al. 1979, Schlafke et al. 1981, Given and Enders 1989). Similar detailed information is lacking for *Martes*; we do, however, know that the uterus of most *Martes* is small and exhibits limited glandular development throughout most of the preimplantation period. This condition suggests that uterine secretory activity is diminished during this time (Danilov and Tumanov 1980*b*, Canivenc et al. 1981). The small size and histological appearance of the corpora lutea in *Martes* suggest that ovarian hormone secretion is reduced during embryonic diapause (Wright 1942, Wright and Coulter 1967, Canivenc et al. 1968, Canivenc et al. 1981). This hypothesis is supported by reports of reduced concentrations of progesterone in plasma and corpora lutea during embryonic diapause in *M. martes*, *M. foina*, and *M. zibellina* (Canivenc and Bonnin 1975, Polyntsev et al. 1975, Bonnin et al. 1977, Canivenc et al. 1981).

Renewed embryonic development and implantation usually occur in late February or March in *M. americana*, *M. foina*, *M. martes*, *M. zibellina*, and *M. pennanti*. One would predict a similar date of implantation in *M. flavigula*, based on known dates of parturition in captive-bred females (Table 30.2). Powell (1982) concluded that implantation could occur as early as January and as late as early April in fishers. Blastocyst implantation is associated with enlargement of the mammae in *M. americana* (Jonkel and Weckwerth 1963) and increased size of the corpora lutea and luteal cells (Canivenc

Table 30.2.   Date of parturition and litter size in *Martes*

Species	Parturition	Litter size		Source
		Range	Mean	
M. americana (American marten)	Mar–Apr	1–5	2.7–3.0	1–3
M. flavigula (yellow-throated marten)	Apr	—	—	4
M. foina (stone marten)	Mar–mid-Apr	1–5	2.5–3.2	5–9
M. martes (Eurasian pine marten)	Late Mar–Apr	2–6	3.0–3.3	5,10–12
M. melampus (Japanese marten)	Mar–mid-May	1–2	—	13,14
M. pennanti (fisher)	Late Mar–early Apr	1–4	2.0–3.0	15–19
M. zibellina (sable)	Late Mar–Apr	1–5	2.8–3.0	5,20,21

Note: — = not reported.
Sources:
1. Ashbrook & Hanson 1927
2. Wright 1963
3. Markley & Bassett 1942
4. A. Andriuskevicius, Kaunus Zoo, Lithuania, pers. commun.
5. Schmidt 1934, 1943
6. Stubbe 1968
7. Jensen & Jensen 1970
8. Canivenc et al. 1981
9. Madsen & Rasmussen 1985
10. Canivenc et al. 1968, 1969
11. Canivenc 1970
12. Danilov & Tumanov 1980*b*
13. Kuroda 1940
14. Imaizumi 1949
15. Lowe 1930
16. Hall 1942
17. Eadie & Hamilton 1958
18. Wright & Coulter 1967
19. Powell 1982
20. Bernatskii et al. 1976
21. Song et al. 1988

et al. 1968). Plasma and luteal progesterone concentrations are significantly elevated during the first half of the postimplantation period (Canivenc and Bonnin 1975, Canivenc et al. 1981). Selås (1990c) suggested that the increased concentrations of steroids in the blood during the time of implantation can be used to determine whether a female is pregnant. But this is not necessarily true, inasmuch as blood levels of these hormones will most likely be the same in pregnant and pseudopregnant females, as is known to be the case in the domestic ferret (*Mustela furo*) (Heap and Hammond 1974). Pseudopregnancy has yet to be confirmed in any *Martes*, however. Duration of the postimplantation period in *M. americana* is 27–28 days (Jonkel and Weckwerth 1963) and about 30–35 days in *M. martes* (Canivenc 1970). Selås (1990c) reported that the postimplantation period ranges between 24 and 30 days in the four most commonly studied species of martens. This period closely approximates that of the mink (27–29 days), ferret (30 days), and western spotted skunk (*Spilogale gracilis*; 28–31 days) (Enders 1952, Foresman and Mead 1973). Since most *Martes* are of similar size, one would predict that the duration of the postimplantation period would be about the same in most species, but perhaps 2–5 days longer in fishers, owing to their larger size. Coulter (1966) estimated the duration of the postimplantation period in fishers to be 30 days; the estimate of 60 days by Hamilton and Cook (1955) seems excessive.

The timing of blastocyst implantation is influenced by the seasonal increase in duration of daylight in spring. Pregnant *M. americana* were subjected to artificial long-day photoperiods in late September. Implantation occurred in late November and parturition followed in late December (Enders and Pearson 1943b, Pearson and Enders 1944). Hastening of blastocyst implantation and reduction in gestation has also been obtained by exposing sables and Eurasian pine martens to long-day photoperiods (Belyaev et al. 1951, Canivenc and Bonnin 1975, Song et al. 1988).

Our knowledge of the hormonal control of pregnancy in the genus *Martes* is fragmentary. Progesterone has been measured during pregnancy in *M. foina* and *M. martes*, and both progesterone and estrogen have been quantified in *M. zibellina* (Canivenc and Bonnin 1975; Polyntsev et al. 1975, 1985; Bonnin et al. 1977; Canivenc et al. 1981). Canivenc et al. (1969) attempted to induce early implantation in the Eurasian pine marten by administering human chorionic gonadotropin. This treatment resulted in formation of new corpora lutea, which assumed an inactive appearance, but implantation was not hastened. Little else is known about the endocrinology of delayed implantation or pregnancy in *Martes*. Despite this lack of specific knowledge, most of what is known about delayed implantation in other species of mustelids is likely applicable to *Martes* (reviewed in Mead 1981,

1989*a*). It seems probable from what we know that increasing day length promotes increased prolactin secretion from the anterior pituitary. This hormonal change in turn leads to increases in luteal cell size, luteal progesterone content, and plasma progesterone levels. Such changes in luteal function precede renewed embryonic development in other mustelids and are believed to be required for the development of a uterine environment conducive to blastocyst implantation (Mead 1981, 1986). Attempts to hasten blastocyst implantation in several species of mustelids by injecting progesterone with or without estrogen have consistently failed (Mead 1981, 1986). Some additional luteal secretion, perhaps a protein, is believed to be required to induce implantation in mustelids (Mead 1986, Mead et al. 1988).

What possible adaptive advantage might have led to delayed implantation in *Martes*? The potential advantages have been discussed by numerous authors (Prell 1927, 1930; Hamlett 1935; Wright 1963; Aitken 1977; Powell 1982; Mead 1989*a*; Sandell 1990). All *Martes* usually litter in late March and April (Table 30.2). This is somewhat earlier than in other similar-sized, temperate-zone mustelids that breed in the spring and lack a prolonged delay of implantation (Table 30.3). Gestation for species listed in Table 30.3 ranges from 41 days in steppe polecats (*Mustela eversmanni*) to 59–77 days in

Table 30.3. Breeding season and month of parturition for several mustelids whose body size is similar to *Martes* and who inhabit the temperate zone but lack a prolonged period of delayed implantation

Species	Breeding season	Parturition	Source
Mustela vison (mink)[a]	Mar–early Apr	Apr–May	1,2
Mustela eversmanni (steppe polecat)	Late Mar–early Apr	Mid-May–early Jun	3,4
Mustela nigripes (black-footed ferret)	Mid-Mar–Apr	May–early Jun	5,6
Mustela lutreola (European mink)	Apr	May–early Jun	7,8
Spilogale putorius interrupta (eastern spotted skunk)	Apr	Late May–early Jun	9,10
Mephitis mephitis (striped skunk)[a]	Late Feb–early Apr	May–early Jun	11

Sources:
1. Hansson 1947
2. Enders 1952
3. Zheng et al. 1983
4. Mead et al. 1990
5. Hillman & Carpenter 1983
6. T. Thorne, Wyoming Game and Fish Dep., pers. commun.
7,8. Moshonkin 1981, 1983
9. Crabb 1944
10. Mead 1968
11. Wade-Smith & Richmond 1978

[a]Has a short, variable period of delayed implantation.

striped skunks (*Mephitis mephitis*). The young of *Martes americana*, *M. martes*, *M. foina*, and *M. zibellina* rapidly attain adult size within 12–16 weeks (Brassard and Bernard 1939, Schmidt 1943) and disperse before mating occurs (Schmidt 1943, Wynne and Sherburne 1984). Fishers require somewhat longer to mature (Coulter 1966, Powell 1982, LaBarge et al. 1990) and the young are not independent until late August or September (Arthur and Krohn 1991). If a marten or fisher were mated on 1 April and had a 40-day gestation, the earliest that birth could occur would be 11 May, or 20–30 days later than parturition in most *Martes*. Therefore, the most plausible selective advantage of delayed implantation for most *Martes* is that it permits matings to occur later in the year and still enables parturition to occur at the earliest possible date. This gives the young of these fast-growing carnivores an additional 20–30 days to perfect their foraging skills, thereby enhancing their chances of survival over winter (Fries 1880, Schmidt 1943). The American mink and striped skunk mate somewhat earlier than most *Martes*, but still do not give birth before late April or May, because females that mate early in the season exhibit a brief delayed implantation whereas those mating late in the season have shorter gestations (Hansson 1947, Wade-Smith et al. 1980).

The preceding discussion does not resolve the question, Why has earlier mating not evolved in *Martes*? Males of most *Martes* appear to be fertile about one month before matings occur, thus suggesting that one must study the females to find the answer (Fig. 30.1). Nor does it explain how or why a summer mating season evolved. Many investigators have presumed that there has been selective pressure to delay onset of the breeding season until environmental conditions were less severe, thereby increasing the probability of males locating widely dispersed females. This could be of particular importance because estrus is thought to be brief in most *Martes*. The amount of energy expended by males of smaller species in search of estrous females in late winter or early spring may have been too great, resulting in selection for mating at a more favorable time of year. This hypothesis is supported by tracking data that suggest that movements of fishers are restricted by soft, thick snow (Raine 1983). One must also consider that delayed implantation might have evolved when climatic conditions were more severe and in response to selective pressures that were stronger than is the case today. Selection for later mating in June–July, as opposed to April, in small-bodied mustelids might have other advantages. Fishers and most martens reportedly spend very little time away from their dens when the young are first born, but are absent for increasing amounts of time as the young get older (Schmidt 1943, Powell and Leonard 1983). The young, which would be left unattended for more than an hour during a copulatory bout, would be less apt to suffer hypothermia if mating occurred during warmer months, thereby increasing selective pressure for late mating. Delaying mating until the young were considerably older might also reduce the potential for adult males to kill or injure the tiny young by stepping on them in their efforts to court females in the den.

Fishers, however, do not delay mating until later in the year, but rather exhibit a postpartum estrus within three to nine days after parturition. Arthur and Krohn (1991) have suggested that this strategy may have evolved to accommodate the longer period of parental care required by fishers and that the energy requirements of adult females with kits may have precluded successful breeding during the summer. This might explain why fishers are the only *Martes* to have evolved a postpartum estrus. Sandell (1990:23) has proposed that a late mating season and subsequent delay of implantation arose because it permitted females to "increase their fitness by choosing their mates" at a time "when the possibilities for female choice or male competition arc grcatest." Malc fishcrs arc known to increase their movements in the spring and to intrude in the home ranges of neighboring males (Leonard 1986, Arthur et al. 1989a). These increased movements have been interpreted as suggesting that males are attempting to locate females (Powell, this volume). It is therefore quite likely that each female could be visited and bred by more than one male. Asocial carnivores like *Martes*, however, are rarely either abundant or concentrated. It therefore seems unlikely that most females would have more than two or three males from which to choose. Some females may have no mate choice, as suggested by Strickland's (1980) report of an increase in barren fishers when the sex ratio was more than three females per male. Mate selection would be further reduced if neighboring males were sexually immature. Moreover, the duration of breeding and estrus are short in *Martes*, thereby decreasing the potential of any single male locating and breeding females beyond its adjacent neighbors.

It has not been possible until recently to propose a plausible physiological explanation for the shift of the breeding season from spring to summer. Such an explanation is based on the following concepts, which have yet to be verified in a variety of species. The first is that seasonal photoperiodic changes merely serve to entrain or synchronize the time at which reproductive events occur (see Herbert 1989). Lacking such environmental cues, reproduction will occur asynchronously owing to endogenous rhythms in individuals (Thorpe 1967, Karsch et al. 1989). Gonadal response in peripubertal animals to a particular photoperiod can be either stimulatory or inhibitory, depending on the photoperiod to which the mother was subjected during pregnancy (Stetson et al. 1986, 1989). Ovulation is usually suppressed while the young are vigorously nursing. And last, starvation or low-protein diets result in decreased gonadotropin secretion from the pituitary, thereby leading to temporary cessation of estrus or menstrual cycles (Fitzgerald et al. 1982, Fitzgerald 1984, Dubey et al. 1986). It is therefore possible that lactation coupled with harsh environmental conditions in the distant past may have prevented some carnivores from obtaining sufficient food to support survival of the young and ovulation, resulting in postponement of the latter until later in the year. When mating was delayed until or after the summer solstice, some females may have produced young without any arrest in embryonic development, but few of these kits would have survived winter. Other late-breeding females may have experienced insufficient secretion of pituitary hormones to promote full luteal function and to induce implantation as the duration of the photoperiod decreased. Embryonic development was arrested until the photoperiod increased in the spring. This change in the light/dark cycle triggered increased prolactin secretion, which in turn increased luteal activity and implantation of the blastocyst. Most adult females that gave birth in early spring may have been unable to breed for another 60–90 days because they were lactating. Lactation would presumably alter pulsatile gonadotropin-releasing hormone secretion. This alteration reduces pituitary secretion of luteinizing hormone and follicle stimulating hormone in several species and prevents an early return to estrus (McNeilly 1988). Depleted energy reserves and possible loss of body weight, which would likely occur at the end of lactation, might also contribute to the reduced ability of the pituitary to secrete sufficient gonadotropins to induce estrus for a few weeks after the kits were weaned. Offspring born in early spring would have a better chance of surviving and thus reproducing to carry on this pattern of reproduction because of their earlier date of birth.

All species of *Martes* give birth in March and April (Fig. 30.1, Table 30.2), fishers and stone martens somewhat earlier than the other species, and Japanese martens possibly somewhat later. Neonates of *M. americana* weigh about 28 g at birth, those of *M. martes* and *M. foina* weigh about 30 g (Brassard and Bernard 1939, Schmidt 1943), and those of fishers weigh about 40 g (Coulter 1966). The eyes of young martens open at 34–38 days (Ashbrook and Hanson 1927, Brassard and Bernard 1939, Schmidt 1943), whereas young fishers do not open their eyes until 48–55 days (Coulter 1966, LaBarge et al. 1990). Young martens begin consuming solid food at 36–45 days. Lactation ceases about six weeks after parturition, and kits begin to emerge from the den at seven to eight weeks (Schmidt 1943, Wynne and Sherburne 1984). Schmidt (1943) believed that family breakup occurred during the breeding season, at which time the young would be 12–16 weeks old. Wynne and Sherburne's 1984 observations, however, suggest that young *M. americana* may not be independent until August, or near the end of the breeding season.

Schmidt (1943) reported that litter size averaged 3.0 for 132 and 38 litters of *M. martes* and *M. foina*, respectively. He noted that litters of 5 or more must be exceptional, considering that females have only four functional mammae (Table 30.2). Strickland and Douglas (1987) summarized litter sizes for 136 pregnancies in captive *M. americana* and reported a mean of 2.9 and a range of 2.2–3.1 for five studies. The sparse data suggest little variation in litter size among species, with most *Martes* averaging slightly over 2.5 per litter. *M. melampus* may be an exception: its litter size reportedly ranges between 1 and 2 and M. Tatara (Kyushu Univ., pers. commun.) reports that the average litter size was 2 (n = 9 litters). Litter size is known to increase and then decline with increasing age in captive American mink and striped skunks (Hansson 1947, Wade-Smith and Richmond 1975). Peak fecundity occurs in the second year and reproductive senescence occurs at about 7 years in mink (Enders 1952). A similar relationship between age and fecundity may exist in captive *M. americana*, with peak fecundity at about 6 years and reproductive senescence at 12 years or more (R. Cochrane, West Virginia Univ., unpubl. data). Several investigators have reported significant increases in numbers of corpora lutea or blastocysts in adult martens and fishers compared with yearlings (Archibald and Jessup 1984, Shea et al. 1985, Douglas and Strickland 1987, Crowley et al. 1990). Strickland and Douglas (1987) cited two reports of captive American martens breeding when 15 years old. Therefore, most species of martens attain sexual maturity, peak fecundity, and reproductive senescence later than other similar-sized mustelids.

Several aspects of reproduction in *Martes* differ from that of other mustelids. A few species, such as the short-tailed weasel (*Mustela erminea*) and western spotted skunk (*Spilogale gracilis*) breed when less than 6 months old (reviewed in Mead 1989a,b), whereas the first mating may be deferred until the second or third year in martens. The unmated domesticated ferret (*Mustela putorius*) remains in constant estrus for up to five months, whereas martens and perhaps fishers exhibit one to three brief periods of sexual receptivity within a 30- to 45-day period each breeding season. All *Martes* exhibit a prolonged period of delayed implantation, during which embryonic development is arrested for 150–300 days. Some mustelids, such as the least weasel (*Mustela nivalis*) and all ferrets (*M. putorius*, *M. eversmanni*, and *M. nigripes*) exhibit direct embryonic development with no known period of embryonic diapause. The mink and striped skunk (*Mephitis mephitis*) have variable gestations with short periods of embryonic diapause lasting several days to a few weeks (reviewed in Mead 1989*a,b*). Some mustelids (e.g., most weasels) have large litters averaging six young or more, whereas litter size in *Martes* is typically two to three.

Wildlife managers commonly need to know annual recruitment to a population. Ideally, one would like to know the number of young weaned by each female, but such data are rarely available. Several reproductive parameters are often used to estimate reproductive success, some of which are less valid than others. The number of corpora lutea indicates only the number of eggs that were ovulated. Inasmuch as ovulation is known to be induced by copulation in sables, the presence of corpora lutea also indicates female sables that have bred. Under favorable ecological conditions one can expect the number of corpora lutea to show relatively good agreement with the number of resulting blastocysts or implanted embryos. Still, agreement cannot be expected to be perfect, for several reasons. In some instances one or more oocytes may fail to be fertilized. For example, one might expect reduced fertilization rates in small isolated populations where low genetic variability has impaired male fertility. In extreme cases this can lead to an increase in structurally abnormal sperm. Based on our knowledge of reproduction in other mustelids, one would also predict that some females that ovulate and form corpora lutea will not have conceived but will become pseudopregnant. The duration of pseudopregnancy in the domestic ferret and steppe polecat is equal to that of pregnancy, and the corpora lutea are retained and continue to secrete progesterone just as in pregnancy (Hammond and Marshall 1930, Heap and Hammond 1974, Mead et al. 1990). It should be noted, however, that pseudopregnancy has yet to be confirmed in any *Martes*. One should also be cautious in estimating the number of ovulated eggs based on the number of luteal structures in ovaries, because all "corpora lutea" may not have been progenitors of ovulated oocytes. Occasionally large antral follicles fail to rupture at ovulation, but still undergo luteinization, and others may undergo an unusual form of atresia and form luteal-like structures known as corpora lutea atretica. The latter are often found in eastern spotted skunks (*Spilogale putorius*) and mink (Mead 1968, Lagerkvist et al. 1992). During years of environmental stress, counting corpora lutea may substantially overestimate actual recruitment rates due to increased intrauterine embryo losses and postnatal mortality.

Blastocyst counts provide a reliable measure of fertilization rates, if one is experienced in flushing the uteri and identifying the embryos and unfertilized eggs. The latter persist in the uterus along with the blastocysts in species that exhibit delayed implantation. Carcasses supplied by trappers, however, are rarely well preserved. In such cases the blastocysts will have died and the blastomeres will likely have undergone autolysis. However, the acellular zonae pellucidae will usually be present in all but the most poorly preserved uteri, and they, along with the unfertilized oocytes, can be flushed and counted. The zonae will be collapsed and difficult to recognize without previous experience. Inasmuch as intrauterine death of unimplanted blastocysts has been estimated at 4% in fishers (Paragi 1990) and likely is about the same in martens, the number of blastocysts and unfertilized oocytes should usually equal the number of corpora lutea vera. Blastocyst counts will give a somewhat better indication of the reproductive state of the population, in that they reflect the percentage of oocytes that were fertilized. This index of fertility still lacks precision in estimating the number of young surviving to weaning.

The number of placental scars, which are formed by accumulation of hemosiderin within macrophages located at previous implantation sites, can accurately indicate the number of implanted embryos surviving through early placentation. They cannot be used to indicate the total number of embryos produced or number of young born, because placental scars are not formed if embryo resorption occurs prior to early placental development and placental scars left by resorbed embryos cannot be distinguished from those of full-term scars (Conaway 1955). Despite these limitations, counts of placental scars provide a more reliable estimate of the number of young born than do counts of blastocysts or corpora lutea. Crowley et al. (1990) reported that counts of placental scars more closely approximated the percentage of radio-collared fishers raising offspring than did counts of corpora lutea or blastocysts. Obtaining this type of data for *Martes* is not without technical difficulty. Several investigators have reported that placental scars disappear soon after birth or that the scars do not always persist until the trapping season (Strickland et al. 1982*b*, Madsen and Rasmussen 1985, Douglas and Strickland 1987). Although placental scars may no longer be readily detected by gross inspection of the untreated uterus, it seems unlikely that they have totally disappeared, since placental scars can be detected for 9–10 months after parturition in cleared uteri of western spotted skunks (Mead, unpubl. data). Wright (1966) reported detecting two generations of placental scars in a badger (*Taxidea taxus*). Thus, there is every reason to expect placental scars of *Martes* to be retained until the time of implantation.

Failure to identify placental scars in uteri of fishers and martens before this time is likely due to improper technique. Placental scars are best detected in reproductive tracts obtained from freshly killed specimens. The uterine horns should be straightened and pinned to cardboard or another material before immersion in 10% formaldehyde or other general fixative for at least 24 hours. The uterus should then be dehydrated in a graded series of alcohols to remove all water and cleared in methyl salicylate or benzyl benzoate. Although fixation may cause external discolorations on the uterine surface to diminish or disappear (Paragi 1990), the clearing process makes the reproductive tract transparent and reveals the reddish brown deposits of hemosiderin. One can also locate unimplanted blastocysts within the lumen of these cleared uteri when viewed through a dissecting microscope with a substage mirror. Failure to fix the reproductive tracts soon after death of the animal, or freezing and thawing of carcasses before fixation, can lead to autolysis of the tissues or rupture of the macrophages and subsequent dispersal of the pigment. The external surface of the uterus also becomes discolored in unfixed specimens and sometimes results in the appearance of dark areas that could easily be mistaken for placental scars. More research will be needed to determine whether uteri from thawed carcasses or those from decomposed specimens can provide meaningful information regarding placental scars in animals collected during the trapping season. Recently, however, placental scars have been detected in uteri of *M. americana* obtained from Alaskan trappers; the uteri were first cleared in methyl salicylate (Johnson and Paragi 1993).

Reproduction in Male *Martes*

Detailed descriptions of the gross and microanatomy of the male reproductive sex organs (excluding the testes) have not been reported for any species of *Martes*. Reproductive tracts of other male mustelids are relatively simple, consisting of paired testes, vasa efferentia, epididymides, vasa deferentia, a prostate, urethra, and a penis with a baculum (Mead 1970). The shape of the baculum is species-specific and its development is stimulated by testosterone in long-tailed weasels (Wright 1950). Consequently, the weight and size of the baculum of fishers and martens significantly increases when males attain sexual maturity (Wright and Coulter 1967, Stubbe 1968, Leonard 1986, Douglas and Strickland 1987, Strickland and Douglas 1987).

Seasonal changes in the testicular cycle have been described in some detail for *M. foina* and *M. martes* (Canivenc et al. 1968; Danilov and Ivanter 1968; Audy 1976*a*,*b*; Audy 1978*a*) and have been partially outlined for *M. americana* (Wright 1942), *M. zibellina* (Song et al. 1988), and *M. pennanti* (Wright and Coulter 1967). All *Martes* for which data are available have permanently scrotal testes and exhibit a distinct seasonal testicular cycle with discontinuous spermatogenesis similar to that of most other male mustelids (Mead 1989*b*). During October-December (the period of sexual inactivity), weight of the paired testes of most martens is about 0.5 g. The diameter of the seminiferous tubules is small (77 μm) and the germinal epithelium consists of Sertoli cells and spermatogonia only (Audy 1976*b*, 1978*a*). The Leydig cells are small (79 μm^2) and contain little or no smooth endoplasmic reticulum or cytoplasmic inclusions (Audy 1978*a*). Plasma testosterone levels are generally less than 1.5 ng/ml (Audy 1976*b*, 1978*b*). In stone martens, there is a transitory increase in plasma testosterone levels in January to 6.0 ng/ml. This is followed by a brief decline in testosterone to 0.3 ng/ml; however, Leydig cell surface area is increased (144 μm^2). Then, there is a steady increase in plasma testosterone which peaks in May at 7.2 ng/ml (Audy 1978*b*). Similar changes in testosterone levels have been reported for *M. martes* (Audy 1976*b*), but such information is lacking for other species of *Martes*. The rise in plasma testosterone levels may be responsible for the increased locomotor activity and aggressive behavior reported by Coulter (1966) and Jensen and Jensen (1970).

All *Martes* complete testicular recrudescence at least one month before estrus (Fig. 30.1). During the breeding season (May–July), the testes of *M. foina* and *M. martes* attain their maximal weight (about 3.2 g, paired weight). The seminiferous tubules are about 220 μm in diameter and contain spermatozoa. The Leydig cells now contain large amounts of smooth endoplasmic reticulum and numerous lipid droplets (Audy 1976*b*, 1978*a*). Testicular regression begins about the time the females enter estrus and probably is completed within 45 days. The annual duration of male fertility is not known with certainty for any *Martes*. This information is vital in that it defines the end of the breeding season regardless of the duration or recurrence of estrus in unmated females. Histological changes in the testes of fishers, which achieve breeding condition earlier in the year, have been incompletely described (Wright and Coulter 1967) and no one has adequately reported on changes in the epididymides, vas deferentia, or prostate of any *Martes*.

Timing of the annual changes in the testicular cycle of *Martes* is probably controlled by seasonal changes in photoperiod, although this has only been documented for stone martens (Audy 1976*b*) and sables (Song et al. 1988). Exposure of male stone martens with regressed testes to long-day photoperiods resulted in elevated plasma testosterone levels during November and December ($\bar{x}$ = 5.3 ± 1.7 [SE] ng/ml) and renewed spermatogenesis in January, whereas testosterone levels in the controls for the comparable time period were considerably lower ($\bar{x}$ = 1.8 ± 1.5 ng/ml).

The age at which males achieve sexual maturity may vary among individuals or populations, but some males of all species listed in Table 30.1 have been reported to achieve sexual maturity at 12–15 months. Schmidt (1943) contended that stone martens were not ready to mate until they were more than 2 years old. Markley and Bassett (1942) reported that the age of the youngest captive male American marten to sire a litter was 3 years. Most captive male sables and Eurasian pine martens do not breed until they are about 27 months old (Table 30.1). After achieving sexual maturity, most, if not all, male mammals continue to produce sperm each breeding season for the duration of their life. A male *M. americana* successfully sired litters at the age of 10. A 15-year-old male was observed to copulate and motile sperm were found in its ejaculate (R. Cochrane, West Virginia Univ., unpubl. data). Semen quality has been reported to vary with age of the male in ranch mink, being somewhat lower in yearlings, maximal during the second and third years, and then lower (Sundqvist 1987). It is not known whether similar age-related changes in semen quality occur in *Martes*.

Semen volume is quite small (<300 μl) in mustelids (Aulerich et al. 1972, Shump et al. 1976, Atherton et al. 1989, Wildt et al. 1989) because these animals lack accessory sex organs other than a prostate. There appear to be no detailed reports of electroejaculation or characterization of the semen for any species of *Martes*.

Research Needs

Although much has been learned about the reproductive biology of martens and fishers, there are many aspects that require further study. More data regarding reproduction in *M. flavigula* and *M. melampus* are clearly needed. Information on the effects of nutrition on puberty and fecundity of all *Martes* would be of great value in helping us to understand why fecundity in yearling martens is so variable. It is also important to ascertain the duration of estrus in unmated females and to determine whether ovulation is spontaneous or induced by copulation. Are multiple bouts of copulation required to induce ovulation or are more eggs ovulated during subsequent copulations? Does pseudopregnancy occur in *Martes* and, if so, what is its duration? Is the hormone profile and uterine physiology the same in pseudopregnancy and pregnancy? How common is pseudopregnancy in wild populations? How long do placental scars persist and what conditions lead to their disappearance?

More complete histological descriptions of seasonal changes in all organs of the male reproductive tract would be beneficial. Seasonal changes in semen characteristics are unknown for all *Martes*. Such information is of critical importance for the most isolated populations, in which high frequencies of abnormal sperm or low viable sperm counts could contribute to impaired reproduction and accelerated extinction. Seasonal changes in hormone profiles in both males and females of more species of *Martes* would also be helpful. Acquisition of this information would help biologists understand the forces that contribute to shifts in breeding populations and would help ensure the long-term survival and genetic diversity of these species in the event that captive breeding programs are required.

Acknowledgments

I thank R. L. Cochrane, K. Foresman, E. W. Jameson, Jr., W. B. Krohn, G. Starypan, and P. L. Wright for reading the manuscript and offering numerous helpful suggestions for revision.

Literature Cited

Abramov, V. K. 1972. Restoration of sable area and numbers in Primorski Territory. Game Biol. Moscow, Lesnaya Promyshlennost 141–161.

Abrams, P. 1983. The theory of limiting similarity. Annu. Rev. Ecol. Syst. 14:359–376.

Adams, C. E. 1981. Observations on the induction of ovulation and expulsion of uterine eggs in the mink, *Mustela vison*. J. Reprod. Fert. 63:241–248.

Adler, G. H. 1988. The role of habitat structure in organizing small mammal populations and communities. Pages 289–299 *in* R. G. Szaro, K. I. Severson, and D. R. Patton, tech. coords. Management of amphibians, reptiles, and small mammals in North America: proceedings of the symposium. USDA For. Serv. Gen. Tech. Rep. RM-166.

Ahti, T., L. Hämet-Ahti, and J. Jalas. 1968. Vegetation zones and their sections in northwestern Europe. Ann. Bot. Fenn. 5:169–211.

Aitken, R. J. 1977. Embryonic diapause. Pages 307–359 *in* M. H. Johnson, ed. Development in mammals. Vol. 1. North Holland, Amsterdam.

Alberta Forestry, Lands, and Wildlife. 1989. Guide to trapping. Alta. For. Lands and Wildl., Edmonton. 12pp.

Alberta Vocational Centre. 1987. Trapping and conservation manual, 1981. Fifth ed. Alta. Advanced Educ. and Alta. For. Lands and Wildl., Lac La Biche. 524pp.

Alcover, J. A., M. Delibes, M. Gosálbez, and J. Nadal. 1986. *Martes martes* Linnaeus, 1758 a les Balears. Misc. Zool. 10:323–333.

Alexander, R. D., J. L. Hoogland, R. D. Howard, K. M. Noonan, and P. W. Sherman. 1979. Sexual dimorphisms and breeding systems in pinnipeds, ungulates, primates, and humans. Pages 402–435 *in* N. A. Chagnon and W. Trons, eds. Evolutionary biology and human social behavior, an anthropological perspective. Duxbury Press, North Scituate, Mass.

Alldredge, J. R., and J. T. Ratti. 1986. Comparison of some statistical techniques for analysis of resource selection. J. Wildl. Manage. 50:157–165.

Allen, A. W. 1982. Habitat suitability index models: marten. U.S. Fish Wildl. Serv., Biol. Serv. Prog. FWS/OBS-82/10.11. 9pp.

423

Allen, A. W. 1983. Habitat suitability index models: fisher. U.S. Fish Wildl. Serv. FWS/OBS-82/10.45. 19pp.

Allen, A. W. 1984. Habitat suitability index models: marten. Revised. U.S. Fish Wildl. Serv. FWS/OBS-82/10.11. 13pp.

Allen, G. 1942. Extinct and vanishing mammals of the Western Hemisphere. Am. Comm. Int. Wildl. Prot., Washington, D.C. 620pp.

Amores, F. 1980. Feeding habits of the stone martens, *Martes foina* (Erxleben, 1777), in south western Spain. Säugetierk. Mitt. 28:316–322.

Anderson, E. 1934. The distribution, abundance and economic importance of the game and fur-bearing mammals of western North America. Proc. Fifth Pac. Sci. Congr., Victoria (1933) 5:4055–4073.

Anderson, E. 1968. Fauna of the Little Box Elder Cave, Converse County, Wyoming: the Carnivora. Univ. of Colo. Stud. Ser., Earth Sci. 6. 71pp.

Anderson, E. 1970. Quaternary evolution of the genus *Martes* (Carnivora, Mustelidae). Acta Zool. Fenn. 130:1–133.

Anderson, E. 1984. Review of the small carnivores of North America during the last 3.5 million years. Pages 257–266 *in* H. H. Genoways and M. R. Dawson, eds. Contributions in Quaternary vertebrate paleontology: a volume in memorial to John E. Guilday. Carnegie Mus. Nat. Hist. Spec. Publ. 8.

Anderson, S. B. 1987. Wild furbearer management in eastern Canada. Pages 1039–1048 *in* M. Novak, J. A. Baker, M. E. Obbard, and B. Malloch, eds. Wild furbearer management and conservation in North America. Ont. Trappers Assoc., North Bay, Ont.

Andrén, H., and P.-A. Lemnell. 1992. Population fluctuations and habitat selection in the Eurasian red squirrel *Sciurus vulgaris*. Ecography 15:303–307.

Anonymous. 1981. Reader's Digest wide world atlas. Reader's Digest Assoc., Pleasantville, N.Y. 240pp.

Anonymous. 1989. Wildlife management plan—Ministik Lake Game Bird Sanctuary. Alta. Fish, Lands and Wildl., and Ducks Unltd., Edmonton. 33pp.

Ansorge, H. 1989*a*. Die Ernährungsökologie des Steinmarders (*Martes foina*) in den Landschaftstypen der Oberlausitz. Pages 473–493 *in* M. Stubbe, ed. Populationsökologie marderartiger Säugetiere. Wiss. Beitr. Univ. of Halle.

Ansorge, H. 1989*b*. Feeding ecological aspects of the pine marten, polecat, and stoat. Pages 494–504 *in* M. Stubbe, ed. Populationsökologie marderartiger Säugetiere. Wiss. Beitr. Univ. of Halle.

Archibald, W. R., and R. H. Jessep. 1984. Population dynamics of the pine marten (*Martes americana*) in the Yukon Territory. Pages 81–97 *in* R. Olson, R. Hastings, and F. Geddes, eds. Northern ecology and resource management. Univ. of Alberta Press, Edmonton.

Arno, S. F., and T. D. Petersen. 1983. Variation in estimates of fire intervals: a closer look at fire history on the Bitterroot National Forest. USDA For. Serv. Res. Pap. INT-301. 8pp.

Arthur, S. M. 1988. An evaluation of techniques for capturing and radio-collaring fishers. Wildl. Soc. Bull. 16:417–421.

Arthur, S. M., R. A. Cross, T. F. Paragi, and W. B. Krohn. 1992. Precision and utility of cementum annuli for estimating ages of fishers. Wildl. Soc. Bull. 20:402–405.

Arthur, S. M., and W. B. Krohn. 1991. Activity patterns, movements, and reproductive ecology of fishers in southcentral Maine. J. Mammal. 72:379–385.

Arthur, S. M., W. B. Krohn, and J. A. Gilbert. 1989*a*. Home range characteristics of adult fishers. J. Wildl. Manage. 53:674–679.

Arthur, S. M., W. B. Krohn, and J. A. Gilbert. 1989*b*. Habitat use and diet of fishers. J. Wildl. Manage. 53:680–688.

Arthur, S. M., T. F. Paragi, and W. B. Krohn. 1993. Dispersal of juvenile fishers in Maine. J. Wildl. Manage. 57:868–874.

Ashbrook, F. G., and K. B. Hanson. 1927. Breeding martens in captivity. J. Hered. 18:499–503.

Ashbrook, F. G., and K. B. Hanson. 1930. The normal breeding season and gestation period of martens. USDA Circ. 107. 6pp.

Ashmole, N. P. 1968. Body size, prey size, and ecological segregation in five sympatric tropical terns (Aves: Laridae). Syst. Zool. 17:292–304.

Aspisov, D. I. 1959. Dinamika populyatsii leskoy kunitsi v volzhsko-kamskom kraye i nekotorie pokazatyeli dlya prognosya izmyenyenii yeye chislyennosti [The population dynamics of the forest marten in the Volga-Kama area with a prognosis of its abundance]. Trudy Vsesoiusnyi Nauchno—issledovatel'skii institut Zhivotn, Syra'ya i Pushniniy 18:29–45 (in Russian).

Atherton, R. W., M. Straley, P. Curry, R. Slaughter, W. Burgess, and R. M. Kitchin. 1989. Electroejaculation and cryopreservation of domestic ferret sperm. Pages 177–187 *in* U. S. Seal, E. T. Thorne, M. A. Bogan, and S. H. Anderson, eds. Conservation biology and the black-footed ferret. Yale Univ. Press, New Haven, Conn.

Aubert, I., and R. Canivenc. 1986. Nidation différée chez les mustélidés: étude ultrastructurale utéro-blastocytaire chez le blaireau européen, la martre et la fouine. Arch. Biol. (Brussels) 97:157–186.

Aubry, K. B., and D. B. Houston. 1993. Distribution and status of the fisher (*Martes pennanti*) in Washington. Northwest. Nat. 73:69–79.

Audy, M. C. 1976*a*. Influence du photopériodisme sur la physiologie testiculaire de la fouine (*Martes foina* Erx.). C.R. Acad. Sci. Paris 283:805–808.

Audy, M. C. 1976*b*. Le cycle sexuel saisonnier du mâle des mustélidés européens. Gen. Comp. Endocrinol. 30:117–127.

Audy, M. C. 1978*a*. Etude ultrastructurale des cellulles de Leydig et de Sertoli au cours du cycle sexuel saisonnier de la fouine (*Martes foina* Erx.). Gen. Comp. Endocrinol. 36:462–476.

Audy, M. C. 1978*b*. Variations saisonnières de la testostérone plasmatique chez la fouine (*Martes foina* Erx.). C.R. Acad. Sci. Paris, Ser. D, 287:721–724.

Aulerich, R. J., R. K. Ringer, and C. S. Sloan. 1972. Electro-ejaculation of mink (*Mustela vison*). J. Anim. Sci. 34:230–233.

Azuma, D. L., J. A. Baldwin, and B. R. Noon. 1990. Estimating the occupancy of spotted owl habitat areas by sampling and adjusting for bias. USDA For. Serv. Gen. Tech. Rep. PSW-124. 9pp.

Baevsky, U. B. 1963. The effect of embryonic diapause on the nuclei and mitotic activity of mink and rat blastocysts. Pages 141–153 *in* A. C. Enders, ed. Delayed implantation. Univ. of Chicago Press, Chicago.

Baevsky, Y. B. 1971. Levels of subcellular differentiation during diapause and activation of the sable embryo *Martes zibellina*. Doklady Akad. Nauk. USSR 197:1458–1460.

Baker, J. M. 1992. Habitat use and spatial organization of pine marten on southern Vancouver Island, British Columbia. M.S. thesis, Simon Fraser Univ., Burnaby, B.C. 119pp.

Baker, R. G. 1983. Holocene vegetational history of the western United States. Pages 109–126 *in* H. E. Wright, Jr., ed. Late-Quaternary environments of the United States. Vol. 2: The Holocene. Univ. of Minnesota Press, Minneapolis.

Balharry, D. 1991. Group stability and intrasexual territoriality in pine martens (*Martes martes*). Page 8 *in* Abstracts of Papers, The biology and management of martens and fishers, Laramie, Wyo.

Balser, D. S. 1960. The comeback of the furbearers. Minn. Volunteer 23(134):57–59.

Balser, D. S., and W. H. Longley. 1966. Increase of the fisher in Minnesota. J. Mammal. 47:547–550.

Banci, V. 1989. A fisher management strategy for British Columbia. Wildl. Bull. B-63, B.C. Minist. Environ., Wildl. Branch, Victoria, B.C. 117pp.

Barkalow, F. S. 1961. The porcupine and fisher in Alabama archaeological sites. J. Mammal. 42:544–545.

Barnes, B. 1988. Converted longspring foothold. Can. Trapper 16(4):9.

Barnes, D. S., and N. Mrosovsky. 1974. Body weight regulation in ground squirrels and hypothalamically lesioned rats: slow and sudden set point changes. Physiol. Behav. 12:251–258.

Barnosky, A. D., and D. L. Rasmussen. 1988. Middle Pleistocene arvicoline rodents and environmental change at 2900 meters elevation, Porcupine Cave, South Park, Colorado. Ann. Carnegie Mus. 57(12):267–292.

Barnosky, C. W., P. M. Anderson, and P. J. Bartlein. 1987. The northwestern U.S. during deglaciation; vegetational history and paleoclimatic implications. Pages 289–321 *in* W. F. Ruddiman and H. E. Wright, Jr., eds. North America and adjacent oceans during the last deglaciation. Geol. Soc. Am., Boulder, Colo.

Barrett, M. W., G. Proulx, D. Hobson, D. Nelson, and J. W. Nolan. 1989. Field evaluation of the C120 Magnum trap for marten. Wildl. Soc. Bull. 17:299–306.

Barrett, M. W., G. Proulx, and N. Jotham. 1988. Wild fur industry under challenge: the Canadian response. Trans. North Am. Wildl. Nat. Resour. Conf. 53:180–190.

Barrett, R. H. 1983. Smoked aluminum track plots for determining furbearer distribution and relative abundance. Calif. Fish Game 69:188–190.

Bartholomew, G. 1977. Body temperature and energy metabolism. Pages 364–449 *in* M. Gordon, ed. Animal physiology: principles and adaptation. Third ed. Macmillan, New York.

Bartness, T. J., and G. N. Wade. 1984. Photoperiodic control of body weight and energy metabolism in Syrian hamster *Mesocricetus auratus*: role of pineal gland, melatonin, gonads and diet. Endocrinology 114:492–498.

Bashanov, V. S. 1943. A new subspecies of sable in Altai. Collections of Khazakst. Branch USSR Acad. Sci., p. 53 (in Russian).

Baskin, L. M. 1990. Conflicts between wildlife and human activity in the forested zone of the USSR. Pages 401–405 *in* S. Myrberget, ed. Trans. Nineteenth Int. Congr. Game Biol., Trondheim, Norway.

Bateman, M. C. 1980. A review of some aspects of the northern American marten. Prepared for Parks Can. by Can. Wildl. Serv., Atlantic Reg. 42pp.

Bateman, M. C. 1982a. Final report on the marten program feasibility phase. Prepared for Parks Can. by Can. Wildl. Serv., Atlantic Reg. 49pp.

Bateman, M. C. 1982b. Progress report, marten re-introduction to Terra Nova National Park. Prepared for Parks Can. by Can. Wildl. Serv., Atlantic Reg. 12pp.

Bateman, M. C. 1984. Progress report, marten re-introduction to Terra Nova National Park. Prepared for Parks Can. by Can. Wildl. Serv., Atlantic Reg. 17pp.

Bateman, M. C. 1986. Winter habitat use, food habits and home range size of the marten, *Martes americana*, in western Newfoundland. Can. Field-Nat. 100:58–62.

Baud, F. J. 1981. Contribution à la connaissance du régime alimentaire hivernal du genre *Martes* en Haute-Savoie. Le Bièvre 3:79–84.

Baudvin, H., J. Dessolin, and C. Riols. 1985. L'utilisation par la martre (*Martes martes*) des nichoirs à choutes dans quelques forêts bourguignonnes. Ciconia 9:61–104.

Beaty, D. W. 1978. Antenna considerations for biomedical telemetry. Telonics, Mesa, Ariz. 17pp.

Bekker, E. F. 1991. A terrestrial furbearer estimator based on probability sampling. J. Wildl. Manage. 55:730–737.

Bekoff, M., and M. C. Wells. 1981. Behavioural budgeting by wild coyotes: the influence of food resources and social organization. Anim. Behav. 29:794–801.

Belyaev, D. K., N. S. Pereldik, and N. T. Portnova. 1951. Experimental reduction of the period of embryonal development in sables (*Martes zibellina* L.). J. Gen. Biol. 12:260–265.

Berg, W. E. 1982. Reintroduction of fisher, pine marten, and river otter. Pages 159–173 *in* G. C. Sanderson, ed. Midwest furbearer management. Proc. Symp. Forty-third Midwest Fish Wildl. Conf., Wichita, Kans.

Bergerud, A. T. 1969. The status of the pine marten in Newfoundland. Can. Field-Nat. 83:128–131.

Bernatskii, V. G., E. G. Snytko, and H. G. Nosova. 1976. Natural and induced ovulation in the sable (*Martes zibellina* L.). Tran. from Dokl. Akad. Nauk. USSR 230:1238–1239, by Consultants Bureau, New York.

Bintz, G. L., and W. W. Mackin. 1980. The effect of water availability on tissue catabolism during starvation in Richardson's ground squirrels. Comp. Biochem. Physiol. 65A: 181–186.

Bintz, G. L., D. L. Palmer, W. W. Mackin, and F. Y. Blanton. 1979. Selective tissue catabolism and water balance during starvation in Richardson's ground squirrels. Comp. Biochem. Physiol. 64A:399–403.

Bissonette, J. A., R. J. Frederickson, and B. J. Tucker. 1988. The effects of forest harvesting on marten and small mammals in western Newfoundland. Report prepared for Nfld. and Laborador Wildl. Div. and Corner Brook Pulp and Paper, Ltd. Utah State Univ., Logan. 109pp.

Bissonette, J. A., R. J. Frederickson, and B. J. Tucker. 1989. Pine marten: a case for landscape level management. Trans. North Am. Wildl. Nat. Resour. Conf. 54:89–101.

Bissonnette, T. H. 1932. Modification of mammalian sexual cycles; reactions of ferrets (*Putorius vulgaris*) of both sexes to electric light added after dark in November and December. Proc. Roy. Soc. Lond., Ser. B, 110:322–336.

Bittner, S. L., and O. J. Rongstad. 1982. Snowshoe hare and allies. Pages 146–163 *in* J. A. Chapman and G. A. Feldhammer, eds. Wild mammals of North America: biology, management, and economics. Johns Hopkins Univ. Press, Baltimore.

Bjärvall, A., E. Nilsson, and L. Norling. 1977. Urskogens betydelse för tjäder och mård. Fauna och Flora 72(1):31–38.

Blackwell, P. 1992. Notes from the trapline. New B.C. Trapper 2(4):14.

Blem, C. R. 1975. Geographic variation in wing-loading of the house sparrow. Wilson Bull. 87:543–549.

Blood, D. A. 1989. Marten. Management guidelines for British Columbia. Wildl. Branch, Minist. Environ. Rep., Victoria, B.C. 6pp.

Blueweiss, L., H. Fox, V. Kudzman, D. Nakashima, R. Peters, and S. Sams. 1978. Relationship between body size and some life history parameters. Oecologia 37: 257–272.

Boggess, E. K., S. B. Linhart, G. R. Batcheller, D. W. Erickson, R. G. Linscombe, A. W. Todd, J. W. Greer, D. C. Juve, M. Novak, and D. A. Wade. 1990. Traps, trapping, and furbearer management. Wildl. Soc. Tech. Rev. 90-1. 31pp.

Bonnin, M., R. Canivenc, and J. Aitken. 1977. Variations saisonnières du taux de la progestérone plasmatique chez la fouine, *Martes foina*, espèce à ovo-implantation différée. C.R. Acad. Sci. Paris 285:1479–1481.

Bookout, T. A. 1965. The snowshoe hare in upper Michigan: its biology and feeding coactions with white-tailed deer. Mich. Dep. Conserv. Res. Dev. Rep. 438. 191pp.

Boss, J., G. Devean, and C. Drysdale. 1987. Kejimkujik National Park—American marten re-introduction program interim report, Feb.–Oct., 1987. Environ. Can., Parks Can., Nat. Resour. Conserv., Kejimkujik Natl. Park. 43pp.

Bourque, B. J. 1976. The Turner Farm site: a preliminary report. Man in the Northeast 11:21–30.

Brainerd, S. M. 1990. The pine marten and forest fragmentation: a review and synthesis. Pages 421–434 *in* S. Myrberget, ed. Trans. Nineteenth Int. Congr. Game Biol., Trondheim, Norway.

Brainerd, S. M., J.-O. Helldin, and E. Lindström. In Prep. Habitat use by Eurasian pine marten (*Martes martes* L.) in the industrial boreal forest.

Braña, F., and J. C. del Campo. 1982. Sobre la alimentación de la marta, *Martes martes* L., en Asturias. Bol. Cienc. Natur. I.D.E.A. 29:131–137.

Brander, R. B. 1973. Life history notes on the porcupine in a hardwood-hemlock forest in upper Michigan. Mich. Acad. 5:425–433.

Brander, R. B., and D. J. Books. 1973. Return of the fisher. Nat. Hist. 82:52–57.

Brassard, J. S., and R. Bernard. 1939. Observations on breeding and development of marten, *Martes a. americana* (Kerr). Can. Field-Nat. 53:15–21.

Broekhuizen, S. 1983. Habitat use of beech marten (*Martes foina*) in relation to landscape elements in a Dutch agricultural area. Pages 614–624 *in* Trans. Sixteenth Int. Congr. Game Biol., High Tatras, Czechoslovakia.

Broekhuizen, S., M. P. A. Lucas, and G. J. D. M. Mueskens. 1989. Behaviour of a young beech marten female (*Martes foina*) during dispersion. Pages 422–432 *in* M. Stubbe, ed. Proc. Mitteleuropäisches Symp. Populationsökologie von Mustelidenarten. Martin-Luther-Universität, Halle-Wittenberg.

Brown, B. 1908. The Conard Fissure, a Pleistocene deposit in northern Arkansas with descriptions of two new genera and twenty new species of mammals. Mem. Am. Mus. Nat. Hist. 9:155–208.

Brown, J. H., and R. C. Lasiewski. 1972. Metabolism of weasels: the cost of being long and thin. Ecology 53:939–943.

Brown, J. H., and B. A. Maurer. 1989. Macroecology: the division of food and space among species on continents. Science 243:1145–1150.

Brown, J. L. 1969. Territorial behavior and population regulation in birds: a review and re-evaluation. Wilson Bull. 81:293–329.

Brown, L. W. 1965. A fisher (*Martes pennanti*) in Sheridan County, Wyoming. Southwest. Nat. 10:143.

Brown, M. K. 1975. The Zimmerman site: further excavations at the Grand Village of Kaskaskia. Ill. State Mus. Rep. Invest. 32:1–67.

Brown, M. K. 1980. The status of the pine marten in New York. Trans. Northeast Sec. Wildl. Soc. 37:217–226.

Brown, M. K., and G. Will. 1979. Food habits of the fisher in northern New York. N.Y. Fish Game J. 26:87–92.

Brown, M. W. 1983. A morphometric analysis of sexual and age variation in the American marten (*Martes americana*). M.S. thesis, Univ. of Toronto, Toronto, Ont. 190pp.

Bryson, R. A., and F. K. Hare. 1974. Climates of North America. Pages 1–47 *in* Climates of North America. Elsevier, New York. 420pp.

Buck, S. 1982. Habitat utilization by fisher (*Martes pennanti*) near Big Bar, California. M.S. thesis, Humboldt State Univ., Arcata, Calif. 85pp.

Buck, S., C. Mullis, and A. Mossman. 1983. Final report: Corral Bottom–Hayfork Bally fisher study. USDA For. Serv. 136pp.

Buker, W. E. 1970. The Drew site (36-Al-62). Pa. Archaeol. 40:21–65.

Bull, E. L., R. S. Holthausen, and L. R. Bright. 1992. Comparison of 3 techniques to monitor marten. Wildl. Soc. Bull. 20:406–410.

Bulmer, M. G. 1974. A statistical analysis of the ten-year cycle in Canada. J. Anim. Ecol. 43:701–718.

Bulmer, M. G. 1975. Phase relations in the ten-year cycle. J. Anim. Ecol. 44:609–622.

Burnett, G. W. 1981. Movements and habitat use of the American marten in Glacier National Park, Montana. M.S. thesis, Univ. of Montana, Missoula. 130pp.

Burnham, K. P., D. R. Anderson, and J. L. Laake. 1981. Line transect estimation of bird population density using a Fourier series. Pages 466–482 *in* C. J. Ralph and J. M. Scott, eds. Estimating numbers of terrestrial birds. Cooper Ornithol. Soc. Studies in Avian Biol. 6.

Burns, J. A. 1984. Late Quaternary palaeoecology and zoogeography of southwestern Alberta: vertebrate and palynological evidence from two Rocky Mountain caves. Ph.D. dissertation, Univ. of Toronto, Toronto, Ont. 306pp.

Burris, O. E., and D. E. McKnight. 1973. Game transplants in Alaska. Alas. Dep. Fish Game., Game Tech. Bull. 4. 57pp.

Buskirk, S. W. 1983. The ecology of marten in southcentral Alaska. Ph.D. dissertation, Univ. of Alaska, Fairbanks. 131pp.

Buskirk, S. W. 1984. Seasonal use of resting sites by marten in southcentral Alaska. J. Wildl. Manage. 48:950–953.

Buskirk, S. W. 1992. Conserving circumboreal forests for martens and fishers. Conserv. Biol. 6:318–320.

Buskirk, S. W., S. C. Forrest, M. G. Raphael, and H. J. Harlow. 1989. Winter resting site ecology of marten in the central Rocky Mountains. J. Wildl. Manage. 53:191–196.

Buskirk, S. W., and H. J. Harlow. 1989. Body-fat dynamics of the American marten *Martes americana* in winter. J. Mammal. 70:191–193.

Buskirk, S. W., H. J. Harlow, and S. C. Forrest. 1988. Temperature regulation in American marten (*Martes americana*) in winter. Natl. Geogr. Res. 4:208–218.

Buskirk, S. W., and S. L. Lindstedt. 1989. Sex biases in trapped samples of Mustelidae. J. Mammal. 70:88–97.

Buskirk, S. W., and L. L. McDonald. 1989. Analysis of variability in home-range size of the American marten. J. Wildl. Manage. 53:997–1004.

Buskirk, S. W., and S. O. MacDonald. 1984. Seasonal food habits of marten in south-central Alaska. Can. J. Zool. 62:944–950.

Buskirk, S. W., P. F. A. Maderson, and R. M. O'Conner. 1986. Plantar glands in North American Mustelidae. Pages 617–622 *in* D. Duvall, D. Müller-Schwarze, and R. M. Silverstein, eds. Chemical signals in vertebrates. Vol. 4. Plenum Press, New York.

Butler, P. M. 1946. The evolution of carnassial dentitions in the Mammalia. Proc. Zool. Soc. Lond. 116:198–220.

Cahalane, V. H. 1961. Mammals of North America. Macmillan, New York. 682pp.

Cameron, A. W. 1958. Mammals of the islands in the Gulf of St. Lawrence. Natl. Mus. Can. Bull. 154. 165pp.

Campbell, T. M. 1979. Short-term effects of timber harvests on pine marten ecology. M.S. thesis, Colorado State Univ., Fort Collins. 71pp.

Canadian General Standards Board. 1984. Animal traps, humane mechanically-powered, trigger activated. CGSB Rep. CAN 2-144.1-M84, Ottawa, Ont. 9pp.

Canadian Trappers Federation. 1984. Canadian trappers manual. Can. Trappers Fed., North Bay, Ont. 227pp.

Canivenc, R. 1970. Contrôle de la biologie lutéale chez les espèces a ovo-implantation différée. Colloque Int. Centre Nat. Recherche Sci. 927:223–233.

Canivenc, R., and M. Bonnin. 1975. Les facteurs écophysiologiques de régulation de la fonction lutéale chez les mammiferes à ovo-implantation différée. J. Physiol. (Paris) 70:533–538.

Canivenc, R., M. Bonnin-Laffargue, and M. Lajus-Boue. 1969. Induction de nouvelles générations lutéales pendant la progestation chez la martre européenne (*Martes martes* L.). C.R. Acad. Sci. Paris, Ser. D, 269:1437–1440.

Canivenc, R., M. Bonnin-Laffargue, and M. C. Relexans. 1968. Cycles génitaux de quelques mustélidés européens. Entretiens de Chizé, Ser. Physiol. 1:85–110.

Canivenc, R., C. Mauget, M. Bonnin, and J. Aitken. 1981. Delayed implantation in the beech marten (*Martes foina*). J. Zool. (Lond.) 193:325–338.

Caro, T. M., C. D. Fitzgibbon, and M. E. Holt. 1989. Physiological costs of behavioural strategies for male cheetahs. Anim. Behav. 39:309–317.

Carpenter, F. L., and R. E. MacMillen. 1976. Threshold model of feeding territoriality and test with a Hawaiian honeycreeper. Science 194:634–642.

Carroll, R. S., M. S. Erskine, P. C. Doherty, L. A. Lundell, and M. J. Baum. 1985. Coital stimuli controlling luteinizing hormone secretion and ovulation in the female ferret. Biol. Reprod. 32:925–933.

Case, T. J. 1978. A general explanation for insular body size trends in terrestrial verte-brates. Ecology 59:1–18.

Casey, T. M., and K. K. Casey. 1979. Thermoregulation of arctic weasels. Physiol. Zool. 52:153–164.

Catlin, G. 1973. Letters and notes on the manners, customs, and conditions of North American Indians. Vol. 1. Dover, New York. 264pp.

Caughley, G. 1977. Analysis of vertebrate populations. John Wiley and Sons, New York. 234pp.

Cederlund, G. 1981. Daily and seasonal activity pattern of roe deer in a boreal habitat. Swed. Wildl. Res. 11:315–353.

Chappel, M. A. 1980. Thermal energetics and thermoregulatory costs of small arctic mammals. J. Mammal. 61:278–291.

Charnov, E. L. 1976. Optimal foraging: attack strategy of a mantid. Am. Nat. 110: 141–151.

Charnov, E. L., G. H. Orians, and K. Hyatt. 1976. Ecological implications of resource depression. Am. Nat. 110:247–259.

Chernikin, Y. M. 1980. Barguzin sable marking. Bull. MOIP Biol. Dep. 85(5):10–24.

Cheylan, G., and P. Bayle. 1988. Le régime alimentaire de quatre espèces de mustélidés en Provence: la fouine *Martes foina*, le blaireau *Meles meles*, la belette *Mustela nivalis* et le putois *Putorius putorius*. Faune de Provence 9:14–26.

Chirkova, A. F. 1967. The relationship between arctic fox and red fox in the far north. Problemy Severa 11:111–113.

Chomko, S. A., and B. M. Gilbert. 1987. The late Pleistocene/Holocene faunal record in the northern Bighorn Mountains, Wyoming. Pages 394–409 *in* R. W. Graham, H. A. Semken, Jr., and M. A. Graham, eds. Late Quaternary mammalian biogeography and environments of the Great Plains and prairies. Ill. State Mus. Sci. Pap. 22, Springfield.

Chotolchu, N., M. Stubbe, and N. Dawaa. 1980. Der Steinmarder *Martes foina* (Erxleben, 1777) in der Mongolei. Acta Theriol. 25:105–114.

Christensen, H. 1985. [Habitat choice, behavior, and diet of the fox (*Vulpes vulpes* L.) during winter in a coniferous forest in southeastern Norway]. M.S. thesis, Univ. of Oslo, Norway. 79pp. (in Norwegian).

Churcher, C. S., P. W. Parmalee, G. L. Bell, and J. P. Lamb. 1989. Caribou from the late Pleistocene of northwestern Alabama. Can. J. Zool. 67:1210–1216.

Churchill, S. J., L. A. Herman, M. F. Herman, and J. P. Ludwig. 1981. Final report on the completion of the Michigan marten reintroduction program. Ecol. Res. Serv., Inc., Iron River, Mich. 132pp.

Clark, A. 1986. Fisher assessment—1985. Pages 406–448 *in* Planning for Maine's inland fish and wildlife, species assessments and strategic plans, fur and game mammals, 1986–1991. Vol. 1: Wildlife, part 1.4. Maine Dep. Inland Fish. Wildl., Augusta. 792pp.

Clark, J. S. 1990. Fire and climate change during the last 750 years in northwestern Minnesota. Ecol. Monogr. 60:135–159.

Clark, T. 1989. Conservation biology of the black-footed ferret. Wildl. Preserv. Trust, Spec. Sci. Rep. 3. 175pp.

Clark, T. W., E. Anderson, C. Douglas, and M. Strickland. 1987. *Martes americana*. Mammalian Species 289:1–8.

Clark, T. W., and T. M. Campbell. 1976. Population organization and regulatory mechanisms of pine martens in Grand Teton National Park, Wyoming. Pages 293–295 *in* R. M. Linn, ed. Conference on scientific research in national parks. Vol. 1. Natl. Park Serv. Trans. Proc. Ser. 5, Washington, D.C.

Clark, T. W., T. M. Campbell III, and T. M. Hauptman. 1989. Demographic characteristics of American marten populations in Jackson Hole, Wyoming. Great Basin Nat. 49:587–596.

Clark, T. W., and M. R. Stromberg. 1987. Mammals in Wyoming. Univ. of Kansas Mus. Nat. Hist., Lawrence. 314pp.

Cleland, C. E. 1966. The prehistoric animal ecology and ethnozoology of the upper Great Lakes region. Univ. of Michigan Mus. Anthropol., Anthropol. Pap. 29. 294pp.

Clem, M. K. 1977a. Food habits, weight changes and habitat selection of fisher during winter. M.S. thesis, Univ. of Guelph, Ont.

Clem, M. K. 1977b. Interspecific relationship of fishers and martens in Ontario during winter. Pages 165–182 *in* R. L. Phillips and C. J. Jonkel, eds. Proc. 1975 Predator Symp., Mont. For. Conserv. Exp. Stn., Univ. of Montana, Missoula.

Clevenger, A. P. 1993a. Spring and summer food habits and habitat use of the European pine marten (*Martes martes*) on the island of Minorca, Spain. J. Zool. (Lond.). 229:153–161.

Clevenger, A. P. 1993b. Pine marten (*Martes martes* Linne, 1758) comparative feeding ecology in an island and mainland population of Spain. Z. Säugetierk. 58:212–224.

Cody, M. L., and J. M. Diamond. 1975. Ecology and evolution of communities. Belknap Press, Cambridge, Mass.

Cohn, C. D. 1963. Feeding frequency and body composition. Ann. N.Y. Acad. Sci. 110:395–409.

Cohn, C., D. Joseph, L. Bell, and M. D. Allweiss. 1965. Studies on the effects of feeding frequency and dietary composition on fat deposition. Ann. N.Y. Acad. Sci. 131: 507–518.

Colbert, E. H. 1935. Siwalik mammals in the American Museum of Natural History. Trans. Am. Philos. Soc. 26:1–401.

Colbert, E. H., and D. A. Hooijer. 1953. Pleistocene mammals from the limestone fissures of Szechwan, China. Bull. Am. Mus. Nat. Hist. 102:1–134.

Cole, C. A., and R. L. Smith. 1983. Habitat suitability indices for monitoring wildlife populations—an evaluation. Trans. North Am. Wildl. Nat. Resour. Conf. 48:367–375.

Collier, D., A. E. Hudson, and A. Ford. 1942. Archeology of the upper Columbian region. Univ. of Wash. Publ. Anthropol. 9:1–178.

Conaway, C. H. 1955. Embryo resorption and placental scar formation in the rat. J. Mammal. 36:516–532.

Connell, J. H. 1980. Diversity and the coevolution of competitors, or the ghosts of competition past. Oikos 35:131–138.

Conover, W. J. 1971. Practical nonparametric statistics. John Wiley and Sons, New York. 462pp.

Cook, A. H. 1949. Furbearer investigations. N.Y. Conserv. Dep. Pittman-Robertson Proj. 1-4, Suppl. G, Final Rep. 57pp.

Cook, S. R., and G. Proulx. 1989. Mechanical evaluation and performance improvement of the rotating jaw Conibear 120 trap. ASTM J. Test. Eval. 17:190–195.

Cooper, S., K. E. Nieman, R. Steele, and D. W. Roberts. 1987. Forest habitat types of northern Idaho: a second approximation. USDA For. Serv. Gen. Tech. Rep. INT-236. 135pp.

Cope, E. D. 1899. Vertebrate remains from Port Kennedy bone deposit. J. Acad. Nat. Sci. Philadelphia 11:193–267.

Corbet, G. B. 1978. The mammals of the Palaearctic region, a taxonomic review. Br. Mus. (Nat. Hist.), London. 314pp.

Corbet, G. B. 1980. The mammals of the Palearctic region. Cornell Univ. Press, Ithaca, N.Y. 314pp.

Corbet, G. B., and J. E. Hill. 1986. A world list of mammalian species. Facts on File, London. 254pp.

Corbett, L. K., and A. E. Newsome. 1987. The feeding ecology of the dingo. III. Dietary relationship of widely fluctuating prey populations in Australia: a hypothesis of alternation of predation. Oecologia 74:215–217.

Corn, J. G., and M. G. Raphael. 1992. Habitat characteristics at marten subnivean access sites. J. Wildl. Manage. 56:442–448.

Cornish, E. R., and N. Mrosovsky. 1965. Activity during food deprivation and starvation of six species of rodent. Anim. Behav. 13:242–248.

Cottrell, W. 1978. The fisher (*Martes pennanti*) in Maryland. J. Mammal. 59:886.

Coulter, M. W. 1960. The status and distribution of fisher in Maine. J. Mammal. 41:1–9.

Coulter, M. W. 1966. Ecology and management of fisher in Maine. Ph.D. dissertation, State Univ. of New York, Syracuse. 183pp.

Cowan, I. McT., and C. J. Guiguet. 1965. The mammals of British Columbia. Third printing, rev. B.C. Prov. Mus. Handb. 11. 414pp.

Cowan, I. McT., and R. H. MacKay. 1950. Food habits of the marten (*Martes americana*) in the Rocky Mountain region of Canada. Can. Field-Nat. 64:100–104.

Crabb, W. D. 1944. Growth, development and seasonal weights of spotted skunks. J. Mammal. 25:213–221.

Criss, S. L., and S. J. Kerns. 1990. Survey of furbearer presence on managed timberlands of interior northern California. Report prepared for Sierra Pacific Indust., Redding, Calif., by Wildland Resource Managers, Round Mountain, Calif. 37pp.

Crowley, S. K., W. B. Krohn, and T. F. Paragi. 1990. Comparison of fisher reproductive estimates. Trans. Northeast Sect. Wildl. Soc. 47:36–42.

Curio, E. 1976. The ethology of predation. Springer-Verlag, New York. 250pp.

Cushing, E. J. 1965. Problems in the Quaternary phytogeography of the Great Lakes region. Pages 403–416 *in* H. E. Wright, Jr., and D. G. Frey, eds. The Quaternary of the United States. Princeton Univ. Press, Princeton, N.J.

Daan, S. 1981. Adaptive daily strategies in behavior. Pages 274–298 *in* J. Aschoff, ed. Handbook of behavioral neurobiology. Vol. 4: Biological rhythms. Plenum Press, New York.

Dagg, A. I., D. Leach, and G. Sumner-Smith. 1975. Fusion of the distal femoral epiphysis in male and female marten and fisher. Can. J. Zool. 53:1514–1518.

Dana, S. T., J. H. Allison, and R. N. Cunningham. 1960. Minnesota lands: ownership, use, and management of forest and related lands. Am. For. Assoc., Washington, D.C. 463pp.

Danilov, P. I., and E. V. Ivanter. 1968. The pine marten in Karelia. Uch zapiski Petrozavodsk ut-ta 15:179–197.

Danilov, P. I., and I. L. Tumanov. 1980*a*. Male reproductive cycles in the Mustelidae. N.Z. Dep. Sci. Ind. Res. Bull. 227:70–80.

Danilov, P. I., and I. L. Tumanov. 1980*b*. Female reproductive cycles in the Mustelidae. N.Z. Dep. Sci. Ind. Res. Bull. 227:81–92.

Dapson, R. W. 1980. Guidelines for statistical usage in age-estimation techniques. J. Wildl. Manage. 44:541–548.

Darwin, C. 1871. The descent of man, and selection in relation to sex. John Murray, London.

Davie, J. W. 1984. Fisher reintroduction. Project Completion Report, Alta. For. Lands and Wildl., Edmonton. 10pp.

Davis, D. E., and R. L. Winstead. 1980. Estimating the numbers of wildlife populations. Pages 221–246 *in* S. D. Schemnitz, ed. Wildlife management techniques manual. Fourth ed. Wildlife Society, Washington, D.C.

Davis, M. B. 1981. Quaternary history and the stability of deciduous forests. Pages 132–177 *in* D. C. West, H. H. Shugart, and D. B. Bodkin, eds. Forest succession. Springer-Verlag, New York.

Davis, M. B. 1983. Holocene vegetational history of the eastern United States. Pages 166–181 *in* H. E. Wright, Jr., ed. Late-Quaternary environments of the United States. Vol. 2: The Holocene. Univ. of Minnesota Press, Minneapolis.

Davis, M. B., and G. L. Jacobson, Jr. 1985. Late-glacial and early Holocene landscapes in northern New England and adjacent areas of Canada. Quat. Res. 23:341–368.

Davis, M. H. 1978. Reintroduction of the pine marten into the Nicolet National Forest, Forest County, Wisconsin. M.S. thesis, Univ. of Wisconsin, Stevens Point. 64pp.

Davis, M. H. 1983. Post-release movements of introduced marten. J. Wildl. Manage. 47:59–66.

Day, M. G. 1966. Identification of hair and feather remains in the gut and faeces of stoats and weasels. J. Zool. (Lond.) 148:201–217.

Dayan, T., D. Simberloff, E. Tchernov, and Y. Yom-Tov. 1989. Inter- and intraspecific character displacement in mustelids. Ecology 70:1526–1539.

Debieve, P., P. Marchesi, and C. Mermod. 1987. Food habits of pine marten (*Martes martes*) and stone marten (*M. foina*) in the Swiss Jura Mountains. Abstract, Proc. Eighteenth Int. Congr. Game Biol., Krakow, Poland.

Debrot, S., and C. Mermot. 1983. The spatial and temporal distribution pattern of the stoat (*Mustela erminea* L.). Oecologia 59:69–73.

Deems, E. F., Jr., and D. Pursley. 1983. North American furbearers: a contemporary reference. Int. Assoc. Fish Wildl. Agencies, in cooperation with Md. Dep. Nat. Resour., Wildl. Adm. 217pp.

Degn, H. J., and B. Jensen. 1977. Skovmaaren (*Martes martes*) i Danmark. Dansk Vildundersolgelser 29:1–20.

Delibes, M. 1978. Feeding habits of the stone marten, *Martes foina* (Erxleben, 1777), in northern Burgos, Spain. Z. Säugetierk. 43:282–288.

Delibes, M. 1983. Interspecific competition and the habitat use of the stone marten *Martes foina* (Erxleben 1777) in Europe. Acta Zool. Fenn. 174:229–231.

de Vos, A. 1951*a*. Overflow and dispersal of marten and fisher in Ontario. J. Wildl. Manage. 15:164–175.

de Vos, A. 1951*b*. Recent findings in fisher and marten ecology and management. Trans. North Am. Wildl. Conf. 16:498–507.

de Vos, A. 1952. Ecology and management of fisher and marten in Ontario. Ont. Dep. Lands For. Tech. Bull., Wildl. Ser., 1. 90pp.

de Vos, A., and S. E. Guenther. 1952. Preliminary live-trapping studies of marten. J. Wildl. Manage. 16:207–214.

DiStefano, J. J. 1987. Wild furbearer management in the northeastern United States. Pages 1077–1090 *in* M. Novak, J. A. Baker, M. E. Obbard, and B. Malloch, eds. Wild furbearer management and conservation in North America. Ont. Trappers Assoc., North Bay, Ont.

DiStefano, J. J., K. J. Royar, D. M. Pence, and J. E. Denoncour. 1989. Marten recovery plan for Vermont. Vt. Dep. Fish Wildl. and USDA For. Serv., Green Mountain Natl. For. 19pp.

Dix, L. M., and M. A. Strickland. 1986*a*. The use of radiographs to classify marten by sex and age. Wildl. Soc. Bull. 14:275–279.

Dix, L. M., and M. A. Strickland. 1986*b*. Sex and age determination for fisher using radiographs of canine teeth: a critique. J. Wildl. Manage. 50:275–276.

Dixon, K. R. 1981. Data requirements for determining the status of furbearer populations. Pages 1360–1373 *in* J. A. Chapman and G. A. Feldhamer, eds. Wild mammals of North America: biology, management, and economics. Johns Hopkins Univ. Press, Baltimore.

Dixon, W. J. 1983. BMDP 83. Biomedical computer programs: P series. Univ. of California Press, Berkeley. 734pp.

Dixon, W. J., and F. J. Massey. 1969. Introduction to statistical analysis. McGraw-Hill, New York. 370pp.

Dodds, D. G., and A. M. Martell. 1971*a*. The recent status of the marten, *Martes americana americana* (Turton), in Nova Scotia. Can. Field-Nat. 85:61–62.

Dodds, D. G., and A. M. Martell. 1971*b*. The recent status of the fisher, *Martes pennanti pennanti* (Erxleben), in Nova Scotia. Can. Field-Nat. 85:63–65.

Douglas, C. W., and M. A. Strickland. 1987. Fisher. Pages 511–529 *in* M. J. Novak, E. Baker, M. E. Obbard, and B. Malloch, eds. Wildlife furbearer management and conservation in North America. Ont. Trappers Assoc., North Bay, Ont.

Douglas, R. J., L. G. Fisher, and M. Mair. 1983. Habitat selection and food habits of marten, *Martes americana*, in the Northwest Territories. Can. Field-Nat. 97:71–74.

Douglas, W. O. 1943. Fisher farming has arrived. Am. Fur Breeder 16:18–20.

Drent, R. H., and S. Daan. 1980. The prudent parent: energetic adjustments in avian breeding. Ardea 68:225–252.

Driver, J. C. 1988. Late Pleistocene and Holocene vertebrates and paleoenvironments from Charlie Lake Cave, northeast British Columbia. Can. J. Earth Sci. 25:1545–1553.

Drysdale, C., and R. Charlton. 1988. American marten re-introduction program progress report, Kejimkujik National Park, November 1987–October 1988. Environ. Can., Parks Can., Nat. Resour. Conserv., Kejimkujik Natl. Park. 12pp.

Dubey, A. K., J. L. Cameron, R. A. Steiner, and T. M. Plant. 1986. Inhibition of gonadotropin secretion in castrated male rhesus monkeys (*Macaca mulatta*) induced by dietary restriction: analogy with the prepubertal hiatus of gonadotropin release. Endocrinology 118:518–525.

Dyck, I. G. 1977. The Harder site: a middle period bison hunter's campsite in the northern Great Plains. Natl. Mus. Man, Mercury Ser., 67:1–339.

Eadie, W. R., and W. J. Hamilton. 1958. Reproduction in the fisher in New York. N.Y. Fish Game J. 5:77–83.

Earle, R. D., and K. R. Kramm. 1982. Correlation between fisher and porcupine abundance in upper Michigan. Am. Midl. Nat. 107:244–249.

Ebersole, S. P. 1980. Food density and territory size: An alternative model and test on the reef fish *Eupomacentrus leocostrictus*. Am. Nat. 115:492–509.

Eger, J. L. 1990. Patterns of geographic variation in the skull of Nearctic ermine (*Mustela erminea*). Can. J. Zool. 68:1241–1249.

Eisenberg, J. F. 1981. The mammalian radiations: an analysis of trends in evolution, adaptation, and behavior. Univ. of Chicago Press, Chicago. 610pp.

Ekman, J. B., and M. K. Hake. 1990. Monitoring starvation risk: adjustment of body reserves in green finches (*Cardvelis chloris L.*) during periods of predictable foraging success. Behav. Ecol. 1:62–67.

Elowe, K. D. 1989. Annual status report—Maine. Pages 11–12 *in* R. Lafond, ed. Proceedings of the northeast fur resources technical committee workshop. Minist. du Loisir, de la Chasse, et de la Pêche, Quebec.

Enders, R. K. 1952. Reproduction in the mink. Proc. Am. Philos. Soc. 96:691–755.

Enders, R. K., and A. C. Enders. 1963. Morphology of the female reproductive tract during delayed implantation in the mink. Pages 129–139 *in* A. C. Enders, ed. Delayed implantation. Univ. of Chicago Press, Chicago.

Enders, R. K., and J. R. Leekley. 1941. Cyclic changes in the vulva of the marten (*Martes americana*). Anat. Rev. 79:1–5.

Enders, R. K., and O. P. Pearson. 1943*a*. The blastocyst of the fisher. Anat. Rec. 85:285–287.

Enders, R. K., and O. P. Pearson. 1943*b*. Shortening gestation by inducing early implantation with increased light in the marten. Am. Fur Breeder 15:18.

Erickson, D. W. 1982. Estimating and using furbearer harvest information. Pages 53–66 *in* G. C. Sanderson, ed. Midwest furbearer management. Proc. Symp. Forty-third Midwest Fish Wildl. Conf., Wichita, Kans.

Erlinge, S. 1967. Home range of the otter (*Lutra lutra* L.) in southern Sweden. Oikos 18:186–209.

Erlinge, S. 1968. Territoriality of the otter (*Lutra lutra* L.). Oikos 19:259–270.

Erlinge, S. 1969. Food habits of the otter *Lutra lutra* L. and the mink *Mustela vison* Schreiber in a trout water in southern Sweden. Oikos 20:1–7.

Erlinge, S. 1972. Interspecific relations between otter *Lutra lutra* and mink *Mustela vison* in Sweden. Oikos 23:327–335.

Erlinge, S. 1974. Distribution, territoriality and numbers of the weasel *Mustela nivalis* in relation to prey abundance. Oikos 25:308–314.

Erlinge, S. 1975. Feeding habits of the weasel *Mustela nivalis* in relation to prey abundance. Oikos 26:378–384.

Erlinge, S. 1977. Spacing strategy in the stoat *Mustela erminea*. Oikos 28:32–42.

Erlinge, S. 1979. Adaptive significance of sexual dimorphism in weasels. Oikos 33: 233–245.

Erlinge, S. 1980. Movements and daily activity patterns of radio-tracked male stoats, *Mustela erminea*. Pages 703–709 *in* C. J. Amlaner and D. MacDonald, eds. A handbook of biotelemetry and radio tracking. Pergamon Press, Oxford.

Erlinge, S. 1981. Food preference, optimal diet and reproductive output in stoats *Mustela erminea* in Sweden. Oikos 36:303–315.

Erlinge, S. 1983. Ecological research on mustelids. Acta Zool. Fenn. 174:169–172.

Erlinge, S. 1986. Specialists and generalists among the mustelids. Lutra 29:5–11.

Erlinge, S. 1987. Why do European stoats *Mustela erminea* not follow Bergmann's rule? Holarctic Ecol. 10:33–39.

Erlinge, S., and M. Sandell. 1986. Seasonal changes in social organization of male stoats, *Mustela erminea*: an effect of shifts between two decisive resources. Oikos 47:57–62.

Errington, P. L. 1967. Of predation and life. Iowa State Univ. Press, Ames. 279pp.

Eshelman, R., and F. Grady. 1986. Quaternary vertebrate localities of Virginia and their avian and mammalian faunas. Pages 43–70 *in* J. N. McDonald and S. O. Bird, eds. The Quaternary of Virginia. Dep. Mines, Minerals, Energy, Charlottesville, Va.

Evans, A. 1986. Feasibility study—American marten *Martes americana americana* (Turton) reintroduction to Kejimkujik National Park. Prepared for Parks Can. by Nesik Biol. Res., Inc. 142pp.

Ewer, R. F. 1973. The carnivores. Cornell Univ. Press, Ithaca, N.Y. 494pp.

Fagen, R. 1988. Population effects of habitat change: a quantitative assessment. J. Wildl. Manage. 52:41–46.

Farrell, D. J., and A. J. Wood. 1968. The nutrition of the mink, *Mustela vision*. I. The metabolic rate of the mink. Can. J. Zool. 46:41–45.

Faulkner, C. H., and J. B. Graham. 1966. Westmoreland-Barber site (40Mi-11) Nickajack Reservoir season II. Natl. Park Serv. Contract, Dep. Anthropol., Univ. of Tennesee.

Fedyk, S., Z. Gegczynska, M. Pucek, J. Raczynski, and M. D. Sikorski. 1984. Winter penetration by mammals of different habitats in the Biebrza Valley. Acta Theriol. 29:317–336.

Fisher, R. A. 1958. The genetical theory of natural selection. Second ed. Dover, New York. 291pp.

Fitzgerald, J. 1984. The effect of castration, estradiol, and LHRH on LH secretion of lambs fed different levels of dietary energy. J. Anim. Sci. 59:460–469.

Fitzgerald, J., F. Michel, and W. R. Butler. 1982. Growth and sexual maturation in ewes: dietary and seasonal effects modulating luteinizing hormone secretion and first ovulation. Biol. Reprod. 27:864–870.

Fitzhugh, E. L., and W. P. Gorenzel. 1985. Design and analysis of mountain lion track surveys. Cal-Neva Wildl. Trans. 1985:78–87.

Floyd, T. J., L. D. Mech, and P. A. Jordan. 1978. Relating wolf scat content to prey consumed. J. Wildl. Manage. 42:528–532.

Foehrenbach, H. 1987. Untersuchungen zur Ökologie des Steinmarders (*Martes foina*, Erxleben 1777) im Alpen- und Nationalpark Berchtesgaden. Dissertation an der Ruprecht-Karls-Universität, Heidelberg. 90pp.

Foresman, K. R., and R. A. Mead. 1973. Duration of postimplantation in a western subspecies of the spotted skunk (*Spilogale putorius*). J. Mammal. 56:521–523.

Fortin, C. 1989. Workshop on marten management. Pages 179–186 *in* Proceedings of the northeast fur resources technical committee workshop. Minist. du Loisir, de la Chasse, et de la Pêche, Quebec.

Fortin, C., and M. Cantin. 1990. Effets du piegeage sur une population nouvellement exploitée de martre d'Amérique, *Martes americana americana*, en milieu boréal. Minist. du Loisir, de la Chasse, et de la Pêche, Quebec, unpub. rep. 63pp.

Fortin, C., M. Cantin, and M. Fortin. 1988. Expérimentation d'une méthode radiographique pour la détermination du sexe et l'estimation de l'âge chez la martre d'Amérique. Minist. du Loisir, de la Chasse, et de la Pêche, Service de l'aménagement et de l'exploitation de la faune, Quebec. 23pp.

Foster, J. B. 1963. The evolution of the native land mammals of the Queen Charlotte Islands and the problem of insularity. Ph.D. dissertation, Univ. of British Columbia, Vancouver. 210pp.

Francis, G. R., and A. B. Stephenson. 1972. Marten ranges and food habits in Algonquin Provincial Park, Ontario. Ontario Minist. Nat. Resour., Res. Rep. 91. 53pp.

Franklin, I. R. 1980. Evolutionary change in small populations. Pages 135–150 *in* M. E. Soulé and B. A. Wilcox, eds. Conservation biology: an evolutionary-ecological perspective. Sinauer, Sunderland, Mass.

Fraser, D. 1984. A simple relationship between removal rate and age-sex composition of removals for certain animal populations. J. Appl. Ecol. 21:97–101.

Fredrickson, L. 1983. Re-introduction of pine marten into the Black Hills of South Dakota, 1977–83. Pages 93–98 *in* Proc. Midwest Furbearer Workshop, March 29–31, 1983, Poynette, Wis.

Fredrickson, L. F. 1989. Pine marten introduction into the Black Hills of South Dakota, 1979–1988. S.D. Dep. Game, Fish Parks, Completion Rep. 90-10. 22pp.

French, A. R. 1988. The patterns of mammalian hibernation. Am. Sci. 76:569–575.

Fretwell, S. D. 1972. Populations in a seasonal environment. Princeton Univ. Press, Princeton, N.J. 217pp.

Fries, S. 1880. Über die Fortpflanzung von *Meles taxus*. Zool. Anz. 3:486–492.

Frison, G. C., and D. N. Walker. 1978. The archaeology of Little Canyon Creek Cave and its associated late Pleistocene fauna. Am. Quat. Assoc. (AMQUA) Abstr. 5:200.

Fur Institute of Canada. 1989. Recommended regulations. Page 190 *in* R. Lafond, ed. Proceedings of the northeast fur resources technical committee workshop. Minist. du Loisir, de la Chasse, et de la Pêche, Quebec.

Gamlin, L. 1988. Sweden's factory forests. New Sci. 117(1597):41–47.

Garcia-Rodriguez, T., M. Ferrell, J. C. Carrillo, and J. Castrovejo. 1987. Metabolic response of *Buteo* to long-term fasting and refeeding. Comp. Biochem. Physiol. 87A:381–386.

Garzon, J., I. Ballarín, L. Cuesta, and F. Palacios. 1980. Datos preliminares sobre la alimentación de la marta comun (*Martes martes martes* Linne, 1758) en España. Actas Reunión Iberoam. Cons. Zool. Vert. 2:323–327.

George, R. L. 1983. The Gnagey site and the Monongahela occupation of the Somerset Plateau. Pa. Archaeol. 53(4):1–97.

Geptner, V. G., N. P. Naumov, and P. B. Yurgenson. 1967. Martes (*Martes zibellina* L.

1758). Pages 507–553 *in* Mammals in the U.S.S.R. Vol. 2. Vysshaya Shkola, Moscow (in Russian).

Gerell, R. 1970. Home range and movements of the mink *Mustela vison* Schreber in southern Sweden. Oikos 21:160–173.

Gerrodette, T. 1987. A power analysis for detecting trends. Ecology 68:1364–1372.

Giannico, G. R. 1986. Geographic and sexual variation of the American pine marten (*Martes americana*) in the Pacific Northwest, with special reference to the Queen Charlotte Islands. M.S. thesis, Univ. of Victoria, Victoria, B.C. 119pp.

Giannico, G. R., and D. W. Nagorsen. 1989. Geographic and sexual variation in the skull of Pacific coast marten (*Martes americana*). Can. J. Zool. 67:1386–1393.

Gibbon, G. 1971. The Bornick site: a Grand River Phase Oneota site in Marquette County. Wis. Archeol. 52(3):85–137.

Gidley, J. W., and C. L. Gazin. 1933. New Mammalia in the Pleistocene fauna from Cumberland Cave. J. Mammal. 14:343- 357.

Gidley, J. W., and C. L. Gazin. 1938. The Pleistocene vertebrate fauna from Cumberland Cave, Maryland. Bull. U.S. Natl. Mus. 171:1–99.

Gieck, E. M. 1986. Wisconsin pine marten recovery plan. Wis. Dep. Nat. Resour., Wis. Endangered Resour. Rep. 22. 24pp.

Gilbert, F. F., and D. G. Dodds. 1987. The philosophy and practice of wildlife management. Robert E. Krieger, Malabar, Fla. 279pp.

Gilbert, J. R. 1979. Techniques and problems of population modeling and analysis. Pages 130–133 *in* Proc. Bobcat Res. Conf., Natl. Wildl. Fed., Washington, D.C.

Gillingham, B. J. 1984. Meal size and feeding rate in the least weasel *Mustela nivalis*. J. Mammal. 65:517–519.

Gittleman, J. L. 1985. Carnivore body size: ecological and taxonomic correlates. Oecologia 67:540–554.

Giulano, W. M., J. A. Litvaitis, and C. L. Stevens. 1989. Prey selection in relation to sexual dimorphism of fishers (*Martes pennanti*) in New Hampshire. J. Mammal. 70:639–641.

Given, R. L., and A. C. Enders. 1989. The endometrium of delayed and early implantation. Pages 175–231 *in* R. M. Wynn and W. P. Jollie, eds. Biology of the uterus. Second ed. Plenum Press, New York.

Gompper, M. E., and J. L. Gittleman. 1991. Home range scaling: intraspecific and comparative trends. Oecologia 87:343–348.

Goodman, M. N., and N. B. Ruderman. 1980. Starvation in the rat. I. Effect of age and obesity on organ weights, RNA, DNA and protein. Am. J. Physiol. 239:E269-E276.

Gordon, C. C. 1986. Winter food habits of the pine marten in Colorado. Great Basin Nat. 46:166–168.

Gordon, K. R. 1986. Insular evolutionary body size trends in *Ursus*. J. Mammal. 67:395–399.

Goszczynski, J. 1976. Composition of the food of martens. Acta Theriol. 21:527–534.

Gould, G. 1987. Nongame wildlife investigations: forest mammal survey and inventory. Calif. Dep. Fish Game, Proj. W-65-R-4. 11pp.

Gould, S. J., and R. F. Johnston. 1972. Geographic variation. Annu. Rev. Ecol. Syst. 3:457–498.

Graham, M. A., and R. W. Graham. 1990. Holocene records of *Martes pennanti* and

Martes americana in Whiteside County, northwestern Illinois. Am. Midl. Nat. 124:81–92.

Graham, R. W. 1976. Late Wisconsin mammal faunas and environmental gradients of the eastern United States. Paleobiology 2:343–350.

Graham, R. W. 1979. Paleoclimates and late Pleistocene faunal provinces in North America. Pages 49–69 *in* R. L. Humphrey and D. J. Stanford, eds. Pre-Llano cultures of the Americas: paradoxes and possibilities. Anthropol. Soc. Wash., Washington, DC.

Graham, R. W. 1986. Response of mammalian communities to environmental changes during the late Quaternary. Pages 300–313 *in* J. Diamond and T. J. Case, eds. Community ecology. Harper and Row, New York.

Graham, R. W. 1987. Late Quaternary mammalian faunas and paleoenvironments of the southwestern plains of the United States. Pages 24–86 *in* R. W. Graham, H. A. Semken, Jr. and M. A. Graham, eds. Late Quaternary mammalian biogeography and environments of the Great Plains and prairies. Ill. State Mus. Sci. Pap. 22, Springfield.

Graham, R. W., and E. C. Grimm. 1990. Effects of global climate change on the patterns of terrestrial biological communities. Trends Ecol. Evol. 5:289–292.

Graham, R. W., and J. I. Mead. 1987. Environmental fluctuations and evolution of mammalian faunas during the last deglaciation in North America. Pages 371–402 *in* W. F. Ruddiman and H. E. Wright, Jr., eds. North America and adjacent oceans during the last deglaciation. Geol. Soc. Am., Boulder, Colo.

Grakov, N. N. 1972. Effect of concentrated clearfellings on the abundance of the pine marten (*Martes martes* L.). Byull. Mosk. o-va Ispyt. Prir. 77:14–23 (in Russian, with English summary).

Grakov, N. N. 1978. Long-term changes in the abundance of the pine marten (*Martes martes*) and some patterns of this process. Byull. Mosk. o-va Ispyt. Prir. 83:46–56.

Grakov, N. N. 1981. Pine marten. Nauka, Moscow. 108pp.

Grayson, D. K. 1984. Time of extinction and nature of adaptation of the noble marten, *Martes nobilis*. Pages 233–240 *in* H. H. Genoways and M. R. Dawson, eds. Contributions in Quaternary vertebrate paleontology: a volume in memorial to John E. Guilday. Carnegie Mus. Nat. Hist. Spec. Publ. 8.

Grayson, D. K. 1985. The paleontology of Hidden Cave: birds and mammals. Pages 125–161 *in* D. H. Thomas, ed. The archaeology of Hidden Cave, Nevada. Anthropol. Pap. Am. Mus. Nat. Hist. 61.

Grayson, D. K. 1987. The biogeographic history of small mammals in the Great Basin: observations on the last 20,000 years. J. Mammal. 68:359–375.

Grenfell, W. E., and M. Fasenfest. 1979. Winter food habits of fishers, *Martes pennanti*, in northwestern California. Calif. Fish Game 65:186–189.

Griffith, B., J. M. Scott, J. W. Carpenter, and C. Reed. 1989. Translocation as a species conservation tool: status and strategy. Science 245:477–480.

Grinnell, J., J. S. Dixon, and J. M. Linsdale. 1937. Furbearing mammals of California. Vol 1. Univ. of California Press, Berkeley. 375pp.

Gross, J. E., J. E. Roelle, and G. L. Williams. 1973. Program ONEPOP and information processor: a systems modeling and communications project. Colo. Coop. Wildl. Res. Unit. Prog. Rep. 327pp.

Grubb, T. G. 1988. Pattern recognition—a simple model for evaluating wildlife habitat.

USDA For. Serv., Rocky Mt. For. Range Exp. Stn., Fort Collins, Colo., Res. Note RM-487. 5pp.

Gruhn, R. 1961. The archeology of Wilson Butte Cave, south-central Idaho. Occas. Pap. Idaho State Coll. Mus. 6. 201pp.

Guilday, J. E. 1956. Archaeological evidence of the fisher in West Virginia. J. Mammal. 37:287.

Guilday, J. E. 1962. The Pleistocene local fauna of the Natural Chimneys, Augusta County, Virginia. Ann. Carnegie Mus. 36:87–122.

Guilday, J. E. 1963. Bone refuse from the Oakfield site, Genesee County, New York. Pa. Archaeol. 33:12–15.

Guilday, J. E. 1967. The climatic significance of the Hosterman's Pit local fauna, Centre County, Pennsylvania. Am. Antiq. 32:231–232.

Guilday, J. E., and H. W. Hamilton. 1973. The late Pleistocene small mammals of Eagle Cave, Pendleton County, West Virginia. Ann. Carnegie Mus. 44:45–58.

Guilday, J. E., H. W. Hamilton, E. Anderson, and P. W. Parmalee. 1978. The Baker Bluff Cave deposit, Tennessee, and the late Pleistocene faunal gradient. Bull. Carnegie Mus. Nat. Hist. 11:1–67.

Guilday, J. E., H. W. Hamilton, and A. D. McCrady. 1969. The Pleistocene vertebrate fauna of Robinson Cave, Overton County, Tennessee. Palaeovertebrata 2:25–75.

Guilday, J. E., P. S. Martin, and A. D. McCrady. 1964. New Paris no. 4: a Pleistocene cave deposit in Bedford County, Pennsylvania. Bull. Natl. Speleol. Soc. 26:121–194.

Guilday, J. E., P. W. Parmalee, and H. W. Hamilton. 1977. The Clark's Cave bone deposit and the late Pleistocene paleoecology of the central Appalachian Mountains of Virginia. Bull. Carnegie Mus. Nat. Hist. 2:1–87.

Guilday, J. E., P. W. Parmalee, and D. P. Tanner. 1962. Aboriginal butchering techniques at the Eschelman site (36LA12), Lancaster County, Pennsylvania. Pa. Archaeol. 32(2):59–83.

Guilday, J. E., P. W. Parmalee, and R. C. Wilson. No date. Vertebrate faunal remains from Meadowcroft Rockshelter (36WH297), Washington County, Pennsylvania. Carnegie Mus. Nat. Hist., unpubl. rep.

Guilday, J. E., and D. P. Tanner. 1965. Vertebrate remains from the Mount Carbon site (46-Fa-7), Fayette County, West Virginia. W. Va. Archaeol. 18:1–14.

Guilday, J. E., and D. P. Tanner. 1966. Animal remains from the Westmoreland-Barber site (40Mi-11), Marion County, Tennessee. Pages 138–140 *in* C. Faulkner and J. B. Graham, eds. Westmoreland-Barber site (40Mi-11) Nickajack Reservoir season II. Natl. Park Serv. Contract, Dep. Anthropol., Univ. of Tennessee.

Gunderson, H. L. 1965. Marten records for Minnesota. J. Mammal. 46:688.

Gunderson, H. L., and J. R. Beer. 1953. The mammals of Minnesota. Univ. of Minnesota Press, Minneapolis. 190pp.

Gustafson, C. E. 1972. Faunal remains from the Marmes Rockshelter and related archeological sites in the Columbia Basin. Ph.D. dissertation, Washington State Univ., Pullman. 183pp.

Hager, M. W. 1972. A late Wisconsinan–Recent vertebrate fauna from Chimney Rock animal trap, Larimer County, Colorado. Univ. of Wyoming Contrib. Geol. 11:63–71.

Hagmeier, E. M. 1955. The genus *Martes* (Mustelidae) in North America: its distribution, variation, classification, phylogeny, and relationship to Old World forms. Ph.D. dissertation, Univ. of British Columbia, Vancouver. 469pp.

Hagmeier, E. M. 1956a. Distribution of marten and fisher in North America. Can. Field-Nat. 70:149–168.

Hagmeier, E. M. 1956b. A numerical analysis of the distributional patterns of North American mammals: a reevaluation of the provinces. Syst. Zool. 5:279–299.

Hagmeier, E. M. 1958. Inapplicability of the subspecies concept to North American marten. Syst. Zool. 7:1–7.

Hagmeier, E. M. 1961. Variation and relationships in North American marten. Can. Field-Nat. 75:122–137.

Hall, E. R. 1926. A new marten from the Pleistocene cave deposits of California. J. Mammal. 7:127–130.

Hall, E. R. 1931. Description of a new mustelid from the later Tertiary of Oregon with assignment of *Parictis primaevus* to the Canidae. J. Mammal. 12:156–158.

Hall, E. R. 1936. Mustelid mammals from the Pleistocene of North America with some systematic notes on some Recent members of the genera *Mustela*, *Taxidea* and *Mephitis*. Carnegie Inst. Wash. Publ. 473:43–119.

Hall, E. R. 1942. Gestation period in the fisher with recommendations for the animal's protection in California. Calif. Fish Game 28:143–147.

Hall, E. R. 1981. The mammals of North America. Second ed. John Wiley and Sons, New York. 1181pp.

Hall, E. R., and K. R. Kelson. 1959. The mammals of North America. Vol. 2. Ronald Press, New York. 536pp.

Hall, M. R. 1987. Nesting success in mallards after partial clutch loss by predators. J. Wildl. Manage. 51:530–533.

Hamilton, D. A., and L. B. Fox. 1987. Wild furbearer management in the midwestern United States. Pages 1100–1116 *in* M. Novak, J. A. Baker, M. E. Obbard, and B. Malloch, eds. Wild furbearer management and conservation in North America. Ont. Trappers Assoc., North Bay, Ont.

Hamilton, W. J., and H. J. Gould. 1940. The normal oestrous cycle of the ferret: the correlation of the vaginal smear and the histology of the genital tract, with notes on the distribution of glycogen, the incidence of growth, and the reaction to intravital staining by trypan blue. Trans. Roy. Soc. Edin. 60:87–105.

Hamilton, W. J., Jr., and A. H. Cook. 1955. The biology and management of the fisher in New York. N.Y. Fish Game J. 2:13–35.

Hamlett, G. W. D. 1935. Delayed implantation and discontinuous development in the mammals. Q. Rev. Biol. 10:432–447.

Hammond, J., Jr. 1951. Effects of androstenedione and progesterone on estrous swelling of the ferret vulva. J. Endocrinol. 7:173–176.

Hammond, J., Jr., and F. H. A. Marshall. 1930. Oestrus and pseudopregnancy in the ferret. J. Exp. Biol. 11:307–319.

Hansson, A. 1947. The physiology of reproduction in mink (*Mustela vison*, Schreb.) with special reference to delayed implantation. Acta Zool. 28:1–136.

Hansson, A. 1978. Small mammal abundance in relation to environmental variables in three Swedish forest phases. Stud. For. Suec. 147. 37pp.

Hardy Associates, Ltd. 1986. Ecological land classification of Elk Island National Park. Report prepared for Parks Can., Calgary, Alta.

Harestad, A. S., and F. L. Bunnel. 1979. Home range and body weight—a reevaluation. Ecology 60:389–404.

Harger, E. M., and D. F. Switzengburg. 1958. Returning the pine marten to Michigan. Mich. Dep. Conserv. Game Div. Rep. 2199. 13pp.

Hargis, C. D., and D. R. McCullough. 1984. Winter diet and habitat selection of marten in Yosemite National Park. J. Wildl. Manage. 48:140–146.

Harington, C. R. 1978. Quaternary vertebrate faunas of Canada and Alaska and their suggested chronological sequence. Syllogeus 15:1–105.

Harlow, H. J. 1981a. Torpor and other physiological adaptations of the badger *Taxidea taxus* to cold environments. Physiol. Zool. 53:267–273.

Harlow, H. J. 1981b. Metabolic adaptations to prolonged food deprivation by the American badger *Taxidea taxus*. Physiol. Zool. 54:276–284.

Harlow, H. J. 1981c. Effect of fasting on rate of food passage and assimilation efficiency in badgers. J. Mammal. 62:173–177.

Harlow, H. J. 1987. Urea-hydrolysis in euthermic hibernators and non-hibernators during periods of food availability and deprivation. J. Therm. Biol. 12:149–154.

Harlow, H. J., and S. W. Buskirk. 1991. Fasting biochemistry of the American marten *Martes americana* and the black-tailed prairie dog *Cynomys lucovicianus*: representative lean and fat animals. Physiol. Zool. 64:1262–1278.

Harlow, H. J., and B. Miller. 1985. Non-shivering thermogenesis in the American badger. Comp. Biochem. Physiol. 80A:159–161.

Harlow, H. J., B. Miller, T. Ryder, and L. Ryder. 1985. Energy requirements for gestation and lactation in a delayed implanter, the American badger. Comp. Biochem. Physiol. 82A:885–889.

Harris, L. D. 1984. An island archipelago model for maintaining biotic diversity in old-growth forests. New forest for a changing world. Proc. 1983 Conven. Soc. Am. For., Portland, Oreg. 640pp.

Harris, L. D., C. Maser, and A. McKee. 1982. Patterns of old growth harvest and implications for Cascades wildlife. Trans. North Am. Wildl. Nat. Resour. Conf. 47:374–392.

Harris, R. B. 1986. Reliability of trend lines obtained from variable counts. J. Wildl. Manage. 50:165–171.

Hart, J. S., H. Pohl, and J. S. Tener. 1965. Seasonal acclimatization in varying hare (*Lepus americanus*). Can. J. Zool. 43:731–744.

Hash, H. S., and M. G. Hornocker. 1978. Range and habitat of male fishers in northwest Montana. Idaho Coop. Wildl. Res. Unit, unpubl. MS. 10pp.

Hash, H., and M. Hornocker. 1980. Immobilizing wolverines with ketamine hydrochloride. J. Wildl. Manage. 44:713–715.

Hauptman, T. N. 1979. Spatial and temporal distribution and feeding ecology of the pine marten. M.S. thesis, Idaho State Univ., Pocatello. 83pp.

Hawksley, O., J. F. Reynolds, and R. L. Foley. 1973. Pleistocene vertebrate fauna of Bat Cave, Pulaski County, Missouri. Bull. Natl. Speleol. Soc. 35:61–87.

Hawley, V. D. 1955. The ecology of the marten in Glacier National Park. M.S. thesis, Montana State Univ., Bozeman. 131pp.

Hawley, V. D., and F. E. Newby. 1957*a*. Marten home ranges and population fluctuations in Montana. J. Mammal. 38:174–184.

Hawley, V. D., and F. E. Newby. 1957*b*. Marten population status. Montana Fish and Game Dep., Wildl. Restor. Div., Completion Rep., Proj. W-49-R-6.

Hayne, D. W. 1949. Calculation of size of home range. J. Mammal. 30:1–18.

Heap, R. B., and J. Hammond, Jr. 1974. Plasma progesterone levels in pregnant and pseudopregnant ferrets. J. Reprod. Fert. 39:149–152.

Heaton, T. H. 1985. Quaternary paleontology and paleoecology of Crystal Ball Cave, Millard County, Utah: with emphasis on mammals and description of a new species of fossil skunk. Great Basin Nat. 45:337–390.

Heckenberger, M. J., J. B. Petersen, and L. A. Basa. 1990. Early Woodland Period ritual use of personal adornment at the Boucher site. Ann. Carnegie Mus. 59:173–217.

Heinselman, M. L. 1973. Fire in the virgin forest of the Boundary Waters Canoe Area, Minnesota. Quat. Res. 3:329–382.

Heinselman, M. L. 1974. Interpretation of F. J. Marshner's map of the original vegetation of Minnesota. (Map.) USDA For. Serv., North Cent. For. Exp. Stn., St. Paul, Minn.

Heisey, D. M., and T. K. Fuller. 1985. Evaluation of survival and cause-specific mortality rates using telemetry data. J. Wildl. Manage. 49:668–674.

Heldmaier, G., S. Steinlechner, J. Rafael, and P. Vsiansky. 1981. Photoperiod control and effects of melatonin on non-shivering thermogenesis and brown adipose tissue. Science 212:917–919.

Helldin, J.-O., and E. Lindström. 1991. [The Swedish marten population during the 1900s]. Viltnytt 30:30–34 (in Swedish).

Helle, P. 1985. Effects of forest fragmentation on bird densities in northern boreal forests. Ornis Fenn. 62:35–41.

Hendrichs, H. 1978. Die soziale Organization von Säugetierpopulationen. Säugetierk. Mitt. 26:81–116.

Henry, S. E., and M. G. Raphael. 1989. Observations of copulation of free-ranging American marten. Northwest. Nat. 70:32–33.

Henry, S. E., M. G. Raphael, and L. F. Ruggiero. 1990. Food caching and handling by marten. Great Basin Nat. 50:381–383.

Henttonen, H. 1989. Metsien rakenteen muutoksen vaikutuksesta myyrräkantoihin ja sitä kautta pikkupetoihin ja kanalintuihin—hypoteesi. Suomen Riista 35:83–90.

Herbert, J. 1969. The pineal gland and light-induced oestrus in ferrets. J. Endocrinol. 43:625–636.

Herbert, J. 1989. Light as a multiple control system on reproduction in mustelids. Pages 138–159 *in* U. S. Seal, E. T. Thorne, M. A. Bogan, and S. H. Anderson, eds. Conservation biology and the black-footed ferret. Yale Univ. Press, New Haven, Conn.

Herman, T., and K. Fuller. 1974. Observations of the marten in the Mackenzie District, NWT. Can. Field-Nat. 88:501–503.

Herreid, C. F., II, and B. Kessel. 1967. Thermal conductance in birds and mammals. Comp. Biochem. Physiol. 21:405–414.

Herrmann, M. 1987. Zum Raum-Zeit-System von Steinmarderrueden (*Martes foina* Erxleben 1777) in unterschiedlichen Lebensräumen des südöstlichen Saarlandes. Diplomarbeit, Universität Bielefeld. 106pp.

Herrmann, M. 1989*a*. Social organization in *Martes foina* and ecological determinants of home range size under urban, agricultural and woodland use of land. Page 996 *in* Abstracts of Papers and Posters, Fifth Int. Theriol. Congr., Rome.

Herrmann, M. 1989*b*. Intra-population variability in the spatial and temporal organization of stone martens (*Martes foina*). Pages 602–603 *in* Abstracts of Papers and Posters, Fifth Int. Theriol. Congr., Rome.

Herrmann, M. 1991. Säugetiere im Saarland. Verbreitung, Gefährdung, Schutz. Michel Verlag, Ottweiler. 166pp.

Herrmann, M., and H. Hendrichs. 1989. Erklärungshypothesen zum Geschlechtsunterschied in der Streifgebietsgrösse beim Steinmarder. Z. Säugetierk. 54, 63. Hauptversammlung 54:21.

Herrmann, M., and J. Knapp. 1984. Vergleichende Freilanduntersuchungen und Aufzuchtbeobachtungen an Stein- und Baummardern. Z. Säugetierk. 49:18, Sonderheft zur 58.

Herscovici, A. 1985. Second nature. The animal rights controversy. CBC Enterprises, Toronto, Ont. 254pp.

Hespenheide, H. A. 1975. Prey characteristics and predator niche width. Pages 158–180 *in* M. L. Cody and J. M. Diamond, eds. Ecology and evolution of communities. Harvard Univ. Press, Cambridge, Mass.

Hesse, S. 1987. Untersuchungen zur Fortpflanzung und Altersstruktur Schleswig-holsteinischer Steinmarder (*Martes foina* Erxleben 1777). Hausarbeit zur wissenschaftlichen Prüfung für das höhere Lehramt, Christian-Albrechts-Universität, Kiel. 156pp.

Heusser, C. J. 1977. Quaternary paleoecology of the Pacific slope of Washington. Quat. Res. 8:282–306.

Heusser, C. J. 1989. North Pacific coastal refugia—the Queen Charlotte Islands in perspective. Pages 91–106 *in* G. G. E. Scudder and N. Gessler, eds. The outer shores. Queen Charlotte Is. Mus. Press, Skidegate, B.C.

Hillman, C. N., and J. W. Carpenter. 1983. Breeding biology and behavior of captive black-footed ferrets. Int. Zoo Yearb. 23:186–191.

Hirsch, E. 1973. Some determinants of intake and patterns of feeding in the guinea pig. Physiol. Behav. 11:687–704.

Hixon, M. A. 1987. Territory area as a determinant of mating systems. Am. Zool. 27:229–247.

Hixon, M. A., F. L. Carpenter, and D. C. Paton. 1983. Territory area, flower density, and time budgeting in hummingbirds: an experimental and theoretical analysis. Am. Nat. 122:366–391.

Hobson, D. F., G. Proulx, and B. L. Dew. 1989. Initial post-release behaviour of marten, *Martes americana*, introduced in Cypress Hills Provincial Park, Saskatchewan. Can. Field-Nat. 103:398–400.

Hodgson, R.G. 1937. Fisher farming. Fur Trade J. Can., Toronto, Ont. 103pp.

Hoffman, J. A. 1983. Progress report, marten re-introduction to Terra Nova National Park. Environ. Can., Parks Can., Nat. Resour. Conserv., Terra Nova Natl. Park. 6pp.

Hoffman, K. 1973. The influence of photoperiod and melatonin on testis size, weight and pelage color in the Djungarian hamster, *Phodopus sungorous*. J. Comp. Physiol. 85:267–282.

Hoffman, R. S., and J. Knox Jones, Jr. 1970. Influence of late-glacial and post-glacial events on the distribution of Recent mammals on the northern Great Plains. Pages 355–

394 *in* W. Dort, Jr., and J. Knox Jones, Jr., eds. Pleistocene and Recent environments of the central Great Plains. Univ. Press of Kansas, Dep. Geology Spec. Publ. 3.

Hoffman, R. S., and D. L. Pattie. 1968. A guide to Montana mammals: identification, habitat, distribution and abundance. Univ. of Montana Found. 129pp.

Hogan, P. 1986. Excavation at Lemoc Shelter (site 5MT2151), a multiple-occupational Anasazi site. Pages 157–277 *in* D. Breternitz, ed. Dolores archeological program: Anasazi communities at Dolores: early small settlements in the Dolores River canyon and western Sagehen Flats area. U.S. Dep. Inter., Bur. Reclamation, Denver, Colo.

Höglund, N. H. 1960. [Studies on the winter diet of pine marten (*Martes martes* L.) in Jämtland province, Sweden]. Viltrevy 1:319–337 (in Swedish).

Holisova, V., and R. Obrtel. 1982. Scat analytical data on the diet of urban stone martens, *Martes foina* (Mustelidae, Mammalia). Folia Zool. 31:21–30.

Holman, J. A. 1985*a*. New evidence on the status of Ladds Quarry. Natl. Geogr. Res. 1:569–570.

Holman, J. A. 1985*b*. Herpetofauna of Ladds Quarry. Natl. Geogr. Res. 1:423–436.

Holmes, T., Jr. 1980. Locomotor adaptations in the limb skeletons of North American mustelids. M.A. thesis, Humboldt State Univ., Arcata, Calif. 160pp.

Holmes, T., Jr. 1987. Sexual dimorphism in North American weasels with a phylogeny of the Mustelidae. Ph.D. dissertation, Univ. of Kansas, Lawrence. 323pp.

Honacki, J. H., K. E. Kinman, and J. W. Koeppl, eds. 1982. Mammal species of the world. Allen Press, Lawrence, Kans. 694pp.

Hoover, R. L., and D. L. Wills, eds. 1984. Managing forested lands for wildlife. Colo. Div. Wildl., in cooperation with USDA For. Serv., Rocky Mt. Reg., Denver, Colo. 459pp.

Hovens, J. P. M., and G. F. E. Janss. 1990. Territorialiteit bij adulte vrouwelijke steen-marters (*Martes foina*). Arnhem, Rijksinstituut voor Natuurbeheer, unpubl. rep. 45pp.

Humason, G. L. 1972. Animal tissue techniques. Third ed. W. H. Freeman, San Francisco. 641pp.

Hunt, D. C., and M. V. Gallagher. 1984. Karl Bodmer's America. Joslyn Art Mus. and Univ. of Nebraska Press, Lincoln. 376pp.

Hunter, M. L., Jr. 1991. Coping with ignorance: the coarse-filter strategy for maintaining biodiversity. Pages 266–281 *in* K. A. Kohm, ed. Balancing on the brink of extinction: the endangered species act and lessons for the future. Island Press, Washington, D.C.

Husar, S. L. 1976. Behavioral character displacement: evidence of food partitioning in insectivorous bats. J. Mammal. 57:331–338.

Imaizumi, Y. 1949. The natural history of Japanese mammals. Yoyo shobo, Tokyo. 348pp.

Ims, R. A. 1987. Male spacing systems in microtine rodents. Am. Nat. 130:475–484.

Ims, R. A. 1988*a*. Spatial clumping of sexually receptive females induces space sharing among male voles. Nature (Lond.) 335:541–543.

Ims, R. A. 1988*b*. The potential for sexual selection in males: effect of sex ratio and spatiotemporal distribution of receptive females. Evol. Ecol. 2:338–352.

Ims, R. A. 1990. Mate detection success of male *Clethrionomys rufocanus* in relation to the spatial distribution of sexually receptive females. Evol. Ecol. 4:57–61.

Ingram, R. 1973. Wolverine, fisher, and marten in central Oregon. Oreg. State Game Comm., Cent. Reg. Admin. Rep. 73–2. 39pp.

Iriarte, J. A., W. L. Franklin, W. E. Johnson, and K. H. Redford. 1990. Biogeographic variation of food habits and body size of the America puma. Oecologia 85:185–190.

Irvine, G. W., L. T. Magnus, and B. J. Bradle. 1964. The restocking of fisher in lake states forests. Trans. North Am. Wildl. Nat. Resour. Conf. 29:307–315.

Irving, L., H. Krog, and M. Monson. 1955. The metabolism of some Alaskan animals in winter and summer. Physiol. Zool. 28:173–185.

IUCN. 1987. The IUCN position statement on translocation of living organisms: introductions, re-introductions, and re-stocking. IUCN, Gland, Switzerland. 20pp.

Iverson, J. A. 1972. Basal energy metabolism of mustelids. J. Comp. Physiol. 81: 341–344.

Jackson, H. H. T. 1961. Mammals of Wisconsin. Univ. of Wisconsin Press, Madison. 504pp.

Jacobson, G. L., Jr., T. Webb III, and E. C. Grimm. 1987. Patterns and rates of vegetation change during the deglaciation of eastern North America. Pages 277–288 *in* W. F. Ruddiman and H. E. Wright, Jr., eds. North America and adjacent oceans during the last deglaciation. Geol. Soc. Am., Boulder, Colo.

Jakes, P. J. 1980. The fourth Minnesota forest inventory: area. USDA For. Serv. Resour. Bull. NC-54. 37pp.

Jaksic, F. M., and H. E. Braker. 1983. Food-niche relationships and guild structure of diurnal birds of prey: competition versus opportunism. Can. J. Zool. 61:2230–2241.

Jedrzejewski, W., and B. Jedrzejewski. 1990. Effect of a predator's visit on the spatial distribution of bank voles: experiments with weasels. Can. J. Zool. 68:660–666.

Jedrzejewski, W., B. Jedrzejewski, and A. Szymura. 1989. Food niche overlaps in a winter community of predators in the Bialowieza primeval forest, Poland. Acta Theriol. 34:487–496.

Jelks, E. B., C. J. Ekberg, and T. J. Martin. 1989. Excavations at the Laurens site. Ill. Hist. Preserv. Agency, Stud. Ill. Archaeol. 5. 136pp.

Jenkins, D., and R. J. Harper. 1980. Ecology of otters in northern Scotland. II. Analyses of otter (*Lutra lutra*) and mink (*Mustela vison*) faeces from Deeside, N.E. Scotland in 1977–78. J. Anim. Ecol. 49:127–160.

Jenks, J. A., R. T. Bowyer, and A. G. Clark. 1984. Sex and age-class determination for fisher using radiographs of canine teeth. J. Wildl. Manage. 48:626–628.

Jenks, J. A., A. G. Clark, and R. T. Bowyer. 1986. Sex and age determination for fisher using radiographs of canine teeth: a response. J. Wildl. Manage. 50:277–278.

Jensen, A., and B. Jensen. 1970. Husmaaren (*Martes foina*) og maarjagten i Danmark 1967–1968. Dansk Undersogelser 15:1–44.

Jewell, P. A. 1966. The concept of home range in mammals. Symp. Zool. Soc. Lond. 18:85–109.

Johnson, D. H. 1980. The comparison of usage and availability measurements for evaluating resource preference. Ecology 61:65–71.

Johnson, D. H., and A. B. Sargeant. 1977. Impact of red fox predation on the sex ratio of prairie mallards. U.S. Fish Wildl. Serv., Wildl. Res. Rep. 6. 56pp.

Johnson, S. A. 1984. Home range, movements, and habitat use of fishers in Wisconsin. M.S. thesis, Univ. of Wisconsin, Stevens Point. 78pp.

Johnson, W. N., and T. F. Paragi. 1993. The relationship of wildfire to lynx and marten populations and habitat in interior Alaska. Ann. Rep., U.S. Fish Wildl. Serv., Koyukuk/Nowitna Ruguge Complex, Galena, Alas. 82pp.

Johnston, D. H., D. G. Joachim, P. Bachmann, K. V. Kardong, R. A. Stewart, L. M. Dix, M. A. Strickland, and I. D. Watts. 1987. Aging furbearers using tooth structure and biomarkers. Pages 228–243 *in* M. Novak, J. A. Baker, M. E. Obbard, and B. Malloch, eds. Wild furbearer management and conservation in North America. Ont. Trappers Assoc., North Bay, Ont.

Jones, J. L. 1991. Habitat use of fishers in northcentral Idaho. M.S. thesis, Univ. of Idaho, Moscow. 147pp.

Jones, L. L. C., and M. G. Raphael. 1991. Ecology and management of marten in fragmented habitats of the Pacific Northwest. Pac. Northwest For. Sci. Lab., Olympia, Wash. 44pp.

Jones, L. L. C., and M. G. Raphael. 1993. Inexpensive camera systems for detecting martens, fishers, and other animals: guidelines for use and standardization. USDA For. Serv. Gen. Tech. Rep. PNW-GTR-306. 22pp.

Jones, L. L. C., L. F. Ruggiero, and J. K. Swingle. 1991. Ecology of marten in the Pacific Northwest: technique evaluation. Page 532 *in* L. F. Ruggiero, K. B. Aubry, A. B. Carey, and M. H. Huff, tech. coords. Wildlife and vegetation of unmanaged Douglas-fir forests. USDA For. Serv. Gen. Tech. Rep. PNW-285.

Jones, P. J. and P. Ward. 1976. The level of reserve protein as the proximate factor controlling the timing of breeding and clutch size in the red-billed quelea *Quelea quelea*. Ibis 118:547–574.

Jonkel, C. J., and R. P. Weckwerth. 1963. Sexual maturity and implantation of blastocysts in the wild pine marten. J. Wildl. Manage. 27:93–98.

Jonsson, H. 1992. [Comparison of pine marten *Martes martes* L. habitat use in fragmented and undisturbed coniferous forests]. Undergr. thesis, Dep. Ecol. Zool., Umeå Univ. 14pp. (in Swedish, with English summary).

Jonsson, S. 1983. [The lynx]. Natur och Kultur, Stockholm. 124pp. (in Swedish).

Joslin, P. 1977. Night stalking—setting a camera trapline for nocturnal carnivores. Photo Life 7:34–35.

Joslin, P. 1988. A photo trapline for cold temperatures. Pages 121–128 *in* H. Freeman, ed. Proc. Fifth Int. Snow Leopard Symp., Wildl. Inst. India, Dehra Dun.

Jounge, J. de. 1981. Predation of the pine marten, *Martes martes* L., in relation to habitat selection and food abundance during winter in central Sweden. Swedish Environ. Prot. Board Rep. 1401.

Kalpers, J. 1983. Contribution à l' étude éco-éthologique de la foine (*Martes foina*): stratégies d'utilisation du domaine vital et des ressources alimentaires. I. Introduction générale et analyse du régime alimentaire. Cahiers d'Ethol. appl. 3:145–163.

Karsch, F. J., J. E. Robinson, C. J. I. Woodfill, and M. B. Brown. 1989. Circannual cycles of luteinizing hormone and prolactin secretion in ewes during prolonged exposure to a fixed photoperiod: evidence for an endogenous reproductive rhythm. Biol. Reprod. 41:1034–1046.

Kartashov, L. M. 1989. Age changes in fertility of sables (*Martes zibellina* L.) in the central OB' region. Sov. J. Ecol. 20:120–124.

Kaufman, D. W. 1974. Differential predation on active and inactive prey by owls. Auk 91:172–173.

Keith, L. B., and L. A. Windberg. 1978. A demographic analysis of the snowshoe hare cycle. Wildl. Monogr. 58. 70pp.

Kelly, G. M. 1977. Fisher (*Martes pennanti*) biology in White Mountain National Forest and adjacent areas. Ph.D. dissertation, Univ. of Massachusetts, Amherst. 178pp.

Kelly, J. P. 1982. Impacts of forest harvesting on the pine marten in the central interior of British Colombia. M.S. thesis, Univ. of Alberta, Edmonton. 54pp.

Kendall, M. P., P. Ward, and S. Bacchus. 1973. A protein reserve in the pectoralis major flight muscle of *Quelea quelea*. Ibis 115:600–601.

Kenward, R. 1987. Wildlife radio tagging: equipment, field techniques and data analysis. Academic Press, London. 222pp.

King, C. M. 1983. The life history strategies of *Mustela nivalis* and *Mustela erminea*. Acta Zool. Fenn. 174:183–184.

King, C. M. 1989*a*. The advantages and disadvantages of small size to weasels, *Mustela* species. Pages 302–334 *in* J. L. Gittleman, ed. Carnivore behavior, ecology, and evolution. Cornell Univ. Press, Ithaca, N.Y.

King, C. M. 1989*b*. The natural history of weasels and stoats. Cornell Univ. Press, Ithaca, N.Y. 253pp.

King, C. M., and J. E. Moody. 1982. The biology of the stoat (*Mustela erminea*) in the national parks of New Zealand. N.Z. J. Zool. 9:49–144.

King, C. M., and P. J. Moors. 1979. On co-existence, foraging strategies and the biogeography of weasels and stoats (*Mustela nivalis* and *Mustela erminea*) in Britain. Oecologia 39:129–150.

King, J. E. 1973. Late Pleistocene palynology and biogeography of the western Ozarks. Ecol. Monogr. 43:539–565.

King, J. E. 1981. Late Quaternary vegetational history of Illinois. Ecology 51:43–62.

King, J. R. 1974. Seasonal allocation of time and energy resources in birds. Pages 4–70 *in* R. A. Paynter, ed. Avian energetics. Nuttal Ornithol. Club, Harvard Univ., Cambridge, Mass.

King, J. R., and D. S. Farner. 1966. The adaptive role of winter fattening in the white-crowned sparrow with comments on its regulation. Am. Nat. 100:403–418.

Kinietz, W. V. 1940. The Indians of the western Great Lakes, 1615–1760. Occas. Contrib. Mus. Anthropol. Univ. of Mich. 10:1–427.

Kirikov, S. V. 1958. Changes in the animal world of our country. Izvestiya USSR Acad. Sci., Geogr. Ser., 1:71–83.

Kirikov, S. V. 1960. Changes in the animal world in natural zones of the USSR. A forest and forest-tundra zone. USSR Acad. Sci. 155pp.

Kirkwood, J. K. 1983. A limit to metabolizable energy intake in mammals and birds. Comp. Biochem. Physiol. 75A:1–3.

Kishida, K. 1927. [Introduced animal research in Japan]. J. Zool. 39:509 (in Japanese).

Klausnitzer, B. 1987. Verstädterung von Tieren (Die neue Brehm-Bücherei). Ziemsen, Wittenberg. 315pp.

Kleiber, M. 1975. The fire of life. Second ed. John Wiley and Sons, New York.

Klein, D. R., M. Meldgaard, and S. G. Fancy. 1987. Factors determining leg length in *Rangifer tarandus*. J. Mammal. 68:642–655.

Klippel, W. E., and P. W. Parmalee. 1982. The paleontology of Cheek Bend Cave: phase II report. Tenn. Valley Authority Contract Rep. 249pp.

Knudsen, K. L., and D. L. Kilgore. 1990. Temperature regulation and basal metabolic rate in the spotted skunk, *Spilogale putorius*. Comp. Biochem. Physiol. 97A:27–33.

Kodric-Brown, A., and J. H. Brown. 1978. Influence of economics, interspecific competition, and sexual dimorphism on territoriality of migrant rufous hummingbirds. Ecology 59:285–296.

Koehler, G. H., and M. G. Hornocker. 1977. Fire effects on marten habitat in the Selway-Bitterroot Wilderness. J. Wildl. Manage. 41:500–505.

Koehler, G. M., J. A. Blakesley, and T. W. Koehler. 1990. Marten use of successional forest stages during winter in north-central Washington. Northwest. Nat. 71:1–4.

Koehler, G. M., W. R. Moore, and A. R. Taylor. 1975. Preserving the pine marten: management guidelines for western forests. Western Wildlands 2:31–36.

Koenig, R., and F. Mueller. 1986*a*. Morphometrische Untersuchungen am mitteleuropäischen Baummarder und Steinmarder I. Jagd und Hege (4):31–33.

Koenig, R., and F. Mueller. 1986*b*. Morphometrische Untersuchungen am mitteleuropäischen Baummarder und Steinmarder II. Jagd und Hege (5):17–19.

Kohn, B. E., and R. G. Eckstein. 1987. Status of marten in Wisconsin, 1985. Wis. Dep. Nat. Resour. Res. Rep. 143. 18pp.

Kohn, B. E., N. F. Payne, J. E. Ashbrenner, and W. A. Creed. 1991. The fisher in Wisconsin. Wis. Dep. Nat. Resour. unpubl. rep. 43pp.

Korpimäki, E., and K. Norrdahl. 1989. Avian predation on mustelids in Europe. 1. Occurrence and effects on body size variation and life traits. Oikos 55:205–215.

Korschgen, L. J. 1980. Procedures for food-habits analyses. Pages 113–127 *in* S. D. Schemnitz, ed. Wildlife management techniques manual. Wildlife Society, Washington, D.C.

Kraft, V. A. 1975. On temperature effect upon the motility of the ermine in winter. Zool. Zhur. 45:148–150.

Krasovsky, L. I. 1970. O poloksityelinoi korrelyatsii myeksdiy chislyennoctioo desnoy kunitsii i sploschchnimi viriybkami lesov na yevropyeskom Syevyerye. Byull. Mosk. o-va Ispyt. Prir. 75:7–15 (in Russian, with English summary).

Krause, T. 1989. NTA trapping handbook—a guide for better trapping. Spearman Publ., Sutton, Nebr. 206pp.

Krebs, J. R., and N. B. Davies. 1984. Behavioral ecology: an evolutionary approach. Second ed. Blackwell, London.

Krohn, W. B., S. M. Arthur, and T. F. Paragi. 1989. Differential vulnerability of fishers to fall fur trapping. Transcript of presentation at 1989 Northeast Fish and Wildl. Conf., Ellenville, N.Y. 21pp.

Krott, P. 1973. Die Fortpflanzung des Edelmarders (*Martes martes* L.) in freier Wildbahn. Z. Jagdwiss. 19:113–117.

Krott, P., and T. Lampio. 1983. Näädän esiintymisestä Suomessa 1900-luvulla. Suomen Riista 30:60–63 (in Finnish, with English summary).

Krüger, H.-H. 1990. Home ranges and patterns of distribution of stone and pine martens. Pages 348–349 *in* S. Myrberget, ed. Trans. Nineteenth Int. Congr. Game Biol., Trondeim, Norway.

Kuehn, D. W. 1989. Winter foods of fishers during a snowshoe hare decline. J. Wildl. Manage. 53:688–692.

Kuehn, D. W., and W. E. Berg. 1981. Use of radiographs to identify age-classes of fisher. J. Wildl. Manage. 45:1009–1010.

Kugelschafter, K. 1988. Untersuchungen zum stoffwechselbedingten Verhalten beim Steinmarder (*Martes foina* Erxleben, 1777) unter Gefangenschaftsbedingungen. Diplomarbeit, Justus Liebig Universität, Giessen. 100pp.

Kugelschafter, K. 1989. First findings as to the influence of tradition in the spreading of the so called car-marten-phenomenon (*Martes foina*). Page 609 *in* Abstracts of Papers and Posters, Fifth Int. Theriol. Congr., Rome.

Kuroda, N. 1940. A monograph of the Japanese mammals. Sanseido, Tokyo. 311pp.

Kurtén, B., and E. Anderson. 1972. The sediments and fauna from Jaguar Cave. II. The fauna. Tebiwa 15:21–45.

Kurtén, B., and E. Anderson. 1980. Pleistocene mammals of North America. Columbia Univ. Press, New York. 442pp.

Kutilek, M. M., R. A. Hopkins, E. W. Clinite, and T. E. Smith. 1983. Monitoring population trends of large carnivores using track transects. Pages 104–106 *in* J. F. Bell and T. Atterbury, eds. Renewable resource inventories for monitoring changes and trends. Oregon State Univ., Corvallis.

Kuzmina, I. Y. 1971. Some data on the middle Urals mammals during the late Anthropogene. Proc. USSR A.C. Zool. Inst. 49:25–37.

Kuzmina, I. Y. 1975. Some data on the middle Urals mammals during the late Pleistocene. Bull. Comm. Quat. Period Stud. (Moscow) 43:44–123.

LaBarge, T., A. Baker, and D. Moore. 1990. Fisher (*Martes*): birth, growth and development in captivity. Mustelid and Viverrid Conserv. Newsl. IUCN/SCC Mustelid and Viverrid Spec. Group (Belgium) 2:1–3.

Laberee, E. E. 1941. Breeding and reproduction in fur bearing animals. Fur Trade J. Can., Toronto, Ont. 166pp.

Labhardt, F. 1991. Der Rotfuchs. Paul Parey, Hamburg.

Lack, D. 1942. Ecological features of the bird faunas of British small islands. J. Anim. Ecol. 11:9–36.

Lagerkvist, G., E. J. Einarsson, M. Forsberg, and H. Gustafson. 1992. Profiles of oestradiol-17β and progesterone and follicular development during the reproductive season in mink (*Mustela vison*). J. Reprod. Fertil. 94:11–21.

Lampe, R. P. 1976. Aspects of the predatory strategy of the North American badger *Taxidea taxus*. Ph.D. dissertation, Univ. of Minnesota, Minneapolis. 103pp.

Lang, R. W. 1968. The natural environment and subsistence economy of the McKees Rocks Village site. Pa. Archaeol. 38:50–80.

Lang, R. W., and A. H. Harris. 1984. The faunal remains from Arroyo Hondo Pueblo, New Mexico. School Am. Res. Press, Santa Fe, N.M. 316pp.

Larson, J. S., and R. D. Taber. 1980. Criteria of sex and age. Pages 143–202 *in* S. D. Schemnitz, ed. Wildlife management techniques manual. Fourth ed. Wildlife Society, Washington, D.C.

Lawlor, T. E. 1982. The evolution of body size in mammals: evidence from insular populations in Mexico. Am. Nat. 119:54–72.

Lea, S. E. G., and R. M. Tarpy. 1986. Hamsters' demand for food to eat and hoard as a function of deprivation and cost. Anim. Behav. 34:1759–1768.

Leach, D., B. K. Hall, and A. I. Dagg. 1982. Aging marten and fisher by development of the suprafabellar tubercle. Can. J. Zool. 56:1180–1911.

Leberg, P. L. 1990. Genetic considerations in the design of introduction programs. Trans. North Am. Wildl. Nat. Resour. Conf. 55:609–619.

Lee, J. E., T. C. White, R. A. Garrott, R. M. Bartmann, and A. W. Alldredge. 1985. Assessing accuracy of a radio-telemetry system for estimating animal locations. J. Wildl. Manage. 49:658–633.

Lee, R. 1978. Forest microclimatology. Columbia Univ. Press, New York. 276pp.

Lehmkuhl, J. F. 1984. Determining size and dispersion of minimum viable populations land management planning and species conservation. Environ. Manage. 8:167–176.

Le Maho, Y., H. Vu Van Kha, H. Koubi, G. Dfwasmes, J. Giraro, P. Ferre, and M. Cagnard. 1981. Body composition, energy expenditure and plasma metabolites in long-term fasting geese. Am. J. Physiol. 241:E342-E354.

Lensink, C. J. 1953. An investigation of the marten in interior Alaska. M.S. thesis, Univ. of Alaska, Fairbanks. 89pp.

Lensink, C. J., R. O. Skoog, and J. L. Buckley. 1955. Food habits of marten in interior Alaska and their significance. J. Wildl. Manage. 19:364–368.

Leonard, R. D. 1980. The winter activity and movements, winter diet and breeding biology of the fisher (*Martes pennanti*) in southeastern Manitoba. M.S. thesis, Univ. of Manitoba, Winnipeg. 181pp.

Leonard, R. D. 1986. Aspects of reproduction in the fisher, *Martes pennanti*, in Manitoba. Can. Field-Nat. 100:32–34.

Leveille, G. A. 1970. Adipose tissue metabolism: influence of periodicity of eating and diet composition. Fed. Proc. 29:1294–1301.

Levins, R. 1968. Evolution in changing environments. Princeton Univ. Press, Princeton, N.J. 120pp.

Lewis, R. B. 1979. Hunter-gatherer foraging: some theoretical explanations and archeological tests. Ph.D. dissertation, Univ. of Illinois, Urbana-Champaign. 255pp.

Li, J. B., and A. L. Goldberg. 1976. Effects of food deprivation on protein synthesis and degradation in rat skeletal muscle. Am. J. Physiol. 231:441–448.

Liat, L. B., N. Sustriayu, T. R. Hadi, and Y. H. Bang. 1980. A study of small mammals in the Ciloto Field Station area, West Java, Indonesia, with special reference to vectors of plague and scrub typhus. Southeast Asian J. Trop. Med. Public Health 11:71–80.

Libois, R. 1991. La fouine (*Martes foina* Erxleben, 1777). *In* M. Artrois and P. Delattre, eds. Encyclopédie des carnivores de France. Soc. française l'étude protection mammifères, Nort sur Erdre. 53pp.

Lilligraven, J. A. 1975. Biological considerations of the marsupial-placental dichotomy. Evolution 29:709–722.

Lima, S. L. 1986. Predation risk and unpredictable feeding conditions: determinants of body mass in wintering birds. Ecology 67:377–385.

Lindlöf, B., and K. Ellström. 1980. Avskjutningsstatistiken ger besked: visst fanns det mer småvilt förr. Svensk Jakt 118:1006–1009.

Lindstedt, S. L., and M. S. Boyce. 1985. Seasonality, fasting endurance and body size in mammals. Am. Nat. 125:879–887.

Lindström, E. 1982. Population ecology of the red fox (*Vulpes vulpes* L.) in relation to food supply. Ph.D. dissertation, Univ. of Stockholm, Sweden. 152pp.

Lindzey, F. G., S. K. Thompson, and J. I. Hodges. 1977. Scent station index of black bear abundance. J. Wildl. Manage. 41:151–153.

Linhart, S. B., and F. F. Knowlton. 1975. Determining the relative abundance of coyotes by scent station lines. Wildl. Soc. Bull. 3:119–124.

Link, W. A., and J. S. Hatfield. 1990. Power calculations and model selection for trend analysis: a comment. Ecology 71:1217–1220.

Litvaitis, J. A., J. A. Sherburne, and J. A. Bissonette. 1985. Influence of understory characteristics on snowshoe hare habitat use and density. J. Wildl. Manage. 49: 866–873.

Liu, K.-B., and N. S. N. Lam. 1985. Paleovegetational reconstruction based on modern and fossil pollen data: an application of discriminant analysis. Ann. Assoc. Am. Geogr. 75:115–130.

Livingston, S. 1984. Faunal analysis. Pages 181–192 *in* E. Lohse, ed. Archeological investigations at site 45-OK-11, Chief Joseph Dam project, Washington. Off. Public Archeol., Univ. of Washington, Seattle.

Livingston, S. 1985. Faunal analysis. Pages 117–133 *in* M. Jaehnig, ed. Archeological investigations at site 45-OK-258, Chief Joseph Dam project, Washington. Off. Public Archeol., Univ. of Washington, Seattle.

Lobachev, I. S. 1973. The ecology of the stone marten in south-east Kazakhstan. Trudy instituta zoologii A.N. Kazakhskoi SSR 34:108–133.

Lockie, J. D. 1961. The food of the pine marten *Martes martes* in west Ross-shire, Scotland. Proc. Zool. Soc. Lond. 136:187–195.

Lockie, J. D. 1964. Distribution and fluctuations of the pine marten, *Martes martes* (L.), in Scotland. J. Anim. Ecol. 33:349–356.

Lockie, J. D. 1966. Territory in small carnivores. Symp. Zool. Soc. Lond. 18:143–165.

Łomnicki, A. 1978. Individual differences between animals and the natural regulation of their numbers. J. Anim. Ecol. 47:461–475.

Łomnicki, A. 1988. Population ecology of individuals. Princeton Univ. Press, Princeton, N.J. 223pp.

Lomolino, M. V. 1985. Body size of mammals on islands: the island rule reexamined. Am. Nat. 125:310–316.

Looman, J., and K. F. Best. 1987. Budd's flora of the Canadian prairie provinces. Res. Branch, Agric. Can. Publ. 1662. 863pp.

Lowe, L. D. 1930. The first authentic report of fisher bred in captivity. Am. Fur Breeder, June:34.

Lower, J. 1989. Thoroughly modern martens. BBC Wildl. 7:326- 329.

Lucas, M. P. A. 1989. Een jonge steenmarter (*Martes foina*) in Nijmegen: opvoeding, ontwikkeling en dispersie. Intern. rapport 89/18 Rijksinstitut voor Natuurbeheer, Arnhem. 28pp.

Lundberg, U. 1987. Soziometrie biologischer Dominanzstrukturen. Biol. Rundschau. 25:355–368.

Lundelius, E. L., Jr., R. W. Graham, E. Anderson, J. Guilday, J. A. Holman, D. W. Steadman, and S. D. Webb. 1983. Terrestrial vertebrate faunas. Pages 311–353 *in* S. C. Porter, ed. Late-Quaternary environments of the United States. Vol. 1: The late Pleistocene. Univ. of Minnesota Press, Minneapolis.

Lyman, C. P., and P. O. Chatfield. 1955. Physiology of hibernation in mammals. Physiol. Rev. 35:403–425.

Lyman, R. L. 1991. Prehistory of the Oregon coast: The effects of excavation strategies and assemblage size on archeological inquiry. Academic Press, New York.

Ma, Y., and J. Wu. 1981. Systematic review of Chinese sables, with description of a new subspecies. Acta Zool. Sinica 27:189–196 (in Chinese with English summary).

MacArthur, R. H. 1972. Patterns of terrestrial bird communities. Pages 189–221 *in* D. S. Farner and J. R. King, eds. Avian biology. Vol. 1. Academic Press, New York.

MacArthur, R. H., and E. R. Pianka. 1966. On optimal use of a patchy environment. Am. Nat. 100:603–609.

Macdonald, D. 1983. The ecology of carnivore social behaviour. Nature (Lond.) 301:379–384.

Mace, R., and T. Manley. 1991. Use of systematically deployed remote cameras to monitor grizzly bears. 1990 report. Montana Fish, Wildl., Parks, Kalispell. 35pp.

MacLean, S. F., B. Fitzgerald, and F. A. Pitelka. 1974. Population cycles in arctic lemmings: winter reproduction and predation by weasels. Arctic and Alpine Res. 6:1–12.

McNab, B. K. 1963. Bioenergetics and the determination of home range size. Am. Nat. 97:133–140.

McNab, B. K. 1969. The economics of temperature regulation in Neotropical bats. Comp. Biochem. Physiol. 31:227–268.

McNab, B. K. 1971. On the ecological significance of Bergmann's rule. Ecology 52: 845–854.

McNab, B. K. 1980. Food habits, energetics, and the population biology of mammals. Am. Nat. 116:106–124.

McNab, B. K. 1983. Ecological and behavioral consequences of adaptation to various food resources. Pages 664–697 *in* J. F. Eisenberg and D. G. Kleiman, eds. Advances in the study of mammalian behavior. Am. Soc. Mammal. Spec. Publ. 7.

McNab, B. K. 1986. The influence of food habits on the energetics of eutherian mammals. Ecol. Monogr. 56:1–19.

McNab, B. K. 1989. Basal rate of metabolism, body size, and food habits in the order Carnivora. Pages 335–354 *in* J. L. Gittleman, ed. Carnivore behavior, ecology and evolution. Cornell Univ. Press, Ithaca, N.Y.

McNamara, J. M., A. I. Houston, and J. R. Krebs. 1990. Why hoard? The economics of food storing in tits *Parvis* spp. Behav. Ecol. 1:12–23.

McNeilly, A. S. 1988. Suckling and the control of gonadotropin secretion. Pages 2323–2349 *in* E. Knobil and J. D. Neal, eds. The physiology of reproduction. Raven Press, New York.

Madsen, A. B., and A. M. Rasmussen. 1985. Reproduction in the stone marten *Martes foina* in Denmark. Natura Jutlandica 21:145–148.

Magoun, A. J., and D. J. Vernam. 1986. An evaluation of the Bear Creek burn as marten (*Martes americana*) habitat in interior Alaska. Final report. U.S. Dep. Inter. Bur. Land Manage. and Alas. Dep. Fish Game Spec. Proj. AK-950-CAH-0. 58pp.

Mahfoud, M., and G. P. Patil. 1982. On weighted distributions. Pages 479–492 *in* G. Kallianpur, P. R. Krishnaiah, and J. K. Ghosh, eds. Statistics and probability: essays in honor of C. R. Rao. North Holland, New York.

Majewski, P., and J. Rolstad. In Prep. Effects of forestry on black woodpecker food supply in the boreal forest.

Major, J. T. 1979. Marten use of habitat in a commercially clear-cut forest during summer. M.S. thesis, Univ. of Maine, Orono. 32pp.

Manville, R. H., and S. P. Young. 1965. Distribution of Alaskan mammals. U.S. Fish Wildl. Serv. Circ. 211. 74pp.

Marchesi, P. 1989. Ecologie et comportement de la martre (*Martes Martes*) dans le Jura Suisse. D.Sc. dissertation, Univ. of Neuchâtel, Switz. 185pp.

Marchesi, P., N. Lachat, R. Lienhard, P. Devieve, and C. Mermod. 1989. Comparaison des régimes alimentaires de la fouine (*Martes foina*) et de la martre (*Martes martes*) dans une région du Jura Suisse. Rev. Suisse Zool. 96:281–296.

Marchesi, P., and C. Mermod. 1989. Regime alimentaire de la martre (*Martes martes* L.) dans le Jura Suisse (Mammalia: Mustelidae). Rev. Suisse Zool. 96:127–146.

Marcot, B. G., M. G. Raphael, and K. H. Berry. 1983. Monitoring wildlife habitat and validation of wildlife-habitat relationships models. Trans. North Am. Wildl. Nat. Resour. Conf. 48:315–329.

Marcum, C. L., and D. O. Loftsgaarden. 1980. A non-mapping technique for studying habitat preferences. J. Wildl. Manage. 44:963–968.

Markley, M. H., and C. F. Bassett. 1942. Habits of captive marten. Am. Midl. Nat. 28:604–616.

Marshall, W. H. 1946. Winter food habits of the pine marten in Montana. J. Mammal. 27:83–84.

Marshall, W. H. 1951*a*. An age determination method for the pine marten in Montana. J. Wildl. Manage. 15:276–283.

Marshall, W. H. 1951*b*. Pine marten as a forest product. J. For. 49:899–905.

Marshall, W. H., and R. K. Enders. 1942. The blastocyst of the marten. Anat. Rec. 84:307–310.

Martell, A. M. 1983. Changes in small mammal communities after logging in north-central Ontario. Can. J. Zool. 61:970–980.

Martin, S. K. 1987. The ecology of the pine marten (*Martes americana*) at Sagehen Creek, California. Ph.D. dissertation, Univ. of California, Berkeley. 223pp.

Martin, T. J. 1986. A faunal analysis of Fort Ouiatenon, an eighteenth century trading post in the Wabash Valley of Indiana. Ph.D. dissertation, Michigan State Univ., East Lansing. 520pp.

Martin, T. J. 1991. Animal remains from the Ogontz Bay site (USFS 09-10-01-328) in Delta County, Michigan. Ill. State Mus. Tech. Rep. 91-000-16.

Martin, T. J., and M. C. Masulis. 1988. Preliminary report on animal remains from Fort De Chartres. Appendix *in* D. Keene, ed. Archeological excavations at Fort De Chartres, 1985–1987. Ill. Hist. Preserv., Springfield.

Maser, C., Z. Maser, J. W. Witt, and G. Hunt. 1986. The northern flying squirrel: a mycophagist in southwestern Oregon. Can. J. Zool. 64:2086–2089.

Maser, C., B. R. Mate, J. F. Franklin, and C. T. Dyrness. 1981. Natural history of Oregon coast mammals. USDA For. Serv. Gen. Tech. Rep. PNW-133. 496pp.

Maser, C., J. M. Trappe, and R. A. Nussbaum. 1978. Fungal–small mammal interrelationships with emphasis on Oregon coniferous forests. Ecology 59:799–809.

Maxham, G. 1970. Return of the marten. Minn. Volunteer 33(189):8–12.

May, R. M. 1973. Stability and complexity in model ecosystems. Princeton Univ. Press, Princeton, N.J. 265pp.

Mayer, M. V. 1957. A method for determining the activity of burrowing mammals. J. Mammal. 38:531.

Mayr, E. 1963. Animal species and evolution. Belknap Press, Cambridge, Mass. 797pp.

Mead, E. M., and J. I. Mead. 1989. Snake Creek Burial Cave and a review of the Quaternary mustelids of the Great Basin. Great Basin Nat. 49:143–154.

Mead, E. M., R. S. Thompson, and T. R. Van Devender. 1982. Late Wisconsinan and Holocene fauna from Smith Creek Canyon, Snake River Range, Nevada. Trans. San Diego Soc. Nat. Hist. 20:1–26.

Mead, R. A. 1968. Reproduction in eastern forms of the spotted skunk (genus *Spilogale*). J. Zool. (Lond.) 156:119–136.

Mead, R. A. 1970. The reproductive organs of the male spotted skunk (*Spilogale putorius*). Anat. Rec. 167:291–302.

Mead, R. A. 1981. Delayed implantation in the Mustelidae with special emphasis on the spotted skunk. J. Reprod. Fert. (suppl.) 29:11–24.

Mead, R. A. 1986. Role of the corpus luteum in controlling implantation in mustelid carnivores. Ann. N.Y. Acad. Sci. 476:25–35.

Mead, R. A. 1989*a*. The physiology and evolution of delayed implantation in carnivores. Pages 437–464 *in* J. L. Gittleman, ed. Carnivore behavior, ecology, and evolution. Cornell Univ. Press, Ithaca, N.Y.

Mead, R. A. 1989*b*. Reproduction in mustelids. Pages 124–137 *in* U. S. Seal, E. T. Thorne, M. A. Bogan, and S. H. Anderson, eds. Conservation biology and the black-footed ferret. Yale Univ. Press, New Haven, Conn.

Mead, R. A., and N. M. Czekala. 1990. Reproductive cycle of the steppe polecat (*Mustela eversmanni*). J. Reprod. Fertil. 88:353–360.

Mead, R. A., M. M. Joseph, S. Neirinckx, and M. Berria. 1988. Partial characterization of a luteal factor that induces implantation in the ferret. Biol. Reprod. 38:798–803.

Mead, R. A., S. Neirinckx, and N. M. Czekala. 1990. Photomanipulation of sexual maturation and breeding cycle of the steppe polecat (*Mustela eversmanni*) and other techniques for more rapid propagation of the species. J. Exp. Zool. 255:232–238.

Mead, R. A., A. W. Rourke, and A. Swannack. 1979. Uterine protein synthesis during delayed implantation in the western spotted skunk and its regulation by hormones. Biol. Reprod. 21:39–46.

Mech, L. D. 1983. Handbook of animal radio-tracking. Univ. of Minnesota Press, Minneapolis. 107pp.

Mech, L. D., and L. L. Rogers. 1977. Status, distribution, and movements of martens in northeastern Minnesota. USDA For. Serv. Res. Pap. NC-143. 7pp.

Melchior, H. R., N. F. Johnson, and J. S. Phelps. 1987. Wild furbearer management in the western United States and Alaska. Pages 1117–1128 *in* M. Novak, J. A. Baker, M. E. Obbard, and B. Malloch, eds. Wild furbearer management and conservation in North America. Ont. Trappers Assoc., North Bay, Ont.

Migula, P. 1969. Bioenergetics of pregnancy and lactation in European common vole. Acta Theriol. 14:167–179.

Miller, B. J., and S. H. Anderson. 1989. Failure of fertilization following abbreviated copulation in the ferret (*Mustela putorius furo*). J. Exp. Zool. 249:85–89.

Miller, D. R. 1961. Marten transplanting in northern Manitoba. Man. Game Branch Biol. Rep. 12pp.

Miller, R. G. 1981. Simultaneous statistical inference. Springer-Verlag, New York. 299pp.

Miller, R. G., R. W. Pitcey, and R. Y. Edwards. 1955. Live trapping marten in British Columbia. Murrelet 36:1–8.

Millward, D. J., and J. C. Waterlow. 1978. Effect of nutrition on protein turnover in skeletal muscle. Fed. Proc. 37:2283–2290.

Mitton, J. B., and M. G. Raphael. 1990. Genetic variation in the marten, *Martes americana*. J. Mammal. 71:195–197.

Mochrie, R. D. 1964. Feeding proteins and efficiency in ruminants. Fed. Proc. 23:85–87.

Moehlman, P. D. 1987. Social organization of jackals. Am. Sci. 75:366–375.

Mohr, C. O. 1947. Table of equivalent populations of North American small mammals. Am. Midl. Nat. 37:223–249.

Mohr, C. O., and W. A. Stumpf. 1966. Comparison of methods for calculating areas of animal activity. J. Wildl. Manage. 30:293–304.

Moors, P. J. 1974. The annual energy budget of a weasel (*Mustela nivalis* L.) population in farmland. Ph.D. dissertation, Univ. of Aberdeen, Scotland. 153pp.

Moors, P. J. 1977. Studies of the metabolism, food consumption and assimilation efficiency of a small carnivore, the weasel (*Mustela nivalis*). Oecologia 27:185–202.

Moors, P. J. 1980. Sexual dimorphism in the body size of mustelids (Carnivora): the roles of food habits and breeding systems. Oikos 34:147–158.

More, G. 1978. Ecological aspects of food selection in pine marten (*Martes americana*). M.S. thesis, Univ. of Alberta, Edmonton. 94pp.

Moreno, S., A. Rodriguez, and M. Delibes. 1988. Summer foods of the pine marten (*Martes martes*) in Majorca and Minorca, Balearic Islands. Mammalia 52:289–291.

Morozov, V. F. 1976. [Feeding habits of *Martes martes* (Carnivora, Mustelidae) in different regions of the north-west of the USSR]. Zool. Zhur. 55:1886–1892 (in Russian, with English summary).

Morris, D. W. 1987. Ecological scale and habitat use. Ecology 68:362–369.

Morrison, P. R. 1960. Some interactions between weight and hibernation function. Harvard Univ. Mus. Comp. Zool. Bull. 124:75–91.

Morrison, P. R., and F. A. Ryser. 1951. Temperature and metabolism in some Wisconsin mammals. Fed. Proc. 10:93–94.

Mortensen, A., and A. S. Blix. 1985. Seasonal changes in the effects of starvation on metabolic rate and regulation of body weight in Svalbard ptarmigan. Ornis Scand. 16:20–24.

Moshonkin, N. N. 1981. Potential polyestricity of the mink (*Lutreola lutreola*). Zool. Zhur. 60:1731–1734.

Moshonkin, N. N. 1983. The reproductive cycle in females of the European mink (*Lutreola lutreola*). Zool. Zhur. 62:3–15.

Mrosovsky, N. 1966. Acceleration of annual cycle to six weeks in captive dormouse. Can. J. Zool. 44:903–910.

Mrosovsky, N., and T. L. Powley. 1977. Set point for body weight and fat. Behav. Biol. 20:205.

Mrosovsky, N., and D. F. Sherry. 1980. Animal anorexia. Science 207:837–842.

Müeskens, G. J. D. M., L. T. J. Meuwiesen, and S. Broekhuizen. 1989. Simultaneus use of day-hides in beech martens (*Martes foina*). Pages 409–421 *in* M. Stubbe, ed. Proc. Mitteleuropäisches Symposium zur Populationsökologie von Mustelidenarten.

Mullis, C. 1985. Habitat utilization by fisher (*Martes pennanti*) near Hayfork Bally, California. M.S. thesis, Humboldt State Univ., Arcata, Calif. 91pp.

Murie, A. 1961. Some food habits of the marten. J. Mammal. 42:516–521.

Murie, O. J. 1974. A field guide to animal tracks. Second ed. Houghton Mifflin, Boston. 375pp.

Myrcha, A. L., L. Ryszkowski, and W. Walkowa. 1969. Bioenergetics of pregnancy and lactation in the white mouse. Acta Theriol. 15:161–166.

Nagorsen, D. W., R. W. Campbell, and G. R. Giannico. 1991. Winter food habits of marten, *Martes americana*, on the Queen Charlotte Islands. Can. Field-Nat. 105: 55–59.

Nagorsen, D. W., J. Forsberg, and G. R. Giannico. 1988. An evaluation of canine radiographs for sexing and aging Pacific coast martens. Wildl. Soc. Bull. 16:421–426.

Nagorsen, D. W., K. F. Morrison, and J. E. Forsberg. 1989. Winter diet of Vancouver Island marten (*Martes americana*). Can. J. Zool. 67:1394–1400.

Nale, R. F. 1963. The salvage excavation of the Boyle site (36-WH-19). Pa. Archaeol. 33:164–194.

Naumov, S. P., and N. N. Rukovsky. 1972. Some aspects of spatial differentiation of mammal populations with special reference to *Martes martes*. Zool. Zhur. 51:1870–1874.

Nelson, K. A. 1979. The occurrence of wolverine (*Gulo luscus*) and other mammals by baited hair traps and snow transects in Six Rivers National Forest. Six Rivers Natl. For., Tech. Rep., Eureka, Calif. 57pp.

Nelson, M. M., and H. M. Evans. 1961. Dietary requirements for lactation in the rat and other laboratory animals. Pages 137–191 *in* S. K. Kon and A. Cowile, eds. Milk: the mammary gland and its secretion. Vol. 2. Academic Press, New York.

Nelson, R. A., J. D. Jones, H. W. Wahner, D. B. McGill, and C. F. Code. 1975. Nitrogen metabolism in bears: urea metabolism in summer starvation and in winter sleep and the role of urinary bladder in water and nitrogen conservation. Mayo Clinic Proc. 50: 141–146.

Nelson, R. D., and H. Salwasser. 1982. The Forest Service wildlife and fish habitat relationship program. Trans. North Am. Wildl. Nat. Resour. Conf. 47:174–183.

Nelson, R. K. 1973. Hunters of the northern forest. Univ. of Chicago Press, Chicago. 339pp.

Nelson, U. C. 1952*a*. Stocking, restocking and introduction of game birds and mammals in Alaska. Q. Rep., Fed. Aid Wildl. Restor. Proj. W-4-D-3. Vol. 3, no. 1. 2pp.

Nelson, U. C. 1952*b*. Stocking, restocking and introduction of game birds and mammals in Alaska. Q. Rep., Fed. Aid Wildl. Restor. Proj. W-4-D-3. Vol. 3, no. 2. 5pp.

Nesvadbova, J., and J. Zedja. 1984. The pine marten (*Martes martes*) in Bohemia and Moravia. Folia. Zool. 33:57–64.

Neu, C. W., C. R. Byers, and J. M. Peek. 1974. A technique for analysis of utilization-availability data. J. Wildl. Manage. 38:541–545.

Neusius, S. W. 1986. Faunal remains from Lemoc Shelter. Pages 279–296 *in* D. Breternitz, ed. Dolores archeological program: Anasazi communities at Dolores: early small settlements in the Dolores River canyon and western Sagehen Flats area. U.S. Dep. Inter., Bur. Reclamation, Denver, Colo.

Neusius, S. W., and M. Gould. 1988. Faunal remains: implications for Dolores Anasazi adaptations. Pages 1049–1143 *in* D. Breternitz, ed. Dolores archeological program: Anasazi communities at Dolores: Grass Mesa Village. U.S. Dep. Inter., Bur. Reclamation, Denver, Colo.

New Brunswick Department of Natural Resources. 1990. New Brunswick land habitat management program. Prog. Rep. to Wildl. Habitat Can. 62pp.

Newby, F. E. 1951. Ecology of the marten in the Twin Lakes area, Chelan County, Washington. M.S. thesis, State College of Washington, Pullman. 38pp.

Newby, F. E., and V. D. Hawley. 1954. Progress on a marten livetrapping study. Trans. North Am. Wildl. Conf. 19:452–460.

Nicht, M. 1969. Ein Beitrag zum Vorkommen des Steinmarders, *Martes foina* (Erxleben, 1777) in der Großstadt (Magdeburg). Z. Jagdwiss. 15:1–6.

Nordquist, G. E., and E. C. Birney. 1988. Mammals: endangered, threatened, and special concern. Pages 294–322 *in* B. Coffin and L. Pfannmuller, eds. Minnesota's endangered flora and fauna. Univ. of Minnesota Press, Minneapolis.

Northwest Territories Renewable Resources. 1992. The Northwest Territories trap exchange program. Northwest Territories Renewable Resources, Yellowknife. 4pp.

Novak, J., M. E. Obbard, J. G. Jones, R. Newmann, A. Booth, A. J. Satterthwaite, and G. Linscombe. 1987. Furbearer harvests in North America. Ont. Trappers Assoc., North Bay, Ont. 270pp.

Novak, M. 1987. Wild furbearer management in Ontario. Pages 1049–1061 *in* M. Novak, J. A. Baker, M. E. Obbard, and B. Malloch, eds. Wild furbearer management and conservation in North America. Ont. Trappers Assoc., North Bay, Ont.

Nyholm, E. S. 1970. On the ecology of the pine marten (*Martes martes*) in eastern and northern Finland. Suomen Riista 22:105–118.

Obbard, M. E. 1987*a*. Red squirrel. Pages 264–281 *in* M. Novak, J. A. Baker, M. E. Obbard, and B. Malloch, eds. Wild furbearer management and conservation in North America. Ont. Trappers Assoc., North Bay, Ont.

Obbard, M. E. 1987*b*. Fur grading and pelt identification. Pages 717–826 *in* M. Novak, J. A. Baker, M. E. Obbard, and B. Malloch, eds. Wild furbearer management and conservation in North America. Ont. Trappers Assoc., North Bay, Ont.

Obbard, M. E., J. G. Jones, R. Newman, A. Booth, A. J. Satterthwaite, and G. Linscombe. 1987. Furbearer harvests in North America. Pages 1007–1038 *in* M. Novak, J. A. Baker, M. E. Obbard, and B. Malloch, eds. Wild furbearer management and conservation in North America. Ont. Trappers Assoc., North Bay, Ont.

Ognev, S. I. 1925. A systematical review of the Russian sables. J. Mammal. 6:276–280.

Ognev, S. I. 1931. Mammals of eastern Europe and northern Asia. 2. Carnivora (Fissipedia). Israel Sci. Transl. Program, Jerusalem, 1962.

Oksanen, T., L. Oksanen, and S. D. Fretwell. 1985. Surplus killing in the hunting strategy of small predators. Am. Nat. 126:328–346.

Olsen, G. H., R. G. Linscombe, V. L. Wright, and R. A. Holmes. 1988. Reducing injuries to terrestrial furbearers by using padded foothold traps. Wildl. Soc. Bull. 16:303–307.

Onderka, D. K., D. L. Skinner, and A. W. Todd. 1990. Injuries to coyotes and other species caused by four models of footholding devices. Wildl. Soc. Bull. 18:175–182.

Orsborn, E. V. 1953. More on marten raising. Fur Trade J. Can. 31(4):14.

Otis, D. L., K. P. Burnham, G. C. White, and D. R. Anderson. 1978. Statistical inference from capture data on closed animal populations. Wildl. Monogr. 62. 135pp.

Pack, J. C., and J. I. Cromer. 1981. Reintroduction of fisher in West Virginia. Pages

1431–1442 *in* J. A. Chapman and D. Pursley, eds. Proc. Worldwide Furbearer Conf., Frostburg, Md.

Pagel, M. D., and P. H. Harvey. 1988. The taxon-level problem in the evolution of mammalian brain size: facts and artifacts. Am. Nat. 132:344–359.

Paget, R. J. 1980. Dormancy of a badger (*Meles meles*) outside the set entrance. J. Zool. (London) 192:558.

Påhlsson, L., R. Frisén, C. H. Ovesen, A. Haapahen, B. Strandli, E. Dahl, and E. Einarsson. 1984. Naturgeogrdfisk regionindelning av Norden. Nordic Advisory Council. 289pp.

Paragi, T. F. 1990. Reproductive biology of female fishers in southcentral Maine. M.S. thesis, Univ. of Maine, Orono. 107pp.

Parmalee, P. W. 1959. Animal remains from the Raddatz Rockshelter, Sk5, Wisconsin. Wis. Archeol. 40:83–90.

Parmalee, P. W. 1960a. Additional fisher records from Illinois. Trans. Ill. Acad. Sci. 53:48–49.

Parmalee, P. W. 1960b. Animal remains from the Aztalan site, Jefferson County, Wisconsin. Wis. Archeol. 41:1–10.

Parmalee, P. W. 1960c. Animal remains from the Durst Rockshelter, Sauk County, Wisconsin. Wis. Archeol. 41:11–17.

Parmalee, P. W. 1960d. A prehistoric record of the fisher in Georgia. J. Mammal. 41:409–410.

Parmalee, P. W. 1963. Vertebrate remains from the Bell site, Winnebago County, Wisconsin. Wis. Archeol. 44:58–69.

Parmalee, P. W. 1964. Vertebrate remains from an historic archaeological site in Rock Island County, Illinois. Trans. Ill. Acad. Sci. 57:167–174.

Parmalee, P. W. 1968. Vertebrate remains from the Mason site (40Fr8), Franklin County, Tennessee. Pages 256–262 *in* C. Faulkner, ed. Archaeological investigations in the Tims Ford Reservoir, Tennessee, 1966. Natl. Park Serv. Contract, Dep. Anthropol., Univ. of Tennessee.

Parmalee, P. W. 1971. Fisher and porcupine remains from cave deposits in Missouri. Trans. Ill. Acad. Sci. 64:225–229.

Parmalee, P. W., and R. D. Oesch. 1972. Pleistocene and Recent faunas from the Brynjulfson caves, Missouri. Ill. State Mus. Rep. Invest. 59pp.

Parsons, G. R., M. K. Brown, and G. B. Will. 1978. Determining the sex of fisher from the lower canine teeth. N.Y. Fish Game J. 25:42–44.

Patterson, B. D. 1984. Mammalian extinction and biogeography in the southern Rocky Mountains. Pages 247–293 *in* M. H. Nitecki, ed. Extinctions. Univ. of Chicago Press, Chicago.

Pauls, R. W. 1978. Behavioral strategies relevant to the energy economy of the red squirrel *Tamiasciurus hudsonicus*. Can. J. Zool. 56:431–435.

Payette, S., C. Morneau, L. Seiois, and M. Desponts. 1989. Recent fire history of the northern Quebec biomes. Ecology 70:656–673.

Pearson, O. P., and R. K. Enders. 1944. Duration of pregnancy in certain mustelids. J. Exp. Zool. 95:21–35.

Pease, J. L., R. H. Vowles, and L. B. Keith. 1979. Interactions of snowshoe hares and woody vegetation. J. Wildl. Manage. 43:43–60.

Petersen, L. R., M. A. Martin, and C. M. Fils. 1977. Status of fishers in Wisconsin, 1975. Wis. Dep. Nat. Resour. Res. Rep. 92. 14pp.

Petterborg, L. 1978. Effect of photoperiod on body weight in the vole *Microtus montanus*. Can. J. Zool. 58:431–435.

Pianka, E. R. 1966. Latitudinal gradients in species diversity: a review of concepts. Am. Nat. 100:33–46.

Picton, H. D. 1979. The application of insular biogeographic theory to the conservation of large mammals in the northern Rocky Mountains. Biol. Conserv. 15:73–79.

Pielou, E. C. 1974. Population and community ecology: principles and methods. Gordon and Breach, New York. 424pp.

Pierce, D. J. 1983. Food habits, movements, habitat use and populations of moose in central Idaho and relationships to forest management. M.S. thesis, Univ. of Idaho, Moscow. 205pp.

Pietz, P. J., and J. R. Tester. 1983. Habitat selection by snowshoe hares in northcentral Minnesota. J. Wildl. Manage. 47:686–696.

Pigozzi, G. 1990. Latrine use and the function of territoriality in the European badger, *Meles meles*, in a Mediterranean coastal habitat. Anim. Behav. 39:1000–1002.

Pittaway, R. J. 1983. Fisher and red fox interactions over food. Ont. Field Biol. 37: 88–90.

Pitts, G. C., and T. R. Bullard. 1968. Some interspecific aspects of body composition in mammals. Pages 45–70 *in* Body composition in animals and man. Nat. Acad. Sci. Proc. 1598.

Pollock, K. H., J. D. Nichols, C. Brownie, and J. E. Hines. 1990. Statistical inference for capture-recapture experiments. Wildl. Monogr. 107. 97pp.

Polyntsev, Y. V., T. G. Novikova, A. G. Volcheck, and V. B. Rozen. 1985. Effect of 17 β-estradiol and progesterone on the processes of mating and ovulation in sable. Page 130 *in* V. G. Safonov, ed. Biology and pathology of farm-bred fur-bearing animals. Oxonian Press, New Delhi.

Polyntsev, Y. V., A. G. Volchek, V. M. Ilyinsky, and V. B. Rozen. 1975. Progesterone and estrogen content during the reproduction of sables. Probl. Endokrinol. 21:86–94.

Ponomarev, A. L. 1938. On the variability and inheritance of colour and pattern in the sable (*Martes zibellina* L.). Zool. Zhur. 17:482–504.

Ponomarev, A. L. 1944. The reaction of some Mustelidae to a gradient in temperature. Zool. Zhur. 38:51–55.

Porter, W. P., and D. M. Gates. 1969. Thermodynamic equilibria of animals with their environment. Ecol. Monogr. 39:227–244.

Powell, R. A. 1975. The misunderstood weasels. Defenders 50:440–441.

Powell, R. A. 1977. Hunting behavior, ecological energetics and predator-prey community stability of the fisher (*Martes pennanti*). Ph.D. dissertation, Univ. of Chicago. 132pp.

Powell, R. A. 1978. A comparison of fisher and weasel hunting behavior. Carnivore 1:28–34.

Powell, R. A. 1979a. Mustelid spacing patterns: variations on a theme by *Mustela*. Z. Tierpsychol. 50:153–165.

Powell, R. A. 1979b. Ecological energetics and foraging strategies of the fisher (*Martes pennanti*). J. Anim. Ecol. 48:195–212.

Powell, R. A. 1979c. Fishers, population models, and trapping. Wildl. Soc. Bull. 7: 149–154.

Powell, R. A. 1980a. Stability in a one-predator–three-prey community. Am. Nat. 115:567–579.

Powell, R. A. 1980b. Fisher arboreal activity. Can. Field-Nat. 94:90–91.

Powell, R. A. 1981a. *Martes pennanti*. Mammalian Species 156:1–6.

Powell, R. A. 1981b. Hunting behavior and food requirements of the fisher (*Martes pennanti*). Pages 883–917 *in* J. A. Chapman and D. Pursley, eds. Proc. Worldwide Furbearer Conf., Frostburg, Md.

Powell, R. A. 1982. The fisher: life history, ecology, and behavior. Univ. of Minnesota Press, Minneapolis. 217pp.

Powell, R. A. 1985. Fisher pelt primeness. Wildl. Soc. Bull. 13:67–70.

Powell, R. A. 1986. Test of an hypothesized relationship between territorial behavior and habitat productivity using black bear (*Ursus americanus*). Abstracts, First Int. Behav. Ecol. Conf., Albany, N.Y.

Powell, R. A. 1989. Effects of resource productivity, patchiness and predictability on mating and dispersal strategies. Pages 101–123 *in* V. Standen and R. A. Foley, eds. Comparative socioecology. Blackwell, Oxford.

Powell, R. A. 1993. The fisher. Second ed. Univ. of Minnesota Press, Minneapolis. 290pp.

Powell, R. A., and R. B. Brander. 1977. Adaptations of fishers and porcupines to their predator-prey system. Pages 45–53 *in* R. L. Phillips and C. Jonkel, eds. Proc. 1975 Predator Symp., Mont. For. Conserv. Exp. Stn., Univ. of Montana, Missoula.

Powell, R. A., and C. M. King. 1989. Effects of food supply during growth on sexual dimorphism of stoats. Am. Soc. Mammal., Thirty-ninth Annu. Meet., Abstr. 166.

Powell, R. A., and R. D. Leonard. 1983. Sexual dimorphism and energy expenditure for reproduction in female fisher *Martes pennanti*. Oikos 40:166–174.

Powell, R. A., and W. J. Zielinski. 1983. Competition and coexistence in mustelid communities. Acta Zool. Fenn. 174:223–227.

Prell, H. 1927. Über doppelte Brunstzeit und verlängerte Tragzeit bei den einheimischen Arten der Mardergattung *Martes* pinel. Zool. Anz. 74:122–128.

Prell, H. 1930. Die verlängerte Tragzeit der einheimischen *Martes* Arten: Ein Erklärungsversuch. Zool. Anz. 88:17–31.

Prest, V. K. 1970. Quaternary geology of Canada. Pages 676–764 *in* R. J. E. Douglas, ed. Geology and economic minerals of Canada. Dep. Energy, Mines, Resour., Ottawa, Ont.

Proulx, G. 1990. Humane trapping program. Annual report 1989/90. Alta. Res. Counc., Edmonton. 15pp.

Proulx, G. 1991. Humane trapping program. Annual report 1990/91. Alta. Res. Counc., Edmonton. 16pp.

Proulx, G., and M. W. Barrett. 1989. Animal welfare concerns and wildlife trapping: ethics, standards and commitments. Trans. Western Sect. Wildl. Soc. 25:1–6.

Proulx, G., and M. W. Barrett. 1991a. Ideological conflict between animal rightists and wildlife professionals over trapping wild furbearers. Trans. North Am. Wildl. Nat. Resourc. Conf. 56:387–399.

Proulx, G., and M. W. Barrett. 1991b. Evaluation of the Bionic trap to quickly kill mink (*Mustela vison*) in simulated natural environments. J. Wildl. Dis. 27:276–280.

Proulx, G., and M. W. Barrett. 1993*a*. Evaluation of mechanically improved Conibear 220™ traps to quickly kill fisher (*Martes pennanti*) in simulated natural environments. J. Wildl. Dis. 29:317–323.

Proulx, G., and M. W. Barrett. 1993*b*. Evaluationof the Bionic^R trap to quickly kill fisher (*Martes pennanti*) in simulated natural environments. J. Wildl. Dis. 29:310–316.

Proulx, G., M. W. Barrett, and S. R. Cook. 1990. The C120 Magnum with pan trigger: a humane trap for mink (*Mustela vison*). J. Wildl. Dis. 26:511–517.

Proulx, G., M. W. Barrett, and S. R. Cook. 1989*a*. The C120 Magnum: an effective quick-kill trap for marten. Wildl. Soc. Bull. 17:294–298.

Proulx, G., S. R. Cook, and M. W. Barrett. 1989*b*. Assessment and preliminary development of the rotating-jaw Conibear 120 trap to effectively kill marten (*Martes americana*). Can. J. Zool. 67:1074–1079.

Proulx, G., I. M. Pawlina, and R. K. Wong. 1993. Re-evaluation of the C120 Magnum and Bionic traps to humanely kill mink. J. Wildl. Dis. 29:184.

Pruitt, W. O. 1957. Observations on the bioclimate of some taiga mammals. Arctic 10:130–138.

Pruitt, W. O., and C. V. Lucier. 1958. Winter activity of red squirrels in interior Alaska. J. Mammal. 39:443–444.

Pulliainen, E. 1981*a*. Winter habitat selection, home range, and movements of the pine marten (*Martes martes*) in a Finnish Lapland forest in winter. Pages 1068–1087 *in* J. A. Chapman and D. Pursley, eds. Proc. Worldwide Furbearer Conf., Frostburg, Md.

Pulliainen, E. 1981*b*. Food and feeding habits of the pine marten in a Finnish Lapland forest in winter. Pages 580–598 *in* J. A. Chapman and D. Pursley, eds. Proc. Worldwide Furbearer Conf., Frostburg, Md.

Pulliainen, E. 1982. Scent-marking in the pine marten (*Martes martes*) in Finnish forest Lapland in winter. Z. Säugetierk. 47:91–99.

Pulliainen, E. 1984. Use of the home range by pine martens (*Martes martes* L.). Acta Zool. Fenn. 171:271–274.

Pulliam, H. R. 1988. Sources, sinks, and population regulation. Am. Nat. 132:652–661.

Punkari, M. 1984. Kunnon metsiä Suomessa vähemmän kuin koskaan. Suomen Luonto 43:12–14, 50 (in Finnish, with English summary).

Purdue, J. R., and B. W. Styles. 1987. Changes in the mammalian fauna of Illinois and Missouri during the late Pleistocene and Holocene. Pages 144–175 *in* R. W. Graham, H. A. Semken, and M. A. Graham, eds. Late Quaternary mammalian biogeography and environments of the Great Plains and prairies. Ill. State Mus. Sci. Pap. 22, Springfield.

Pyke, G. H., H. R. Pulliam, and E. L. Charnov. 1977. Optimal foraging: a selective review of theory and tests. Q. Rev. Biol. 52:137–154.

Pyne, S. J. 1982. Fire in America: a cultural history of wildland and rural fire. Princeton Univ. Press, Princeton, N.J. 654pp.

Quann, J. D. 1985. Progress report, Fundy National Park American Marten Reintroduction Program, 1985. Environ. Can., Parks Can., Nat. Resour. Conserv., Fundy Natl. Park. 34pp.

Quick, H. F. 1953. Wolverine, fisher and marten studies in a wilderness region. Trans. North Am. Wildl. Conf. 18:512–533.

Quick, H. F. 1955. Food habits of marten (*Martes americana*) in northern British Columbia. Can. Field-Nat. 69:144–147.

Quick, H. F. 1956. Effects of exploitation on a marten population. J. Wildl. Manage. 20:267–274.

Radinsky, L. B. 1984. Basicranial axis length v. skull length in analysis of carnivore skull shape. Biol. J. Linn. Soc. 22:31–41.

Rahel, F. J. 1990. The hierarchical nature of community persistence: a problem of scale. Am. Nat. 136:328–344.

Raine, R. M. 1981. Winter food habits, responses to snow cover and movements of fisher (*Martes pennanti*) and marten (*Martes americana*) in southeastern Manitoba. M.S. thesis, Univ. of Manitoba, Winnipeg. 145pp.

Raine, R. M. 1982. Range of juvenile fisher, *Martes pennanti*, and marten, *Martes americana*, in southeastern Manitoba. Can. Field-Nat. 96:431–438.

Raine, R. M. 1983. Winter habitat use and responses to snow cover of fisher (*Martes pennanti*) and marten (*Martes americana*) in southeastern Manitoba. Can. J. Zool. 61:25–34.

Raine, R. M. 1987. Winter food habits and foraging behaviour of fishers (*Martes pennanti*) and martens (*Martes americana*) in southeastern Manitoba. Can. J. Zool. 65:745–747.

Ralls, K. 1976. Mammals in which females are larger than males. Q. Rev. Biol. 51:245–276.

Ralls, K. 1977. Sexual dimorphism in mammals: avian models and unanswered questions. Am. Nat. 111:917–938.

Ralls, K., and P. H. Harvey. 1985. Geographic variation in size and sexual dimorphism of North American weasels. Biol. J. Linn. Soc. 25:119–167.

Rand, A. L. 1944. The status of the fisher (*Martes pennanti*) in Canada. Can. Field-Nat. 58:77–81.

Raphael, M. G. 1984. Wildlife populations in relation to stand age and area in Douglas-fir forests of northwestern California. Pages 259–274 *in* W. R. Meehan, T. R. Merrill, Jr., and T. A. Hanley, eds. Fish and wildlife relationships in old-growth forests: proceedings of a symposium. Am. Inst. Fish. Res. Biol., Juneau, Alaska.

Raphael, M. G. 1988. Long-term trends in abundance of amphibians, reptiles, and mammals in Douglas-fir forests of northwestern California. Pages 23–31 *in* R. C. Szaro, K. E. Severson, and D. Patton, tech. coords. Management of amphibians, reptiles and small mammals in North America. USDA For. Serv. Gen. Tech. Rep. RM-166

Raphael, M. G., and R. H. Barrett. 1981. Methodologies for a comprehensive wildlife survey and habitat analysis in old-growth Douglas-fir forests. Cal-Neva Wildl. Trans. 1981:106–121.

Raphael, M. G., and R. H. Barrett. 1984. Diversity and abundance of wildlife in late successional Douglas-fir forests. Pages 34–42 *in* New forests for a changing world. Proc. 1983 Soc. Am. For. Natl. Conven., Portland Ore.

Raphael, M. G., and S. E. Henry. 1990. Preliminary suggestions for monitoring marten in the Rocky Mountain region. USDA For. Serv., Pac. Northwest For. Sci. Lab., Olympia, Wash. 6pp.

Raphael, M. G., and K. V. Rosenberg. 1983. An integrated approach to wildlife inventories in forested habitats. Pages 219–222 *in* J. F. Bell and T. Atterbury, eds. Renewable resource inventories for monitoring changes and trends. Oregon State Univ., Corvallis.

Raphael, M. G., C. A. Taylor, and R. H. Barrett. 1986. Smoked aluminum track stations record flying squirrel occurrence. USDA For. Serv. Res. Note PSW-384. 3pp.

Rasmussen, A. M., and A. B. Madsen. 1985. The diet of the stone marten *Martes foina* in Denmark. Natura Jutlandica 21:141–144.

Ray, C. E. 1967. Pleistocene mammals from Ladds, Bartow County, Georgia. Ga. Acad. Sci. Bull. 25:120–150.

Rayevski, V. V. 1947. The life of Kondo-Sosva sable. Preserv. Dep., Counc. Minist. RSFSR, Moscow. 220pp.

Real, L. A. 1980. On uncertainty and the law of diminishing returns in evolution and behavior. Pages 37–64 *in* J. E. R. Staddon, ed. Limits to action. Academic Press, New York.

Reed, T. R. 1982. Interspecific territoriality in the chaffinch and great tit on islands and the mainland of Scotland: playbacks and removal experiments. Anim. Behav. 30: 171–181.

Rego, P. W. 1984. Factors influencing harvest levels of fisher in southcentral and southeastern Maine. M.S. thesis, Univ. of Maine, Orono. 54pp.

Rego, P. 1989. Annual status report—Connecticut. Pages 7–9 *in* R. Lafond, ed. Proceedings of the northeast fur resources technical committee workshop. Minist. du Loisir, de la Chasse, et de la Pêche, Quebec.

Reig, S., and W. Jedrzejewski. 1988. Winter and early spring food of some carnivores in the Bialowieza National Park, eastern Poland. Acta Theriol. 33:57–65.

Repenning, C. A., and F. Grady. 1988. The microtine rodents of the Cheetah Room fauna, Hamilton Cave, West Virginia, and the spontaneous origin of *Synaptomys*. U.S. Geol. Surv. Bull. 1853:1–32.

Rettie, A. 1971. Summary of marten and fisher transplanting in Parry Sound Forest District in 1956–63. Ont. Minist. Nat. Resour., final rep. 4pp.

Reynolds, J. C., and N. J. Aebischer. 1991. Comparison and quantification of carnivore diet by faecal analysis: a critique, with recommendations, based on a study of the fox *Vulpes vulpes*. Mammal Rev. 21:97–122.

Rhodes, R. S. 1984. Paleoecology and regional paleoclimatic implications of the Farmdalian Craigmile and Woodfordian Waubonsie mammalian local faunas, southwestern Iowa. Ill. State Mus. Rep. Invest. 40:1–51.

Riabov, L. S. 1982. The stone marten in the wild and in captivity. Zhur. Okhota i okhotnich'e khoziaistvo 3:18–20.

Richardson, L., T. W. Clark, S. C. Forest, and T. M. Campbell III. 1986. Black-footed ferret recovery: a discussion of some options and considerations. Great Basin Nat. Mem. 8:169–184.

Ricker, W. E. 1975. Computation and interpretation of biological statistics of fish populations. Inf. Can. Bull. 191, Ottawa, Ont. 382pp.

Rising, J. D. 1987. Geographic variation of sexual dimorphism in size of savannah sparrows (*Passerculus sandwichensis*): a test of hypotheses. Evolution 41:514–524.

Ritchie, J. C. 1976. The late-Quaternary vegetational history of the western interior of Canada. Can. J. Bot. 54:1793–1818.

Ritchie, J. C. 1987. Postglacial vegetation of Canada. Cambridge Univ. Press, New York. 178pp.

Ritter, A. 1986. Marten assessment. Pages 564–601 *in* Planning for Maine's inland fish and wildlife. Species assessments and strategic plans, fur and game mammals 1986–1991. Maine Dep. Inland Fish. Wildl. Vol. 1, part 1.4.

Roberts, R. C. 1979. The evolution of avian food-storing behavior. Am. Nat. 114:418–438.

Robinson, A. 1918. The formation, rupture and closure of ovarian follicles in ferrets and ferret-polecat hybrids and some associated phenomena. Trans. Roy. Soc. Edin. 52:303–362.

Robinson, W. B. 1953. Coyote control with Compound 1080 stations in national forests. J. For. 51:880–885.

Robson, D. S., and D. G. Chapman. 1961. Catch curves and mortality rates. Trans. Am. Fish. Soc. 90:181–189.

Rogers, C. M. 1987. Predation risk and fasting capacity: do wintering birds maintain body mass? Ecology 68:1051–1061.

Rogers, K. 1975. Faunal remains from the Zimmerman site—1971. Pages 80–91 *in* M. K. Brown, ed. The Zimmerman site: further excavations at the Grand Village of Kaskaskia. Ill. State Mus. Rep. Invest. 32.

Rognrud, M. 1983. General wildlife restocking in Montana, 1941–1982. Final Rep., Fed. Aid Wildl. Restor. Proj. W-5-D.

Rollins, R. 1989. Abundance of furbearers in British Columbia: a survey of trappers. R. B. Rollins Assoc., unpubl. rep., Victoria, B.C. 22pp.

Rolstad, J. 1991. Consequences of forest fragmentation for the dynamics of bird populations: conceptual issues and the evidence. Biol. J. Linn. Soc. 42:149–163.

Rolstad, J., and P. Wegge. 1987. Habitat characteristics of capercaillie *Tetrao urogallus* display grounds in southeastern Norway. Holarctic Ecol. 10:219–229.

Rolstad, J., and P. Wegge. 1989. Effects of logging on capercaillie (*Tetrao urogallus*) leks. III. Extinction and recolonization of lek populations in relation to clearfelling and fragmentation of old forest. Scand. J. For. Res. 4:129–135.

Rolstad, J., P. Wegge, and B. B. Larsen. 1988. Spacing and habitat use of capercaillie during summer. Can. J. Zool. 66:670–679.

Romanowski, J. 1989. Diet of the stone marten in urban habitats. Page 974 *in* Abstracts of Papers and Posters, Fifth Int. Theriol. Congr., Rome.

Romanowski, J., and G. Lesinski. 1991. A note on the diet of stone marten in southeastern Romania. Acta Theriol. 36:201–204.

Roscoe, J. T., and J. A. Byars. 1971. An investigation of the restraints with respect to sample size commonly imposed on the use of the chi-square statistic. J. Am. Stat. Assoc. 66:755–759.

Rosenberg, K. V., and M. G. Raphael. 1986. Effects of forest fragmentation on vertebrates in Douglas-fir forests. Pages 263–272 *in* J. Verner, M. L. Morrison, and C. J. Ralphs, eds. Wildlife 2000: modeling habitat relationships of terrestrial vertebrates. Univ. of Wisconsin Press, Madison.

Rosenzweig, M. L. 1966. Community structure in sympatric carnivora. J. Mammal. 47:602–612.

Rosenzweig, M. L. 1968. The strategy of body size in mammalian carnivores. Am. Midl. Nat. 80:299–315.

Rossolimo, O. L., and I. J. Pavlinov. 1974. Sexual dimorphism in the development, size, and proportions of the skull in the pine marten (*Martes martes* L.: Mammalia, Mustelidae). Byull. Mosk. o-va Ispyt. Prir. Otd. Biol. 79:23–35. (Translation *in* C. M.

King, editor. 1980. Biology of mustelids: some Soviet research. Vol. 2. Sci. Inf. Div. Dep. Sci. Indust. Res., Wellington, N.Z. Pp. 180–191.)

Roughgarden, J. 1972. Evolution of niche width. Am. Nat. 106:683–718.

Roughton, R. D., and M. W. Sweeny. 1982. Refinements in scent-station methodology for assessing trends in carnivore populations. J. Wildl. Manage. 46:217–229.

Rowe, J. S. 1972. Forest regions of Canada. Dep. Environ., Can. For. Serv. Publ. 1300. 172pp.

Roy, K. 1990. Cabinet Mountains fisher reintroduction study. Interim progress report. Montana Coop. Wildl. Res. Unit, Univ. of Montana, Missoula.

Roy, K. D. 1991. Ecology of reintroduced fishers in the Cabinet Mountains of northwest Montana. M.S. thesis, Univ. of Montana, Missoula. 102pp.

Roze, U. 1989. The North American porcupine. Smithsonian Inst. Press, Washington, D.C. 261pp.

Ruggiero, L. F., R. F. Holthausen, B. G. Marcot, K. B. Aubry, J. W. Thomas, and E. C. Meslow. 1988. Ecological dependency: the concept and its implications for research and management. Trans. N. Am. Wildl. Nat. Resour. Conf. 53:115–126.

Ruiz-Olmo, J., and J. M. Lopez-Martín. In Press. Seasonal food of pine marten (*Martes martes* L., 1758) in a fir forest of Pyrenean Mountains (northeastern Spain). Proc. First European Congr. Mammal., Lisbon, Portugal.

Ryan, T. A., Jr., B. L. Joiner, and B. F. Ryan. 1980. Minitab II reference manual. Pennsylvania State Univ., University Park. 138pp.

Sacher, G. A., and E. F. Staffeld. 1974. Relation of gestation time to brain weight for placental mammals: implication for the theory of vertebrate growth. Am. Nat. 108: 593–615.

Salwasser, H., C. K. Hamilton, W. B. Krohn, J. Lipscomb, and C. H. Thomas. 1983. Monitoring wildlife and fish: mandates and their implications. Trans. North Am. Wildl. Nat. Resour. Conf. 48:297–307.

Sandell, M. 1986. Movement patterns of male stoats, *Mustela erminea*, during the mating season: differences in relation to social status. Oikos 47:63–70.

Sandell, M. 1989a. The mating tactics and spacing patterns of solitary carnivores. Pages 164–182 *in* J. L. Gittleman, ed. Carnivore behavior, ecology, and evolution. Cornell Univ. Press, Ithaca, N.Y.

Sandell, M. 1989b. Ecological energetics, optimal body size and sexual dimorphism: a model applied to the stoat, *Mustela erminea* L. Func. Ecol. 3:315–324.

Sandell, M. 1990. The evolution of seasonal delayed implantation. Q. Rev. Biol. 65: 23–42.

Sasaki, H., and Y. Ono. 1989. Spacing patterns of Siberian weasel on a small island in Japan. Page 616 *in* Abstracts of Papers and Posters, Fifth Int. Theriol. Congr., Rome.

SAS Institute, Inc. 1985a. SAS user's guide: basics. Version 5. SAS Inst., Cary, N.C. 1290pp.

SAS Institute, Inc. 1985b. SAS user's guide: statistics. Version 5. SAS Inst., Cary, N.C. 956pp.

SAS Institute, Inc. 1988. SAS/STAT user guide. Release 6.03. SAS Institute, Cary, N.C. 1028pp.

Schaffer, M. L. 1983. Determining minimum viable population sizes for the grizzly bear. Int. Conf. Bear Res. Manage. 5:133–139.

Schamberger, M., and W. B. Krohn. 1982. Status of the habitat evaluation procedures. Trans. North Am. Wildl. Nat. Resour. Conf. 47:154–164.

Schamel, D., and D. M. Tracy. 1986. Encounters between arctic foxes, *Alopex lagopus*, and red foxes, *Vulpes vulpes*. Can. Field-Nat. 100:562–563.

Schemmel, R. 1976. Physiological consideration of lipid storage and utilization. Am. Zool. 16:661–670.

Schempf, P. F., and M. White. 1977. Status of six furbearer populations in the mountains of northern California. USDA For. Serv., Calif. Reg., unpubl. rep. 51pp.

Schlafke, S., A. C. Enders, and R. L. Given. 1981. Cytology of the endometrium of delayed and early implantation with special reference to mice and mustelids. J. Reprod. Fert. (suppl.) 29:135–141.

Schmidt, F. 1934. Über die Fortpflanzungsbiologie vom sibirischen Zobel (*Martes zibellina* L.) und europäischen Baummarder (*Martes martes*). Z. Säugetierk. 9:392–403.

Schmidt, F. 1943. Naturgeschichte des Baum- und des Steinmarders mit vergleichenden Betrachtungen ihrer nächsten Verwandten, besonders des sibirischen Zobels und des amerikanischen Fichtenmarders. *In* D. Mueller-Using, ed. Monographien der Wildsäugetiere. Vol. 10. Inst. für Jagdkunde, Univ. of Göttingen, Leipzig. 258pp.

Schoener, T. W. 1967. The ecological significance of sexual dimorphism in size in the lizard *Anolis conspersus*. Science 155:474–477.

Schoener, T. W. 1968. Sizes of feeding territories among birds. Ecology 49:124–141.

Schoener, T. W. 1974. Resource partitioning in ecological communities. Science 185: 27–39.

Scholander, P. F., R. Hock, V. Walters, and L. Irving. 1950. Adaptation to cold in arctic and tropical mammals and birds in relation to body temperature, insulation and basal metabolic rate. Biol. Bull. 99:259–271.

Schorger, A. W. 1942. Extinct and endangered mammals and birds of the Great Lakes region. Trans. Wis. Acad. Sci., Arts Letters 34:24–57.

Schröpfer, R., W. Biedermann, and H. Szczesniak. 1989. Saisonale Aktionsraumveränderungen beim Baummarder *Martes martes* L. 1758. Pages 433–442 *in* M. Stubbe, ed. Populationsökologie Marderartiger Säugetiere. Wiss. Beitr. Univ. of Halle.

Schupbach, T. A. 1977. History, status, and management of the pine marten in the upper peninsula of Michigan. M.S.F. thesis, Michigan Tech Univ., Houghton. 70pp.

Scotts, D. J., and S. A. Craig. 1988. Improved hair-sampling tube for the detection of rare mammals. Aust. Wildl. Res. 15:469–472.

Seber, G. A. F. 1973. The estimation of animal abundance. Charles Griffin, London. 506pp.

Seber, G. A. F. 1984. Multivariate observations. John Wiley and Sons, New York. 678pp.

Selander, R. K. 1966. Sexual dimorphism and differential niche utilization in birds. Condor 68:113–151.

Selås, V. 1990a. Måren. Norges dyr 1:142–151. Cappellens Forlag A/S, Oslo.

Selås, V. 1990b. Hiplassering hos mår (*Martes martes*). Fauna (Oslo) 43:27–35 (in Norwegian, with English summary).

Selås, V. 1990c. Marens reproduksjonsbiologi. Fauna (Oslo) 43:19–26.

Semken, H. A., Jr. 1983. Holocene mammalian biogeography and climatic change in the eastern and central United States. Pages 182–207 *in* H. E. Wright, Jr., ed. Late-Quaternary environments of the United States. Vol. 2: The Holocene. Univ. of Minnesota Press, Minneapolis.

Semken, H. A., Jr. 1984. Paleoecology of a late Wisconsinan/Holocene micromammal sequence in Peccary Cave, northwestern Arkansas. Pages 405–431 *in* H. H. Genoways and M. R. Dawson, eds. Contributions in Quaternary vertebrate paleontology: a volume in memorial to John E. Guilday. Carnegie Mus. Nat. Hist. Spec. Publ. 8.

Semken, H. A., Jr., and C. R. Falk. 1987. Late Pleistocene/Holocene mammalian faunas and environmental changes on the northern plains of the United States. Pages 176–313 *in* R. W. Graham, H. A. Semken, and M. A. Graham, eds. Late Quaternary mammalian biogeography and environments of the Great Plains and prairies. Ill. State Mus. Sci. Pap. 21, Springfield.

Serafini, P., P. Valier, and S. Lovari. 1992. Food habits of sympatric red foxes and stone martens in a rural area. Abstracts, Fifty-fourth Congr. Unione Zool. Ital., Perugia, Italy.

Seton, E. T. 1925. Lives of game animals. Doubleday, Page, Garden City, N.Y.

Seton, E. T. 1929. Lives of game animals. Vol. 2, Part 2. Doubleday, Doran, Garden City, N.Y.

Shackelford, R. M. 1952. Superfetation in the ranch mink. Am. Nat. 86:311–319.

Shaffer, M. L. 1981. Minimum population sizes for species conservation. Bioscience 31:131–134.

Shannon, C. E., and W. Weaver. 1949. The mathematical theory of communication. Univ. of Illinois Press, Urbana-Champaign. 117pp.

Shea, M. E., N. L. Rollins, R. T. Bowyer, and A. G. Clark. 1985. Corpora lutea number as related to fisher age and distribution in Maine. J. Wildl. Manage. 49:37–40.

Shepherd, D. S., and J. H. Greaves. 1984. A weather-resistant tracking board. Pages 112–113 *in* Proc. Eleventh Vertebr. Pest Conf., Univ. of California, Davis.

Shubin, I. G., and H. G. Shubin. 1975. Sexual dimorphism in mustelids (Mustelidae, Carnivora). Zhur. obshchei Biol. 36:283–290. (Translation *in* C. M. King, editor. 1980. Biology of mustelids: some Soviet research. Vol. 2. Sci. Inf. Div. Dep. Sci. Indust. Res., Wellington, N.Z. Pp. 197–205.)

Shump, A. U., R. J. Aulerich, and R. K. Ringer. 1976. Semen volume and sperm concentration in the ferret (*Mustela putorius*). Lab. Anim. Sci. 26:913–916.

Sibbald, I. R., D. G. Sinclair, E. V. Evans, and D. L. T. Smith. 1962. The rate of passage of feed through the digestive tract of the mink. J. Biochem. Physiol. 40:1391–1394.

Siegel, S. 1956. Nonparametric statistics for the behavioral sciences. McGraw-Hill, New York. 312pp.

Silver, H. 1957. Marten. Pages 258–262 *in* A history of New Hampshire game and furbearers. N.H. Fish Game Dep. Surv. Rep. 6. 466pp.

Silverman, B. W. 1986. Density estimation for statistics and data analysis. Chapman and Hall, London. 175pp.

Simberloff, D. 1987. The spotted owl fracas: mixing academic, applied, and political ecology. Ecology 68:766–772.

Simberloff, D., and W. Boecklen. 1981. Santa Rosalia reconsidered: size ratios and competition. Evolution 1206–1228.

Simms, D. A. 1979. North American weasels: resource utilization and distribution. Can. J. Zool. 57:504–520.

Simon, T. 1980. An ecological study of the pine marten in the Tahoe National Forest. M.S. thesis, California State Univ., Sacramento. 143pp.

Simpson, G. G. 1945. Principles of classification and a classification of mammals. Bull. Am. Mus. Nat. Hist. 85:1- 350.

Sinclair, G. 1986. American marten live trapping results for re-introductions to Fundy and Kejimkujik national parks. Environ. Can., Parks Can., Nat. Resour. Conserv., Fundy Natl. Park. 9pp.

Sinclair, G. 1987. Plan update for the reintroduction of the American marten, Fundy National Park. Environ. Can., Parks Can., Nat. Resour. Conserv., Fundy Natl. Park. 9pp.

Sjörs, H. 1965. Forest regions. Acta Phytogeogr. Suec. 50:48–63.

Skinner, D. L., and A. W. Todd. 1988. Distribution and status of selected mammals in Alberta as indicated by trapper questionnaires in 1987. Alta. For. Lands and Wildl., Fish and Wildl. Occas. Pap. 4, Edmonton.

Skirnisson, K. 1983. Aktivität und Home-Range-Grössen von Steinmardern (*Martes foina*) in Norddeutschland. Pages 1–5 *in* Trans. Sixteenth Int. Congr. Game Biol., High Tatras, Czechoslovakia.

Skirnisson, K. 1986. Untersuchungen zum Raum-Zeit-System freilebender Steinmarder (*Martes foina* Erxleben, 1777). Beitr. zur Wildbiol. 6:1–200.

Slough, B. G. 1989. Movements and habitat use by transplanted marten in the Yukon Territory. J. Wildl. Manage. 53:991–997.

Slough, B. G., R. H. Jessup, D. I. McKay, and A. B. Stephenson. 1987. Wild furbearer management in western and northern Canada. Pages 1062–1076 *in* M. Novak, J. A. Baker, M. E. Obbard, and B. Malloch, eds. Wild furbearer management and conservation in North America. Ont. Trappers Assoc., North Bay, Ont.

Slough, B. G., and C. M. Smits. 1985. Yukon marten management, progress to August, 1985. Yukon Fish Wildl. Branch, Dep. Renewable Resour., unpubl. rep., Whitehorse, Yukon. 57pp.

Smith, C. C. 1968. The adaptive nature of social organization in the genus of tree squirrels, *Tamiasciurus*. Ecol. Monogr. 39:31–63.

Smith, R. E., and R. C. Hock. 1963. Brown fat: thermogenic effector of arousal in hibernators. Science 140:199–200.

Snedecor, G. W., and W. G. Cochran. 1967. Statistical methods. Sixth ed. Iowa State Univ. Press, Ames. 593pp.

Snyder, J. E. 1984. Marten use of clearcuts and residual stands in western Newfoundland. M.S. thesis, Univ. of Maine, Orono. 31pp.

Snyder, J. E., and J. A. Bissonette. 1987. Marten use of clear-cuttings and residual forest in western Newfoundland. Can. J. Zool. 65:169–174.

Snyder, N. F. R., and J. W. Wiley. 1976. Sexual size dimorphism in hawks and owls of North America. Ornithol. Monogr. 20:1–96.

Sokal, R. R., and F. J. Rohlf. 1981. Biometry. W. H. Freeman, San Francisco. 859pp.

Sondaar, P. Y. 1977. Insularity and its effect on mammal evolution. Pages 671–707 *in* M. K. Hecht, P. C. Goody, and B. M. Hecht, eds. Major patterns in vertebrate evolution. Plenum Press, New York.

Sonerud, G. A. 1985*a*. Nest hole shift in Tengmalm's owl *Aegolius funereus* as defense against nest predation involving long-term memory in the predator. J. Anim. Ecol. 54:179–192.

Sonerud, G. A. 1985*b*. Risk of nest predation in three species of hole nesting owls:

influence on choice of nesting habitat and incubation behavior. Ornis Scand. 16: 261–269.

Sonerud, G. 1986. Effect of snow cover on seasonal changes in diet, habitat, and regional distribution of raptors that prey on small mammals in boreal zones of Fennoscandia. Holarctic Ecol. 9:33–47.

Song, J. H., Y. Tong, and Y. Xiao. 1988. Effects of light on reproduction and molting of *Martes zibellina*. J. Ecol. China 7:17–29.

Soper, J. D. 1970. The mammals of Jasper National Park, Alberta. Can. Wildl. Serv., Rep. Ser. 10. 80pp.

Soukkala, A. M. 1983. The effects of trapping on marten populations in Maine. M.S. thesis, Univ. of Maine, Orono. 40pp.

Soutiere, E. C. 1978. The effects of timber harvesting on the marten. Ph.D. dissertation, Univ. of Maine, Orono. 60pp.

Soutiere, E. C. 1979. Effects of timber harvesting on marten in Maine. J. Wildl. Manage. 43:850–860.

Soutiere, E., and M. W. Coulter. 1975. Interim report, re-introduction of marten to the White Mountain National Forest, New Hampshire, 1975. Univ. of Maine, Orono. Unpubl. 3pp.

Soutiere, E. C., and J. D. Steventon. 1981. Seasonal pelage change of the marten (*Martes americana*) in Maine. Can. Field-Nat. 95:356.

Spaulding, W. G., E. B. Leopold, and T. R. Van Devender. 1983. Late Wisconsin paleoecology of the American Southwest. Pages 259–293 *in* S. C. Porter, ed. Late-Quaternary environments of the United States. Vol. 1: The late Pleistocene. Univ. of Minnesota Press, Minneapolis.

Spencer, W. D. 1981. Pine marten habitat preferences at Sagehen Creek, California. M.S. thesis, Univ. of California, Berkeley. 121pp.

Spencer, W. D., R. H. Barrett, and W. J. Zielinski. 1983. Marten habitat preferences in the northern Sierra Nevada. J. Wildl. Manage. 47:1181–1186.

Spencer, W. D., and W. J. Zielinski. 1983. Predatory behavior of pine martens. J. Mammal. 64:715–717.

Stach, J. 1959. On some Mustelinae from the Pliocene bone breccia of Weze. Acta Palaeontol. Polonica 4:101–118.

Stafford, T. W., and H. A. Semken. 1990. Accelerator ^{14}C dating of two micromammal species representative of the late Pleistocene disharmonious fauna from Peccary Cave, Newton County, Arkansas. Curr. Res. Pleistocene 7:129–132.

Stamps, J. A., and K. Tollestrup. 1984. Prospective resource defense in a territorial species. Am. Nat. 123:417–427.

Standing Committee on Aboriginal Affairs and Northern Development. 1986. The fur issue. Can. House of Commons, Issue 1, Ottawa, Ont. 81pp.

Stenlund, M. H. 1955. A recent record of the marten in Minnesota. J. Mammal. 36:133.

Stenson, G. B. 1988. Oestrus and the vaginal smear cycle of the river otter, *Lutra canadensis*. J. Reprod. Fert. 83:605–610.

Stetson, M. H., J. A. Elliott, and B. D. Goldman. 1986. Maternal transfer of photoperiodic information influences the photoperiodic response of prepubertal Djungarian hamsters. Biol. Reprod. 34:664–670.

Stetson, M. H., S. N. Ray, N. Creyaufmiller, and T. H. Horton. 1989. Maternal transfer of photoperiodic information in Siberian hamsters. II. The nature of the maternal signal, time of signal transfer, and the effect of the maternal signal on peripubertal reproductive development in the absence of photoperiod input. Biol. Reprod. 40: 458–466.

Stevens, C. L. 1968. The food of fisher in New Hampshire. N.H. Dep. Fish Game, unpubl. rep. (Cited in R. A. Powell. 1982. The fisher. Univ. of Minnesota Press, Minneapolis. 217pp.)

Steventon, J. D. 1979. Influence of timber harvesting upon winter habitat use by marten. M.S. thesis, Univ. of Maine, Orono. 24pp.

Steventon, J. D., and J. T. Major. 1982. Marten use of habitat in a commercially clear-cut forest. J. Wildl. Manage. 46:175–182.

Stewart, J. D. 1987. Prehistoric and historic cultural resources of selected sites at Harlan County Lake, Harlan County, Nebraska. U.S. Army Corps of Engineers, Final Rep., Kansas City, Mo.

Stone, W. B., A. S. Clauson, D. E. Slingerlands, and B. L. Weber. 1975. Use of Romanosky stains to prepare tooth sections for aging mammals. N.Y. Fish Game J. 22:156–158.

Storch, I. 1988. [Home range utilization by pine martens]. Z. Jagdwiss. 34:115–119 (in German, with English and French summaries).

Storch, I., E. Lindström, and J. de Jonge. 1990. Habitat selection and food habits of the pine marten in relation to competition with the red fox. Acta Theriol. 35:311–320.

Streeter, R. G., and C. E. Braun. 1968. Occurrence of pine marten, *Martes americana* (Carnivora: Mustelidae) in Colorado alpine areas. Southwest. Nat. 13:449–451.

Strickland, M. A. 1981. Fisher and marten study 1979–80 and 1980–81. Ont. Minist. Nat. Resour., Algonquin Reg. Prog. Rep. 7. 172pp.

Strickland, M. A. 1989. Marten management in Ontario. Pages 155–174 in R. Lafond, ed. Proceedings of the Northeast fur resources technical committee workshop. Minist. du Loisir, de la Chasse, et de la Pêche, Quebec.

Strickland, M. A., and C. W. Douglas. 1981. The status of fisher in North America and its management in southern Ontario. Pages 1443–1458 in J. A. Chapman and D. Pursley, eds. Proc. Worldwide Furbearer Conf., Frostburg, Md.

Strickland, M. A., and C. W. Douglas. 1983. The marten. Ont. Minist. Nat. Resour., unpubl. rep., Parry Sound. 14pp.

Strickland, M. A., and C. W. Douglas. 1984. Results of questionnaires sent to trappers of fisher and marten in the Algonquin region (Ontario) in five consecutive years 1979 to 1983. Ont. Minist. Nat. Resour. Rep., Toronto. 49pp.

Strickland, M. A., and C. W. Douglas. 1987. Marten. Pages 530–546 in M. Novak, J. A. Baker, M. E. Obbard, and B. Malloch, eds. Wild furbearer management and conservation in North America. Ont. Trappers Assoc., North Bay, Ont.

Strickland, M. A., C. W. Douglas, M. Novak, and N. P. Hunziger. 1982a. Marten. Pages 599–612 in J. A. Chapman and G. A. Feldhamer, eds. Wild mammals of North America: biology, management, economics. Johns Hopkins Univ. Press, Baltimore.

Strickland, M. A., C. W. Douglas, M. Novak, and N. P. Hunziger. 1982b. Fisher. Pages 586–598 in J. A. Chapman and G. A. Feldhamer, eds. Wild mammals of North America: biology, management, economics. Johns Hopkins Univ. Press, Baltimore.

Strickland, M. A., C. W. Douglas, M. K. Brown, and G. R. Parsons. 1982c. Determining the age of fisher from cementum annuli of the teeth. N.Y. Fish Game J. 29:90–94.

Stroganov, S. U. 1969. Carnivorous mammals of Siberia. Israel Sci. Transl. Program, Jerusalem. 522pp.

Stubbe, M. 1968. Zur Populationsbiologie der *Martes*-Arten. Beitr. Jagd.- und Wild-forsch. 104:195–203.

Stuebe, M. M., and E. D. Ketterson. 1982. A study of fasting in tree sparrows (*Spizella arborea*) and dark-eyed juncos (*Junco hyemalis*): ecological implications. Auk 99: 299–308.

Suckling, G. C. 1978. A hair sampling tube for the detection of small mammals in trees. Aust. Wildl. Res. 5:249–252.

Sullivan, M. J. 1984. American marten re-introduction progress report, Fundy National Park, 1984. Environ. Can., Parks Can., Nat. Resour. Conserv., Fundy Natl. Park. 30pp.

Sundqvist, C. 1987. Male infertility in mink breeding. Ph.D. dissertation, Abo Akademi, Abo, Finland. 578pp.

Swanson, G. A., T. Surber, and T. S. Roberts. 1945. The mammals of Minnesota. Minn. Dep. Conserv. Tech. Bull. 2. 108pp.

Swihart, R. K., and N. A. Slade. 1985. Testing for independence of observations in animal movements. Ecology 66:1176–1184.

Syrjanov, A. 1989. Reproduktion und Anzahl des Zobels *Martes zibellina* (L., 1758) im Gebeit Krasnoharsk. Pages 461–465 *in* M. Stubbe, ed. Populationsökologie marderar-tiger Säugetiere. Wiss. Beitr. Univ. of Halle.

Tapper, S. C. 1976. The diet of weasels, *Mustela nivalis*, and stoats, *Mustela erminea*, during early summer, in relation to predation on gamebirds. J. Zool. (Lond.) 179: 219–224.

Tapper, S. C. 1979. The effect of fluctuating vole numbers (*Microtus agrestis*) on a population of weasels (*Mustela nivalis*) on farmland. J. Anim. Ecol. 48:603–617.

Taylor, C. A., and M. G. Raphael. 1988. Identification of mammal tracks from sooted track stations in the Pacific Northwest. Calif. Fish Game 74:4–15.

Taylor, C. R., K. Schmidt-Nielsen, and J. L. Raab. 1970. Scaling the energetic cost of running to body size in mammals. Am. J. Physiol. 219:1104–1107.

Taylor, M. E., and N. Abrey. 1982. Marten, *Martes americana*, movements and habitat use in Algonquin Provincial Park, Ontario. Can. Field-Nat. 96:439–447.

Taylor, S. L. 1993. Thermodynamics and energetics of resting site use by the American marten (*Martes americana*). M.S. thesis, Univ. Wyoming, Laramie. 89pp.

Teilhard de Chardin, P., and P. Leroy. 1945. Les mustelides de Chine. Institut de Geo-biologie (Beijing) 12:1–56.

Tester, U. 1986. Vergleichende Nahrungsuntersuchung beim Steinmarder (*Martes foina* Erxleben 1777) in grosstädtischem und ländlichem Habitat. Säugetierk. Mitt. 33: 37–52.

Thomas, J. W., R. G. Anderson, C. Maser, and E. L. Bull. 1979. Snags. Pages 60–77 *in* J. W. Thomas, ed. Wildlife habitats in managed forests—the Blue Mountains of Oregon and Washington. USDA For. Serv. Agric. Handb. 553.

Thomas, J. W., L. F. Ruggiero, R. W. Mannan, J. W. Schoen, and R. A. Lancia. 1988. Management and conservation of old-growth forests in the United States. Wildl. Soc. Bull. 16:252–262.

Thomasma, L. E., T. D. Drummer, and R. O. Peterson. 1991. Testing the habitat suit-ability index model for the fisher. Wildl. Soc. Bull. 19:291–297.

Thompson, I. D. 1986. Diet choice, hunting behavior, activity patterns, and ecological energetics of marten in natural and logged areas. Ph.D. dissertation, Queen's Univ., Kingston, Ont. 179pp.

Thompson, I. D. 1991. Will marten become the spotted owl of the east? For. Chron. 67:136–140.

Thompson, I. D. 1994. Marten populations in uncut and logged boreal forest in Ontario. J. Wildl. Manage. 57:000–000.

Thompson, I. D., and P. W. Colgan. 1987*a*. Numerical responses of martens to a food shortage in northcentral Ontario. J. Wildl. Manage. 51:824–835.

Thompson, I. D., and P. W. Colgan. 1987*b*. Effects of logging on home range characteristics and hunting activity of marten in Ontario. *In* B. Bobek, K. Perzanowski, and W. L. Regelin, eds. Global trends in wildlife management. Swiat Press, Krakow, Poland.

Thompson, I. D., and P. W. Colgan. 1990. Prey choice by marten during a decline in prey abundance. Oecologia 83:443–451.

Thompson, I. D., I. J. Davidson, S. O'Donnell, and F. Brazeau. 1989. Use of track transects to measure the relative occurrence of some boreal mammals in uncut forest and regeneration stands. Can. J. Zool. 67:1816–1823.

Thompson, W. K. 1949. A study of marten in Montana. Pages 181–188 *in* Proc. Twenty-ninth Ann. Conf. Western Assoc. State Game and Fish Comm., Seattle, Wash.

Thorpe, D. H. 1967. Basic parameters in the reaction of ferrets to light. Pages 53–70 *in* G. W. Wolstenholme and M. O'Connor, eds. Effects of external stimuli on reproduction. Ciba Found. Stud. Group. Little, Brown, Boston.

Timofeev, V. V., and V. N. Nadeev. 1955. Sobol'. Teknicheskoi i ekonomicheskoi lit-ry po voposam zagotovok, Moscow.

Todd, A. W., and E. K. Boggess. 1987. Characteristics, activities, lifestyles, and attitudes of trappers in North America. Pages 59–76 *in* M. Novak, J. A. Baker, M. E. Obbard, and B. Malloch, eds. Wild furbearer management and conservation in North America. Ont. Trappers Assoc., North Bay, Ont.

Toft, C. A., and P. J. Shea. 1983. Detecting community-wide patterns: estimating power strengthens statistical inference. Am. Nat. 122:618–625.

Tomak, C. H. 1974. Prairie Creek: a stratified site in southwestern Indiana. Proc. Ind. Acad. Sci. 84:65–68.

Torbit, S. C., L. H. Carpenter, D. M. Swift, and A. W. Alldredge. 1985. Differential loss of fat and protein by mule deer during winter. J. Wildl. Manage. 49:80–85.

Travis, H., W. G. Pilbeam, and R. Cole. 1978. Relationship of vulvar swelling to estrus in mink. J. Anim. Sci. 46:219–224.

Trillmich, F. 1986. Are endotherms emancipated? Some considerations on the cost of reproduction. Oecologia 69:631–633.

Trivers, R. L. 1972. Parental investment and sexual selection. Pages 136–179 *in* B. G. Campbell, ed. Sexual selection and the descent of man, 1871–1971. Aldine, Chicago.

Underwood, L. S. 1971. The bioenergetics of the arctic fox *Alopex lagopus* L. Ph.D. dissertation, Pennsylvania State Univ., State College.

USDA Forest Service. 1986. Lewis and Clark National Forest, forest plan.

USDA Forest Service. 1987. Gallatin National Forest, forest plan.

U.S. Fish and Wildlife Service. 1981. Standards for the development of suitability index models. Ecol. Serv. Man. 103. U.S. Fish Wildl. Serv., Div. Ecol. Serv. U.S. Gov. Printing Office, Washington, D.C. 68pp.

Van Devender, T. R., R. S. Thompson, and J. L. Betancourt. 1987. Vegetation history of the deserts of southwestern North America; the nature and timing of the late Wisconsin–Holocene transition. Pages 323–362 *in* W. F. Ruddiman and H. E. Wright, Jr., eds. North America and adjacent oceans during the last deglaciation. Geol. Soc. Am., Boulder, Colo.

Van Dyke, F. G., R. H. Brocke, and H. G. Shaw. 1986. Use of road track counts as indices of mountain lion presence. J. Wildl. Manage. 50:102–109.

Van Sickle, W. D. 1990. Methods for estimating cougar numbers in southern Utah. M.S. thesis, Univ. of Wyoming, Laramie. 46pp.

Van Valen, L. 1965. Morphological variation and width of ecological niche. Am. Nat. 99:377–390.

Van Zant, K. L., G. R. Hallberg, and R. G. Baker. 1980. A Farmdalian pollen diagram from east-central Iowa. Proc. Iowa Acad. Sci. 87:52–55.

van Zyll de Jong, C. G. 1969. The restoration of marten in Manitoba, an evaluation. Man. Wildl. Branch Biol. Rep. 26pp.

Vaughn, M. R., and L. B. Keith. 1981. Demographic responses of experimental snow-shoe hare populations to overwinter food shortage. J. Mammal. 45:354–380.

Venge, O. 1956. Experiments on forced interruption of the copulation in mink. Acta Zool. Stockholm 37:287–304.

Volodin, V. I., V. V. Bobrov, and V. N. Ravnushkin. 1980. The Aidash Cave. Nauka, Novosibirsk 97–127.

Voorhies, N. R. 1990. Vertebrate paleontology of the proposed Norden Reservoir area, Brown, Cherry and Keya Paha counties, Nebraska. Div. Archeol. Res., Tech. Rep. 82-09. Univ. of Nebraska, Lincoln. 733pp.

Wabakken, P. 1985. Vinternæring, habitatbruk og jaktatferd hos mår (*Martes martes*) i sørøst-norsk barskog. M.S. thesis, Univ. of Oslo, Norway. 85pp.

Wade-Smith, J., and M. E. Richmond. 1975. Care, management, and biology of captive striped skunks (*Mephitis mephitis*). Lab. Anim. Sci. 25:575–584.

Wade-Smith, J., and M. E. Richmond. 1978. Reproduction in captive striped skunks (*Mephitis mephitis*). Am. Midl. Nat. 100:452–455.

Wade-Smith, J., M. E. Richmond, R. A. Mead, and H. Taylor. 1980. Hormonal and gestational evidence for delayed implantation in the striped skunk, *Mephitis mephitis*. Gen. Comp. Endocrinol. 42:509–515.

Waechter, A. 1975. Ecologie de la fouine en Alsace. La Terre et la Vie 3:399–457.

Walker, D. N. 1987. Late Pleistocene/Holocene environmental changes in Wyoming: the mammalian record. Pages 334–393 *in* R. W. Graham, H. A. Semken, Jr., and M. A. Graham, eds. Late Quaternary mammalian biogeography and environments of the Great Plains and prairies. Ill. State Mus. Sci. Pap. 22, Springfield.

Wallace, K., and R. Henry. 1985. Return of a Catskill native. Conservationist (New York) 40(3):16–19.

Wang, L. C. H. 1978. Energetic and field aspects of mammalian torpor: the Richardson's ground squirrel. Pages 109–145 *in* L. C. H. Wang and J. W. Hudson, eds. Strategies in the cold-natural torpidity and thermogenesis. Academic Press, New York.

Warner, P., and P. O'Sullivan. 1982. The food of the pine marten *Martes martes* in

County Claire. Pages 323–330 *in* F. O'Gorman and J. Rockford, eds. Trans. Fourteenth Int. Congr. Game Biol., Dublin, Ireland.

Watts, W. A. 1983. Vegetational history of the eastern United States 25,000 to 10,000 years ago. Pages 294–310 *in* S. C. Porter, ed. Late-Quaternary environments of the United States. Vol. 1: The late Pleistocene. Univ. of Minnesota Press, Minneapolis.

Watts, W. A., and R. C. Bright. 1968. Pollen, seed and mollusk analysis of a sediment core from Pickerel Lake, northeastern South Dakota. Geol. Soc. Am. Bull. 79:855–876.

Webb, T., III, E. J. Cushing, and H. E. Wright, Jr. 1983. Holocene changes in the vegetation of the Midwest. Pages 142–165 *in* H. E. Wright, Jr., ed. Late-Quaternary environments of the United States. Vol. 2: The Holocene. Univ. of Minnesota Press, Minneapolis.

Webb, W. S. 1974. Indian Knoll. Univ. of Tennessee Press, Knoxville. 365pp.

Weber, D. 1987. Zur Biologie des Iltisses (*Mustela putorius* L.) und den Ursachen seines Rückgangs in der Schweiz. Inaugural dissertation, Naturhistorisches Mus., Basel. 194pp.

Webster, G. S. 1978. Dry Creek Rockshelter: cultural chronology in the western Snake River region of Idaho. Tebiwa 15:1–35.

Weckwerth, R. P. 1957. The relationship between the marten population and the abundance of small mammals in Glacier National Park. M.S. thesis, Montana State Univ., Missoula. 76pp.

Weckwerth, R. P., and V. D. Hawley. 1962. Marten food habits and population fluctuations in Montana. J. Wildl. Manage. 26:55–74.

Weckwerth, R. P., and P. L. Wright. 1968. Results of transplanting fishers in Montana. J. Wildl. Manage. 32:977–980.

Weems, R. E., and B. B. Higgins. 1977. Post-Wisconsinan vertebrate remains from a fissure deposit near Ripplemead, Virginia. Bull. Natl. Speleol. Soc. 39:106–108.

Weishampel, J. F. 1990. Maintaining genetic variation in a one-way, two-island model. J. Wildl. Manage. 54:676–682.

White, J. A., H. G. McDonald, E. Anderson, and J. M. Soiset. 1984. Lava blisters as carnivore traps. Pages 241–256 *in* H. H. Genoways and M. R. Dawson, eds. Contributions in Quaternary vertebrate paleontology: a volume in memorial to John E. Guilday. Carnegie Mus. Nat. Hist. Spec. Publ. 8.

Whitney, L. F. 1948. Feeding our dog. Van Nostrand, New York. 243pp.

Wilbert, C. J. 1992. Spatial scale and seasonality of habitat selection by martens in southeastern Wyoming. M.S. thesis, Univ. of Wyoming, Laramie. 91pp.

Wilcox, B. A. 1978. Insular ecology and conservation. Pages 95–119 *in* M. E. Soulé and B. A. Wilcox, eds. Conservation biology: an evolutionary-ecological perspective. Sinauer, Sunderland, Mass.

Wildt, D. E., M. Bush, C. Morton, F. Morton, and J. D. Howard. 1989. Semen characteristics and testosterone profiles in ferrets kept in a long-day photoperiod, and the influence of HCG timing and sperm dilution medium on pregnancy rate after laproscopic insemination. J. Reprod. Fert. 86:349–358.

Williams, G. L., D. R. Russell, and W. K. Seitz. 1977. Pattern recognition as a tool in the ecological analysis of habitat. Pages 521–531 *in* Classification, inventory, and analysis of fish and wildlife habitat. U.S. Fish Wildl. Serv. FWS/OBS-78/76.

Wilson, D. S. 1975. The adequacy of size as a niche difference. Am. Nat. 109:769–784.

Wilson, R. L. 1968. Systematics and faunal analysis of a lower Pliocene vertebrate assemblage from Trego County, Kansas. Contrib. Mus. Paleont. Univ. of Mich. 22: 75–126.

Winnett, G., and R. DeGabriele. 1982. A hair sampling tube for the detection of small and medium-sized mammals. Aust. Mammal. 5:143–145.

Wintemberg, W. J. 1928. Uren prehistoric village site, Oxford County, Ontario. Mus. Can. Bull. 51. 97pp.

Wood, J. E. 1959. Relative estimates of fox population levels. J. Wildl. Manage. 23: 53–63.

Worthen, G. L., and D. L. Kilgore. 1981. Metabolic rate of pine marten in relation to air temperature. J. Mammal. 62:624–628.

Wozencraft, W. C. 1989. The phylogeny of the Recent Carnivora. Pages 495–535 *in* J. L. Gittleman, ed. Carnivore behavior, ecology and evolution. Cornell Univ. Press, Ithaca, N.Y.

Wright, H. E., Jr. 1981. Vegetation east of the Rocky Mountains 18,000 years ago. Quat. Res. 15:113–125.

Wright, P. L. 1942. Delayed implantation in the long-tailed weasel (*Mustela frenata*), the short-tailed weasel (*Mustela cicognani*), and the marten (*Martes americana*). Anat. Rec. 83:341–353.

Wright, P. L. 1948. Preimplantation stages in the long-tailed weasel (*Mustela frenata*). Anat. Rec. 100:593–607.

Wright, P. L. 1950. Development of the baculum of the long-tailed weasel. Proc. Soc. Exp. Biol. Med. 75:820–822.

Wright, P. L. 1963. Variations in reproductive cycles in North American mustelids. Pages 77–97 *in* A. C. Enders, ed. Delayed implantation. Univ. of Chicago Press, Chicago.

Wright, P. L. 1966. Observations on the reproductive cycle of the American badger (*Taxidea taxus*). Pages 27–45 *in* I. W. Rowlands, ed. Comparative biology of reproduction in mammals. Academic Press, New York.

Wright, P. L., and M. W. Coulter. 1967. Reproduction and growth in Maine fishers. J. Wildl. Manage. 31:70–87.

Wunder, B. A. 1975. A model for estimating metabolic rate of active or resting mammals. J. Theor. Biol. 49:345–354.

Wynne, K. M., and J. A. Sherburne. 1984. Summer home range use by adult marten in northwestern Maine. Can. J. Zool. 62:941–943.

Yacoe, M. E. 1983. Maintenance of the pectoralis muscle during hibernation in the big brown bat *Eptesicus fuscus*. J. Comp. Physiol. 152:97–104.

Yeager, L. E. 1950. Implications of some harvest and habitat factors on pine marten management. Trans. North. Am. Wildl. Conf. 15:319–334.

Yerbury, H. 1947. Raising marten in captivity. Fur Trade J. Can. 25:14.

Yermolova, N. M. 1978. Theriofauna of the Angara Valley during the late Anthropogene. Nauka, Novosibirsk. 220pp.

Ylönen, H. 1989. Zum einfluss der Musteliden *Mustela nivalis* und *M. erminea* auf zyklische Kleinnager am Beispiel von *Clethrionomys*-Populationen in Mittelfinland. Pages 553–562 *in* M. Stubbe, ed. Populationsökologie marderartiger Säugetiere. Wiss. Beitr. Univ. of Halle.

Young, S. P. 1944. The wolves of North America, part I. Am. Wildl. Inst., Washington, D.C. 385pp.

Youngman, P. M., and F. W. Schueler. 1991. *Martes nobilis* is a synonym of *Martes americana*, not an extinct Pleistocene-Holocene species. J. Mammal. 72:567–577.

Yurgensen, P. B. 1947. Sexual dimorphism in feeding as an ecological adaptation of a species. Pages 79–83 *in* C. M. King, ed. 1975. Biology of mustelids: some Soviet research. Br. Libr. Lending Div., Boston Spa, Yorkshire.

Zar, J. H. 1984. Biostatistical analysis. Second ed. Prentice-Hall, Englewood Cliffs, N.J.

Zdansky, O. 1924. Jungtertiare carnivoren China. Palaeontol. Sinica 2:1–149.

Zeimans, G., and D. N. Walker. 1974. Bell Cave, Wyoming: preliminary archaeological and paleontological investigations. Wyo. Geol. Surv. Rep. Invest. 10:88–90.

Zheng, S., J. Zeng, and R. Cui. 1983. On ecology and energy dynamics of masked polecat (*Mustela eversmanni*) in Haibei, Qinghai province. Acta Theriol. Sinica 3: 35–46.

Ziegler, A. C. 1963. Unmodified mammal and bird remains from Deer Creek Cave, Elko County, Nevada. Pages 15–24 *in* M. E. Shutler and R. Shutler, Jr., eds. Deer Creek Cave, Nevada. Nev. State Mus. Anthropol. Pap. 11, Carson City.

Zielinski, W. J. 1986. Relating marten scat contents to prey consumed. Calif. Fish Game 72:110–116.

Zielinski, W. J., W. D. Spencer, and R. D. Barrett. 1983. Relationship between food habits and activity patterns of pine martens. J. Mammal. 64:387–396.

Zumeta, D. C. 1990. Public meetings held on Minnesota's timber harvesting/generic EIS. Minn. For. 3(3):7.

Index

Most main entries apply to *Martes* in general. For additional page references, see entries under individual species.

479